AGRI-ENVIRONMENTAL MANAGEMENT IN EUROPE

AGRI-ENVIRONMENTAL MANAGEMENT IN EUROPE

Sustainable Challenges and Solutions – From Policy Interventions to Practical Farm Management

Prof Kathy Lewis
Dr John Tzilivakis
Dr Doug Warner
Dr Andy Green

 5m Publishing

First published 2018

Published by
5M Publishing Ltd,
Benchmark House,
8 Smithy Wood Drive,
Sheffield, S35 1QN, UK
Tel: +44 (0) 1234 81 81 80
www.5mpublishing.com

A Catalogue record for this book is available from the British Library

ISBN 9781912178445

Book layout by Servis Filmsetting Ltd, Stockport, Cheshire
Printed by CPI Anthony Rowe Ltd, UK
Photos by the authors unless otherwise indicated
Back cover photo courtesy of: Dr John Tzilivakis

Contents

List of tables

List of figures

Acronyms and abbreviations

AD	Anaerobic Digestion
AEI	Agri-Environmental Indicator
BAT	Best Available Technique (or Technology)
C	Carbon
CAD	Centralised Anaerobic Digestion
CAP	Common Agricultural Policy
CBD	Convention on Biological Diversity
CE	Circular Economy
CoE	Council of Europe
CFCs	Chlorofluorocarbons
CH_4	Methane
CO	Carbon monoxide
CO_2	Carbon dioxide
CO_2e	Carbon dioxide equivalent
EAFRD	European Agriculture Fund for Rural Development
EAP	Environmental Action Programme
EC	European Commission/European Community
ECCP	European Climate Change Programme
EEA	European Environment Agency
EEP	European Environment Programme
EIA	Environmental Impact Assessment
EMAS	Eco-Management and Auditing Scheme
ERA	Environmental Risk Assessment
ETS	Emissions Trading System
EU	European Union
EU-28	The 28 Member States of the European Union
EUROSTAT	EU statistical data service
FAO	Food and Agriculture Organisation
GAEC	Good Agricultural and Environment Conditions
GDD	Groundwater Daughter Directive

GDP	Gross Domestic Product
GE	Green Economy
GES	Good Environmental Status
GIS	Geographical Information Systems
GM	Genetic Modification
GPS	Global Positioning System
GWP	Global Warming Potential
HCFC	Hydrochlorofluorocarbons
IEA	International Energy Agency
IED	Industrial Emissions Directive
IFM	Integrated Farm Management
IFWM	Integrated Farm Waste Management
IPCC	International Panel on Climate Change
IPM	Integrated Pest Management
ISM	Integrated Soil Management
IUCN	International Union for Conservation of Nature
IWM	Integrated Water Management
JRC	EC's Joint Research Centre
K	Potassium
LCA	Life Cycle Assessment
LCT	Life Cycle Thinking
LEAF	Linking Environment And Farming
LFA	Less Favoured Area
MEA	Millenium Ecosystem Assessment
MRL	Maximum Residue Limit
MS	Member State
N	Nitrogen
NH_3	Ammonia
N_2O	Nitrous oxide
NERT	National Emissions Reduction Targets
NO_3^-	Nitrate
NGO	Non-Governmental Body
NPK	Nitrogen, Phosphorus, Potassium
NUTS	Nomenclature of Territorial Units for Statistics
NVZ	Nitrate Vulnerable Zones
ODS	Ozone Depleting Substance
OECD	Organisation for Economic Co-operation and Development
P	Phosphorus
PES	Payments for Ecosystem Services
PM	Particulate Matter
PO_4^{2-}	Phosphate
PRD	Partial Rootzone Drying
PSR	Pressure–State–Response framework
RDP	Rural Development Programme
SAC	Special Area of Conservation

SO$_2$	Sulphur dioxide
SEA	Strategic Environmental Assessment
SOC	Soil Organic Carbon
SOM	Soil Organic Matter
SPA	Special Protection Area
SPR	Source–Pathway–Receptor
SRC	Short Rotation Coppice
TAN	Total Ammoniacal Nitrogen
TER	Toxicity–Exposure Ratio
TFP	Total Factor Productivity
TRWR	Total Renewable Water Resources
UN	United Nations
UNEP	United Nation's Environment Programme
UNESCO	United Nations Educational, Scientific and Cultural Organization
UV	Ultraviolet
VOC	Volatile Organic Compound
VRT	Variable Rate Technology
WEI	Water Exploitation Index
WFD	Water Framework Directive
WFPS	Water Filled Pore Space
WMO	World Meteorological Organisation

CHAPTER 1

Introduction

1.1. Introduction to the book

This book endeavours to explore the complex agri-environmental challenges European society faces and highlights some of the tools and techniques available to tackle these challenges, ranging from the strategic level of European policy making and formulation of interventions, down to practical on-farm solutions and mitigation options. It is not possible in a book such as this to delve deep into the science of each issue; instead it provides a comprehensive overview seeking more to provide an insight into the integrated nature and complexity of the agri-environment. The book seeks to cover much of western and middle Europe including Iceland and Cyprus but excluding Russia and seeks to consider all the cultural and political elements that affect the agri-environment. It should be noted that a significant proportion of the EU's agriculture and environment policy and regulatory framework has been adopted by European countries not in the EU, such as Iceland and Switzerland.

The book has five themes through which these challenges are explored:

- Chapter 2: Atmospheric pollution and climate change
- Chapter 3: Biodiversity
- Chapter 4: Water
- Chapter 5: Resources
- Chapter 6: Cultural heritage.

Each chapter has broadly the same structure. The first part considers the background and nature of the issues and challenges, looking at the fundamental causes, the key threats and the current scientific understanding. Having set the scene, the second part proceeds to explain how the problems are being tackled across Europe by governments and major organisations, that is, taking a 'top-down' perspective. This part explores how policy, legislation, instruments and interventions have developed over time and how current frameworks work. The final part looks at the issues taking the opposite 'bottom-up' perspective and discusses on-the-ground activities that can be taken by farmers, land owners and rural communities to manage and mitigate the problems. This part includes descriptions of practical on-farm management options including management planning; land, soil, water, livestock, crop, infrastructure and habitat management; and technological and biotechnology solutions.

Many of the five themes are inherently connected, thus a critical and emerging challenge is to develop integrated solutions and approaches, where multiple challenges and objectives are simultaneously addressed. This forms the topic of Chapter 7. The final chapter (Chapter 8) will explore the knowledge and technology that may be developed in the near future that could be utilised to help tackle current issues; and will scan the horizon for other agri-environmental issues and challenges that may lie ahead. It also briefly considers how policy might adapt and develop to address emerging issues.

This first chapter introduces and provides a background to the agri-environmental topic, including the issues and challenges faced by society. It then provides an overview of some elements that are common to multiple issues and thus chapters in this book, so that they do not need to be repeated within each chapter. This includes key policies and interventions such as the European Union's Common Agricultural Policy and various other EU regulations and directives, and key biological, geological, chemical and physical cycles and processes, such as the carbon and nitrogen cycles.

1.2. Background

Although many parts of Europe, particularly the north-west may be perceived as being highly urbanised, in reality 95 per cent of the region (409 Mha; Hart *et al.*, 2013) is defined as being rural. Of this, 165 Mha (38% of the EU; Hart *et al.*, 2013) may be forest, but 191 Mha (45% of the EU; Hart *et al.*, 2013) is under agricultural production of one sort or another, 107 Mha (25% of the EU) being cropland and 84 Mha (20% of the EU), grassland (Hart *et al.*, 2013). The historical dominance of this land use means that much of Europe's countryside has been shaped by agriculture over many centuries, and as a result the industry has a central role to play in both the maintenance of the

rural environment (particularly less intensive forms of agriculture), and in achieving the broader environmental objectives of European governments, including the EU as a whole and its individual Member States (MSs) (OECD, 2008; EC, 2006; EC, 2013a). The agricultural industry is essential for the delivery not only of a wide range of vital and/or desirable ecosystem services, not least of which of course is food, but also fibre and energy products, recreational facilities, water resources, flood regulation and many others (Hart *et al.*, 2013). In addition, although farming generally only accounts for a small proportion of most European countries' economies, for example approximately 1.7 per cent of the EU's GVA (gross value added; Eurostat, 2014), it is the foundation of many rural economies and communities, and therefore essential in maintaining social cohesion (EC, 2013a).

The environmental impacts of agriculture have, for a number of years, been the focus of considerable public attention and regulatory control. Some of these have been widely reported (EEA, 2007; Zalidis *et al.*, 2004), including the pollution of surface and groundwaters resulting from the use of nitrogen fertilisers (Skinner *et al.*, 1997) and pesticides (Warren *et al.*, 2003), the declines in biodiversity such as farmland bird populations (Chamberlain *et al.*, 2000) and honeybees (Goulson *et al.*, 2008), and the emission of greenhouse gases (GHGs) from ruminant livestock (O'Mara, 2011), whilst others such as pollution from the use of veterinary products (Kay *et al.*, 2005) are less widely understood outside the scientific community. There are, therefore, a variety of linked systems at work in such environments, some of which may be complementary whilst others are contradictory, and not always in a consistent way. Intensive agriculture for example, is often questioned due to the negative environmental impacts which may result, including pollution of surface and groundwaters, increased GHG emissions (Cooper *et al.*, 2009; Woods *et al.*, 2010; Rey Benayas and Bullock, 2012), damage to soils with the subsequent potential for increased erosion (Pimentel and Kounang, 1998; Louwagie *et al.*, 2009), increased demand for water for irrigation (Hart *et al.*, 2013) and the reduced biodiversity which often seems to accompany specialisation in a single form of production, for example arable (Stoate *et al.*, 2001). Nevertheless, it cannot be denied that intensification of production has led to significant increases in the output (food, fuel and fibre) produced by the agricultural industry (Donald *et al.*, 2001; EC, 2006), indeed, intensification is generally defined as being an increase in productivity per unit area of land with 'sustainable intensification' referring to the goal of achieving 'more from less', that is, increasing productivity without (or at least not to the same degree) increasing the use of natural resources. Thus, agricultural intensification has an essential role to play in maintaining and/or improving food security (both within Europe and beyond) in a world in which the global population is expected to reach 9.6 billion by 2050 (UN, 2014). In addition, intensification may also result in a number of environmental benefits, including (in some cases) reductions in GHG emissions and efficiencies in the use of natural resources such as fossil fuels and water, and it is even argued that increased production in one area of land may free up land for biodiversity through the process of 'sparing' (Burney *et al.*, 2010; Garnett, 2010; Phalan *et al.*, 2011), although the evidence for this is inconclusive (Rudel *et al.*, 2009).

This serves to illustrate the complexity of the interactions at work between agricultural production and the environment in which it operates, something which is further added to by the heterogeneous nature of European agriculture. Although EU cropland is dominated by arable agriculture (which includes cereals, pulses, oilseeds, and root and leafy vegetables, amongst others), in Mediterranean countries permanent crops may be important, and similarly grassland may be anything from intensive lowland pasture to unimproved and upland grazing (Hart *et al.*, 2013). In addition, the regional make-up of the industry can vary significantly both between and within European countries. The proportion of land given over to crops, for example, ranges from 48 per cent of land in Denmark (2.1 Mha) to as little as 4 per cent in the Republic of Ireland (350,000 ha), which in contrast has the highest proportion of grassland (64% or 4.5 Mha). It is, therefore, the goal of agricultural and agri-environment policies, and their delivery on the ground, to strike the most advantageous balance between the various elements of this multifaceted system – albeit what constitutes the 'most advantageous balance' may be interpreted differently by different people (farmers, environmentalist, consumers, regulators, and so on).

1.3. The policy landscape

The agri-environment is a highly complex entity with mankind being not only a creator of the problems but also having the power to prevent, manage and mitigate them. The complexity of the system means that a problem in one area is rarely, if ever, an isolated issue. It invariably influences the quality and/or function of other areas. Air pollution, for example, does not just affect air quality but can cause water and soil pollution via pollutant deposition, damage habitats and biodiversity and exacerbate climate change, as well as potentially affecting crop productivity. If the impact is not direct, it can be secondary or even tertiary. A single plant species may provide food or habitat for a 'seemingly' insignificant insect, but if its populations diminish then the insect's predators will have a shortage of food and their populations will also drop, potentially causing a chain reaction. If the insect provided other ecosystem services then these will also be less productive. For example, if the insect was a pollinator, then other plant species pollinated by that insect may also fail and it soon becomes obvious that the insect is far from insignificant.

As is the case with many policy areas, there are many players involved. This is, perhaps, more so when it comes to protecting the environment as pollution does not remain in one location and does not recognise national boundaries. Airborne pollution can travel extremely long distances when in the lower atmosphere and become deposited in locations far beyond its source point. Therefore, several international organisations, such as the United Nations, have debated and promoted multilateral conventions and treaties to tackle a particular issue such as climate change and the loss of biodiversity. Signatories to these agreements, be they individual countries or whole communities such as the EU, must then take action to ensure that their obligations are met and so these conventions and treaties subsequently become drivers for policies, legislation and other interventions and initiatives in a particular area to be developed and enacted.

Policies to manage and mitigate these complex and integrated issues are, out of necessity, also complex and integrated, requiring a whole-system, holistic perspective. This complexity is even more apparent considering environmental and ecosystem services protection must be achieved whilst also ensuring we can deliver a plentiful supply of affordable, nutritious food as well as the other agricultural goods and services we need (green fuel, fibre, pharmaceuticals, biochemicals, leisure facilities, etc.). Therefore, policy development is difficult to get right, especially in a large area such as Europe, where much of it (i.e. the EU) is covered by a single, harmonised approach but where the geology, landscapes, climate, soils, habitats and wildlife are hugely diverse. Indeed, it is only recently that an integrated approach to agri-environmental policy has begun to emerge.

With respect to the agri-environment in Europe there are many important policy instruments, several of which cover more than one of the themes addressed within this book (see Table 1.1) such as the Nitrates Directive and the Water Framework Directive. Although the scope of this book is Europe, this continent is dominated by the EU and so much of the policy is EU-centric; this is especially the case as those countries outside the EU but in the European Economic Area (i.e. Iceland, Liechtenstein and Norway) are obliged to adopt the EU's directives. The EU's Common Agricultural Policy (CAP) is undoubtedly the most influential and most complex of them all with repect to both the land area to which it applies and the financial cost of its implementation.

The influence of the CAP cannot be underestimated. Introduced in 1962 by the Treaty of Rome (European Community, 1957), it has been the single most important element of common policy. It is the cornerstone of policy relating to agriculture and rural areas and accounts for over 40 per cent of the total EU budget. Initially, the main objectives, as defined in Article 39, were to increase agricultural productivity, improve the standard of living for the agricultural community and address various market concerns (prices, stability and product availability). These objectives were achieved by implementing a system of agricultural subsidies (farm payments) and other incentive programmes and initiatives largely linked to productivity.

However, during the intervening decades it has evolved significantly from this initial

Table 1.1: Overview of key EU policies, directives and strategies

Year	Common name	Atmospheric pollution and climate change	Biodiversity	Water	Resources	Cultural heritage
1962	Common Agricultural Policy	✓	✓	✓	✓	✓
1979	Birds Directive		✓			
1991	Nitrates Directive	✓	✓	✓	✓	
1992	Habitats Directive		✓			
1992	Natura 2000		✓			✓
2000	Water Framework Directive		✓	✓	✓	
2007	2020 Climate and Energy Package	✓			✓	
2011	Biodiversity Strategy		✓			
2013	Strategy on Adaptation to Climate Change	✓			✓	
2014	2030 Climate and Energy Framework	✓			✓	

emphasis to encompass a much broader range of policy objectives (Latacz-Lohmann and Hodge, 2003). In the 1970s, there was increasing recognition of the environmental problems caused by agriculture including pollution from the excessive use of fertilisers and pesticides and damage to the country-side and biodiversity resulting from changes in the structure of the farming industry (e.g. hedge removal). Initially, such issues were addressed by specific legislation prohibiting certain activities, for example in relation to nitrate pollution (Latacz-Lohmann and Hodge, 2003), and often at a national level until made mandatory across the EU through legislation such as the Drinking Water Directive (European Council, 1998) and the Nitrates Directive (European Council, 1991) (Latacz-Lohmann and Hodge, 2003); however, attempts to produce such 'draconian' legislation covering other environmental issues met with considerable resistance from farmers, who saw it as an unwanted interference in what they could do with

their own land (EC, 2013a). Consequently, it was identified that broad environmental issues would need to be addressed through voluntary, incentive-led approaches, of the sort now seen in Europe's agri-environment schemes (Latacz-Lohmann and Hodge, 2003).

By the 1980s, the CAP had become a victim of its own success. Although it had been outstandingly successful in terms of increasing food production, it was now viewed as being one of the key drivers for environmental degradation in rural areas (Latacz-Lohmann and Hodge, 2003), and at the same time, the high cost of supporting overproduction, which led to the so-called food 'mountains' and 'lakes', was seen as excessive. As a result, integration of environmental goals into agricultural policy began with the pace of change accelerating into the 1990s such that the CAP gradually evolved to include an ever greater agri-environmental component (EC, 2013a). The CAP reforms of 1992 marked a major shift in emphasis from the price support mechanisms which had dominated up to that point, to a greater concentration on direct aid payments, in a process known as 'decoupling', and the introduction of a number of specific measures intended to encourage 'environmentally-friendly farming', including the Agri-Environmental Regulation, which made it mandatory for EU MSs to have an agri-environment scheme (EEC, 1992; Zalidis *et al.*, 2004; Latacz-Lohmann and Hodge, 2003).

The CAP's approach to encourage the use of more environmentally sound farming practices via farm subsidies has been adopted by other, non-EU European countries, albeit in a less complex way, to the extent that it is largely the norm across the European area. For example, Norway places significant importance on ensuring the well-being of rural communities, particularly because many of these are in remote locations with harsh and challenging environments. According to the OECD (OECD, 2015) Norway's agricultural sector is one of the most heavily subsidised of all the OECD areas. This is not just through farm payments but also through special tax arrangements. However, in order to receive these benefits farmers must comply with good agricultural practice and participate in subsidy schemes that are not linked to production but, rather, have environmental objectives such as the National Environment Scheme (Nasjonalt miljøprogram) administered by the Norwegian Agricultural Authority. The main difference between this scheme and that of the EU agri-environment schemes is that it does not prescribe certain practices and activities but, instead, prohibits practices seen as environmentally damaging such as removing habitats, for example water bodies and stone walls, and using pesticides on field margins. In Switzerland, since 1999, all direct payments have been based on stringent proof of ecological performance

(cross compliance). This ensures that ecological methods are used throughout the country. Iceland's approach is quite different although its environment is equally harsh, if not more so. Farm subsidies are not linked directly to environmental objectives but part of the payment is tied to 'quality management' which includes sustainable land use. This is mainly to help control the country's severe soil erosion and land degradation problems.

The CAP has continued to evolve since the 1992 reforms. The Treaty of Amsterdam (1997) reaffirmed the EU's commitment to sustainable development (Article 6), and was soon followed, at the Cardiff European Council in 1998, by the start of what became known as the Cardiff integration process, which required the various arms of the EU to develop comprehensive strategies to integrate environmental issues into their respective areas of concern. A requirement which the European Commission responded to in its communication 'Directions towards sustainable agriculture' (EC, 1999) that identified the reforms to be made to the CAP as part of the EU-wide Agenda 2000 legislative package. This was the first major period of reform since 1992, and established that the CAP should not only improve the competitiveness of EU agriculture, guarantee food safety and quality, stabilise farm incomes and support the competitiveness of rural areas across the European Union, but also provide environmental benefits and enhance the rural landscape (Zalidis *et al.*, 2004). At this time, the CAP was reorganised into two distinct 'pillars', the first dealing with market-related policy and the second sustainable development policy, and although agri-environment schemes fall under Pillar 2, there is also a requirement to take environmental concerns into account in the implementation of Pillar 1 measures, and as such they work together

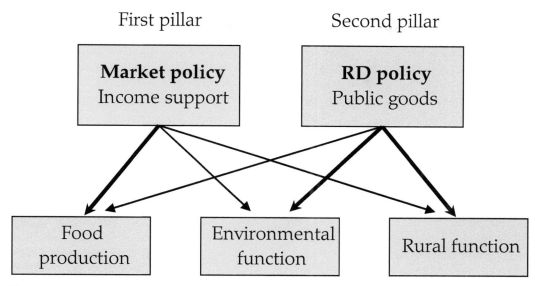

Figure 1.1: Complementary nature of the two pillars of the CAP
(Source: European Communities, 2008)

to provide the environmental improvements required in order to meet a range of policy objectives including those relating to climate change, water management and use, and biodiversity (see Figure 1.1). Under Pillar 1, payments were increasingly decoupled from production, and the requirements of cross compliance meant that in order to obtain the full rate of direct income support, a basic level of environmental performance had to be delivered (based on the 'polluter pays' principle); whilst under Pillar 2, producers were rewarded for voluntarily delivering a higher level of performance (the 'provider gets' principle) and EU MSs were required to establish 'Codes of good agricultural practice' (Choe and Fraser, 1999).

In 2003, an agreement on a marked reform of the CAP (EC, 2003) saw the transfer of funding from Pillar 1 to Pillar 2, and it was at this point that the decoupling of production and payment became a significant element of policy, and in so doing many of the incentives for intensifying production were reduced or removed (EC, 2006;

EC, 2013b). In addition, cross compliance (requiring beneficiaries of direct payments to maintain all agricultural land in good agricultural and environmental condition) and modulation became compulsory, with the latter increasing the budget available for measures under Pillar 2 (EC, 2000). EU rural development policy, including that relating to agri-environmental policy, was further developed though Council Regulation (EC) No 1698/2005 (European Council, 2005), which established policy for the 2007–2013 period. Although this continued in much the same vein, it changed the overall approach to developing Pillar 2 – rural development programmes (see Figure 1.1). First, it established a single fund to support rural development, the European Agriculture Fund for Rural Development (EAFRD), with a single set of financial rules and constraints, streamlining administration. It also focused attention on a set of three core policy objectives, with a thematic axis dedicated to each, of which Axis 2 is of most relevance to agri-environmental policy, together with an overarching methodological axis dedicated to the LEADER approach, namely:

- Axis 1: to improve the competitiveness of the agricultural and forestry industries – which receive a minimum of 10 per cent of overall funding
- Axis 2: to support land management and improving the environment – minimum 1.25 per cent
- Axis 3: to improve the quality of life of rural populations and support economic diversification – minimum 10 per cent.

This system was further refined by the subsequent CAP Health Check that was agreed in 2008 and which introduced a number of amendments with the aim of simplifying it and freeing up further resources for addressing environmental challenges. In particular, funds were moved from direct payments in favour of rural development, including support for environmental improvements, such that today Pillar 2 of the CAP accounts for just over 20 per cent of the total CAP budget (€12.4 billion in 2007), which in turn absorbs just under half of the European Union's total annual budget (€55.1 billion of €126.5 billion in 2007; European Communities, 2008).

CAP reform for the period 2014–2020 continues the process of reform started back in the early 1990s. In its communication entitled 'The CAP towards 2020: Meeting the food, natural resources and territorial challenges of the future' (EC, 2010a), a period which ties in with Europe 2020, the EU's growth strategy for that period (EC, 2010b), the Commission set out its vision for the 2013 to 2020 period, with a number of implications for the future of rural development policy, and this has formed the basis for the agreed policy. Under this, the overall two-Pillar structure is maintained, as (in broad terms) are the three main axes of rural development (see Figure 1.2). However, the

links between Pillars are strengthened with the aim of producing a more holistic policy (EC, 2013b), and the planned changes have been heralded by some as the 'greening' of the CAP. In particular, as well as cross compliance (the compulsory basic level of environmental provision), under Pillar 1 a 'greening payment' has been introduced in which 30 per cent of direct payments are linked to the provision of certain environmental services (maintenance of permanent grassland, ecological focus areas and crop diversification). This too is compulsory, and therefore is intended to introduce practices that are beneficial for the environment and climate on most of the utilised agricultural area (EC, 2013b). Under Pillar 2, 30 per cent of the budget for each rural development programme must be reserved for voluntary measures that are beneficial for the environment and climate change including agri-environmental/climate measures (the new CAP specifically refers to agri-environment climate rather than just agri-environment

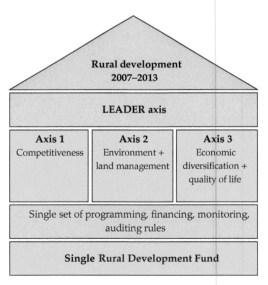

Figure 1.2: The three thematic axes and the methodological (LEADER) axis of the RDP (Source: European Communities, 2008)

payments; European Parliament and Council, 2013), organic farming, Areas of Natural Constraint, Natura 2000 areas, forestry measures and other investments which are beneficial for the environment or climate, and as such the extent to which agri-environment and other related measures are to contribute to farm incomes is to be considerable.

1.4. Key biological, geological, chemical and physical cycles and processes

1.4.1. Introduction

Within agri-environmental science there are a number of fundamental processes concerned with the cycling of biological, chemical and geological elements through the Earth and its atmosphere. Cycles can be gaseous and/or sedimentary and include the principal components of life and all ecosystems – nitrogen, oxygen, carbon and water.

1.4.2. Carbon cycle

The carbon cycle is summarised in Figure 1.3 (data derived from NASA Earth Observatory, 2011). It is a complex process which describes how carbon moves from the atmosphere, through animals and plants and then returns to the atmosphere. From the perspective of agricultural production, fossil fuels and land use change are key mechanisms in determining net carbon emissions to the atmosphere.

Carbon is the fourth most abundant element in the universe, and is essential to life on Earth. All living cells contain carbon compounds such as fats, carbohydrates and proteins. Additionally, carbon is part of the ocean, our atmosphere as the gas, carbon

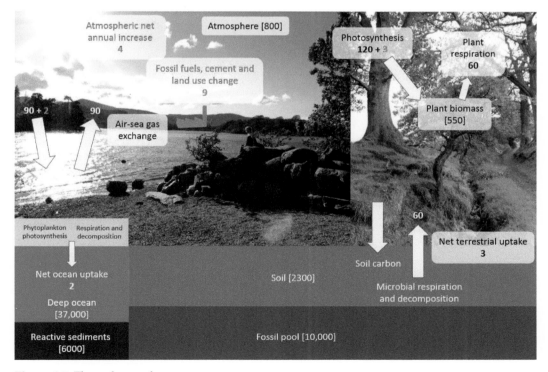

Figure 1.3 The carbon cycle
Key: [ ] = stored carbon as (Gt); red text = anthropogenic emissions
(Photos courtesy of: Dr Doug Warner)

dioxide, and is found in solid materials as diverse as limestone (also known as calcite, calcium carbonate, $CaCO_3$), wood, plastic, diamonds and graphite.

Carbon does not remain in one place but moves, in its many forms, between the atmosphere, oceans, biosphere and geosphere as described by the carbon cycle, illustrated in Figure 1.3. This cycle consists of several storage carbon reservoirs and the processes by which the carbon moves between reservoirs. Carbon reservoirs include the atmosphere, the oceans, vegetation, rocks and soil. If more carbon enters a pool than leaves it, that pool is considered a *net carbon sink*. If more carbon leaves a pool than enters it, that pool is considered *net carbon source*. The global carbon cycle, one of the major biogeochemical cycles, can be divided into geological and biological components. The geological carbon cycle operates on a timescale of millions of years, whereas the biological carbon cycle operates on a timescale of days to thousands of years.

The geological component of the carbon cycle refers to processes related to carbon within minerals and rocks and includes weathering, dissolution, precipitation of minerals, burial and subduction, and volcanic eruptions. In the atmosphere, carbonic acid forms by a reaction with atmospheric carbon dioxide and water. The newly formed carbonic acid falls to Earth as rain where it reacts with minerals present on the surface, slowly dissolving them into their component materials and molecules through the process of chemical weathering. These component materials are then transported by surface waters eventually to the ocean, where they precipitate out as minerals like limestone to eventually create rock formations. Carbon materials deposited on the seabed are pushed deeper into the Earth by tectonic forces where they heat up and eventually melt and rise back up to the surface, where they are

released as carbon dioxide and returned to the atmosphere. This return to the atmosphere can occur violently through volcanic eruptions, or more gradually in seeps, vents and carbon dioxide-rich hot springs.

The biological part of the carbon cycle is, perhaps, the most important aspect of the cycle for agri-environmental science. The key processes that govern the movement of carbon between land, ocean and atmosphere are photosynthesis and respiration. Virtually all multicellular life on Earth depends on the production of sugars from sunlight and carbon dioxide (photosynthesis), and the metabolic breakdown (respiration) of those sugars to produce the energy needed for movement, growth and reproduction. Plants take in carbon dioxide from the atmosphere during photosynthesis, and release it back into the atmosphere during respiration.

On land, the major exchange of carbon with the atmosphere results from photosynthesis and respiration. During daytime in the growing season, leaves absorb sunlight and take up carbon dioxide from the atmosphere. At the same time, plants, animals and soil microbes consume the carbon in organic matter and return carbon dioxide to the atmosphere. Photosynthesis stops at night when the sun cannot provide the driving energy for the reaction, though respiration continues. This kind of imbalance between these two processes is reflected in seasonal changes in the atmospheric carbon dioxide concentrations. During winter in the northern hemisphere, photosynthesis ceases when many plants lose their leaves, but respiration continues. This condition leads to an increase in atmospheric carbon dioxide concentrations during the northern hemisphere winter. With the onset of spring, however, photosynthesis resumes and atmospheric carbon dioxide concentrations are reduced.

In the oceans, phytoplankton use carbon to make shells of calcium carbonate. The shells settle to the bottom of the ocean when phytoplankton die and are buried in the sediments. The shells of phytoplankton and other creatures can become compressed over time as they are buried and are often eventually transformed into limestone. Additionally, under certain geological conditions, organic matter can be buried and over time form deposits of the carbon-containing fuels, coal and oil. It is the non-calcium-containing organic matter that is transformed into fossil fuel. Both limestone formation and fossil fuel formation are biologically controlled processes and represent long-term sinks for atmospheric carbon dioxide.

As discussed in Chapter 2 and in section 1.4.6 below, human activities are significantly altering the natural carbon cycle. Activities such as the burning of fossil fuels, deforestation and intensive farming have contributed to a long-term rise in atmospheric carbon dioxide leading to climate change. Agricultural practices and the way land is used also affect the amount of carbon stored in plant matter and soil and consequently the amount of carbon in the atmosphere.

1.4.3. Nitrogen cycle

The second key cycle with respect to agricultural sources is the nitrogen cycle (Figure 1.4).

Nitrogen, like carbon, is one of the main chemical elements that all living organisms depend upon. It is a major component of chlorophyll, the most important pigment needed for photosynthesis, as well as amino

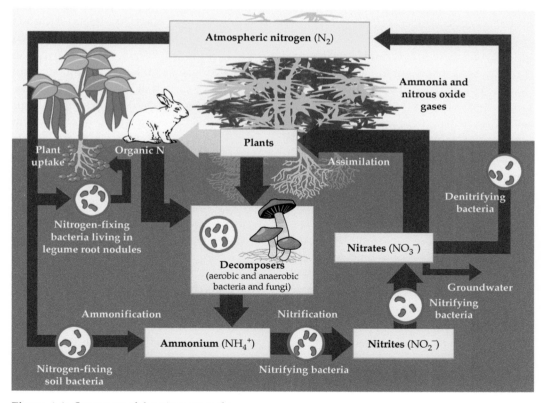

Figure 1.4: Summary of the nitrogen cycle
(Image courtesy of: Johann Dréo [adapted])

acids, the key building blocks of proteins. Although nitrogen is very abundant in the atmosphere, accounting for 78.09 per cent of dry air by volume, in its atmospheric form (dinitrogen gas) it is not accessible to most biological species and ecosystems. Nitrogen does exist in other forms available to plants and ecosystems but these are much less abundant and often in short supply; indeed, a shortage of nitrogen will often restrict productivity in many ecosystems, particularly plants/crops. It is only when nitrogen is converted from the atmospheric dinitrogen gas into ammonia (NH_3) that it becomes available to plants.

As well as dinitrogen gas and ammonia, nitrogen exists in many different forms, including both inorganic (e.g. ammonia, nitrate) and organic (e.g. amino and nucleic acids) forms. Nitrogen undergoes many different transformations in the ecosystem, changing from one form to another as organisms use it for growth and, in some cases, energy. The natural cycle of death and decay of plant and animal tissue, combined with the excretion of metabolites by animals liberates nitrogen contained within the species into the environment. This occurs in all ecosystems. The major transformations of nitrogen are nitrogen fixation, nitrification, denitrification and ammonification.

Nitrogen fixation is the process of converting dinitrogen gas into biologically available nitrogen. Dinitrogen gas is a very stable compound and a large amount of energy is required to convert it to plant available forms. As a result, only a select group of single-celled organisms called prokaryotes are able to carry out this energetically demanding process. These prokaryotes include aquatic organisms, such as cyanobacteria, free-living soil bacteria, such as *Azotobacter*, bacteria that form associative relationships with plants, such as *Azospirillum*, and most

importantly, bacteria, such as *Rhizobium* and *Bradyrhizobium*, that form symbioses with legumes and other plants (Postgate, 1982). Prokaryotes utilise the enzyme nitrogenase to catalyse the conversion of atmospheric dinitrogen to ammonia which plants then use to produce nitrogenous biomolecules. Although most nitrogen fixation is carried out by prokaryotes, some nitrogen can be fixed abiotically by lightning or certain industrial processes, including the combustion of fossil fuels.

Nitrification is another important step in the global nitrogen cycle that converts fixed nitrogen to nitrate such that it can be absorbed by plants and used to produce amino acids, nucleic acids and chlorophyll. Most nitrification occurs aerobically and is also carried out exclusively by prokaryotes. Nitrification has two discrete steps, each carried out by a specific type of microorganism. The first step is the oxidation of ammonia to nitrite (NO_2^-), which is carried out by microbes known as ammonia-oxidisers. The second step in nitrification is the oxidation of nitrite to nitrate (NO_3^-). This step is carried out by a completely separate group of prokaryotes, known as nitrite-oxidising bacteria. Some of the genera involved in nitrite oxidation include *Nitrospira*, *Nitrobacter*, *Nitrococcus* and *Nitrospina*. For complete nitrification, both ammonia oxidation and nitrite oxidation must occur.

Denitrification is the process that converts nitrate to nitrogen gas, thus removing bioavailable nitrogen and returning it to the atmosphere. Dinitrogen gas is the ultimate end product of denitrification, but other intermediate gaseous forms of nitrogen exist such as nitrous oxide (N_2O), a greenhouse gas. Denitrification is an anaerobic process that occurs mostly within soils, sediments and anoxic zones (i.e. areas depleted of dissolved oxygen) in large water bodies (lakes and oceans). Like nitrogen fixation, denitrification is carried out by prokaryotes. Some denitrifying bacteria include species in the genera *Bacillus, Paracoccus* and *Pseudomonas.* Denitrification is a significant process as it removes fixed nitrogen from the ecosystem and returns it to the atmosphere in a biologically unavailable form. This is particularly important in agriculture where the loss of nitrates in fertiliser is detrimental and costly. However, denitrification in wastewater treatment plays a very beneficial role by removing unwanted nitrates from the wastewater effluent, thereby reducing the chances that the water discharged from the treatment plants will cause undesirable consequences such as algal blooms.

Ammonification is the process by which organic nitrogen (e.g. in amino acids, DNA) in animal waste, dead animals and plants is returned to the nitrogen cycle. Various fungi and prokaryotes decompose this material and release inorganic nitrogen back into the ecosystem as ammonia. The ammonia then becomes available for uptake by plants and other microorganisms for growth.

Although a requirement of plant growth, any surplus nitrogen within the soil will proceed along the pathways highlighted above (Machefert *et al.,* 2002; Oenema *et al.,* 2005; Smith and Conen, 2004; Smith *et al.,* 2008). Two important pollutants of air are potentially derived from the nitrogen cycle. These are nitrous oxide (N_2O) and ammonia (NH_3) (see Chapter 2).

Within the agricultural ecosystem, to ensure yields and crop quality, there is the frequent addition of supplementary nitrogen, in either inorganic or organic form as fertiliser, and increased inputs from animals via potentially high stocking rates coupled with supplementary feeding. Further, these additional inputs may be concentrated within relatively small areas depending on the farming system. Thus, farm activities have been exerting an ever-increasing

impact on the global nitrogen cycle. Human activities, such as fertiliser manufacturing and the use of fossil fuels, have also significantly altered the amount of fixed nitrogen in the Earth's ecosystems. In fact, some predict that by 2030, the amount of nitrogen fixed by human activities will exceed that fixed by microbial processes (Vitousek *et al.,* 1997). Increases in available nitrogen can alter ecosystems by increasing primary productivity and impacting carbon storage (Galloway *et al.,* 1994).

1.4.4. Phosphorous cycle

The phosphorous cycle is a little less complex, in comparison to the nitrogen and carbon cycles. Nevertheless, phosphorus is vital as it is needed by every living organism to grow and function. It is a major component of adenosine triphosphate (ATP), a small molecule that is responsible for energy management in cells, and is also one of the main components of DNA and RNA. Unlike carbon and nitrogen, most of the phosphorous on Earth is stored in soil and rocks in the form of phosphate minerals. Weathering causes it to break free from the rocks and wash away where it becomes mixed in the soil or dissolved in water. Plants then absorb the phosphorus from the ground or water and animals obtain the phosphorus by eating the plants and drinking natural waters. Phosphorus is returned to the environment when organisms die and decay. Organic phosphate is gradually released as inorganic phosphate or becomes incorporated into more stable organic materials and becomes part of the soil organic matter. The release of inorganic phosphate from organic phosphates is called mineralisation and is caused by microorganisms breaking down organic compounds. The activity of microorganisms is highly influenced by soil temperature and soil moisture. The process is most rapid when soils are warm and moist

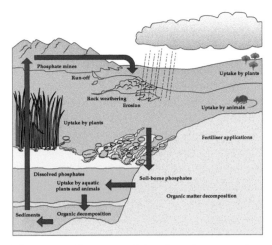

Figure 1.5: Summary of the phosphorous cycle

but well drained. Phosphate can potentially be lost through soil erosion and to a lesser extent to water running over or through the soil.

As is the case with the carbon and nitrogen cycles, farming practices influence the global phosphate cycle. As crop growth and productivity are dependent on phosphate, it is added to soils as fertiliser, often in combination with nitrogen. Phosphate fertiliser is produced from mined phosphate rock (Figure 1.5). That applied is taken up by plants or lost to the environment, for example via run-off.

1.4.5. Water cycle

Water exists in numerous places across the Earth, these primarily being oceans and seas, surface water bodies (lakes, rivers, ponds, etc.), frozen water bodies (glaciers, ice-sheets, ice-caps), below ground resources (aquifers) and as water vapour in the atmosphere. It also accumulates in plants and soil matrices. However, water is not static and there is continuous movement and exchange between these deposits governed by a number of processes that are collectively known as the water (or hydrological) cycle (Figure 1.6).

There are three main exchange mechanisms: evaporation, condensation and precipitation. During the evaporation process, water in the liquid form is converted to water vapour. Evaporation is driven by temperature (i.e. the process is faster at higher air temperatures) and to a lesser extent by wind. Water lost by evaporation from plants and soils is termed evapotranspiration. During the condensation process, the reverse happens and water vapour is converted to liquid water. High in the atmosphere, very small particulate matter acts as nuclei upon which water condenses and forms clouds. When condensation happens at ground level, fog and mists are formed. Condensation is also temperature-dependent. As water vapour cools, it reaches a limit, (also known as the dew point) and conversion to a liquid occurs.

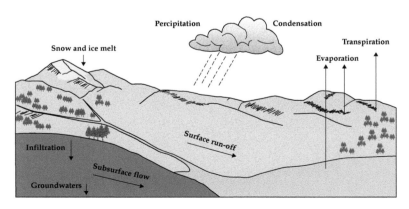

Figure 1.6: Summary of the water cycle

Air pressure will also influence the rate of condensation. The third process, 'precipitation', occurs when water condensate in the atmosphere accumulates to the point when it is too heavy and unstable and so falls to Earth as rain, snow or hail. This point occurs when the pressure caused by the upward air movement is less than that of the downward pressure exerted by the cloud.

In practice, a cycle of movement occurs whereby the sun heats up liquid water on the earth which evaporates and moves to the vapour state. Water vapour is carried on air currents to the atmosphere where the air is much cooler, causing it to condensate into clouds. Cloud movement and changing conditions eventually cause the clouds to release the liquid water they hold which is then deposited back to earth. Some of this precipitation will fall as snow and accumulate in the colder regions of the planet, forming the ice packs and glaciers. However, most precipitation falls back into oceans or onto land. Some of the land deposited water will run off and flow into surface water bodies or infiltrate into the ground to be absorbed by soils and plant material, subsequently to be lost by evaporation. That not absorbed

in this way will eventually drain into and replenish aquifers.

1.4.6. The greenhouse effect

The greenhouse effect is named due to its perceived similarity to the effect of solar radiation passing through glass and warming a greenhouse. This is actually poorly named, as the physical processes inside a greenhouse are fundamentally different to the processes in the atmosphere. A greenhouse works by reducing airflow, isolating the warm air inside the structure so that heat is not lost by convection. However, with regard to our atmosphere, visible light, as well as ultraviolet light, travels from the Sun to the Earth and some of this energy (about one third) is reflected back to space by the upper atmosphere, and the remainder travels to the surface of the Earth where it is absorbed. The Earth then radiates infrared energy out into space but some of this energy gets 'captured' by a layer of gaseous compounds known as greenhouse gases (GHGs). Some of the energy captured by these gases is sent into space, but some of it gets deflected back to Earth. This process is illustrated in Figure 1.7. Solar radiation

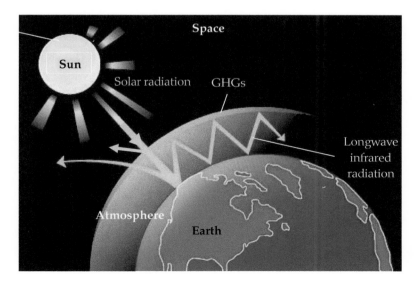

Figure 1.7: The greenhouse effect

passes through the atmosphere (~343 Watts per m^2); some of this is reflected by the atmosphere and the Earth's surface (~103 Watts per m^2), resulting in a net incoming radiation of 240 Watts per m^2. Solar energy is absorbed by the Earth's surface, warms it, and is converted into heat resulting in the emission of longwave infrared radiation back to the atmosphere. Some of the infrared radiation passes through the atmosphere and is lost into space, and some is absorbed by GHG molecules and re-emitted, warming the Earth's surface and the troposphere.

1.5. Impact assessment frameworks, concepts and techniques

1.5.1. Introduction

In addition to understanding some of the key processes, agri-environmental management also draws upon numerous environmental impact assessment frameworks and techniques, such as those used in environmental risk assessment (ERA) and life cycle assessment (LCA). The following sections within this chapter provide an overview of these frameworks and a description of the concepts and techniques that are utilised within agri-environmental management, and thus referred to in numerous places throughout this book.

1.5.2. Environmental risk assessment (ERA)

1.5.2.1. Overview

Environmental Risk Assessment (ERA) is an 'objective' and scientific process to identify, characterise and quantify (where possible) risks to the environment. Generally, these risks are negative effects and impacts, but in theory the process can be equally applied to positive effects and impacts. It tends be part of a broader framework of environmental risk

management (ERM). ERA is often context specific and approaches may vary between industry sectors and between industry and regulators, and there are probably as many different models, diagrams, concepts and ideas as there are risk assessments. However, there are some fundamental aspects that are common to all.

First, risk assessment is generally associated with a hazard, something negative to be avoided if possible. However, a hazard can only be defined in the context of a receptor, for example some chemicals are only toxic to some organisms, and it is exposure that determines whether the hazard is realised, that is, does an organism become exposed to the chemical? This process is often referred to as the source–pathway–receptor model (see Chapter 1.5.2.2), where there is the 'source' of the hazard, and a 'pathway' for a 'receptor' to become exposed. If we then calculate probable exposure (pathway) and assess potential impact (e.g. toxicity), then an assessment of the risks can be made.

ERA is generally split into qualitative and quantitative risk assessment, with the latter being the more complex and costly to undertake, but providing a more accurate assessment. ERA also tends to be a tiered and iterative process, usually starting with simple qualitative approaches as a means of scoping, and then moving onto more complex quantitative processes. The process may then be repeated to fine-tune the assessment.

Although there is a tiered approach to risk assessment, there can be some common steps within each tier:

- **Hazard identification and analysis:** used to identify all activities within the specified boundaries that may pose hazards to humans, animals or the environment and, therefore, need to be included in the assessment. Hazards can only be defined within the context of an end receptor (something that is harmed by being exposed to the hazard). Event tree or fault tree analysis is a common and accepted means of identifying hazards.

- **Exposure assessment:** used to identify the likely pathways and extent of exposure to the previously identified hazards. There are a number of aspects to consider: a clear definition of the hazard; the characteristics of the local environment, both in terms of how this affects the hazard and also what specific receptors may be exposed and their sensitivity to the hazard; the behaviour of the hazard; and specific 'dose–response' relationships. The effect or consequences of the hazard/ exposure also needs to be assessed, for example does it affect the health of the receptor, that is, affect its reproductive capability (use of 'No Observable Effect Concentration'; NOEC), or does it kill the receptor and if so, what proportion of the population might it affect (use of 'lethal dose' or 'concentration'; LD_{50} or LC_{50})?

- **Risk estimation:** combines the outputs from the hazard and exposure assessments to provide an indication of the likely risk. There are many different techniques ranging from simple matrices of hazard magnitude and likelihood, through to more sophisticated approaches using scoring, ranking and weighting methods and/or techniques to quantify risks.

- **Risk communication:** uses techniques to convey the risks that have been estimated to various stakeholders. In many respects this is a field of study of its own, especially with respect to risk perception and/or communicating risks to the public. However, there are a number of basic aspects to consider including: the target audience/purpose of the risk assessment; avoiding unrealistic levels of precision in risk estimates; communication of data quality; and communication of uncertainty or unknown risks.

- **Risk management/options appraisal:** Having identified and communicated the risks, this step involves identifying options to reduce or mitigate the risks. It is usually combined with a cost–benefit analysis in order to identify options that have the highest mitigation potential for the least cost.

1.5.2.2. Source–pathway–receptor (SPR) model

The Source–Pathway–Receptor (SPR) model (Figure 1.8) is a conceptual model used in environmental management and environmental risk assessments. It describes the causal links between an event that leads to a polluting emission and the receptor that is ultimately impacted upon.

The *source* part of the model refers to the identification of the origin of the pollutant. This could be, for example, an oil spill, vehicle exhaust emissions or application of

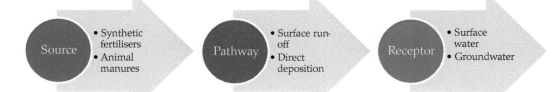

Figure 1.8: Source–Pathway–Receptor model for phosphorus pollutant

a nitrogen fertiliser. There may also be more than one source. For example, phosphorus pollution could come from animal manures, excess chemical fertiliser run-off, waste water, animal feeds and/or sewage.

The second part of the model, *pathway*, considers the various pathways the pollutant can use to move in the environment. This could be simply air, water or soil. It could move through one environmental media to another in a sequential manner or move through more than one media simultaneously. The pathway could also be described in more scientific terms such as, with respect to nitrogen pollution, leaching, volatilisation or denitrification. Other examples are movement with air currents, drainage and distribution via animals or birds.

The final step of the model, *receptor*, considers what and/or who might be affected. This could be described in simple terms as where the pollutant ends up, such as air or soil, but pollutants rarely remain in one place within the environment and so multiple receptors are common. For example, the nitrogen pollution for fertiliser applications might end up in the atmosphere, in ground- and surface waters and, in consequence, aquatic ecosystems will also be receptors.

Some applications of the model also add a further step and consider the *consequences* of the receptor receiving the pollutant. This can refer to the damage that may occur such as climate change, loss of biodiversity and increased incidence of respiratory problems. It may involve an assessment of how sensitive the receptor is to the pollutant.

The approach provided by the SPR model allows conclusions to be reached about the potential risks of the pollutant and it is also useful to aid the identification of intervention points – points where controls can be implemented effectively.

1.5.2.3. Toxicity: exposure ratios in risk assessment

Environmental risk assessments often involve an initial risk characterisation process whereby the predicted exposure (E) of an organism (e.g. human, animal, bird, fish, plant, beneficial insect, etc.) is estimated and compared to an appropriate toxicological end point (T, e.g. the LD_{50}, LC_{50}, NOEL, etc. – see Table 1.2) to give a Toxicity–Exposure Ratio (TER). The TER value is then compared with various levels of concern, often referred to as 'trigger values' which identify when further studies or regulatory actions are required. The basic process is shown in Figure 1.9. In practice, this will be modified depending on the organism and exposure route (e.g. oral, dermal, inhalation, etc.).

TERs are a relatively crude measure of risk and their use has been criticised (Calow, 1998; Bennett *et al.*, 2005; Mineau, 2005) but nevertheless they are still used extensively in regulatory first-tier environmental risk assessments.

Exposure is the most complex part of the assessment and this is often done using mathematical models. It requires a knowledge of the chemical's physico-chemical properties (e.g. solubility, volatility) and how it behaves in the environment. Data are required that will provide information on, for example, how quickly a chemical degrades in soils and water, and if it will bind with other substances (e.g. organic carbon) such that it is locked and cannot move freely to contaminate other environmental media such as groundwater. This type of information is then used as input data into various mathematical models to determine estimates of environmental concentrations, usually as time-series data.

There are many different mathematical models available for predicting the fate and environmental concentrations of chemicals, such as pesticides, in the environment.

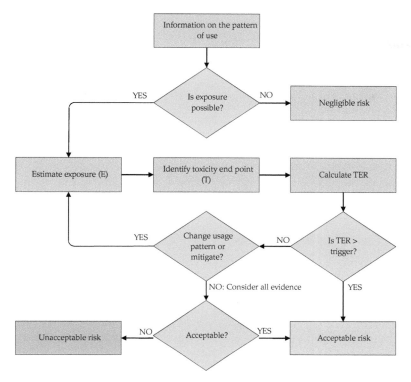

Figure 1.9: Basic flow chart of the TER risk assessment process

For example, the MACRO model has been used since the early 1990s to investigate the effects of macropore flow on soil hydrology and contaminant transport under transient field conditions. It is widely used in pesticide exposure and risk assessments for groundwater and surface water, for example in European pesticide registration procedures (Jarvis and Larsbo, 2012). The Pesticide Root Zone Model (PRZM) is a field-scale model that represents the environment as a single spatial compartment and estimates pesticide concentrations in run-off and soil (Carsel *et al.*, 1998). AgDrift is a model for estimating pesticide spray drift from aerial, ground boom and airblast applications (Teske *et al.*, 2002; Bird *et al.*, 2002).

The toxicity/ecotoxicity end-point data is normally taken from scientific literature, regulatory documents or databases (e.g. Lewis *et al.*, 2016). Data will vary according to species and often also according to gender and age. In regulatory processes it is not possible to assess every type of organism and so usually an indicator species or that which is the most sensitive is used as a surrogate. For example, the mallard duck (*Anas platyrhynchos*) or Japanese quail (*Coturnix japonica*) are often used as indicator species for birds, the rainbow trout (*Oncorhynchus mykiss*) or bluegill sunfish (*Lepomis macrochirus*) for fish and the water flea (*Daphnia magna*) for aquatic invertebrates.

Toxicological/ecotoxicological end points are defined as the concentration of a substance that causes an effect observed in a toxicity study. The effect is selected based on the health concern and so could be related to mortality, reproduction, growth, onset of a particular disease, such as cancer, or failure of a particular organ, such as the liver. Equally, it could be the highest concentration at which no effect is observed. A summary of the common end points is

Table 1.2: Common toxicological end points

Endpoint	Description
LD_{50}	Lethal Dose 50: The lethal dose for 50% of the tested population.
LC_{50}	Lethal Concentration 50: The lethal concentration for 50% of the tested population.
EC_{50}	Effective Concentration 50: The concentration at which 50% of its maximum effect is observed in the tested population.
NOEC	No Observed Effect Concentration: The concentration at which no adverse effect is observed in the test population.
NOEL	No Observed Effect Level: The level of exposure at which no adverse effect is observed in the tested population.
LOEC	Lowest Observed Effect Concentration: The lowest concentration found to cause an effect in the tested population.

given in Table 1.2 and illustrated graphically in Figure 1.10.

1.5.3. Life cycle assessment (LCA)

1.5.3.1.   Overview

Life Cycle Assessment (LCA) is a tool that can be used to logically and systematically evaluate the environmental effects of a product, process, or activity. More specifically, it aims to take a holistic approach covering all stages of the life cycle of a product, process or activity to develop optimal solutions. When a system is changed to reduce the impacts from single production processes, there is a danger that environmental burdens are shifted from one part of the system to another or one impact is improved at the expense of another – this is also known as displacement or pollution swapping. LCA attempts to address this issue by taking a holistic systems perspective. This inherently

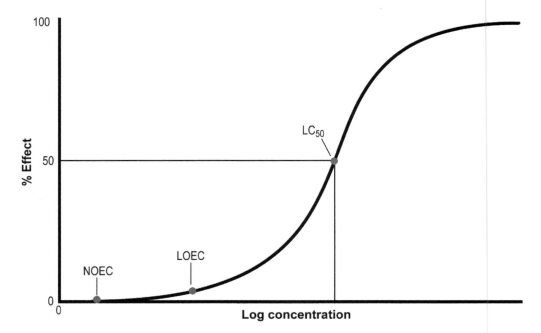

Figure 1.10: Toxicological end points

means that the assessment process is more complex, thus LCA also provides a logical and systematic framework for managing this complexity.

LCA was developed from energy analysis in the early 1960s and environmental criteria added during the 70s and 80s. The technique has had other names including life cycle analysis; resource and environmental profile analysis; product life cycle assessment; eco-balancing; substance flow analysis; and cradle-to-grave assessments. It is not ecological footprinting or whole-life costing – these are different tools using different techniques for different contexts. The first code of practice was developed by SETAC in 1993 and the ISO standards emerged in 2006 (the main ones being ISO 14040 and 44:2006).

There are four key stages in an LCA:

1. Goal and scope definition: clearly defining the purpose and objectives of the LCA; the functional unit, system boundaries; data requirements; impact assessment methods and any assumptions. This is somewhat of an obvious stage, but it is a critical one. It should clearly define the purpose of the LCA which consequently governs everything else from boundary setting through to communication of the end results. The key elements are outlined below:

    *   Functional unit: is a measure of the function of the system being studied. It provides a reference to which the inputs and outputs can be related and enables comparison of two or more different systems. Examples include: per kg of product; per production of a hot cup of coffee or the production of 1000 hot cups of coffee; for a paint system, per unit surface protected for ten years.

    *   System boundaries: determine which unit processes are to be included

in the LCA study. Defining system boundaries can be subjective; it is important not to exclude any important processes. The following boundaries can be considered: boundaries between the production and natural systems; boundaries between different production systems; geographical area (spatial boundaries); and time horizon (temporal boundaries).

*   Data requirements: The reliability of the results from LCA studies strongly depends on the extent to which data quality requirements are met. The following parameters should be taken into account: time-related coverage; geographical coverage; technology coverage; precision, completeness and representativeness of the data; consistency and reproducibility of the methods used throughout the data collection; and uncertainty of the information and data gaps. It is important to document the data as much as possible.

*   Assumptions: The goal and scope definition process will involve making decisions which include a number of assumptions. It is important that all assumptions are clearly stated so that they can be taken into account when the results of the study are utilised.

2. Inventory: the data collection phase – quantifying all the inputs and outputs from each life cycle stage, including all emissions to air, land and water. The most important steps of the inventory are: description of system in flow diagrams; identification of unit processes to be modelled separately; qualitative determination of inputs and outputs; quantitative determination of inputs and outputs; inventory data collection; documentation of the data (metadata); and calculation of the inventory, including

allocations and covering the inventories of the background data sets. Data are usually categorised into system flows: elementary flows (resource use and emissions); product flows, that come from or go into the production process (goods and services); waste flows (a sub-type of product flows). In the impact assessment, it is the quantitative infor-mation on the elementary flows (and in some methods the waste flows) that is used to analyse impacts.

3. Impact assessment: classifying and characterising the emissions from each life-cycle stage in terms of their envi-ronmental effects and impacts. Typically, this is for midpoint impact categories such as climate change, acidification, ecotoxicity, eutrophication, and so on, but in some instances, it may include an assessment of the end point impact cat-egories, such as impacts on ecosystems and human health. The latter are more difficult to determine, so are considered optional (under the ISO standards), whereas the midpoint categories are compulsory. Impacts are characterised using equivalency factors (see Chapters 1.5.3.2 and 1.5.3.3 below) to convert the data from the inventory into quantified impacts. In some instances, the data can be normalised, usually by expressing them in relation to a recognised refer-ence point, for example expressing emis-sions as a percentage of total emissions for a region or per head of population.

4. Interpretation: evaluating the findings of the inventory and impact assess-ment in phases in relation to the goal and scope of the LCA. This may also include hotspot identification; data quality assessments; sensitivity analy-sis; scenario analysis; and limitations and reflections on study. It is important that: conclusions and recommendations

should relate to the goal and scope of the study; any limitations of the study within the given goal and scope of the LCA study are listed; any comparisons, for example to LCAs of similar products, are based on common approaches and data; and inappropriate generalisations of findings are avoided.

In terms of agri-envirommental manage-ment, undertaking full LCAs is usually confined to specific one-off projects, usu-ally undertaken by a company for one or more of its products or by government to gain an understanding of the effects of poli-cies. It is generally not a technique that is applied at the level of farm management. However, there are techniques used within LCA that are or can be used at the farm level, especially techniques from the impact assessment stage, such as Global Warming Potential (GWP) for carbon footprinting. The concepts and techniques are described in the following sections within this chapter.

1.5.3.2.   Global warming potential (GWP)
The relative contribution each greenhouse gas makes towards climate change is com-monly expressed as their GWP, where the GWP for each gas is calculated relative to the effect of carbon dioxide and are thus expressed as carbon dioxide equivalents (CO_2e). GWP for a gas is calculated by taking into account its estimated duration in the atmosphere and its ability to absorb outgoing infrared radiation. Consequently, there can be different values for each gas for different time horizons, as shown in Table 1.3.

Table 1.3 shows how the GWP for methane, for example, decreases from 72 to 25 as we go from a 20-year to 100-year time horizon because methane has an average lifetime of 12 years, and as the time horizon increases beyond its average lifetime, an

Table 1.3: GWP values and lifetimes

Gas	Lifetime (years)	GWP time horizon		
		20 years	100 years	500 years
Methane	12	72	25	7.6
Nitrous Oxide	114	310	298	153
HFC-23 (hydrofluorocarbon)	270	1200	14800	12200
HFC-134a (hydrofluorocarbon)	14	3830	1430	435
Sulphur Hexafluoride	3200	16300	22800	32600

increasing amount of methane is destroyed by interacting with other gases. In contrast, sulphur hexafluoride (a man-made chemical with no agricultural sources) has a much longer lifetime of 3200 years on average, and consequently its GWP increases as the time horizon increases from 20, 100 to 500 years (over a 500-year period, the emission of 1 tonne of sulphur hexafluoride is the same as the emission of 32,600 tons of carbon dioxide). Therefore, even a tiny amount of sulphur hexafluoride emissions can make a large contribution to climate change. The Intergovernmental Panel for Climate Change (IPCC) propose a time horizon of 100 years, GWP_{100}. The GWPs are revised occasionally by the IPCC, but the current GWP_{100} values for methane and nitrous oxide, for example, are 25 and 298 respectively. The calculation of GWP is a well-established technique, but it does have some uncertainties attached to it. For example, it is difficult to predict the lifetime of a gas that is highly reactive to other gases in the atmosphere, because the gas's lifetime is likely to be hard to quantify and have a changing lifetime, over the time period of measurement. Also, because current measurement tools are imprecise, it is difficult to measure the composition of the atmosphere accurately as it changes. Finally, as substances interact and change in the atmosphere, their ability to absorb outgoing radiation can also change. As a result, its ability to absorb outgoing infrared radiation may also change over time.

1.5.3.3. Other equivalency factors

As is the case with GWP, different substances contribute towards a specific environmental impact to differing degrees. Therefore, it is not scientifically sound to simply add concentrations/masses of different chemicals together within LCA, hence the 'equivalency' approach is quite often used. Table 1.4 provides a summary of the key approaches. There are undoubtedly many critics of this approach and, indeed, LCA more broadly (e.g. Klinglmair *et al.*, 2014; Ayres, 1995) with the key issues being variability and high levels of uncertainty as well as problems associated with data quality and gaps. Nevertheless, the approach is widely used, although in terms of human and ecological toxicity other approaches such as Toxicity–Exposure Ratios (see Chapter 1.5.2.3) are more usually used within risk assessments but are fundamentally more complex.

1.5.3.4. Water footprints

The concept of water footprinting was introduced by Hoekstra and Hung (2002) with the aim of introducing an effective tool to aid improvements in sustainable water use. It is analogous to ecological footprinting (Wackernagel and Rees, 1998) and also has resonance with carbon footprinting (Wiedmann and Minx, 2008; Pandey *et al.*, 2011) seeking to provide a sound basis for an assessment of local environmental, social and economic risks and impacts associated

Table 1.4: Summary of LCA techniques that use the 'equivalency approach'

Environmental impact (potential)	Equivalency factors	Description
Acidification	kg SO_2 eq.; or mol H^+ eq.	Acidification can be caused by a number of different gases including sulphur dioxide, nitrogen oxides and ammonia. Within LCA, they are expressed as kg of sulphur dioxide equivalents (kg SO_2 eq.) or equivalent molar concentration of the hydrogen ion (mol H^+ eq.). For example, the acidification potential of ammonia is 1.88 kg SO_2 eq[1].
Eutrophication	kg PO_4^{2-} eq; kg NO_3 eq; kg P eq.; or mol N eq.	A build-up of nutrients, i.e. ammonia, nitrogen oxides and phosphates in the environment can lead to eutrophication. Within LCA, the mass is expressed as either kg of nitrate, nitrogen, molar concentration of the nitrogen; phosphate or phosphorus equivalents, depending on the LCA model being used. For example, the eutrophication potential of ammonia is 3.64 kg NO_3-eq or 0.35 kg PO_4^{2-}eq[2]. The different equivalency factors are often used for different media. For example, mol N eq. for terrestrial eutrophication; kg P eq. for freshwater eutrophication; and kg N eq. for marine eutrophication.
Human and ecological toxicity	Kg 1,4-DB eq; CTUe; or CTUh	This is used as a measure of the inherent toxicity of a chemical released into the environment to either humans, or one of three environmental compartments (freshwater, marine or terrestrial). It is expressed as kg equivalents of 1,4-dichlorobenzene. Comparative toxic units for human (CTUh) can also be used for human toxicity, cancer and non-cancer effects; and comparative toxic units for ecosystem (CTUe) can be used for freshwater ecotoxicity.
Ozone depletion	kg CFC-11 eq.	Depletion of the stratospheric ozone layer (ODP) due to anthropogenic emissions of ozone-depleting substances (CFCs in particular) can affect human health and ecosystem functions. Within LCA, the ODP potential of chemicals is expressed as kg of trichlorofluoromethane-11 equivalents. For example, tetrachloromethane has an ODP value of 1.08 kg CFC-11-eq[2].
Particulate matter	kg PM_{10} eq; or kg PM2.5 eq.	Assessment of the impact of fine solid and liquid particulate matter (<PM_{10} μm, <$PM_{2.5}$ μm, <$PM_{0.1}$ μm) such as dust, pollen and mould spores, primarily, on human health. The smaller the particulate size, the greater it's potential to cause harm. Particle mass is expressed as kg of particulate matter <10 μm in size (PM_{10}) or <2.5 μm in size ($PM_{2.5}$) equivalents.
Photochemical ozone formation	kg C_2H_4 eq; or kg NMVOC eq.	This is used to assess the potential for smog formation and its impacts on human health and ecosystem quality. The most important chemicals are carbon monoxide, nitrogen oxides and volatile organic compounds. It is expressed as kg of ethylene equivalents or equivalent of non-methane volatile organic compounds. For example, carbon monoxide has an equivalency factor of 0.027 kg C_2H_4-eq[3].

Table 1.4: (continued)

Environmental impact (potential)	Equivalency factors	Description
Abiotic resource depletion	Kg Sb eq.	Used to assess the consumption of non-biological resources such as fossil fuels, minerals, metals and water and is a measure of their scarcity. It is expressed as kg of antimony equivalents.
Resource depletion water	m^3 water eq.	Characterisation factors for water resource depletion are calculated using a reference water resource flow, based on an EU consumption weighted average, and water flows are related to this reference flow. This enables impacts in water consumption to be expressed as cubic metres of water equivalent.
Ionising radiation, human health	kBq U^{235} eq. (to air)	Elementary flows of radionucleides are assessed using equivalent uranium radiation measured in kilo Becquerel.

1: Data suggested by Hauschild and Wenzel, 1998
2: Data suggested by Heijungs *et al.*, 1992
3: Data suggested by Jenkin and Hayman, 1999

with water use. A water footprint is a measure of the total net water consumed by a region (e.g. a country, catchment or farm), a company (e.g. a business as a whole, a whole supply chain or component of that supply chain such as a farm or process) or an individual (Chapagain and Orr, 2009) both directly and indirectly. The water footprint can also be defined as the cumulative virtual water content of all goods and services consumed by one individual or by the individuals of one country (Hoekstra and Hung, 2002), where the definition of 'virtual water' was previously given by Allan (1994) as the total volume of water needed to produce a product or service. Whilst the ecological footprint illustrates the area of land needed to sustain a population, the water footprint represents the volume of water required. However, whilst the ecological footprint does not take into account where such land is used nor how it is used, the water footprint is a much more explicit indicator taking into consideration both the geographical location where the water is used (UNESCO, 2006) and the type of water consumed, classifying water

into three broad types. Blue water within the footprint refers to the consumption of surface and groundwaters. The green water footprint refers to the consumption of rainwater stored in the soil as soil moisture which mostly occurs during the growing cycle of crops and forest. The grey water footprint is defined as the volume of water required to assimilate waste and it is quantified as the volume of water needed to dilute pollutants to such an extent that the quality of the ambient water remains above the agreed water quality standards (Chapagain and Hoekstra, 2011).

From a business or supply chain perspective the water footprint offers a number of opportunities. It can help identify the most water-intensive processes and activities which can then be explored in more detail regarding potential efficiency savings through, for example, water-efficient technologies, water conservation measures and wastewater treatments (Hoekstra, 2008). In addition, businesses can reduce their water footprint not only by reducing the water consumption and pollution in their own operations, but through engaging with their suppliers or transforming their business model in order to better control their supply chain (Hoekstra *et al.*, 2012). The water footprint can also help companies better understand water issues and their relative impact to local situations. Thus, the approach can be extremely useful to assess water use within a business as it informs better decision-making.

References

Allan, J.A. (1994) Overall perspectives on countries and regions. In: Rogers P, Lydon P (eds.) *Water in the Arab World: Perspectives and Prognoses*. Harvard University Press, Cambridge.

Ayres, R.U. (1995) Life cycle analysis: a critique. *Resources, Conservation and Recycling*, 14(3–4), 199–223. DOI: 10.1016/0921-3449(95)00017-D

Bennett, R.S., Dewhurst, I., Fairbrother, A., Hooper, M., Leopold, A., Mineau, P., Mortensen, S., Shore, R.F. and Springer, T. (2005) A new interpretation of avian and mammalian reproduction toxicity test data in ecological risk assessment. *Ecotoxicology*, 14(8), 801–815. DOI: 10.1007/s10646-005-0029-1

Bird, S.L., Perry, S.G., Ray, S.L. and Teske, M.E. (2002) Evaluation of the AGDISP aerial spray algorithms in the AgDRIFT model. *Environmental Toxicology and Chemistry*, 21(3), 672–681. DOI: 10.1002/etc.5620210328

Burney, J.A., Davis, S.J. and Lobell, D.B. (2010) Greenhouse gas mitigation by agricultural intensification. *Proceedings of the National Academy of Science USA*, 107(26), 12052–12057. DOI: 10.1073/pnas.0914216107

Calow, P. (1998) Ecological risk assessment: risk for what? How do we decide? *Ecotoxicology and Environmental Safety*, 40(1–2), 15–18. DOI: 10.1006/eesa.1998.1636

Carsel, R.F., Imhoff, J.C., Hummel, P.R., Cheplick, J.M. and Donigan, A.S. (1998) PRZM-3. *A model for predicting pesticide and nitrogen fate in the crop root and unsaturated soil zones*. Users Manual for Release 3.0. US Environmental Protection Agency.

Chamberlain, D.E., Fuller, R.J., Bunce, R.G.H., Duckworth, J.C. and Shrubb, M. (2000) Changes in the abundance of farmland birds in relation to the timing of agricultural intensification in England and Wales. *Journal of Applied Ecology*, 37(5), 771–788. DOI: 10.1046/j.1365-2664.2000.00548.x

Chapagain, A.K. and Hoekstra, A.Y. (2011) The blue, green and grey water footprint of rice from production and consumption perspectives. *Ecological Economics*, 70(4), 749–758. DOI: 10.1016/j.ecolecon.2010.11.012

Chapagain, A.K. and Orr, S. (2009) An improved water footprint methodology linking global consumption to local water resources: a case of Spanish tomatoes. *Journal of Environmental Management*, 90(2), 1219–1228. DOI: 10.1016/j.jenvman.2008.06.006

Choe, C. and Fraser, I. (1999) Compliance monitoring and agri-environmental policy. *Journal of Agricultural Economics*, 50(3), 468–487. DOI: 10.1111/j.1477-9552.1999.tb00894.x

Cooper, T., Hart, K. and Baldock, D. (2009) *The provision of public goods through agriculture in the European Union - Report prepared for DG agriculture and rural development, Contract No 30-CE-1233091/00-28*. Institute for European Environmental Policy, London.

Donald, P.F., Green, R.E. and Heath, M.F. (2001) Agricultural intensification and the collapse of Europe's farmland bird populations. *Proceedings of the Royal Society, London B.*, 268(1462), 25–29. DOI: 10.1098/rspb.2000.1325

EC – European Commission (1999) *Communication from the Commission to the Council; the European Parliament; the Economic and Social Committee and the Committee of the Regions – Directions towards sustainable agriculture (COM(1999) 22 final)*. European Commission (EC), Brussels.

EC – European Commission (2000) Agriculture and the Environment. European Commission (EC), DG Agriculture, Brussels.

EC – European Commission (2003) A Long-Term Policy Perspective for Sustainable Agriculture (COM(2003) 23 final). European Commission (EC), Brussels.

EC – European Commission (2006) *Communication from the Commission to the Council and the European Parliament*

–*Development of agri-environmental indicators for monitoring the integration of environmental concerns into the common agricultural policy (COM(2006) 508)*. European Commission (EC), Brussels.

EC – European Commission (2010a) *Communication from the Commission to the European Parliament, The Council, The European Economic and Social Committee and the Committee of the Regions – The CAP towards 2020: meeting the food, natural resources and territorial challenges of the future (COM(2010) 672 final).* European Commission (EC), Brussels.

EC – European Commission (2010b) *Europe 2020: a European strategy for smart, sustainable and inclusive growth (3.3.2010 COM(2010)2020).* European Commission (EC), Brussels.

EC – European Commission (2013a) *The European Union explained: Agriculture – a partnership between Europe and farmers.* Publications Office of the European Union, European Commission (EC), Luxembourg.

EC – European Commission (2013b) *Agricultural policy perspectives brief no. 5 – Overview of CAP reform 2014–2020.* European Commission (EC), Brussels.

EEA – European Environment Agency (2007) *Europe's environment: the fourth assessment.* European Environment Agency (EEA), Copenhagen.

EEC – European Ecomomic Communities (1992) Council Regulation (EEC) No 2078/92 of 30 June 1992 on agricultural production methods compatible with the requirements of the protection of the environment and maintenance of the countryside. *Official Journal of the European Communities,* L 215, 85–90.

European Communities (2008) *EU rural development policy 2007–2013: Factsheet.* Directorate General for Agriculture and Rural Development, European Commission (EC), Brussels.

European Community (1957) *The Treaty of Rome – 25 March 1957.* European Community, Brussels.

European Council (1991) Council Directive of 12 December 1991 concerning the protection of waters against pollution caused by nitrates from agricultural sources (91/676/EEC). *Official Journal of the European Communities,* L 375, 1–8.

European Council (1998) Council Directive 98/83/EC of 3 November 1998 on the quality of water intended for human consumption. *Official Journal of the European Communities,* L 330, 32–54.

European Council (2005) Council Regulation (EC) No 1698/2005 of 20 September 2005 on support for rural development by the European Agricultural Fund for Rural Development (EAFRD). *Official Journal of the European Union,* L 277, 1–40.

European Parliament and Council (2013) Regulation (EU) No 1305/2013 of the European Parliament and of the Council of 17 December 2013 on support for rural development by the European Agricultural Fund for Rural Development (EAFRD) and repealing Council Regulation (EC) No 1698/2005. *Official Journal of the European Union,* L 347, 487–548.

Eurostat (2014) *Europe in Figures –- Eurostat Yearbook.* Eurostat, Luxembourg, Luxembourg.

Galloway, J.N., Hiram Levy, H. and Kasibhatla, P.S. (1994) Year 2020: consequences of population growth and development on deposition of oxidized nitrogen. *Ambio,* 23(2), 120–123.

Garnett, T. (2010) *Intensive Versus Extensive Livestock Systems and Greenhouse Gas Emissions.* Food Climate Research Network, Oxford.

Goulson, D., Lye, G.C. and Darvill, B. (2008) Decline and conservation of bumble bees. *Annual Review of Entomology,* 53, 191–208. DOI: 10.1146/annurev.ento.53.103106.093454

Hart, K., Allen, B., Lindner, M., Keenleyside, C., Burgess, P., Eggers, J. and Buckwell, A. (2013) *Land as an environmental resource. Report prepared for DG environment, Contract No ENV.B.1/ETU/2011/0029.* Institute for European Environmental Policy, London.

Hauschild, M., Wenzel, H. and Technical University of Denmark Institute for Product Development (Copenhagen) (1998) *Environmental Assessment of Products: Scientific Background* (Vol. 2). London: Chapman and Hall.

Heijungs, R., Guinée, J.B., Huppes, G., Lankreijer, R.M., Udo de Haes, H.A., Wegener Sleeswijk, A., Ansems, A.M.M., Eggels, P.G., Duin, R.V. and De Goede, H.P. (1992) Environmental Life Cycle Assessment of Products: Guide and Backgrounds (Part 1). CLM, Leiden, Netherlands.

Hoekstra, A.Y. (2008) *Globalization of Water*. John Wiley and Sons, Inc.

Hoekstra, A.Y and Hung, P.Q. (2002) *Virtual water trade: a quantification of virtual water flows between nations in relation to international crop trade*, Value of water: Research Report Series No. 11. IHE Delft, Netherlands. Available at. http://waterfootprint.org/media/downloads/Report11.pdf

Hoekstra, A.Y., Chapagain, A.K., Aldaya, M.M., and Mekonnen, M.M. (2012) The Water Footprint Assessment Manual: Setting the Global Standard. Earthscan, London.

Jarvis, N. and Larsbo, M. (2012) MACRO (v 5. 2): model use, calibration, and validation. *Transactions of the ASABE*, 55(4), 1413–1423.

Jenkin, M.E. and Hayman, G.D. (1999) Photochemical ozone creation potentials for oxygenated volatile organic compounds: sensitivity to variations in kinetic and mechanistic parameters. *Atmospheric Environment*, 33(8), 1275–1293.

Kay, P., Blackwell, P.A. and Boxall, A.B.A. (2005) A lysimeter experiment to investigate the leaching of veterinary antibiotics through a clay soil and comparison with field data. *Environmental Pollution*, 134(2), 333–341.

Klinglmair, M., Sala, S., and Brandão, M. (2014) Assessing resource depletion in LCA: A review of methods and methodological issues. *The International Journal of Life Cycle Assessment*, 19(3), 580–592.

Latacz-Lohmann, U. and Hodge, I. (2003) European agri-environmental policy for the 21st century. *The Australian Journal of Agricultural and Resource Economics*, 47(1), 123–139.

Lewis, K.A., Tzilivakis, J., Warner, D. and Green, A. (2016) An international database for pesticide risk assessments and management.

Human and Ecological Risk Assessment: An International Journal, 22(4), 1050–1064.

Louwagie, G., Gay, S.H. and Burrell, A. (2009) *Final report on the project 'Sustainable Agriculture and Soil Conservation (SoCo)', JRC Scientific and Technical report*. Publications Office of the European Union, Luxembourg.

Machefert, S.E., Dise, N.B., Goulding, K.W.T. and Whitehead, P.G. (2002) Nitrous oxide emission from a range of land uses across Europe. *Hydrology and Earth System Sciences*, 6, 325–337.

Mineau, P. (2005) A review and analysis of study endpoints relevant to the assessment of 'long term' pesticide toxicity in avian and mammalian wildlife. *Ecotoxicology*, 14(8), 775–799.

NASA Earth Observatory (2011) *The Carbon Cycle*. NASA (National Aeronautics and Space Administration). Available at: https://earthobservatory.nasa.gov/Features/CarbonCycle/

OECD – Organisation for Economic Co-operation and Development (2008) *Environmental performance of agriculture in OECD countries since 1990*. Paris, France.

OECD – Organisation for Economic Co-operation and Development (2015) *Producer and consumer support estimates*, OECD Agricultural Statistics Database, Paris, France.

Oenema, O., van Liere, L. and Schoumans, O. (2005) Effects of lowering nitrogen and phosphorus surpluses in agriculture on the quality of groundwater and surface water in the Netherlands. *Journal of Hydrology*, 304(1), 289–301.

O'Mara, F.P. (2011) The significance of livestock as a contributor to global greenhouse gas emissions today and in the near future. *Animal Feed Science and Technology*, 166–167, 7–15. DOI: 10.1016/j.anifeedsci.2011.04.074

Pandey, D., Agrawal, M. and Pandey, J.S. (2011) Carbon footprint: current methods of estimation. *Environmental Monitoring and Assessment*, 178(1), 135–160.

Phalan, B., Balmford, A., Green, R.E. and Scharlemann, J.P.W. (2011) Minimising the harm to biodiversity of producing more food globally. *Food Policy*, 36, S62–71.

Pimentel, D. and Kounang, N. (1998) Ecology of soil erosion in ecosystems. *Ecosystems*, 1, 416–426.

Postgate, J. R. (1982) *The Fundamentals of Nitrogen Fixation*. CUP Archive, ISBN 978 05 2128 4943.

Rey Benayas, J.M. and Bullock, J.M. (2012) Restoration of biodiversity and ecosystem services on agricultural land. *Ecosystems*, 15(6), 883–899. DOI: 10.1007/s10021-012-9552-0

Rudel, T.K., Schneider, L., Uriarte, M., Turner II, B.L., DeFries, R., Lawrence, D., Geoghegan, J., Hecht, S., Ickowitz, A., Lambin, E.F., Birkenholtz, T., Baptista, S. and Grau, R. (2009) Agricultural intensification and changes in cultivated areas, 1970–2005. *Proceedings of the National Academy of Sciences, USA*, 106(49), 20675–20680. DOI: 10.1073/pnas.0812540106

Skinner, J.A., Lewis, K.A., Bardon, K.S., Tucker, P., Catt, J.A. and Chambers, B.J. (1997) An overview of the environmental impact of agriculture in the U.K. *Journal of Environmental Management*, 50(2), 111–128. DOI: 10.1006/jema.1996.0103

Smith, K.A. and Conen, F. (2004) Impacts of land management on fluxes of trace greenhouse gases. *Soil Use Management*, 20(2), 255–263. DOI: 10.1111/j.1475-2743.2004.tb00366.x

Smith, P., Martino, D., Cai, Z., Gwary, D., Janzen, H., Kumar, P., McCarl, B., Ogle, S., O'Mara, F., Rice, C., Scholes, B., Sirotenko, O., Howden, M., McAllister, T., Pan, G., Romanenkov, V., Schneider, U., Towprayoon, S., Wattenbach, M. and Smith, J. (2008) Greenhouse gas mitigation in agriculture. *Philosophical Transactions Royal Society of London Biological Sciences*, 363(1492), 789–813. DOI: 10.1098/rstb.2007.2184

Stoate, C., Boatman, N.D., Borralho, R.J., Carvalho, C.R., de Snooc, G.R. and Eden, P. (2001) Ecological impacts of arable intensification in Europe. *Journal of Environmental Management*, 63(4), 337–365. DOI: 10.1006/jema.2001.0473

Teske, M.E., Bird, S.L., Esterly, D.M., Curbishley, T.B., Ray, S.L. and Perry, S.G. (2002) AgDrift®: a model for estimating near-field spray drift from aerial applications. *Environmental Toxicology and Chemistry*, 21(3), 659–671. DOI: 10.1002/etc.5620210327

UN – United Nations (2014) *Concise report on the world population situation in 2014, United Nations, Department of Economic and Social Affairs Population Division*, Report number: ST/ESA/SER.A/354. United Nations, New York, ISBN 978-92-1-151518-3.

UNESCO (2006) *Water, a shared responsibility: the United Nations world water development report 2*. UNESCO Publishing, Paris, France/Berghahn Books, Oxford, UK.

Vitousek, P.M., Aber, J.D., Howarth, R.W., Likens, G.E., Matson, P.A., Schindler, D.W., Schlesinger, W.H. and Tilman, D.G. (1997) Human alteration of the global nitrogen cycle: Sources and consequences. *Ecological Applications*, 7(3), 737–750. DOI: 10.1890/1051-0761(1997)007[0737:HAOTGN]2.0.CO;2

Wackernagel, M. and Rees, W. (1998) *Our Ecological Footprint: Reducing Human Impact on the Earth* (No. 9). New Society Publishers, Canada.

Warren, N., Allan, I.J., Carter, J.E., House, W.A. and Parker, A. (2003) Pesticides and other micro-organic contaminants in freshwater sedimentary environments: a review. *Applied Geochemistry*, 18(2), 159–194. DOI: 10.1016/S0883-2927(02)00159-2

Wiedmann, T. and Minx, J. (2008) *A definition of 'carbon footprint'*. Chapter 1 in Pertsova, C.C. (Ed.) Ecological Economics Research Trends. Nova Science. pp1–11.

Woods, J., Williams, A., Hughes, J.K., Black, M. and Murphy, R. (2010) Energy and the food system. *Philosophical Transactions of the Royal Society B*, 365(1554), 2991–3006. DOI: 10.1098/rstb.2010.0172

Zalidis, G.C., Tsiafouli, M.A., Takavakoglou, V., Bilas, G. and Misopolinos, N. (2004) Selecting agri-environmental indicators to facilitate monitoring and assessment of EU agri-environmental measures effectiveness. *Journal of Environmental Management*, 70(4), 315–321. DOI: 10.1016/j.jenvman.2003.12.006

Atmospheric pollution and its consequences

2.1 Setting the scene

2.1.1 Introduction

This chapter seeks to explore the nature, sources, concentrations and effects of atmospheric pollution arising from agricultural practices. Air pollutants affect the concentration of the planet's atmosphere, altering its composition and its behaviour, which has the potential to damage all life on Earth. The atmosphere has a direct interface to other environmental media, that is, soil/land and all surface water bodies. As described in Chapter 1, agricultural production involves the manipulation of various natural cycles including those of nitrogen, carbon and water. Inevitably, therefore, there is interchange of agricultural pollutants between the different media and the potential for natural cycles, on which agriculture depends, to become polluted. In order to understand how this can be managed and mitigated, it is necessary to understand their sources. Due to the interconnection between the different environmental media and the interchange of pollutants, it is inevitable that there will be overlaps with other topics, such as biodiversity and water and where this occurs reference will be to the relevant chapters where further information can be found.

The main agricultural emissions to the atmosphere are carbon dioxide (CO_2), methane (CH_4), nitrous oxide (N_2O) and ammonia (NH_3). However, the range of pollutants that can arise from agriculture is very broad, and includes pesticides, smoke and particulates, volatile organic compounds (VOCs), bioaerosols and noxious odours. Although these latter mentioned substances normally occur in smaller concentrations, they still have the potential to impact on the environment, damage human health and negatively affect the quality of life of individuals, in particular those within rural communities. All air pollutants have the potential to cause a number of different direct and indirect impacts. Direct impacts can include toxic effects arising from the contact or inhalation of the pollutants by humans and biodiversity, whilst the indirect impacts include climate change from emissions of greenhouse gases (GHGs) and deposition of harmful substances into aquatic and terrestrial ecosystems.

Table 2.1 provides an overview of the relative contribution by agriculture and land use to emissions of these pollutants. The sources of emissions are inherently linked to agricultural practices and land management. Figure 2.1 provides a holistic perspective of

Table 2.1: Agricultural contribution to atmospheric emissions

Emission	Estimated contribution	Scale	References
Carbon dioxide (CO_2)	10–21%	Global	Calverd, 2005; WRI, 2011
Methane (CH_4)	43%	Global	WRI, 2011
Nitrous oxide (N_2O)	70%	Global	WRI, 2011
Ammonia (NH_3)	93–94%	Europe	EEA, 2016; Eurostat, 2015
Pesticides	99%	Europe	Webb *et al.*, 2016
Bioaerosols	8.6% (based on PM_{10} emissions)	Europe	Seedorf, 2004
NMVOCs*	12.25%	Europe	EEA, 2016

* NMVOCs = Non-methane volatile organic compounds

the main pollutants and the key activities from which they arise.

Figure 2.1 shows that a number of air pollutants are caused by the manufacture of the inputs that farming relies upon. These emissions are in addition to those that occur when these inputs are used. Figure 2.1 also shows that the management of biomass and soils can also act as a sink for carbon.

The following sections within this chapter provide an overview of emission sources (Chapter 2.1.2) and carbon sinks (Chapter 2.1.3), which are then explored in detail in

Chapter 2.3 due to the inherent influence of agricultural practices on emissions and sinks.

2.1.2. An overview of emission sources

One of the most important agricultural inputs is fertiliser. These are chemical compounds formulated to provide plants with a range of essential nutrients including nitrogen, phosphorus and potassium as well as a number of minerals and trace elements. With respect to nitrogen fertilisers, the majority of those applied, in the agricultural context, are ammonium compounds

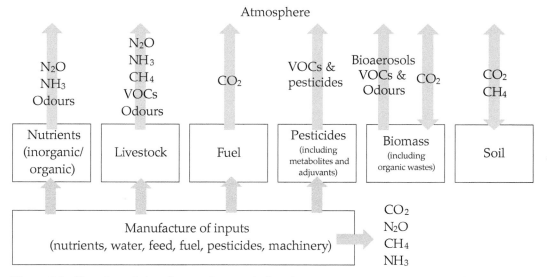

Figure 2.1: Overview of air pollutants from agricultural systems

(e.g. ammonium nitrate (NO_3^-), ammonium sulphate ((NH_4)$_2$$SO_4$), ammonium phosphate (($NH_4$)$_3$$PO_4$), urea ($CH_4N_2O$)) and nitrate compounds (e.g. ammonium nitrate (NH_4NO_3), calcium nitrate ($Ca(NO_3)_2$), potassium nitrate (KNO_3)). The manufacture and use of these compounds can contribute to N_2O, NH_3 and CO_2 emissions.

Nitrous oxide can be generated during fertiliser manufacture, spreading of inorganic fertilisers and livestock manures. It is a naturally occurring trace gas but like CO_2 its atmospheric concentrations have increased significantly in recent years. Globally, around 60 per cent of anthropogenic emissions of N_2O originate from agriculture (IPPC, 2001a), but it does vary from one country to the next. For example, in 2008, the per cent of the total N_2O emissions from the agricultural sector was 31, 41, 55, 58, 61, 71 and 80 for Belgium, Finland, Italy, Germany, UK, France and Hungary, respectively (World Bank, 2016).

Inorganic nitrogen fertiliser is also responsible for emissions of ammonia, a pungent, corrosive gas which is emitted from terrestrial ecosystems as part of the natural functioning of the nitrogen cycle (Chapter 1, Figure 1.4). However, natural levels are significantly less than anthropogenic emissions (for example, Bouwman *et al.*, 1997; Watts, 2000). To illustrate, air concentrations reported by Sutton *et al.* (2008) range from 0.11 μg m^{-3} NH_3 in a purely rural location to 5.1 μg m^{-3} NH_3 for a kerbside location in a large city centre, but have been reported as high as 8.1 μg m^{-3} NH_3 for an intensive livestock area in a lowland rural location. The 2013 data for Europe attributes 93.3 per cent of ammonia emissions to agriculture (Eurostat, 2015) although variation exists depending on Member State (MS). In the UK for example, agriculture is responsible for 83 per cent of NH_3 emissions, of which most arises from livestock production, with an estimated 19 per cent, 7 per cent, 20 per cent and 7 per cent originating

from housing, storage, spreading and grazing deposition, respectively (Misselbrook *et al.*, 2000). A further 20 per cent is emitted from the application of ammonium and urea containing inorganic nitrogen fertiliser. The Dutch National Emissions Model for Ammonia showed that the total NH_3 emission from agriculture in the Netherlands in 2009 was 88.8 Gg NH_3–N, of which 50 per cent arose from housing, 37 per cent from manure application, 9 per cent from mineral N fertilizer, 3 per cent from outside manure storage, and 1 per cent from grazing (Velthof *et al.*, 2012).

A key component of the carbon cycle (Chapter 1.4.2), and to a lesser extent the nitrogen cycle (Chapter 1.4.3), is the combustion of fossil fuels which results in emissions of CO_2. The combustion of fossil fuels does not arise from the agricultural sector in isolation but will occur from many other sectors including industry and transport. However, agriculture and food production does make a contribution (directly and indirectly) including emissions arising from the energy/fuel (and associated emissions) embedded in farm inputs, those used on-farm and those used beyond the farm gate. The use of energy in the agricultural and horticultural sector is highly variable between enterprises, including between different crops. For example, in the UK, roughly 3 million ha of cereals were grown in 2007 and just 2000 ha of glasshouse crops (Defra, 2012). However, the direct energy consumption was roughly the same for both, that is, ~5000 GWh (Defra, 2007), which equates to ~0.0017 GWh ha^{-1} for cereals and 2.5 GWh for glasshouse crops. Clearly, the energy intensity of glasshouse production is a lot higher, with the bulk of the energy being used for heating. Thus, energy efficiency in some enterprises is more significant than others.

With regard to embedded emissions, the energy use and associated emissions from farm inputs can be considerable. For example, the manufacture of inorganic fertilisers requires significant amounts of energy resulting in CO_2 emissions. One of the causes of this increase of atmospheric CO_2 is the growth in the use of nitrogen-based fertilisers. These compounds are produced at a low cost by the Haber-Bosch process (Sutton *et al.*, 2008) and so cost has not hindered their widespread use (indeed, it has been a key driver for agricultural intensification) and hence there has been an associated increase in N_2O, NH_3 and CO_2 emissions (Sutton and Fowler, 2002; Erisman, 2004; Erisman *et al.*, 2007).

Pesticides also require energy during their manufacture, albeit to a much lesser extent than inorganic fertiliser. Once on-farm, some are prone to being lost to the atmosphere via volatilisation and/or drift during and after application, and thus can become airborne pollutants. Once airborne, regardless of their product formulation, they have the potential to be transported by air as fine liquid particles, dusts and in the vapour phase. When conditions are suitable they will be deposited back to earth, either as the parent substance or as a metabolite, potentially contaminating non-target areas such as neighbouring crops, water bodies, woodlands and other habitats.

Livestock can be a source of a number of air pollutants, in particular methane. Enteric methane is produced as a result of microbial fermentation of feed. Methane is produced predominately in the rumen (~85–90%) and to a much lesser extent in the large intestines (~10–15%). The ruminal methane production process begins with digestive microorganisms hydrolysing proteins, starches and plant cell wall polymers into amino acids and sugars. These products are then fermented into volatile fatty acids (including acetate, propionate and butyrate), hydrogen gas and CO_2. In the rumen, methanogenic

bacteria use hydrogen gas, CO_2 and formic acid as substrates to generate energy for growth. Nitrous oxide is also a product of the ruminant fermentation process, albeit to a lesser extent than from manures and urine (described above; IPPC, 2001b; Hamilton *et al.*, 2007).

Carbon dioxide from livestock respiration has been estimated as accounting for 21 per cent of anthropogenic GHGs worldwide (Calverd, 2005). However, the FAO (2006) excludes livestock respiration from its global estimate of greenhouse gases arising from livestock agriculture as it considers that it is not a net source of CO_2 as it is part of a rapidly cycling biological system, where the plant matter consumed was itself created through the conversion of atmospheric CO_2 into organic compounds (Hamilton *et al.*, 2007). Since the emitted and absorbed quantities are considered to be equivalent, livestock respiration was not included. Some sources suggest this argument is flawed (Worldwatch, 2009) and that any opportunity for reduced CO_2 should be considered. Nevertheless, CO_2 arising from livestock respiration appears to receive little attention.

The livestock industry is also a major emitter of ammonia, which as well as being environmentally damaging, is a precursor of fine particulate matter, a serious public health threat. Ammonia is also one of the main causes of nuisance odours (Hales *et al.*, 2012). The usual concern with odours emanating from livestock enterprises is that they are unpleasant and offensive, however, they can also exacerbate breathing disorders such as asthma. Odours are considered a form of air pollution and contribute towards poor air quality.

As well as CH_4, N_2O, NH_3 and odour, there are emerging concerns with regard to emissions of volatile organic compounds (VOCs) and bioaerosols. VOCs from livestock enterprises, particularly ethanol and methanol produced by ruminal bacteria such as *Streptococcus bovis*, may contribute towards photochemical smog and present a risk to human health. Bioaerosols are small airborne particulates that contain living biological matter. Common components include skin particles, feed, faecal bacteria (such as those like *Staphylococcae* and *Streptococcae*), fungi, moulds, viruses, yeasts and a range of organic compounds. Bioaerosols are most commonly associated with intensive enterprises and their concentrations can grow disproportionately with stocking rate, the highest concentrations being seen where large numbers of animals are kept in confined spaces. They are particularly associated with intensive pig and poultry units. Like odours, VOCs and bioaerosols can antagonise respiratory issues in both humans and animals; these pollutants can help spread disease-causing bacteria and viruses (Environment Agency, 2008).

Soil and its management can affect the emission and transport of many pollutants, including those emitted to air. It can also be a source of emissions itself with respect to the carbon that is stored within the soil, and consequently it can also be carbon sink helping to mitigate climate change (see Chapter 2.1.3). Carbon is present in soils as Soil Organic Carbon (SOC), where, according to the IPCC (2006) it comprises 58 per cent of Soil Organic Matter (SOM), the lignin, protein and cellulose component of soils. In terms of GHG emissions, increases in SOM enhance SOC, thus sequestering atmospheric C within the soil.

The decomposition of organic material under anaerobic conditions results in the production of methane (Freibauer, 2003; Smith *et al.*, 2008). Conditions conducive with this environment include waterlogged soils (for example, rice crops) and livestock

rumen during the fermentation of ingested plant material or during manure storage (Freibauer, 2003; Mosier *et al.*, 1998; IPCC, 2006). Several species of methanogenic bacteria exist, for example *Syntrophomonas wolfei*, and the most frequent substrates utilised are either acetate (CH_3COO^-) during acetoclastic methanogenesis or hydrogen (H_2) and CO_2 during hydrogenotrophic methanogenesis (McInerney *et al.*, 1981; Thauer, 1977). Methanotrophic bacteria can also remove CH_4 from the atmosphere.

As with soil, biomass can be a store of carbon, and as such this can become a source of CO_2 for example via combustion. Biomass and organic wastes, can also become a source of bioaerosols, for example from the composting process.

There are many activities on the farm that can lead to potentially damaging concentrations of dust and fine particulate matter in the farm atmosphere. Agricultural dusts, such as that from grain and crop handling and storage operations, seed dressing and handling, and livestock, can be a complex mixture of organic and inorganic materials which often contain fungal spores, bacteria, endotoxins, mites and insect debris, animal faeces, plant dust, soil, bedding, feed and feed components, pesticides and other chemicals, and so on. Chronic exposure to agriculture-related organic dusts is associated with an increased risk of developing respiratory diseases such as rhinosinusitis, asthma, chronic bronchitis, chronic obstructive pulmonary disease (COPD) and hypersensitivity pneumonitis (Poole and Romberger, 2012). Indeed, farmer's lung and organic dust toxicity syndrome (ODTS) are names given to two farm occupational diseases caused by inhaling airborne mould spores. Organic dust has also been associated with conjunctivitis and various skin allergic reactions. As well as negative impacts on farm workers and others in the vicinity of agricultural operations, there is also evidence that the health of livestock can also be affected and it is thought to affect animal performance and the efficiency of livestock operations (Cambra-López *et al.*, 2010).

2.1.3. *An overview of carbon sinks*

The carbon sequestered in soils, referred to as Soil Organic Carbon (SOC), is present as a component (approximately 58%) of Soil Organic Matter (SOM), and decaying plant and animal tissue. A critical point to consider with C sequestration is first, accumulation after a change of land use does not continue indefinitely and second, any reversion to the original land use or management practice will reverse any accumulation of C back to the original levels. An increase in SOC or biomass C proceeds until either the vegetation reaches maturity, or the soil has increased in C such that the rate of C accumulation equals the rate of CO_2 emission. At this point, no more C is gained within the system which is said to have reached equilibrium. The land use must be maintained in order to preserve this C that has been accumulated. Where reversion to the original land use occurs, CO_2 will be emitted in quantities greater than it is sequestered, decreasing the overall C sequestered. The rate of SOC loss typically proceeds more rapidly than it is accumulated (Smith *et al.*, 2000a, b). The time taken to reach equilibrium varies depending on the potential equilibrium that may be reached, and the rate of C accumulation. The quantity of SOM and hence C sequestered in soils at equilibrium for a given land use is influenced by climatic variables, for example annual precipitation and mean annual temperature (Ganuza and Almendros, 2003; Verheijen *et al.*, 2005), the dominant soil type, type of land use (e.g. annual cultivation, permanent or temporary grassland, woodland) and management

practice (e.g. reduced or zero tillage; Dawson and Smith, 2007; Schils *et al.*, 2008).

Biomass refers to living plant tissue, and includes stem, roots and leaves. Although the roots are present in the soil they are not categorised as soil organic carbon because they are living. The proportion of C in biomass is approximately 48 per cent (IPCC, 2006). Biomass can be a carbon sink, via the process of photosynthesis and the absorption of CO_2 from the atmosphere (Freibauer *et al.*, 2004; Smith *et al.*, 2008). A worst-case scenario is represented by bare soil, where minimum plant biomass is present. Consequently, there will be negligible return of plant residue to the soil and accumulation of SOC (Chapter 2.3.8). Although potentially beneficial for rare arable flora or ground-nesting birds, such management is not conducive with climate change mitigation. It increases the risk of loss of residual N in surface run-off (increased risk of indirect N_2O emission) especially when combined with water (Freibauer *et al.*, 2004; Smith et al., 2008) or wind erosion (Janzen, 1987). Cultivated land and grassland typically achieve full biomass potential within one year and because no further C accumulation occurs after year 1, the C at equilibrium is relatively low in comparison to land covered by woody plant species, such as hedgerows or trees.

2.1.4. *Consequences and impacts*

2.1.4.1. *Overview*

There are a number of consequences and impacts that arise due to atmospheric pollution. These include climate change, due to emission of greenhouse gases, and exposure of humans and other organisms to harmful substances in the air (either by contact or inhalation) and/or by deposition into other media, such as water and biodiversity habitats. Table 2.2 summarises the main effects

Table 2.2: Emissions to the atmosphere and their main effects and impacts

Emission	Effects and impacts
Carbon dioxide (CO_2)	• Greenhouse effect and climate change
Methane (CH_4)	• Greenhouse effect and climate change
Nitrous oxide (N_2O)	• Greenhouse effect and climate change
Ammonia (NH_3)	• Contributes to acid rain and deposition can result in acidification of habitats • Deposition can result in eutrophication of aquatic and terrestrial habitats • N_2O emissions can occur following ammonia deposition • Combines with other pollutants to form particulates and cause respiratory problems in humans and animals
Pesticides	• Deposited pesticides (from spray drift) can result in eco-toxic effects on non-target organisms that are exposed • Can exacerbate human and livestock respiratory problems • Can include VOCs
Bioaerosols	• Can cause respiratory problems, nausea, headache and fatigue in humans • Can assist the spread of human and animal disease-causing bacteria and viruses
VOCs	• Precursors to ground level ozone • Will contribute to smog • Can exacerbate respiratory diseases
Noxious odours	• Can impact on the quality of life • Can cause respiratory problems, nausea and headache
Dusts	• Will contribute to smog • Can impact on the quality of life • Can cause respiratory problems, nausea and headache

and impacts that arise from emissions of pollutants to the atmosphere.

2.1.4.2. The greenhouse effect and climate change

The greenhouse effect (see Chapter 1.4.6) is influenced by atmospheric concentrations of a number of different substances, known as greenhouse gases (GHGs). The concentrations of GHGs in the atmosphere have increased gradually since the early 1800s which coincides with the peak of the industrial revolution (around 1760–1830). Figure 2.2 shows the increase in concentration for CO_2, CH_4 and N_2O which are the most important greenhouse gases. It can be seen that the concentration of CO_2 has increased from ~275 to 375 ppm (~36 per cent increase); CH_4 has increased from ~700 to 1950 ppb (~180 per cent increase);

and N_2O has increased from ~270 to 320 ppm (~18 per cent increase). It is the continued accumulation of these gases that is increasing the greenhouse effect, and thus raising concerns with respect to the global impacts of this change.

'Global warming' is the term that is most often used in association with the impact of the greenhouse effect. There are various predictions, but increases in temperature of 2–7°C are estimated (IPCC, 2007). However, perhaps the more important issue is the consequential change in climate and the environmental and human impacts that are associated with these changes. Changes in climate are not just about an increase in temperature. Indeed, some places may get warmer, but some places may get cooler. There are also likely to be changes in precipitation in terms of amounts, frequency,

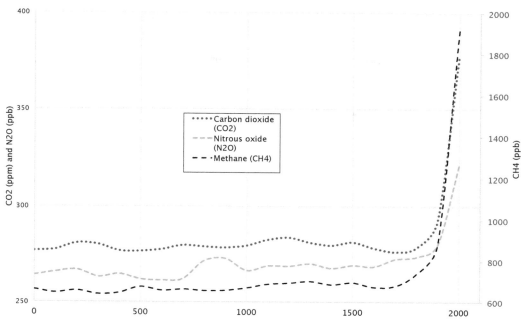

Figure 2.2: Atmospheric concentrations of greenhouse gases
(Created from: IPCC, 2007)

intensity and patterns, and more frequent and/or more extreme weather events.

Such changes in the Earth's climate will inevitably have a number of consequences and these can vary from one location to another. Figure 2.3 provides a summary of the potential implications of climate change in the context of rural development in Europe. These include some of the more obvious effects such as melting of glaciers, sea ice and permafrost, and changes in sea levels due to thermal expansion and ice melt. There may also be more floods and droughts, due to the precipitation changes. However, there are also a number of, perhaps, less obvious impacts but which, nevertheless could become serious particularly for food and water security. For example, as crops respond directly to climate (temperature and rainfall), and to CO_2 enrichment of the atmosphere, we may expect to see increases or decreases in crop yields depending on geographical

location and crop type. Research has shown that global wheat yields are likely to fall by up to 6 per cent for every 1°C rise in temperature (Asseng *et al.*, 2015). Wheat is one of the world's most important crops and such a loss would have serious consequences for food security. However, depending on location, some crops may benefit from the lengthening of growing seasons. CO_2 also has a fertilisation effect in some crops such as wheat, rice and many fruit and vegetables which consequently could improve yields. Consequently, potato and fruit growers in northern Europe may expect to see increased yields but those in southern Europe may not benefit due to drier soils.

Pest and disease pressure can also be expected to change, for example there may be species migration and it is generally predicted that temperature increases may extend the geographical range of some pests simply because environmental conditions

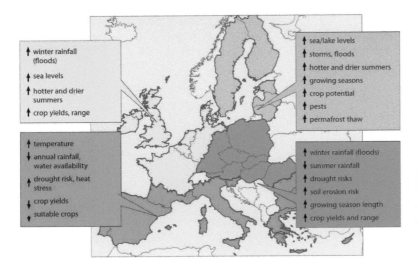

Figure 2.3:
Implications for rural areas from climate change in Europe
(Source: European Union, 2010)

become favourable. However, there are also other effects that may increase pest pressure. Elevated CO_2 can increase the levels of simple sugars in crop foliage and lower their nitrogen content which can result in increased pest damage as insects may need to increase feeding to meet their metabolic nitrogen requirements (Venkataraman, 2016; Yin *et al.*, 2010; Zavala *et al.*, 2013). Warmer temperatures can reduce the efficacy of some pesticides by encouraging them to break down quicker. Climate change can also impact on species populations such as pollinators which may also reduce yields (see Chapter 3.1).

2.1.4.3. Acid rain and acidification of habitats

The potential significant environmental damage caused by deposition of acidic material has been recognised since the late 1980s. It was first identified in Sweden in 1872 and the first scientific studies can be traced back to the 1950s (Seinfeld and Pandis, 1998). The issue did not begin to get public attention until the 1970s when the effects were visually evident. Whilst acidic deposition is commonly referred to as 'acid rain', it actually refers to both wet and dry deposition.

Wet deposition refers to rain, hail, snow, dew and fog that are polluted with sulphur dioxide and oxides of nitrogen, these pollutants being emitted as by-products from industrial processes and the burning of fossil fuels. Some natural processes also generate acidic molecules. For example, nitrogen oxides are formed when a thunderstorm produces lightning and various sulphurous gases are discharged from erupted volcanoes and rotting vegetation. Acidic deposition can have significant harmful effects on the environment, damaging trees and other vegetation and causing the acidification of soils and water bodies. However, policy and regulatory interventions (see Chapter 2.2) have greatly reduced sulphur emissions such that these are no longer the main cause of acidification, however, there may be a legacy effect. Nevertheless, as the emission of other acidic molecules such as nitrogen oxides and ammonium are declining at a much slower rate, acidification is still a cause for concern, albeit much reduced from the days of high SO_2 emissions, which peaked around the mid 1970s.

Acid deposition has multiple effects on trees and vegetation. First, there is the direct effect of acid damaging foliage by chemically burning it. This can eventually kill the tree but also weakens it and makes it much more vulnerable to insect and disease attack resulting in dieback. Conifers are often more affected than broad-leaved species as their needles are vulnerable to acid deposition all year around. There are also significant secondary effects on biodiversity that relies on the trees and other affected vegetation as habitat and a source of food.

Some of the most dramatic effects of acid deposition were seen on Europe's forests in the early to mid 1980s. In 1983, a survey in West Germany showed that 34 per cent of the country's total forest area had been damaged by air pollution including 50 per cent of the Black Forest (Hairsine, 2013). Around the same time, Switzerland also recorded significant damage to its alpine forests with the worst affected areas near the Gotthard Pass in the centre of the country, where many trees had been killed and up to 65 per cent of trees showed some vegetative damage. The 'Black Triangle' region – the border area shared between Germany, Poland and Czech Republic – has over 80 per cent forest cover, and acidification, due

to acid rain. It was first observed there in the early 1950s peaking in the mid 1980s. For example, the Jizera Mountains experienced large-scale dieback (40 to 80 per cent) of spruce stands, a decrease in pH of surface waters and decline of life in streams and reservoirs (see Figure 2.4). However, due to the many EU policy interventions to curb and mitigate air pollution some recovery of Europe's affected forests has now been observed.

Acidic deposition enters aquatic bodies either directly as polluted precipitation or indirectly through the catchment system. The significance of the damage often depends on the pH of the surrounding soils and sediments. If these are alkaline then this can neutralise the acidity and reduce, if not completely mitigate, any damage. However, if the surrounding soils and sediments are not alkaline then the water body will gradually become acidic. This can then have a major impact on the aquatic species that the water body can support. Fish communities dwindle due to high mortality, a reduced growth rate, skeletal deformities and failed reproduction. Subsequently,

Figure 2.4: Forest in the 'Black Triangle' damaged by acid rain (Photo courtesy of: Lovecz)

acidified water bodies become home only to species that can tolerate high-acid conditions. Populations of high-value fish such as trout and salmon tend not to be supported as these species are particularly sensitive to acidic water conditions. Eventually, even species that can tolerate acid environments can be lost due to food shortages. Soft-bodied creatures such as snails, leeches and crayfish are also highly susceptible. Some insect species demonstrate growth effects under acidic conditions. For example, the larvae of dragonflies and some water beetles can become exceptionally large. The first signs of acidification were observed in vulnerable lake ecosystems in Scandinavia (Odén, 1968) but it was not long after before other areas of Europe noted damage. For example, in Norway, acidic deposition caused the loss of fish populations in thousands of lakes until relatively recently (Hesthagen *et al.*, 1999).

Aquatic plant life populations can change quite dramatically under acidified conditions. Many algal species are the first to disappear but some of the filamentous species can thrive in these conditions and rapidly populate aquatic bodies. Where the water's pH drops below 6.0, floating mats of filamentous algae, sometimes referred to as 'flab' can develop; these mats are often dominated by Zygnemataceans algae (species of *Mougeotia, Zygogonium* and *Spirogyra;* Muniz, 1990). Macrophytes, aquatic plants that grow in or near water as either emergent, submergent or floating vegetation, also gradually disappear. Loss of these species can have significant secondary effects on the insect and fish life that depend on them for survival.

As well as damaging freshwater environments, acidic deposition can have serious consequences for the world's marine life, especially those that have calcium carbonate-based shells and skeletons or which depend upon calcium carbonate habitats such as coral reefs. Sea grass meadows and kelp forests are also highly susceptible.

Like water bodies, the significance of the effects of acidic deposition on soils depends on the soil composition and the bedrock lithology. Soils rich in chalk and limestone have the capacity to neutralise the pollutants, thus the damage that does occur depends on the soil's buffering capacity, and often no serious consequences occur. Where the soil is not alkaline, such as that on a bedrock of aluminium-based minerals such as feldspar and muscovite, it is more susceptible and acidic deposition can significantly alter the soil chemistry and thus, the species that the soil supports.

Soil pH is an important variable in determining soil quality. It influences the retention of nutrient cations in soils (e.g. calcium and potassium are lost from acid soils by leaching), the release of toxic metals (phytotoxic aluminium is mobilised in acid soils as well) and general microbial activity. Indeed, microorganism populations can change significantly with the loss of those that break down organic matter to produce available nitrogen needed for plant growth, thus acidic deposition can impact on the efficient functioning of the nitrogen cycle (see Chapter 1.4.3), potentially reducing agricultural production. As the chemical properties of the soil change, the rate of microbial processes slow and so the amount of available plant nutrients reduces. At the same time, the changing chemistry can have other effects, for example interfering with the phosphorus cycle (see Chapter 1.4.4) and locking phosphate such that it is unavailable to plants. A disturbance in the supply of nutrients as well as the acid-rich environment can seriously damage tree and plant roots, in particular the fine, smaller ones which the plants use for absorbing nutrients. In consequence, plants suffer stunted growth, are more vulnerable

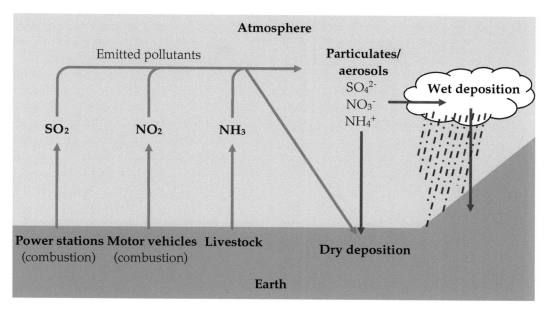

Figure 2.5: Ammonia emission and deposition

to pests and disease and often die. This can also affect the wildlife which feeds and lives on the plants and trees. Thus, the overall variety of fauna and flora able to grow in a specific area will alter.

2.1.4.4. Eutrophication of aquatic and terrestrial habitats

If nitrogen and phosphorus pollution reaches water bodies it can result in a series of adverse effects known as eutrophication. Eutrophication, also referred to as hyper-trophication, is the depletion of oxygen in a water body, which kills aquatic life. It is a response to the addition of excess nutrients, mainly phosphates but also nitrogen which induces extremely rapid growth of plants and algae, the decaying of which consumes oxygen from the water. The unbalanced ecosystem and changed chemical composition makes the water body unsuitable for recreational and other uses, and the water becomes unacceptable for human consumption. In extreme cases, massive algal blooms can cover the water surface and some forms

of these are toxic to humans, domestic pets and wildlife. In many cases, eutrophication is caused by pollution from terrestrial sources (see Chapter 1.4) but atmospheric deposition of nitrogen has its role to play. Nitrogen from the atmosphere can arise from a number of different sources. Predominately, this is as nitrogen oxide emitted from coal- or oil-burning electric utilities or other industries and from petrol- and diesel-fuelled vehicles. However, nitrogen also enters the atmosphere as it volatilises from manure or fertiliser (see Nitrogen cycle, Chapter 1.4.3). The consequences of this are described in detail in the subsequent chapters on biodiversity (Chapter 3) and water (Chapter 4).

The deposition of atmospheric N may be termed wet or dry. Wet deposition arises in response to incorporation within water vapour and deposition within rainfall, typically at a long distance from the source (Figure 2.5). Dry deposition results from increased concentration of NH_3 within the air, with a greater tendency to be more local in distribution, and concentration gradients evident in relation to the source. The nitrogen status of vegetation, for example, may decrease directly with distance from a source of NH_3, such as an area of intensive livestock housing, in relation to the prevailing wind direction. Indeed, ammonia is now recognised as the dominant factor behind deposition of nitrogen, and the eutrophication and depletion of nitrogen-sensitive habitats (Cocks, 1993; Erisman and Wyers, 1993; Sutton *et al.*, 1993, 1995, 2001). The molecular structure (NH_3 or NO_3) and nature (wet or dry) of the nitrogen deposited has been correlated with the magnitude of environmental impact on sensitive terrestrial habitats.

2.1.4.5.   Ecotoxic effects

In 1962, in the book by Rachel Carson, *Silent Spring* (Carson, 1962), the various toxic effects of the indiscriminate use of pesticides were bought to the world's attention including their impact on biodiversity, environmental pollution, extensive contamination of water bodies, the ecological imbalances these effects introduced and their route into the food chain. The book, and Ms. Carson's public engagements, highlighted the important aspects of pesticide pollution alerting the world's political, business, scientific and public sectors to the dangers. As a result, pesticides began to receive much closer scientific and political scrutiny. The public debate moved quickly from *whether* pesticides were dangerous to *which* ones were dangerous, and the burden of proof shifted from the opponents of unrestrained pesticide use to the manufacturers.

Nowadays, we know categorically that chemical pesticides can have a range of toxic effects on both humans and biodiversity. Eventually, airborne pesticide pollution will be deposited to earth, either being washed out by rain ('wet deposition') or falling as dry particles and thus non-target areas, people,

fauna and flora can become exposed and suffer damage as a result. There is significant scientific evidence to show that some airborne pesticide residues can be deposited considerable distances away from the site of application. Some, particularly those that are hydrolytically stable, such as the organochlorines, have been detected in remote areas such as the Arctic and Antarctic regions. Residues have also been found in air, rain, cloud water, fog and snow. The behaviour of pesticides in the atmosphere is complex and the concentrations found depend on several variables such as their volatility, photostability, method of application and the extent of use (Unsworth *et al.*, 1999; Oehme, 1991; Shen *et al.*, 2005). Nevertheless, it is local areas that are most at risk.

First, non-target agricultural areas (including those on neighbouring farms) can be receptors and thus there can be impacts on production. As described above, acid deposition and terrestrial eutrophication can directly damage crops causing a range of problems. Visible markings on crop leaves and produce can quickly appear as dead tissue or it can develop slowly as a yellowing of the leaf. Both airborne pollution and that deposited into the soil can also cause seriously stunted growth and reduced yields. In extreme cases, atmospheric deposition of pesticides can also lead to premature crop death but this does not usually occur unless there has been recurrent or long-term exposure. How quickly the problem develops and how economically damaging the injury becomes is dependent on a number of factors including: the specific nature and concentration of the pollutant; the length of exposure to that pollutant as well as the crop species; and possibly the particular variety and the growth stage at which it is exposed, with seedlings and young plants being more susceptible. In the case of livestock production, air pollution can lead to respiratory issues and even infection as a result of eating polluted feed and water.

Second, terrestrial habitats such as woodlands, flower meadows and other natural areas can be affected by pesticide deposition such that they suffer similar damage to that described above for crops. However, there can also be secondary effects on biodiversity due to the damage to habitats and, potentially, the loss of food sources for a range of species including pollinators (see Chapter 3). Aquatic ecosystems can be equally affected if pesticide spray drift is deposited into surface waters. Deposited herbicides, for example, can kill aquatic plants either growing in the margins or covering the water surface. This can have a range of effects in addition to the loss of the plants. Aquatic habitats will be damaged and their physical area reduced. It can alter the light levels in the water body and so alter the range of species that find it a suitable breeding place and a mass of rotting vegetation can lead to a serious degradation of water quality due to elevated nitrogen levels (see Chapter 3) and releases of methane to the atmosphere.

2.1.4.6.   *Human health*

The effects and impacts outlined above, such as climate change and acid rain can result in impacts on human health, directly and indirectly. However, this section within this chapter focuses more on the direct effects of air pollutants on human health, especially with respect to ammonia, particulate matter, pesticides and bioaerosols.

Ammonia is relatively short-lived within the atmosphere, with a typical lifespan maximum numbering a few days (Dentener and Crutzen, 1994). Upon emission, it reacts with sulphur or oxides of nitrogen to form fine particulate matter, typically within the $PM_{2.5}$ range (less than 2.5 microns diameter; Anderson *et al.*, 2003) that is also derived from the combustion of fossil fuels (Chapter 2.1.2). Particles within this size range are responsible for adverse health effects, particularly respiratory conditions (EEA, 2015).

The term 'particulate matter' (PM) usually refers to very small-sized particles typically between 0.01 and 1000 microns in size. PM size is important as the smaller the size the more inhalable the particles are and so more likely to damage human health. It is normal to include size when PM is being referred to. PM_{10} refers to material 10 microns or smaller, PM_{50} refers to particles smaller than 50 microns, and so on. Fine particulates are usually $PM_{2.5}$. PM is usually composed of substances such as pollen, spores, soil particles, combustion products from fossil fuels (e.g. carbon soots) and crop dust. The environmental impacts of PM include poor air quality, smogs and reduced visibility but it is the health effects that are the most concerning. They have been shown to cause a range of respiratory effects such as irritation of the airways, decreased lung function and aggravated asthma and bronchitis (Jansen *et al.*, 2005; Burnett *et al.*, 2014). They have also been linked with various cardiovascular events including irregular heartbeat and non-fatal heart attacks (Burnett *et al.*, 2014; Martinelli *et al.*, 2013).

Related to PMs are bioaerosols, which are also very fine particles but their composition is somewhat different to general PM matter as they include a significant proportion of living organisms such as bacteria, fungal spores, mycotoxins, moulds and virus particles. Research tends to suggest that these airborne organisms are generally not viable as they are killed by the high humidity of the atmosphere (Sattar *et al.*, 2016; Sattler *et al.*, 2001). However, this is not always the case and spores, moulds and yeasts can and do survive (Pan *et al.*, 2016). These substances can cause respiratory problems, nausea, headache and fatigue in vulnerable individuals and have been linked with the spread of viruses and bacterial infections (Sattar *et al.*, 2016).

Many farming activities can generate large quantities of dust. For example, grain dust can be produced in significant quantities when certain crops (e.g. cereals, maize) are harvested, dried, transported and processed. These particles are rarely pure, often being contaminated with bacteria, fungi, moulds, insects and pesticide residues. If inhaled they can aggravate respiratory illnesses including asthma and bronchitis. Exposure to agricultural pesticides can cause a broad range of different health effects potentially affecting every part of the body, being absorbed via the inhalation of aerosols, dusts and vapours. The type of effect will vary with the specific chemical, for example organophosphates can damage an enzyme known as acetylcholinesterase and cause problems for the central nervous system manifesting as memory loss, loss of coordination, reduced speed of response to stimuli, reduced visual ability, altered or uncontrollable mood and general behaviour, and reduced motor skills (Jaga and Dharmani, 2003), whereas other pesticides can bind to hormones such as oestrogen and disrupt the endocrine system which regulates growth, metabolism and reproduction. Other pesticides can cause cancer or can induce mutations of genetic material. Other possible health effects include asthma, allergies, and hypersensitivity, as well as impacts on fertility, reproduction and foetal development (Mostafalou and Abdollahi, 2013; Osman, 2011).

2.1.4.7. Soil

An overview of the carbon cycle, including the role of soils, is illustrated earlier in Figure 1.3. To recap, carbon may be stored within carbon reservoirs. These reservoirs are numerous but include soils. The soil functions as a sink if the carbon that enters it

exceeds the carbon that leaves. Where more carbon leaves, it functions as a *source*. As can be seen in Figure 1.3, microbial respiration and decomposition within soil is a critical component of the carbon cycle and its functioning as a source or sink. The carbon that is stored within soils can be lost to the atmosphere as CO_2 (under aerobic conditions) and CH_4 (under anaerobic conditions). The precise pathway that the carbon follows, and importantly whether it is lost to the atmosphere or remains within the soil, is determined by the localised environment within the soil itself. These emissions are an issue in themselves, but can also have further impacts. For example, the cultivation and drainage (i.e. the creation of an aerobic soil environment from what was an anaerobic environment) of soils with a high SOC can lead to emissions of CO_2 via oxidation. This can be a significant issue, as large amounts of soil and CO_2 can be lost in a relatively short space of time. Figure 2.6 shows an iron post located at Holme Fen in Cambridgeshire in the UK. The post was driven into the peat

Figure 2.6: The Holme Fen post

soil here in 1852 so that it was flush with the ground. As the soil was drained to make way for intensive agriculture the peat has shrunk and oxidised, releasing the carbon as CO_2, and now the post stands 4 metres high above the ground – so a soil 4 metres deep has been lost in just 160 years. This equates to the loss of a significant carbon sink, one that had accumulated over many thousands of years but that was lost in a matter of decades due to a change in land use and management. The importance of land use and the impact of agricultural management practices on soil carbon is discussed in greater detail in Chapter 2.3.8.

2.1.4.8.   *Noxious odours*

Noxious and offensive odours arising from agricultural holdings is a common problem that can have serious impacts on local residents, impacting on their quality of life and their ability to fully enjoy the countryside. There are numerous sources including the storage and spreading of livestock manures, slurries and sewage sludge, and from intensive livestock enterprises, particularly pigs and poultry. An odour is produced by gaseous chemicals which are detected by olfaction as 'smells' or by a pungent sensation such as a burning or tingling sensation in the nose or throat. The type and intensity of the odour largely depends on the chemicals present and the degree of offensiveness can be very subjective. Analysis of livestock excreta identified a large number of different malodorous compounds. Eriksen *et al.* (2010) divided these chemicals into four main groups: (i) volatile fatty acids; (ii) indoles and phenols; (iii) ammonia and volatile amines; and (iv) sulphur-containing compounds. The latter group is considered to be responsible for many of the worst offending odours such as hydrogen sulphide, methanethiol (methyl mercaptan), dimethyl sulphide and dimethyl disulphide. Many of these sulphur compounds only need to be present in very small concentrations for the human nose to detect them.

2.2.   *Policy and interventions*

2.2.1.   *Introduction*

Current scientific understanding advocates that there are a number of benefits to be realised if policy and regulations related to air pollution, energy efficiency and climate change are closely integrated. Climate change mitigation will help reduce air pollutants, and measures to deliver clean air will help reduce greenhouse gas emissions. As well as the obvious advantages of promoting sustainable development and moving towards low carbon societies there are also significant cost savings and benefits for both the environment, ecosystem well-being and human health. However, historically these issues were addressed separately within policy and although a much greater level of integration is now seen, the process is far from complete. Therefore, policies and interventions relating to air pollution can be broadly split into three areas:

- **Clear air:** A considerable number of policies relating to clean air relate to the urban environment and have been introduced to mitigate pollution from the transport and industry sectors. Indeed, many people believe that rural areas are free from air pollution. This is far from the case and most clean air policies are now sufficiently broad in their scope to address air pollution from agriculture. Policies and interventions include those that are concerned with ensuring that air is free from potentially harmful substances (with respect to humans) as well as those relating to dark smoke and noxious odours. The main focus is on human health although there will be additional benefits for biodiversity, environmental quality and agricultural productivity.

- **Energy and climate change:** Agriculture is a significant energy user and a contributor towards climate change, however, it can also create energy on-farm. In terms of climate change, farming practices can contribute to the problem but can also help mitigate it via carbon sequestration and GHG emission management. Climate change predictions for farming are not all negative; whilst some farms will undoubtedly have problems relating to poor yields, increased pest pressures, and so on, others may see benefits from increased crop yields and new cropping opportunities. Policies and interventions recognise these complexities. They focus on encouraging energy efficiency as well as on the management and mitigation of GHGs emissions and their impact on climate, but also on encouraging farmers to recognise the opportunities that climate change might offer.

- **Chemicals emissions:** This relates to policies and interventions that seek to tackle chemical emissions to the atmosphere. In the context of the agri-environment, the focus is on more localised and specific issues such as pesticide drift, VOCs and bioaerosols and the impacts these have on the health and welfare of workers, residents and bystanders as well as non-target species and the environment more generally.

There are a significant number of policies, European-wide strategies and legal statutes that relate to air quality; the most important ones, in the context of the agri-environment, are summarised in Table 2.3.

2.2.2. Clean air

Air pollution has been a major concern of European governments since the late 1800s.

The issue of poor air quality was first tackled by the European Union in the early 1970s following the first United Nations conference on the environment in Stockholm in 1972. As a consequence of this conference, air pollution was one of the priority areas of the first European Environment Programme in 1973, recognising the importance of air quality to the economic development and prosperity of the Community. A major issue was the common occurrence of smog in cities and industrialised areas and by the mid 1900s there was a growing awareness of the environmental damage caused by acid rain. By the mid 1900s, many countries had introduced legislation to tackle air pollutants. For example, in the UK the Clean Air Acts of 1956 and 1968 introduced measures to improve air quality by, for example, introducing smoke control areas and prohibiting the emission of dark smoke from chimneys. Bulgaria was one of the last European nations to introduce air quality regulations. Its 'Law on the Purity of the Atmospheric Air' was introduced in 1996 and introduced environmental standards for air quality and was mainly targeted towards industry and transport. Unlike that of many other European nations, this law applied from its inception to communal agriculture.

With very few exceptions such as Bulgaria, the focus at this time was on the control of emissions from industrial processes, transport and the energy sector, and not from agriculture. Within the EU, various directives (e.g. 70/220/EEC; 72/306/EEC) introduced a common approach for managing air-polluting emissions from motor vehicles and in 1975 a procedure for exchanging air quality information between EU MSs was established (75/441/EEC). Following this, there were several other air-quality related directives including, for example, limiting lead in air (82/884/EEC), combating pollution from industry (84/360/EEC), limiting

Table 2.3: The main policies and legal instruments relating to air pollution

Policies and instruments	Description	Relevant areas		
		CA[1]	E & CC[1]	CE[1]
UN Earth Summit, 1992	A major UN conference held in Rio de Janerio that addressed sustainable development and covered, for example, environmental concerns, climate change, consumption rates of natural resources and population growth. Led to the adoption of Agenda 21 and, over time, the integration of sustainability policies in almost every aspect of life.	✓	✓	✓
1979 Convention on Long Range Transboundary Air Pollution (CLRTAP)	A UN convention that aimed to limit and, as far as possible, gradually reduce and prevent air pollution including long-range transboundary air pollution. In the context of the agri-environment this is predominately airborne pesticide residues. Under CLRTAP, a number of major environmental air pollution problems have been addressed via scientific and policy collaboration leading to a number of protocols which identify a range of signatory obligations.	✓		✓
Gothenburg Protocol	One of the protocols of CLRTAP concerned with abating acidification, eutrophication and ground-level ozone.	✓		
Industrial Emissions Directive (IED) (2010/75/EU)	Replaces the Integrated Pollution Prevention and Control Directive along with several other related instruments. It establishes a framework for the control of the main industrial activities, giving priority to intervention at source, ensuring prudent management of natural resources and taking into account, when necessary, the economic situation and specific local characteristics of the place in which the industrial activity is taking place. Regulations apply, particularly to intensive pig and poultry farms.	✓		✓
Best Available Technology (BAT)	A general term with a literal meaning but in this context is linked to regulations on limiting polluting emissions from industry (i.e. the IED). BAT is industry-specific and the associated reference documents provide guidelines on emission limits. However, BAT can also be site-specific as the emission guidelines can be modified according to cost–benefit analysis and local economic or geographical constraints.	✓		✓
2001 National Emissions Ceilings Directive (2001/81/EC)	This Directive sets emission limits, or 'ceilings', for each member state, and the European Union as a whole. Ceilings are specified for four important air pollutants: nitrogen oxides, non-methane volatile organic compounds, sulphur dioxide and ammonia. Implements the EU's obligations under the Gothenburg Protocol.	✓	✓	
The Clean Air for Europe Programme (CAfE)	CAfE was launched in 2001 with the aim to develop long-term, strategic and integrated policy advice to protect against significant negative effects of air pollution on human health and the environment. Linked closely to the Thematic Strategy on Air Pollution.	✓		✓

Policies and instruments	Description	Relevant areas		
		CA[1]	E & CC[1]	CE[1]
EU Clean Air Policy Package	Aims to substantially reduce air pollution across the EU. The strategy sets out objectives for reducing the health and environmental impacts of air pollution by 2030, and contains legislative proposals to implement stricter standards for emissions and air pollution	✓		✓
Ambient Air Quality Directive (2008/50/ EC)	Cornerstone of current EU air quality legislation, updated and merged previous legal instruments. It sets legally binding limits for concentrations in outdoor air of major air pollutants that impact public health, such as particulate matter and nitrogen dioxide. As well as having direct effects, these pollutants can combine in the atmosphere to form ozone, a harmful air pollutant (and potent greenhouse gas) which can be transported great distances by weather systems.	✓	✓	
Thematic Strategy on Air pollution	Linked to CAfE this is the EU's long-term, strategic and integrated policy to protect against the effects of air pollution on human health and the environment.	✓		
Strategy on Adaptation of Climate Change	Adopted in 2013, this Strategy sets out a framework for encouraging Europe to adopt adaptation measures to deal with unavoidable impacts of anticipated changes to the climate. It has three objectives: (1) to promote and encourage action by MSs, (2) to enable better informed decision-making by acting on, increasing and sharing scientific knowledge and (3) focusing particularly on vulnerable sectors such as agriculture.		✓	
UN Kyoto Protocol	A multilateral international agreement linked to the United Nations Framework Convention on Climate Change, which commits signatories to binding emission reduction targets for reducing the emissions of greenhouse gases.		✓	
European Climate Change Programme (ECCP)	The aim of the ECCP is to identify and develop all the necessary elements of an EU strategy to implement the Kyoto Protocol. It includes a comprehensive package of policy measures and initiatives to reduce GHGs.		✓	
2020 Climate and Energy Package	Brings together a number of binding legal instruments to ensure the EU meets its climate and energy targets. These are currently (1) 20% reductions in GHG emissions based on 1990 levels, (2) 20% of EU energy derived from renewable sources and (3) a 20% improvement in the EU's energy efficiency. Includes the Emissions Trading System which is the EU's main measure for cutting GHGs from large-scale industrial installations.		✓	
2030 Climate and Energy Framework	A forward-looking and planning strategy for climate and energy. It includes EU-wide targets and policy objectives for the period between 2020 and 2030. These targets aim to help the EU achieve a more competitive, secure and sustainable energy system and to meet its long-term 2050 greenhouse gas reductions target.		✓	

Policies and instruments	Description	Relevant areas		
		CA[1]	E & CC[1]	CE[1]
Energy Roadmap	This sets out four main routes to a more sustainable, competitive and secure energy system in 2050 for Europe: (1) energy efficiency, (2) renewable energy, (3) nuclear energy and (4) carbon capture and storage. It combines these routes in different ways to create and provide an analysis of seven possible scenarios for 2050.		✓	
Thematic Strategy on the Sustainable Use of Pesticides	A thematic Strategy to reduce the impacts of pesticides on human health and the environment, and more generally, to achieve a more sustainable use of pesticides. Embedded within the associated regulations is a robust regulatory approvals process that includes assessment of the risks from airborne pesticide drift.			✓

[1] Key: CA – clean air, E & CC – energy and climate change, CE – chemical emissions

the emissions of sulphur dioxide and oxides of nitrogen from power plants (88/609/EEC) and, later, controlling air pollution from municipal waste incinerators (98/369/EEC). Various conventions and international agreements enabled a largely common approach across Europe and, indeed, the entire western world, recognising that air pollution, like many environmental issues, does not recognise national borders or geophysical boundaries. These initiatives included the 1979 Convention on Long Range Transboundary Air Pollution and the 1985 Convention for the Protection of the Ozone Layer as well as the 1992 United Nations Framework Convention on Climate Change.

The first EC directive was the 1996 Air Quality Framework Directive (Council Directive on Ambient Air Quality Assessment and Management). This described the processes and techniques that were to be used to evaluate and report air quality across the EU. It also provided a list of specific pollutants of concern. Linked to this directive were four 'daughter' directives concerned with specific pollutants. The first addressed the more common pollutants: SO_2, NO_2, NO_x,

PM_{10}, $PM_{2.5}$ and lead. The other three were much more specific, the second concerned with benzene and carbon monoxide (CO), the third, ozone and the fourth, various metals and polycyclic aromatic hydrocarbons (PAHs). In addition, various other legislative actions were taken to reduce polluting emissions, including those emitted to air, from key sectors.

At this time, the agricultural pollutants were still not a major focus for policy initiatives even though there was growing awareness of problems, especially those concerned with ammonia and noxious odours. This began to change when the Integrated Pollution Prevention and Control (IPPC) Directive was adopted (96/61/EC now revised and updated as 2010/75/EU – Industrial Emissions Directive [IED]). Industrial production processes accounted for a considerable share of the greenhouse gases, acidifying substances, waste water pollutants and solid waste pollution in Europe. To reduce these emissions, a set of common rules for permitting and controlling industrial installations have been adopted by EU MSs. Operators of industrial installations undertaking activities covered by the IED are required to obtain an environmental permit from the specific authority in that MS. Many different types of installations are covered by the IED Directive including large, industrial-scale pig and poultry farms where the issue of ammonia emissions and noxious odours are a major concern. The IED Directive requires that the Best Available Techniques (BAT) are used in that operation to ensure that emissions to air, land and water (including noxious odours) are minimised and kept below limits prescribed during the authorisation and licensing process (see also Chapter 5). BATs are described in published reference documents (BREFs) which include details of the associated emission levels (AEL). BREFs are industry-specific and provide guidelines to MSs for setting limits within operational permits, and BREFs for the intensive rearing of poultry and pigs (BREF 07.2003) and for slaughterhouses and animal by-products industries (BREF 05.2005) have been published. The IED, in certain instances, allows MS competent authorities some flexibility to set less strict emission limit values. This is possible only where an assessment shows that achieving the emission levels associated with BAT would lead to disproportionately higher costs compared to the environmental benefits that would be realised.

The 1999 Gothenburg Protocol set targets for reducing four major air pollutants by 2010; these being sulphur dioxide, nitrogen oxides, volatile organic compounds (VOCs) and ammonia. This also saw agriculture take on a greater focus, the industry being a major emitter of ammonia. This protocol lead to the introduction of the 2001 National Emissions Ceilings Directive (2001/81/EC) and the start of the current policy approach. In addition, the EU's 6th Environment Action Programme (EAP) introduced in 2002 included various air quality objectives. However, these were, perhaps,

overambitious and by the end of the period of the EAP (2010) these objectives had not been fully achieved and so these were reviewed and revised and taken forward to the 7th EAP that runs to 2020.

The Clean Air for Europe Programme (CAfE) was introduced in 2003 with the aim to establish a long-term, integrated strategy to tackle air pollution and to protect against its effects on human health and the environment. Part of this programme was to review current policies and targets with the findings to feed into the development of the new forward-looking strategy to extend until 2020. In consequence, the Thematic Strategy on Air Pollution was introduced in 2005 and in 2008 the Air Quality Directive (2008/50/EC) set strict targets for EU MSs on the atmospheric concentrations of the most harmful substances including fine particulate matter. This updated the 1996 Air Quality Framework Directive and merged much of the previous legislation into the one directive. This is still the cornerstone of EU air quality policy and has introduced a much closer integration between the policies relating to air quality and climate change management and mitigation. In the context of the agri-environment, air quality is of most concern to biodiversity and maintaining eco-system services and so this thematic strategy is described in detail in Chapter 3.

During the period 2011–2013, the European Commission conducted a further review of EU air quality policy, which resulted in the adoption of the Clean Air Policy Package. As part of the package, the Commission proposed a Clean Air Programme for Europe, updating the 2005 Thematic Strategy in order to set new objectives for EU air policy for 2020 and 2030.

A major pillar of this Thematic Strategy and at the heart of current EU air pollution legislation is an updated and revised

National Emissions Ceilings Directive (2016 2016/2284/EU) repealing and replacing the 2001 legislation. This directive sets national reduction commitments for five pollutants (sulphur dioxide, nitrogen oxides, volatile organic compounds, ammonia and fine particulate matter) responsible for acidification, eutrophication and ground-level ozone pollution which leads to significant negative impacts on human health and the environment.

The 1999 Gothenburg Protocol was the start of the target-setting policy approach and this agreement was updated in 2012, introducing new targets for signatory nations. While the original protocol set national emission ceilings for 2010 for each targeted pollutant, the revised protocol specifies emission reduction commitments in terms of percentage reductions from base 2005 to 2020. It was also extended to cover one additional air pollutant, namely particulate matter ($PM_{2.5}$), and thereby also black carbon as a component of $PM_{2.5}$.

The IED discussed above is the main regulatory control mechanism for offensive odours at EU level. However, most European nations also have national level control mechanisms. Germany has, probably, the most developed regulatory process in terms of odour. The federal government's main instrument is the '1999 Guideline on Odour in Ambient Air (GOAA)' which describes an assessment process for odour nuisance and sets minimum impact standards. Within many nations, odour management is undertaken as part of planning process but emission levels or impact standards are not often set. For example, in the UK, the Town and Country Planning Act 1990 sets out the regulatory framework for land use planning and advises local authorities that they should undertake an assessment and take into account the impacts that a new development will have on the quality of air,

including odour, but the regulation does not set specific odour limits.

2.2.3. Energy and climate change

In contrast to clean air policies, awareness of climate change is a relatively recent phenomenon. Although there had been scientific debate previously, it was not until the late 1970s that the World Meteorological Organisation (WMO) began to express concern that anthropogenic activities that lead to the emission of CO_2 might lead to a serious warming of the Earth's lower atmosphere. Scientific concerns grew over the next decade and the issue began to climb political agendas. The first notable initiative was the United Nation's (UN) Environment Programme (UNEP) established the International Panel on Climate Change (IPCC) to investigate and report on the scientific evidence on climate change. The IPCC's first report supported the development of the UN Framework Convention on Climate Change Policies in 1991 which was, subsequently, signed by 166 nations at the Earth Summit in Rio de Janeiro in 1992, coming into force in 1994. This framework included the major objective of seeking to stabilise the climate by controlling damaging emissions and introduced a GHG emission monitoring programme for all signatory nations. In 1997, the UN Kyoto Protocol was adopted, entering into force in February 2005. This international agreement established targets for reducing GHG emissions, from a number of EU MSs (and the EU as a whole). The first commitment period started in 2008 and finished in 2012 following which the 'Doha Amendment to the Kyoto Protocol' was adopted that introduced a second commitment period up to 2020 and that included new targets and revised the list of GHGs to be reported.

As a consequence of the Kyoto Protocol, the European Commission launched the

European Climate Change programme in 2000, in order to set out how it would meet the target of reducing emissions by 8% by 2012. It was widely acknowledged that agriculture can aid the process in a number of ways by reducing its own emissions of CO_2, CH_4 and N_2O, increasing carbon sequestration, and by producing biomass for energy/biofuels. It is also highly incentivised to do so as agriculture is probably the industry most vulnerable to the negative impacts of climate change as farming activities directly depend on climatic conditions. Although the Kyoto Protocol targets for the EU as a whole were seen by some as quite challenging, following debate amongst MSs the target was tightened still further such that the aim is to reduce GHG emissions by 20 per cent by 2020 (EEA, 2014), and some EU MSs have gone even further, setting tighter national targets (Lewis *et al.*, 2012). Indeed, in the recently published Climate 2030 white paper, the Commission has indicated a desire to reduce greenhouse gas emissions by 40 per cent (below the 1990 levels), and to increase the share of renewable energy to at least 27 per cent, by 2030, both of which are likely to have an influence on future agricultural and agri-environmental policy.

Currently, it is the EU 2020 Climate and Energy Package policy initiative that sets these legally-binding targets for greenhouse gas emissions as well as targets for a number of other areas related to energy efficiency and climate change. These targets were agreed by EU MSs in 2007 and enacted in EU law in 2009. The aim is to ensure that the EU's energy and climate policy objectives are delivered such that the EU overall reduces its dependency on imported energy, increases its energy security, creates jobs, boosts competitiveness and ensures smart, sustainable growth. Thus, it closely integrates climate change and energy policy.

The plan is that these targets will be revisited from time to time and tightened further post 2020. The Package has three main targets:

1. To achieve a 20 per cent reduction (based on 1990 levels) of greenhouse gas emissions by 2020;
2. To ensure that, by 2020, at least 20 per cent of all the EU's energy comes from renewable sources;
3. To improve the EU's energy efficiency by 20 per cent by 2020.

The Package includes a number of closely related activities including regulations, research and investment activities, monitoring programmes and awareness building. For example:

- The Emissions Trading System (ETS) is the primary means by which greenhouse gas emissions from large-scale premises in the industry, power and aviation sectors are managed. It works by placing a cap on the quantity of emissions these large-scale premises may emit. Within this capped limit, companies receive or can buy emissions allowances. These allowances can be traded between companies as needed but by the end of an accounting period each company must surrender sufficient allowances to cover all its emissions. Failure to do so results in significant financial penalties being imposed. Over time, the CAP will gradually be reduced to mitigate greenhouse gas emissions. It has been estimated that these sectors, collectively, account for around 45 per cent of the EU's greenhouse gas emissions. As well as EU MSs, Iceland, Liechtenstein and Norway participate in the system.
- The National Emissions Reduction Targets (NERT) applies to sectors not

covered by ETS such as public buildings and domestic housing, agriculture, waste and transport (excluding aviation). Targets vary from one MS to another in order to take into account a MS's wealth. This is known as the 'effort-sharing decision' and allows some of the poorest MSs to actually increase emissions, as economic growth is inevitably accompanied by increased emissions, whilst the richest must reduce the most. Each MS is required to report their emissions annually. Activities vary by MS and include, for example, promotion of public transport, use of electric vehicles, improving housing stock and encouraging climate-friendly farming and the generation of biogas by agricultural holdings.

- Binding national targets for renewable energy (i.e. that from wind, solar, hydro, tidal, biomass and geothermal) have been set for each MS such that they must, by 2020, raise the amount of renewable energy consumed. Again, targets vary by MS (between 10 to 49 per cent) depending upon the MS's current achievements in this area. Each MS is required to develop national action plans describing how they will meet their obligations including the establishment of sectorial targets and planned policy initiatives. Biannual progress reports must be published. Post 2020, the target for renewable energy will be raised such that an increase of 27 per cent by 2030 will be sought.
- Energy efficiency is a key area of activity with actions taken to promote innovation in low carbon technologies, research through, for example, the EU's Horizon 2020 research funding initiative, funding programmes for carbon capture and storage, and from regulatory drivers such as the Energy Efficiency Directive

that sets out various rules and obligations aimed at ensuring the EU reaches its targets for improving energy efficiency. Measures include the requirement for mandatory energy efficiency certificates to accompany the sale or rental of property, the establishment of energy efficiency standards for various products such as central heating boilers, white electrical goods and other large household appliances, the introduction of smart energy consumption meters in domestic properties and mandatory energy audits for large companies.

In order to develop the EU such that it is as resilient to the impact of climate change as possible, in 2013 EU MSs collectively adopted a strategy known as the 'Strategy on Adaptation to Climate Change'. This is comprised of a package of mechanisms, binding energy and climate targets and regulations to ensure that greenhouse gas emissions are minimised and reduced significantly against the 1990 baseline levels. Linked closely to the 2020 Climate and Energy Package and the more forward-looking 2030 Climate and Energy Framework, this strategy has three core objectives:

1. Ensuring that appropriate adaptation action is taken by all MSs. This includes each MS developing their own national adaptation strategy. Indicators of progress have also been developed and these are being used to monitor activity at MS level.

2. Making sure that all decision-making is science-based and to this end further scientific research is being undertaken to plug gaps in knowledge and mechanisms put in place to provide knowledge and data sharing. The results and findings of research, funded under various EU funding mechanisms such as Horizon

2020, will be used to directly inform the strategy and steer its future direction.

3. Focusing additional attention on those sectors, for example, agriculture, fisheries, water and energy, seen as particularly vulnerable to the impact of climate change to encourage them to become more resilient. This includes ensuring that current policies in these areas, such as the Common Agricultural Policy and the Common Fisheries Policy, are 'climate-proofed', that is, adaptation becomes central to these policies to ensure they are better able to cope with current and future climate change impacts. Other activities include looking at ways to use scarce water resources more efficiently, improving flood defences and investing in better forestry practices to guard against the impact of storms and fires. Similarly, 'climate-proofing' is seen as vital for infrastructure sector development such as those related to energy, transport and buildings. The Strategy also seeks to encourage the use of insurance and other financial products to protect infrastructure assets and development against natural disasters and other climate-related risks.

This EU-wide policy is seen as particularly important as cooperation between MSs is vital considering that often vulnerable features such as river basins, groundwater supplies and forests transcend national borders. At times of natural disasters, it is often direct neighbours that are asked for humanitarian assistance and good relationships can help deliver this with speed.

Not only does the Strategy seek to protect against the adverse effects of climate change but it also highlights the opportunities it could bring. Within the agricultural sector, there are indications that for some crops, northern European MSs will see crop yields increase and opportunities for growing different crops and new varieties, more tolerant of drought, for example, will arise. Responsive farmers across Europe will be able to benefit economically from their ability to store carbon in the soil or directly reduce their greenhouse gas emissions. Many farmers are saving money by improving the energy efficiency of their operations and exploring alternatives to traditional fossil fuels such as wind, solar (see Figure 2.7) and biofuel crops such that wind and solar farms are now a common sight in many parts of Europe (see Chapter 5.3). The Strategy on Adaptation to Climate Change seeks to ensure that whilst adaption is paramount, the arising opportunities should be accepted as this will ensure that the EU's competitiveness improves.

The EU's longer-term policy objectives are to move the EU to a low carbon economy by 2050 whilst achieving other, often conflicting objectives, of increasing competitiveness and ensuring that the EU is energy secure. To this end, the Energy Roadmap 2050 looks to identify and implement ways to deliver this objective by further changes in legislation, and by making investments in, for example, areas related to:

- New technologies, such as those related to carbon capture and storage
- Increasing the EU's consumption of renewable energy
- Improving energy efficiency
- Improving grid infrastructure by, for example, replacing that which is old and outdated with low-carbon alternatives.

Moving along this Roadmap and building on the 2020 Climate and Energy Package, the 2030 Climate and Energy Framework tightens the current 2020 targets and seeks to reduce greenhouse gas emissions still further, achieving at least a 40 per cent

Figure 2.7: Wood storage barn with solar panels
(Photo courtesy of: Pixabay.com)

reduction on the 1990 levels. In addition, the aim is to increase the amount of renewable energy consumed to at least 27 per cent and also improve energy efficiency across the EU such that at least a 27 per cent improvement is seen.

Greening of the CAP has helped ensure that agriculture plays its part in reaching the GHG emission targets. Article 28 of Regulation (EU) No 1305/2013 concerned with providing support to MSs on rural development relates directly to measures for 'agri-environment-climate'.

2.2.4. Chemical emissions

Chemicals are a part of modern-day society. Whilst they offer a wide range of benefits, many chemicals are also toxic and can harm human health and the environment when they are emitted to air. In the context of the agri-environment, the main chemicals used are those used in crop protection (i.e. chemical pesticides and biocides), however, there are also various chemical reactions that occur in the natural environment that can cause other toxic substances to form and cause air pollution, for example volatile organic compounds.

Pesticides have an important role to play in ensuring that populations have access to an abundant supply of safe and healthy food (see also Chapter 5). However, their use does involve risks to human health and the environment and so EU pesticide policy advocates the sustainable use of these chemicals to minimise the risks and maximise the benefits. This is delivered via its Thematic Strategy on the Sustainable Use of Pesticides and the Common Agricultural Policy. Over the last 30 years or so, there has been considerable global concern regarding the contamination of the environment by pesticides.

Whilst water contamination from run-off and sub-surface flow is a dominant pollution pathway (see also Chapter 4), pesticide residue concentrations resulting from spray drift and volatilisation are also a concern (Dabrowski and Shulz, 2003). Pesticide residues generated during the application process can move (drift) beyond the target area (e.g. foliage or soil in the case of pre-emergent herbicides or fumigants) to other non-target receptors as well as water such as plants, farmed and domestic animals, wildlife, and operators, workers, bystanders and local residents in the vicinity of the treatment area. In addition, some studies have shown that airborne pesticides can transport very long distances (Unsworth *et al.*, 1999). Whilst all these issues have for many years been addressed within regulatory risk assessments (to a greater or lesser extent dependent on the sophistication of the science and data availability), public concerns regarding their own exposure to agricultural pesticides and the impacts on their health, particularly from spray drift, began to gain momentum around the early 2000s. For example, in the UK, public concern prompted a judicial review by the Royal Commission on Environmental Pollution (RCEP) regarding the risks to residents and bystanders from agricultural crop spraying (RCEP, 2005). Consequently, whilst the actual policies (e.g. the Thematic Strategy for the Sustainable use of Pesticides) have not changed, the regulatory processes that underpin them have, such that the risk assessments performed under EC Regulation 1107/2009 concerned with the placing of plant protection products on the market, used to guard against exposure from pesticide drift, have been improved following a number of research projects and data collation activities.

Biocides are similar in many respects to pesticides; indeed, the two terms are often interchanged. However, biocides are more commonly used to refer to substances such as disinfectants rather than specific crop pests. Biocides are regulated in a similar manner to pesticides via the Biocidal Product Regulation (Regulation (EU) 528/2012) which concerns the placing on the market and use of biocidal products. Biocides are not used in the same way within agriculture as pesticides and the risk of drift is not such a concern. Nevertheless, there are risks to workers, operators and indeed some bystanders from volatilisation in particular and these are considered in the regulatory approvals process.

Volatile Organic Compounds (VOCs) generally refer to organic compounds that are highly volatile such that they have a strong tendency to move from the liquid to the gaseous phase. Their formation is sometimes a natural process, such as the burning of crop stubbles or other biomass. Some pesticide active substances are classed as VOCs and some product formulations include VOCs as solvents. VOCs can also be formed by the anaerobic digestion of livestock wastes and feed materials. In some instances, VOCs can be beneficial to plants by attracting pollinating insects.

The VOC Solvents Emissions Directive is the main policy instrument for the reduction of industrial emissions of volatile organic compounds (VOCs) in the European Union but it has little relevance to the agri-environment, being more applicable to industry, particularly those using large quantities of solvents such as the printing and pharmaceutical industries. Emissions from agriculture are largely unregulated other than by the enforcement of good agricultural practice and the regulations relating to pesticides and biocides.

2.3. Farm level management and protection

2.3.1. Introduction

Chapters 2.1 and 2.2 outlined the key air pollutants, their effects and impacts and the policy responses that have been implemented to tackle them. This section focuses on the farm level and how emissions can be reduced. This does not mean totally eliminating emissions, for example by taking land out of production (albeit this may be an option in some instances), but rather finding solutions that are optimal, meaning maintaining production while reducing emissions (sustainable intensification). As such, this section is structured based on farm management components, rather than management of individual air pollutants.

A related issue, of a more strategic nature, is one of displacement. Some options to reduce emissions may result in a loss of production and consequently this may result in that production (and its associated emissions) taking place elsewhere. Thus, it is important to ensure that any reductions achieved are not simply displaced (or even increased) by production at a different location. This can depend on site-specific factors, such as soil type or soil organic carbon content, and consequently can be a case of avoiding undertaking certain practices or not growing some crops in some locations. This is sometimes beyond the influence of an individual farm (albeit it can be considered at individual farm scale) and so it can overlap with policies aimed at tackling the issue. Thus, it is important that solutions at the farm level are synergistic with policy responses, so that emissions are reduced overall at regional, national and global levels.

The farm activities discussed below are all components of a single system and are extensively interconnected, however, they can be considered subcomponents of the wider system. In so doing, this provides a logical and systematic structure to examine the reduction of emissions from agricultural production, and the agri-environmental management options that could be applied in different circumstances, accounting for the sustainable intensification and displacement issues outlined above.

2.3.2. Inorganic nutrients

2.3.2.1. Introduction

Inorganic nutrients and their management are a key issue with respect to air pollution, particularly regarding N_2O and NH_3. They are a fundamental component of cropping enterprises as well as for livestock, both in terms of fertiliser for grass and forage crops and embedded emissions in feeds. The issue is inherently linked with the nitrogen cycle and is connected to other pollutants, such as the loss of nitrate via leaching (see Chapter 1.4.3). Therefore, improvements in management of inorganic nutrients could significantly reduce the environmental pressures associated with nutrient losses whilst also reducing production costs and mitigating the nitrogen contamination of air, surface and groundwaters.

Options to reduce emissions often involve techniques to control or limit the processes that lead to their production, especially the avoidance of excess nitrogen which can then be lost to the environment. Some of these options involve different techniques, while others involve alternative inputs. Each option can also impact upon production and/or the profitability of the enterprise, hence a balance needs to be struck to find an optimal solution, albeit in some instances there are win–win solutions.

Emissions of air pollutants can arise from several subcomponents including

Table 2.4: Fertiliser manufacture: GHG emissions

Fertiliser	Production technique	Nutrient content	CO_2 (kg CO_2e/ kg)	N_2O (kg CO_2e/ kg)	Other (kg CO_2e/ kg)	Total (kg CO_2e/ kg)
Urea	Europe average	46% N	0.64	0	0.08	0.73
	Best Available Technology	46% N	0.45	0	0.07	0.52
Urea ammonium nitrate	Europe average	32% N	0.88	0.59	0.05	1.53
	Best Available Technology	32% N	0.72	0.14	0.04	0.9
Calcium ammonium nitrate	Europe average	26.5% N	0.66	0.97	0.04	1.68
	Best Available Technology	26.5% N	0.5	0.22	0.03	0.75
Ammonium nitrate	Europe average	35% N	0.82	1.29	0.05	2.17
	Best Available Technology	35% N	0.62	0.29	0.04	0.96
Calcium nitrate	Europe average	15.5% N	0.4	1.06	0.02	1.49
	Best Available Technology	15.5% N	0.3	0.24	0.02	0.56
Ammonium phosphate	Europe average	18% N 46% P_2O_5	0.67	0	0.03	0.7
NPK 15-15-15	Europe average	15% N 15% P_2O_5 15% K_2O	0.56	0.47	0.03	1.06
Triple superphosphate	Europe average	48% P_2O_5	0.34	0	0.01	0.35
Muriate of potash	Europe average	60% K_2O	0.29	0	0.01	0.3

Source: Brentrup and Pallière (2008)

manufacture; storage and handling; application; nutrient uptake and use efficiency and fate.

2.3.2.2. Manufacture

Fertilisers are considered as an indirect energy use on-farm. There are significant energy use and emissions issues associated with the production of N-fertiliser, and the mining and processing of phosphate fertilisers. Most ammonium-based inorganic nitrogen fertilisers are manufactured via the Haber or Haber-Bosch process (Brentrup and Pallière, 2008). It is energy-intensive due to the high temperature and pressure required to drive the conversion of atmospheric nitrogen (N_2) to ammonia (NH_3) in the presence of hydrogen (H_2). The manufacturing process and hence the energy used is dependent on the type of fertiliser. Table 2.4 shows the GHG emissions associated with the production of a range of different fertilisers.

Table 2.4 shows that the nitrates have the highest GHG emissions per kg of product, with ammonium nitrate having the highest per kg of product based on European average data. However, if we take into account the per cent nitrogen, then it's a slightly different picture, as calcium nitrate has the highest GHG emissions per per cent N. We can also see that with modern technology the GHG emissions are significantly lower, almost half in some instances and that the differences between the different fertilisers is less. This reduction observed in 'best available technology' manufacturing facilities results from the removal and prevention of N_2O emitted during the process from entering the atmosphere (Brentrup and Pallière, 2008). Consequently, there is scope to reduce emissions before products reach the farm, by purchasing those which have lower emissions during their manufacture, albeit taking into account the needs of the crop, in terms of the fertiliser (and nutrient content) needed.

2.3.2.3.   Storage and handling

Once the fertiliser has been purchased and delivered to the farm, it then needs to be stored and handled – this is another point where it can be lost as air pollution, although the amount lost is relatively small compared with manufacturer and application.

Poor storage and handling techniques resulting in leaks and spills can result in gaseous emissions of air pollutants. Therefore, it is essential that good practice is adhered to. Considerable amounts of energy are used, and therefore indirect greenhouse gases and other air emissions are generated, during the manufacture and subsequent transport of inorganic fertilisers and so general good storage and housekeeping practice to avoid wastage is essential. This includes taking care to store in a way not to cause damage to containers and, if this occurs, to promptly make these safe. Once the fertiliser is exposed to the atmosphere air emissions are likely to occur; an event that will also diminish the quality of the product. Good planning will help avoid the accumulation of outdated materials leading to excessive quantity of fertiliser in-store and so reduce environmental risk as well as waste.

Ideally, inorganic fertilisers should be stored inside a dedicated building constructed of material that will not easily burn, such as concrete, brick or steel. However, in some circumstances, especially in densely populated areas, outside storage may offer benefits as it can help reduce the risk of fire due to electrical wiring and equipment. However, if ammonium nitrate fertilisers are stored outdoors, precautions should be taken to prevent packaging from deteriorating due to sunlight or water (e.g. covering it with plastic). Care must also be taken in this situation to collect or divert to a safe place any contaminated run-off caused by rainwater.

Most inorganic fertilisers will decompose when heated, releasing gases, some of which may be toxic, hazardous and/or harmful. Fertilisers based on ammonia (e.g. ammonium nitrate, ammonium phosphate as well as urea or urea-based fertilisers) are more susceptible to this than others. NPK fertilisers can give off a number of gases when decomposing, for example ammonia, oxides of nitrogen and hydrogen chloride depending on the source material. Whilst ammonium nitrate has a melting point of 170°C and decomposes above 210°C, decomposition can occur if store heaters are used too close to fertiliser containers or packaging. Direct electrical heaters, and indeed other electrical equipment, should not be used inside fertiliser stores. All ammonia-based fertilisers will emit ammonia gas when decomposing and those containing ammonium nitrate will also release oxides of nitrogen. Ammonium nitrate is not in itself combustible but, as it is an oxidising agent, it can assist other materials to burn, even if air is excluded. It is therefore important that precautions are taken and emergency procedures are in place to prevent a disaster event.

2.3.2.4.   Application

Application of fertiliser is the point at which it is deliberately released into the environment in order to provide nutrients to crops. However, this release can also result in pollution if not appropriately managed. First, if fertiliser is not applied within the target area (i.e. the cropped areas) it is instantly a potential pollutant – the nutrients could end up in terrestrial or aquatic habitats and/or be lost as gaseous emissions. Second, excess nutrients within the cropped area are at risk of being lost – the right amount needs to be applied at the right time in order to minimise any excess. Excess nutrients, that the plant does not uptake, are at risk of being lost as NO_3^-, NH_3 or N_2O pollution. There are two key aspects to help understand and reduce emissions.

1. AVOIDING EXCESS NITROGEN

When nitrogen is in the soil at levels in excess of crop requirements it can be lost via a number of different pathways (i.e. volatilisation; denitrification; and leaching), some of which may result in emissions of N_2O and NH_3. The management of the conditions that affect these pathways is important, and is covered in the next section within this chapter, but avoiding excess nitrogen in the soil is an important first step towards reducing potential emissions.

Reducing or not applying nitrogen fertiliser is an obvious option that can minimise excess nitrogen and thus reduce N_2O emissions. Measures that reduce the application of inorganic nitrogen fertiliser and the number of livestock can also result in a direct reduction of NH_3, and, according to Freibauer *et al.* (2004), may also reduce atmospheric CH_4. However, this can also reduce or eliminate productivity from agricultural land and so requires careful targeting to optimize mitigation. In cases where cropping ceases and the application of nitrogen fertiliser stops, ideally this should be undertaken in combination with the maintenance of plant cover. This allows the assimilation and reduction of residual soil nitrogen and prevents further nitrogen loss to the environment through, for example, leaching.

Other, more subtle, strategies may reduce emissions without compromising agricultural yield. As highlighted in Figure 2.5, the quantity of nitrogen released via ammonia deposition is variable depending on location, proximity to the source of ammonia emission and whether wet or dry deposition has occurred. Accounting for this deposition in grassland fertiliser recommendations presents an opportunity to reduce the application of ammonia- or urea-containing compounds, and further emissions to the atmosphere. Establishing the quantity of residual nitrogen within soils via soil testing, for example, prior to fertiliser application is a further method to reduce the quantity of supplementary nitrogen applied. The residual soil nitrogen is included within the overall fertiliser recommendations and the amount to be applied is adjusted as necessary. A high level of residual soil nitrogen means that the quantity of inorganic nitrogen to be applied can be decreased in response. Although ammonia is reduced due to a reduction in the quantity of N applied, the measures do not differentiate between nitrogen fertiliser formulations, that is, quantities of nitrogen as both ammonium nitrate and urea will be reduced.

Other potential mitigation strategies are nitrification inhibitors, which slow the formation of nitrate within the soil; and slow-release fertiliser products or coated fertilisers which deliver nitrogen to the crop root zone more in tandem with crop growth. Slow-release fertiliser formulations and nitrification inhibitors (Smith *et al.*, 2008; Sylvester-Bradley and Withers, 2012) both reduce the rate at which nitrogen is released into the environment, reducing the risk of leaching into watercourses. The rate of nitrogen release is decreased simultaneously to crop uptake, similar to the gradual fixation of nitrogen by clover. Surplus NO_3^- does not accumulate within the soil, decreasing the emission of N_2O from leachate or surface run-off. Nitrification inhibitors decrease the rate of NH_4^+ conversion to NO_3^-, again slowing the rate of NO_3^- formation and leaching risk. Trials of such products have found emission reductions of between 9 per cent (neem coating applied to urea in wheat) and 89 per cent (average 65%) for dicyianimide applied to urea in spring barley (Smith *et al.*, 2008). Although financial savings may be derived from a reduced environmental loss of nutrient, an additional and net cost overall may be incurred by the farmer when switching to alternative products.

When there is excess nitrogen and the crop has been harvested and/or the land is fallow, the prevention of leaching on vulnerable sandy soils may be achieved by the presence of a cover or catch crop during the winter before a spring-sown crop, and this should not compromise crop yield. A cover crop temporarily removes the residual soil nitrogen present as NO_3^- although the precise impact on N loss by leaching, and the indirect N_2O emissions that result from leachate (IPCC, 2006) is difficult to quantify. Cover crops are reported to reduce N leaching typically by between 25 and 50 kg N ha^{-1} (Silgram and Harrison, 1998). Warner *et al.* (2016) estimate this value to be 4–88 kg N ha^{-1} as a function of annual rainfall and dominant soil texture. The presence of ground cover has other benefits. It prevents soil loss from wind erosion (Duncan, 2008), and any nutrients and C contained within that soil. C and N are returned to the soil within the crop residues, however, this is simultaneous to the crop actively growing and utilising that nitrogen. Species that are able to grow and utilise nitrogen at relatively low soil temperatures and that establish extensive ground cover and rooting systems, for example barley, are considered to be among the most effective (Tzilivakis *et al.*, 2015). Field beans (*Phabia*) are criticised as

cover crops due to a reduced surface cover combined with rooting systems that do not branch extensively laterally (Tzilivakis *et al.*, 2015). The growing of a cover crop requires fuel to power machinery (Hülsbergen and Kalk 2001; Williams *et al.*, 2009) to both plant the seed then either perform a shallow cultivation or drive a sprayer to apply a herbicide in order to remove it once its cover crop function is no longer needed. The reduction in GHG emissions (i.e. N_2O) due to the decrease in nitrate leaching needs to exceed the increase in CO_2e that results from fuel consumption and depreciation associated with the additional machinery use. Warner *et al.* (2016) report that the additional GHG emissions from this fuel and from the production of seed are eliminated when leaching is reduced by 15 kg N ha^{-1}yr^{-1} or more. From an overall GHG mitigation perspective, not solely N_2O mitigation, it is important that cover crops are not used in situations where leaching is low risk, for example on heavier clay soils, where the reduction may typically be below this 15 kg N ha^{-1}yr^{-1} threshold.

Similar to cover crops, measures such as the undersowing of cultivated crops, such as maize with a legume, permits the synchronised release of nitrogen during crop growth and offers the potential to substitute inorganic supplementary nitrogen, providing that this nitrogen is deducted from the fertiliser recommendations. The potential quantity of nitrogen substituted is dependent on the legume species and, in the case of clover, the proportion within the sward (Cuttle *et al.*, 2003; Rochette and Janzen, 2005).

2. UNDERSTANDING THE CONDITIONS THAT LEAD TO EMISSIONS

Where large applications of fertiliser are combined with soil conditions favourable to denitrification, large amounts of N_2O can be produced and emitted to the atmosphere. Some additional N_2O is thought to arise in agricultural soils through the process of nitrogen fixation. The nitrogen cycle (Chapter 1.4.3) illustrates how two key processes are responsible for the majority of N_2O emissions within soils. The first, microbial nitrification, sees ammonium (NH_4^+) oxidised to nitrate (NO_3^-) under aerobic conditions (Machefert *et al.*, 2002). Ammonium is derived initially from the decomposition of plant or animal tissue as part of the decay process. The second process is denitrification in which nitrate (NO_3^-) is converted to mainly dinitrogen gas (N_2) under anaerobic conditions. In both processes, a proportion of the nitrogen forms N_2O (Machefert *et al.*, 2002), however, this is highly variable both temporally and spatially. It is influenced by a combination of local site conditions (soil texture, soil drainage and annual rainfall), method of land management (nitrogen fertiliser regime in particular), and the quantity of plant biomass within the soil (enhanced by the incorporation of crop residues). It is also stimulated in response to an increase in soil temperature because it is a microbial process (Dobbie *et al.*, 1999; Dobbie and Smith, 2003; Smith *et al.*, 1996). The fraction of N released during nitrification and denitrification that forms N_2O is variable depending on factors such as soil type and land use. Average proportions are cited by De Vries *et al.* (2003) as being 0.0125 and 0.035, respectively, on mineral soils, that is, for every kg of nitrogen released from nitrification, 0.0125 kg forms N_2O. For nitrogen released from denitrification, 0.035 kg forms N_2O. On peat soils this increases to 0.02 and 0.06, respectively, due to the high organic matter content of the soil. Critically, a greater quantity of N_2O is emitted via denitrification compared to nitrification. Denitrification also renders nitrogen that is available to the plant as NO_3^- unavailable in the form of N_2. Since emissions vary considerably due to spatial factors, different agroclimatic regions within Europe

with different annual precipitation patterns and variations in soil type, will have variable emissions. The key point to remember is that it will be highly site-specific.

Soil texture, drainage and rainfall, and their impact on the soil Water Filled Pore Space (WFPS) are concepts key to understanding N_2O emissions from soils and which pathway (nitrification or denitrification) will proceed. The WFPS determines whether aerobic or anaerobic conditions prevail within the soil. In general, when soils are below 55 per cent, WFPS N_2O emissions are predominantly from nitrification, although both processes may occur simultaneously at WFPS above this percentage. Focusing on drainage, another key determinant of WFPS is soil compaction, the loss of soil micro- and macroaggregate structure due to the exertion of pressure (Mudgal and Turbé, 2010). If we picture a wall constructed of small, regular-shaped interlocking bricks, this represents a small particulate clay soil. If those bricks, while wet and pliable, have a force applied to them so that they become tightly packed together, water will be unable to drain freely through that wall creating anaerobic conditions suitable for denitrification to proceed. This represents a compacted clay soil. Compaction may result due to the high-axle loads of large agricultural machinery (EC, 2006). It is a particular risk on wet soils in combination with soils that have a high per cent clay content (Louwagie *et al.*, 2009). Soil compaction prevents the movement of plant roots and soil fauna such as earthworms, inhibiting the benefit of natural drainage through tunnels, especially from anecic species (deep earth dwellers) that burrow vertically into the deeper soil profiles (Mudgal and Turbé, 2010).

Ammonia (NH_3) is emitted to the atmosphere by a process known as volatilisation, a process comparable to evaporation. In a similar manner to moisture evaporating under environmental stresses such as heat and wind, so does NH_3. Oxides of nitrogen (NO_x) are also released during volatilisation of inorganic and organic nitrogen, although this is predominantly NH_3 (IPCC, 2006).

Waterlogged peat soils or watercourses adjacent to agricultural land will not be fertilised deliberately, but the run-off of nitrogen from surrounding agricultural land may result in potentially high emissions of N_2O due to a high risk of denitrification (De Vries *et al.*, 2003). Measures that prevent this process, such as zero input grass buffer strips, offer a significant mitigation role.

Another pathway whereby N_2O may be formed is the leaching of nitrate (NO_3^-). Plants utilise N as NH_4^+ and NO_3^- although preference tends to be temperature-dependent (Abberton *et al.*, 2008). Nitrate does not readily bind with soil colloids and is easily removed (leached) from the soil profile by excess flow of water from, for example, heavy rainfall. A mean 1 per cent of N leached forms N_2O (IPCC, 2006), that is, for every 1 kg of nitrogen that is leached as NO_3^-, 0.01 kg will form N_2O. Another pathway is the production of ammonia. Ammonium nitrate fertiliser is a key source of atmospheric ammonia; this is, however, smaller than the contribution made by urea. Sommer and Hutchings (2001) report that surface-applied urea contributes an estimated 15 per cent of the total NH_3 emissions, compared to 2 per cent for ammonium nitrate and 5 per cent for ammonium sulphate. The emission of NH_3, in addition to the type of fertiliser, is further dependent on whether applied to arable land or grassland (Misselbrook *et al.*, 2000). When urea is applied to grassland, 23 per cent of the N forms NH_3 compared to 1.6 per cent for ammonium nitrate (van der Weerden and Jarvis, 1997). A similar pattern is observed on arable land, with emissions of 11.5 and 0.8 per cent, respectively.

2.3.3. Machinery and field operations

With respect to machinery and field operations, the main concern for air quality are those that are related to energy/fuel use and GHG emissions both directly and indirectly. As with other on-farm activities, emissions occur each time machinery is used as this effectively reduces a part of its working life (depreciates). Once the working life of the machine ceases, it is scrapped. In addition, the machine requires maintenance and repair, again something to which each unit of use will contribute. The energy and GHGs associated with all components of the machine's life cycle, its manufacture, maintenance and eventual scrapping, can be calculated and then apportioned per unit of time used for a given operation. Machine components subject to potentially high risk of mechanical damage, such as tillage implements, have proportionally higher associated indirect emissions (Hülsbergen and Kalk, 2001; Williams *et al.*, 2009).

Once machinery and inputs are on the farm, fuel is combusted to utilise them, in terms of the energy required for farm infrastructure, field operations, and so on. Fossil fuels are used directly to power a wide range of land management machinery, particularly that used in conducting operations such as soil tillage and agrochemical application (Hülsbergen and Kalk, 2001; and Kalk and Hulsbergen, 1999; Williams *et al.*, 2006, 2009). Direct fossil fuel consumption depends on the type (e.g. depth of tillage, number of passes, engine size and forward speed of the machine) and number of land management operations that take place, which are of course themselves, crop, site and production system dependent; and as a result different products (crop or livestock based) from different locations will have different emissions associated with the fuel consumed by tillage operations in producing them. In many instances the underlying fuel consumption from tillage, and therefore the quantity by which it is reduced, will vary with location. For example, the removal of ploughing in a reduction in CO_2e emissions of 0.099 tCO_2e ha^{-1}, compared to 0.05 and 0.036 on medium or coarse soils respectively, or a reduced cultivation depth reduces energy consumption, particularly on heavy soils (Kalk and Hülsbergen, 1999). The key challenge is to achieve reductions without an associated reduction in yield and/or shifting of the burden to elsewhere (displacement).

2.3.4. Water use

First, water use often requires the supply of mains water, which also has embedded emissions. Mains water treatment is an energy intensive process. According to the Environment Agency (2008), the carbon equivalent (CO_2e) of mains-treated water lies between 0.25–0.29 t CO_2e per mega litre (or kg CO_2e L^{-1}). For a crop irrigated with 350 mm ha^{-1} (or 3.5 mega litres ha^{-1}) this equates to just over 1 t CO_2e ha^{-1}. Second, the application of water to crops via irrigation is a further energy intensive process (Mosier *et al.*, 2006). The energy consumed by irrigation is determined principally by the source and requirement for pumping coupled with the pressure at which it is delivered (Soto-García *et al.*, 2013). High-pressure systems such as rainguns, or the requirement to extract groundwater from a greater depth increases the fuel required to deliver the water to the crop. A low-pressure system, where water is extracted from a surface reservoir and delivered via a low-pressure trickle system, tends to be the least energy intensive. Soto-García *et al.* (2013) cite 0.06 kWh m^3 of water extracted from an on-farm reservoir storage facility compared to 0.95 kWh m^3 where the source is groundwater.

Measures that include strategies to reduce the need for irrigation (see Chapter 4.3) will be particularly relevant to mitigating fossil fuel consumption in the southern agro-climatic regions, although are applicable to localised regions of northern European countries, for example the East of England. Fundamental maintenance of machinery in good working condition (e.g. using the correct size pump and hose length for the depth of water to be abstracted) is one way to reduce the fuel consumed and emissions associated with the application of irrigation water. Other methods to optimise application include scheduling and systems with greater efficiency (e.g. trickle irrigation; Sakellariou-Makrantonaki *et al.*, 2007). Trickle irrigation delivers water at a lower pressure, consuming smaller quantities of energy, and using smaller volumes of water because evaporation is reduced and delivery is direct to the crop roots (Sakellariou-Makrantonaki *et al.* 2007). Water sensors may be used to control irrigation such that the desired soil water capacity (usually field capacity) is not exceeded, also reducing the risk of nitrate leaching and denitrification (see also Chapter 4.3).

The use of harvested rainwater (i.e. untreated and previously unused water) as an alternative for irrigation or pesticide application eliminates energy and CO_2 emissions during its treatment. Mitigation strategies therefore include means by which to collect rainwater such as the construction of on-farm reservoirs (Warner *et al.*, 2016).

The soil management strategies (Chapter 2.3.8) that enhance SOM are of relevance to water use and they can improve the water-holding capacity of soils in addition to reducing soil evaporation, thus decreasing the requirement for irrigation and the associated energy and emissions.

2.3.5. Pest and disease management

MANUFACTURE

Crop protection makes relatively minor contributions to the overall GHG balance in many crops due to small quantities of active ingredient (Hülsbergen and Kalk 2001; Tzilivakis *et al.*, 2005a, b). Overall, this is equivalent to between 4 per cent (spring barley) and 12.5 per cent (potatoes) of the overall energy costs in UK arable crop production (Audsley *et al.* 2009), so this is not a major source of emissions, unlike supplementary nitrogen fertiliser.

Use

In order to understand how pesticide residues in air can be mitigated, an understanding of how they become airborne and the factors affecting air concentrations and transport is required. The technique chosen to apply pesticides depends largely on the product formulation (e.g. liquids, solids, dusts, fogs) but other factors also play a part including the crop to be treated (tall crops such as fruit trees and vines will need an orchard – air blast – sprayer whereas glasshouse crops can be treated with handheld devices or with those mounted on the overhead gantry). In a significant proportion of field treatments, a liquid pesticide is applied under pressure such that the liquid is forced through a small aperture (i.e. the spray nozzle). In consequence, the liquid is atomised into fine liquid particles and aerosols, forming a mist. This mist has the potential to remain airborne for a period of time and can, therefore, move with air currents and wind. Generally, the larger (and so heavier) the spray droplets the quicker they will be deposited to the crop/soil. Pesticide drift is usually associated with the movement of liquid particles with the corresponding movement of air during the application process (primary drift). However, post-application or 'secondary' drift can also occur due to volatilisation of the pesticide from crop foliage or soil surface. Factors that affect the amount of spray drift and its direction of travel are governed by environmental and meteorological conditions, application

technique and product choice. There are four main approaches to its mitigation.

1. **Application technique:** The equipment and technique used to apply the pesticide to the crop will have considerable influence on how much drift is created. The main parameter is the size of the droplets produced by the sprayer nozzle. Most spray nozzles are classified according to the mean particle size produced (e.g. very fine <100 µm; fine 100–175 µm, medium 175–250 µm, etc.). The amount of spray drift is usually related to the proportion of fine spray droplets created. However, spray pressure is also a factor with lower pressures creating less drift. In addition, the height of application can also influence drift. Ideally, the sprayer boom height should be as low as possible. How far any generated drift travels and the direction of movement will be governed by nozzle spacing and angle and the forward speed of the sprayer as well as meteorological conditions (Hilz and Vermeer, 2013). Under the EC Directive 2009/128/EC on the sustainable use of pesticides, aerial spraying is prohibited except in emergency situations. Therefore, it is not expected that this application route will add to atmospheric concentrations of pesticides in Europe.

2. **Responding to weather conditions:** Wind speed will affect the distance the drift travels and it is generally considered best practice to avoid spraying when the wind speed is greater than 6.5 km/h. Wind direction will govern the direction of movement and so if the wind is moving towards sensitive features such as water courses, then spraying should be avoided. Air temperature and relative humidity are also important as these can encourage evaporation, decreasing the

droplet size (Hobson *et al.*, 1993) and so promoting drift. High air temperature and low relative humidity will encourage drift. Spraying is best avoided when the relative humidity is below 50 per cent and when temperatures are high. Air temperature can also influence atmospheric stability and off-target movement of spray droplets. Therefore, in the summer months, the best conditions for spraying will be early or late in the day when it is generally cooler.

3. **Pesticide and product choice:** The physical–chemical properties of the pesticide active substance and the product formulation (including both the presence of other substances such as adjuvants and the physical state, i.e. solid, liquid, etc.) will influence the pesticide's tendency to drift.

   The volatility of the active substance will affect the rate and tendency for a pesticide to move into the vapour phase. This will be affected by temperature and humidity. However, many modern-day pesticides have low vapour pressures and so are less likely to be problematic in this respect. The physical–chemical properties of the chemical will also influence how it is formulated into useable products. For any specific pesticide active substance there may be a variety of formulation types available. A summary of the common ones is given in Table 2.5. The physical properties of these specific formulations will influence the risk of drift (Crush, 2006; De Schampheleire *et al.*, 2009; Hilz and Vermeer, 2013).

   Some pesticide formulations and mixes include adjuvants that are designed to alter the way the pesticide droplets behave. Adjuvants can offer a range of benefits including improving mixing, antifoam agents, buffering agents, surfactants and drift control agents. The correct choice of adjuvants can significantly affect the amount of spray drift that is generated and so these substances can be used for mitigation as well as allowing greater safety and efficacy in applications in less favourable operating conditions and environments (Oliveira *et al.*, 2013). There are two main types of adjuvants that can reduce drift, both of which modify the spray characteristics. First, there are a range of 'drift control agents which tend to be thickeners and second, there are products known as 'drift inhibitors' (or retardants) that are usually polyacrylamide or polyvinyl polymers. Thickeners work by reducing the tendency for small droplets to form by increasing the viscosity of the spray liquid. The inhibitors work by increasing droplet size improving deposition and on-target placement. For example, some oils (e.g. vegetable oils) have been shown to reduce drift more than others (mineral oils; Western *et al.*, 1999).

   Dressed seeds are usually considered a sound environmental choice on the basis that they should reduce the amount of pesticides used during the growing seasons. This is somewhat simplistic as there are additional risks including those to seed-eating birds and mammals and those relating to seed dust drift. The action of sowing treated seeds can cause the surface coating to abrade, generating, in some cases, significant dust clouds of seed dust containing small amounts of pesticides. Factors affecting this type of drift include the quality of the treated seed, the seed drilling technology and meteorological conditions (Nuyttens *et al.*, 2013).

4. **Site management:** Whilst site management does not reduce the amount of airborne pesticides, it can prevent its

Table 2.5: Some pesticide formulations and their characteristics

Formulation type	Code	Characteristics
Emulsifiable concentrate	EC	Solvent (usually organic oil) based system. Forms an oil-in-water emulsion when diluted in the sprayer tank. Solvent type will affect viscosity. May reduce drift when used with a flan-fan nozzle but this may depend on the surfactant concentration.
Flowable or suspension concentrate	SC	Tiny solid particles suspended in a liquid – usually water. Often includes wetting agents. High viscosity and so often quite thick in undiluted state. When diluted in sprayer tank the solids disperse. They can settle and so they ideally need to be kept agitated. These formulations have been shown to decrease spray drift in wind tunnel experiments more efficiently than emulsified oils.
Soluble liquid concentrate	SL	These are usually water-based products that contain a dissolved active ingredient (often a salt). This is one of the formulation types that actually contains dissolved molecules, not suspended particles.
Oil dispersion	OD	An OD is a non-aqueous SC. It combines a very good biological efficacy with an environmental friendly formulation. The active ingredient is dispersed in organic oils or methylated crop oils. Can be expected to reduce spray drift as the oil will reduce the fine droplet fraction.
Suspo-emulsions	SE	A water-based formulation that contains both suspended solids, similar to an SC, and emulsion droplets. Usually has high viscosity. Requires good agitation after dilution to maintain a homogenous dispersion. Can be expected to reduce spray drift as the oil will reduce the fine droplet fraction.
Granules	GR	A solid product applied dry and spread over the surface using spreading equipment. This formulation, especially when the granules are fine, can cause dust drift.
Wettable powders	WP	Dry formulation of fine, solid particles often mixed with a carrier. After dilution, they are similar to flowables in that they exist as solids suspended in water. Can be very dusty. These formulations are not expected to affect drift.
Water dispersible granules	WG	Dry formulations that are similar to WPs except they consist of larger particles and are typically much less dusty. After dispersion in water, they form a suspension of solids. These formulations are not expected to affect drift.

(Crush, 2006; De Schampheleire *et al.*, 2009; Hilz and Vermeer, 2013)

transport and help to keep it in the target area such that it does not contaminate non-target habitats, biodiversity and local populations. Windbreaks, which are often rows of trees or high hedgerows, are used to protect crops from wind damage and reduce soil wind-erosion. However, they have also been shown to reduce the risk of pesticide drift entering sensitive areas and surface waters. Their effectiveness depends on their siting in relation to wind direction and the vegetation density. A dense leaf canopy will block the movement of the pesticide. Aerodynamics of the windbreak shape and the local site topography will also be a factor in their effectiveness. It should be realised, however, that the trees themselves and any wildlife within them will be exposed to the pesticide and may be impacted upon (Wenneker *et al.*, 2004). Similarly, artificial barriers, which are vertical structures made from a variety of materials including metal and plastic cloth, can perform a similar function and have the advantage of not being adversely affected by the pesticide. One of the primary economic advantages of a living (tree, hedge) windbreak is that it is a cheap and cost-effective solution due to low establishment and maintenance costs. The primary economic disadvantage is that it may take several years to develop; therefore, the economic benefit is not immediate.

2.3.6. Livestock

2.3.6.1. Introduction

Livestock can be a source of a number of air pollutants including CH_4, N_2O and NH_3. Emissions can occur directly from the animal itself, from excreta inside housing, deposition in the field and slurry and manure handling and storage systems (the latter are covered in organic nutrients, Chapter 2.3.7). A key challenge is tackling the issue of pollution or 'nitrogen swapping', that is for example, in an effort to reduce NH_3 emissions, greenhouse gases in the form of N_2O are

increased, or vice versa (Sutton *et al.*, 2007). Emissions vary based on the type of animal, their diet and associated farm management and animal husbandry practices, including housing and grazing.

2.3.6.2. Livestock types

In terms of types of livestock, there are three key types which differ in how they affect emissions: ruminants (such as cattle and sheep), monogastrics (such as pigs) and poultry (such as chickens). Figures 2.8–2.10 illustrate the key differences between them. Ruminants effectively have four stomachs, and the rumen contains a number of flora and fauna, including bacteria and protozoa, to help break down feed, including cellulose – in the process this can release CH_4. Then at the other end, urine and faeces are excreted and more CH_4 is released along with nitrogen compounds. Pigs have almost the same digestive system as humans, known as monogastrics, as there is only one stomach. Methane is produced from excreta, but it is relatively low. The bigger issue is nitrogen compounds, especially ammonia. In poultry, as with pigs, the main issue is nitrogen. Excreta from poultry is a lot dryer as poultry

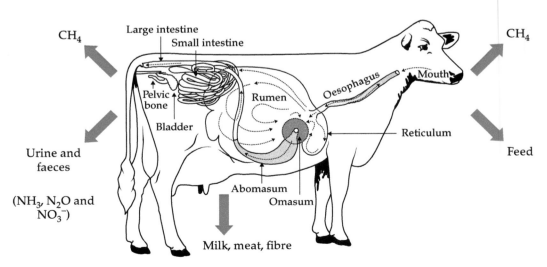

Figure 2.8: Ruminant livestock: sources of pollutants

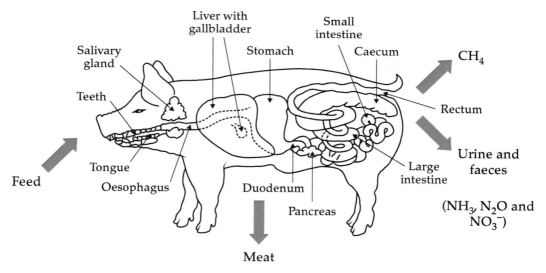

Figure 2.9: Monograstric livestock: sources of pollutants

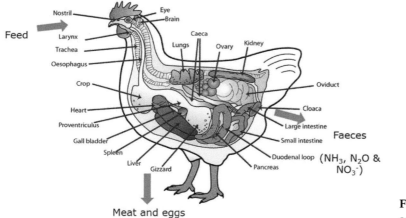

Figure 2.10: Poultry: sources of pollutants

do not actually urinate (it is combined in guano), but excreta can have a high N content, so N pollution is an issue, especially ammonia.

CH$_4$ production via fermentation under anaerobic conditions also proceeds within the stomach of animals as part of digestion. The process is, however, particularly prolific in ruminant livestock such as cattle and sheep (IPCC, 2006). Larger animals, for example dairy cows, produce more CH$_4$ which, as a general rule, increases with body weight. Diets that contain fibrous components more difficult to digest and that have a high volatile solid content, for example straw (Thomas, 2004), also increase CH$_4$ production within the rumen (Duncan, 2008). In contrast, feed with a greater starch content (e.g. maize derivatives) is noted by certain authors to decrease enteric CH$_4$ production per MJ of dietary energy, per kg of feed intake and per kg of product (Beauchemin *et al.*, 2008; Johnson and Johnson, 1995; Lovett *et al.*, 2006; Mills *et al.*, 2003; Smith *et al.*,

2008; Yan *et al.*, 2000). An increase in the formation of propionate ions ($C_2H_5COO^-$) reduces the availability of hydrogen that would otherwise form CH_4 (Monteny *et al.*, 2006).

2.3.6.3. Diet

With regard to diet, CH_4 emissions may be decreased through enhancement of the nutritional quality of grazing land by including high protein forage such as legumes (Alcock and Hegarty 2005; Waghorn *et al.* 2002) summarised in Table 2.6. It is essential, however, that this is derived simultaneously to the animal's protein needs in order to prevent excretion of excess nitrogen that has implications for N_2O emissions.

The grazing of forage crops (e.g. fodder beet or fodder radish) by livestock, and the greater ME per kg of dry matter (Table 2.6) and the associated increased digestibility, has the potential to decrease CH_4. Inclusion of high-starch feed products derived from maize provides a further option to reduce CH_4 production in the livestock rumen. However, the recommendation of maize as a means to mitigate GHG emissions from livestock comes with a caveat: first, any increase in the area of maize grown that requires a change in land use from previously uncultivated land such as grassland to one of annual cultivation is likely to result in a loss of SOC (Lovett *et al.*, 2006; Warner *et al.*, 2008a, b; Williams *et al.*, 2006); and second, maize cropping can increase soil erosion which, as highlighted in Chapter 5.1.4.4, potentially results in emission of CO_2. Therefore, a holistic perspective is needed to optimise emissions.

Dietary additives are also cited as offering potential to reduce enteric fermentation in livestock (Beauchemin *et al.*, 2008; Monteny *et al.*, 2006; Smith *et al.*, 2008). A recently completed systematic review of the direct environmental benefits of feed additives (Lewis *et al.*, 2013) identified 130 compounds that have the potential to reduce emissions including methane. Examples include lipids (oils) to increase digestion rate or halogenated compounds to prevent CH_4 production by methanogenic bacteria (van Nevel and Demeyer, 1996; Wolin *et al.*, 1964; Smith *et al.*, 2008). More than 40 of these substances showed potential to reduce emissions by more than 40 per cent. However, many of the compounds can be costly (e.g. essential

Table 2.6: Properties of selected feed types of relevance to greenhouse gas emissions

Feed type	Crude protein N (kg/kgDM)	Metabolisable energy (MJ kgDM)	Starch (g/kgDM)	Volatile solids (kg/kgDM)
Clover (red) aerial part (fresh)	0.0392	10.5	0	0.254
Dairy concentrates[1]	0.0357	11.8	267.8	0.209
Fodder beet	0.0096	12.0	0	0.358
Grass hay (average)	0.0182	8.6	0	0.363
Grass silage (average)	0.0338	10.8	0	0.346
Grazing	0.0248	11.2	0	0.359
Kale	0.0256	12.0	0	0.343
Lucerne (fresh)	0.0282	9.8	0	0.252
Lucerne silage	0.0304	8.5	0	0.246
Maize silage	0.0144	11.0	250	0.367
Wheat whole crop fermented	0.0152	10.5	200	0.364

[1]60% wheatfeed; 20% barley; 20% rapeseed meal
(from Thomas, 2004)

Figure 2.11: Cattle grazing
(Photo courtesy of: Dr John Tzilivakis)

oils) and do not have benefit for yield, thus would incur a cost to the farmer.

2.3.6.4. Grazing

Many livestock systems utilise grass and source of forage by allowing animals to directly graze it (see Figure 2.11). Urine and faeces from livestock contain nitrogen. If we refer back to the nitrogen cycle (Chapter 1.4.3) it can be seen that the deposition of nitrogen containing material onto soils in the form of excretory products will result in the emission of N_2O. The quantity of nitrogen deposited is dependent on the type of stock, the proportion of the year the animal remains outside, stocking rate and diet (Abberton *et al.*, 2008; Moorby *et al.*, 2007). For a given type of animal, it is further determined by the crude protein content of the diet relative to the protein incorporated into the animal as body growth, milk production in the case of dairy cows and during pregnancy (Abberton *et al.*,

2008; Freibauer, 2003; IPCC, 2006; Moorby *et al.*, 2007; Williams *et al.*, 2006). According to Misselbrook *et al.* (2000), 7 per cent of NH_3 emissions result from grazing deposition. This correlates with the quantity of nitrogen excreted coupled with local site-specific variables described in the previous paragraph. The quantity of nitrogen excreted by livestock is a product of age, type of animal (cattle, sheep, pigs and poultry), stocking rate and diet (see Chapter 2.3.6.3). Inorganic nitrogen applied to grassland in combination with grazing deposition, further increases the quantity of NH_3 derived from grazing cattle (Jarvis and Bussink, 1990).

As described above (see Chapters 2.1.4 and 2.3.2), when nutrients are applied to the soil they are prone to being lost as pollutants, including gaseous emissions to the air. Management of the soil can affect these processes, especially managing the level of compaction. For example, large numbers of livestock grazed in a small area for a prolonged period, or during the winter are also cited as a cause of topsoil compaction (Louwagie *et al.*, 2009) which, coupled with the N contained within grazing deposition, increases N_2O emission rates further. Lower stocking rates may still result in localised areas of soil compaction around, for example, feeding troughs (Moorby *et al.*, 2007).

Measures on grassland include the removal of livestock during the winter, the implementation of extensive grazing systems and the creation of input-free grass buffer strips on temporary grassland. These options reduce nitrogen inputs at vulnerable times of the year and reduce the impact of supplementary organic nutrients on sensitive habitat features such as watercourses. Fencing will prevent direct access by livestock to watercourses and direct deposition into them.

The removal of livestock during the winter from soils vulnerable to compaction (Chapter 5) or high-carbon organic soils prevents the deposition of nitrogen onto potentially waterlogged and anaerobic soils. The removal of livestock from grazing land during the winter coincides with limited grass growth and nutrient assimilation; the nitrogen deposited does not then become an environmental burden. Less intensive grazing systems result in a reduced rate of N ha^{-1} applied as grazing deposition and the risk of overlap between urine patches. This results in decreased N leached, decreased risk of poaching and denitrification, and a decline in the emission of NH$_3$ due to the greater N efficiency of farms (ADAS, 2007).

The emission of NH$_3$ is not influenced by the presence of high organic matter peat soils in the same manner as N$_2$O. Any reduction results from decreased risk of volatilization if deposition occurs on frozen soils. These livestock may be moved to alternative grazing areas, woodchip corrals or housed indoors. Where the latter results, manure storage will increase, and the NH$_3$ emissions further influenced.

Sheep grazing tends to be undertaken most of the year outside, although in upland areas removal from moorland areas to areas closer to farm dwellings is likely during the winter.

2.3.6.5. Housing
Much of the nitrogen consumed in livestock diets is excreted in manure as urea (uric acid for poultry) which is readily broken down to ammonia, thus, livestock housing can be a significant source of emissions (ISUS, 2004; Koerkamp, 1994; Koerkamp *et al.*, 1998). Ammonia results primarily from the breakdown of urea (present in urine) by the enzyme urease (excreted in faeces). Undigested feed protein and wasted feed are additional sources of ammonia in animal production systems. Emissions increase as a function of the total ammoniacal nitrogen

(TAN) content of the material (the quantity of nitrogen that exists as NH$_4^+$), with high NH$_4^+$ containing manures or slurries at greater risk of a higher rate of release. The design of livestock housing, especially with respect to the surface area exposed, is a key factor that can influence emissions and can overlap with slurry and manure collection systems (see Chapter 2.3.7.2) with respect to the duration of exposure (Webb *et al.*, 2005).

Reducing emissions from naturally-ventilated buildings where the manure is handled as slurry is difficult to achieve, however emissions may be reduced by 25 per cent by the addition of straw litter, as TAN is immobilised in the straw (AHDB, 2009; Webb *et al.*, 2005). However, straw-bedded housing systems may require building conversion, have higher manure management costs and may present potential animal hygiene/milk quality concerns. Reducing the area available to the animals, hence reducing the floor area contaminated by excreta, can also reduce emissions, but this may be at the expense of animal welfare. Emissions from poultry buildings can be reduced if the dry matter of the manure or litter is 60 per cent or more, thus minimising water spillages can help reduce emissions (Webb *et al.*, 2005). Finally, reducing the period of time livestock are housed will reduce emissions from housing – this could displace emissions to the field, but generally emissions from livestock grazing in the field are much lower than from housing (AHDB, 2009). However, extending the grazing season needs to be carefully managed in order to avoid soil compaction, sward damage and associated water pollution risks (see Chapter 4).

2.3.7. Organic nutrients

2.3.7.1. Introduction
What goes into livestock, and does not get metabolised, must come out. These animal

wastes are organic nutrients and thus there is scope for them to be lost as air pollutants: NH_3, CH_4 and N_2O. This can occur after deposition in the field or in housing; during collection and storage; during application (as a fertiliser); and after application. The first of these, deposition, is covered under grazing (see Chapter 2.3.6.4) – deposition could be classed under organic nutrients, but it is generally not feasible to collect, store and utilise it as an organic fertiliser, thus in this instance it is considered part of the livestock sub-system. However, collection, storage and application are covered in this section within this chapter, as they are all points where there is scope for emissions reduction.

2.3.7.2.   Collection

As mentioned in Chapter 2.3.6.5, emissions from collection systems overlap with housing to some extent, as generally collection takes place within housing and/or the farmyard. As explained in Chapter 2.3.6.5, emissions tend to correlate with the surface area exposed and the duration of exposure (Webb *et al.*, 2005), thus these elements can be controlled to some extent by the design and operation of collection systems.

As a starting point, more frequent removal of slurry can help reduce emissions. For dairy cubicle housing, increasing the frequency of scraping the fouled floor areas from once up to four times per day has been shown to reduce ammonia (by up to 20%), where this was combined with a floor design which leads to the rapid draining of urine. Similarly, research has shown that cleaning dairy cow collecting yards through pressure washing can be very effective in reducing ammonia emissions (by 90% compared with daily scraping), although this requires additional labour and increases slurry/dirty water volumes (AHDB, 2009). NH_3 emissions from pig housing can also be reduced if the surface area of the exposed slurry or manures is reduced and/or it is frequently removed and placed in covered storage outside the building (UNECE, 1999). In buildings housing laying hens, NH_3 emissions can be reduced by drying manure on belts fitted beneath cages, which remove manure. Some reduction is also possible by frequently removing the droppings using belts.

2.3.7.3.   Storage

Manure and slurries stored in tanks or lagoons gradually decompose and pollute the air with hundreds of different gases including oxides of nitrogen and ammonia. The amount of N_2O emitted depends upon the type of manure storage system, the manure composition, the length of the storage period and the conditions under which it is stored (e.g. temperature, oxygen availability). Where stored anaerobically it creates conditions suitable for denitrification. If, however, the conditions are strictly anaerobic, the lack of oxygen prevents the formation of NO_3^- to denitrify in the first place. Storage that contains a combination of aerobic and anaerobic microsites is generally the most prolific in terms of N_2O production.

Factors that determine the rate of NH_3 emission during manure storage include the physical characteristics of the material and the method of storage, that is, whether solid or slurry, the amount of ventilation within the store (rate of air movement and exchange), air temperature and relative humidity. Slurry-based and solid manure systems both risk NH_3 emission during the storage stage itself, the rate being increased further by the surface area of the manure exposed to the atmosphere, and the type of storage vessel (for example, circular store, weeping wall or lagoon). A potentially negative impact of slurry-based systems is that emissions of NH_3 are increased in addition to those of CH_4 (Moorby *et al.*, 2007).

Manures produced from diets of high volatile solid content will potentially produce more CH_4 during storage (Chadwick, 2005; IPCC, 2006; Thomas, 2004; Williams *et al.*, 2009). CH_4 emissions increase in response to higher temperatures, in addition to the storage method and notably, its impact on the presence of aerobic or anaerobic conditions (Monteny *et al.*, 2006; Sommer *et al.*, 2007). If manure is stored as slurry, the prevalent anaerobic conditions combined with the high proportion of organic compounds renders this suitable for CH_4 production (Monteny *et al.*, 2006). A cooler ambient temperature decreases the rate of CH_4 production, especially temperatures of 15°C or below (Monteny *et al.*, 2006).

The development of a crust in liquid slurries reduces emissions from an estimated 4.4 to 2.1 g N m^{-2}d^{-1}, as does covering of the store (Misselbrook *et al.*, 2000). Covering of storage tanks and lagoons (Paustian *et al.*, 2004) is a potential mitigation strategy. In addition to covering the store, if it is completely sealed shut and the gas harvested, then emission to the atmosphere is reduced even more. Two GHG mitigation parameters exist in this situation: the prevention of the release of CH_4 from the manure or slurry to the atmosphere; and the substitution of fossil fuel from the combustion of the CH_4 collected (Clemens *et al.*, 2006; Holm-Nielsen *et al.*, 2009; Junior *et al.*, 2015). This process is known as anaerobic digestion and, in terms of manure storage, is where the impact is most notable in mitigating CH_4 emission from livestock systems. Anaerobic digestion has the potential to also reduce NH_3 emissions (and therefore indirect N_2O emission) during storage, providing that the facility is entirely sealed (Clemens *et al.*, 2006; Holm-Nielsen *et al.*, 2009). The emission of N_2O from material applied after treatment has been found to be not significantly different to slurry not processed by anaerobic

digestion (Clemens *et al.*, 2006; Junior *et al.*, 2015). The mitigation component is therefore restricted solely to the storage phase. It does, however, mean that the process does not increase emissions in other parts of the N cycle (Chapter 1, Figure 1.4).

2.3.7.4. Application

Finally, when slurry and manure are applied to land, there is scope for emissions. Variables that influence NH_3 emission from material spread on soils include local site-specific spatial factors such as the soil type, the soil cation exchange capacity (the quantity of negative charges on the soil particles) and the soil pH (Bussink and Oenema, 1998). Emissions are further influenced by anthropogenic factors such as the type, method and timing of manure application coupled with climatic variables of increased wind speed and soil temperature during and immediately post application (Defra, 2010).

A surface application of liquid slurry during hot summer weather provides optimal conditions for ammonia volatilisation to proceed. The ammoniacal nitrogen content of the material is also a key variable. Misselbrook *et al.* (2000) cite emission factors of 15–59 per cent of the total ammoniacal N (TAN) applied for slurries of less than 4 per cent and greater than 8 per cent dry matter, respectively (Smith and Chambers, 1995), if undertaken between August and April. This value increases to 60 per cent for all dry matter contents if applied between May and July. For solid manure, this figure is greater still, at 76 per cent. Where a crust forms under hot, dry conditions the material becomes effectively encapsulated, reducing the further release of ammonia (Anderson *et al.*, 2003). A direct heat relationship is not strongly obvious from field measurements due to this process (Menzi *et al.*, 1998), although a correlation does exist. On frozen ground, slurry remains on the surface where volatilisation may continue, albeit slowly.

In the case of urea or high-nitrogen live-stock slurry (dairy cow or pig), the method of application is a significant factor. The shallow injection of slurry (the delivery of nitrogen containing liquid beneath the soil surface as opposed to on it) may reduce emissions by up to 80 per cent, while soil incorporation within 24 hours gives a reduction of around 30 per cent (Misselbrook *et al.*, 2000). The surface incorporation of urea will effectively reduce ammonia emissions to those similar to where ammonium nitrate is applied. The equipment required to apply supplementary nitrogen in the form of slurry more accurately is more expensive with no discernible yield increase.

Measures to mitigate volatilisation (e.g. avoidance of surface application of manures during the summer) also reduce NO_x (Hodgkinson *et al.*, 2002), decreasing green-house gas emissions in the form of N_2O.

2.3.8. Soil management

Soil and its management can be both a source and sink of pollutants, especially CO_2 and CH_4. It is a complex system to understand and manage, with many interrelated components and there are often trade-offs and synergies to consider. Soil management can also affect the emission of other pollutants to air (see above) and to water (see Chapter 4). This section within this chapter focuses more on the soil itself and its constituents, especially carbon.

Soils are a store of carbon and the enhancement of Soil Organic Matter (SOM) and Soil Organic Carbon (SOC) is targeted as a priority to improve the condition of soils throughout Europe. Different soil types can hold different quantities of SOC before reaching capacity or equilibrium. The quantity of SOM and SOC at a given moment in time (i.e. when at equilibrium) for a given land use is influenced by climatic variables, for example annual precipitation and mean annual temperature (Ganuza and Almendros, 2003; Verheijen, *et al.*, 2005), the dominant soil type, type of land use (e.g. annual cultivation, permanent or temporary grassland, woodland) and management practice (e.g. reduced or zero tillage; Dawson and Smith, 2007; Schils *et al.*, 2008). Typical values are summarised in Table 2.7 (based on Bradley *et al.*, 2005).

Changing the soil type is not an easy undertaking, therefore the land manager is restricted within the confines of the local environmental variables. An awareness of the soil type means that SOC may be enhanced to optimal levels (Bending and Turner, 2009; Johnston, 2008). The quantities of SOC in Table 2.7 allow estimation of the potential SOC accumulated for a given land use change (e.g. cropland to grassland where accumulation of SOC ceases when the grassland value is attained). It does not allow prediction of the change in SOC within the same land use although, for example, for cropland and improved grassland, the figure will be somewhere between the two values. Cultivated agricultural land typically has smaller quantities of carbon in soils and biomass than other forms of land use such as permanent grassland or woodland (Bradley *et al.*, 2005; Dyson *et al.*, 2009; Smith *et*

Table 2.7: Mean SOC (t CO_2e/ha) in England (0–30 cm) categorised by land use and soil type

Land use	Organic	Organo-mineral	Mineral	Other	All
Forestland	839.7	447.3	392.3	128.3	337.3
Cropland	623.3	429.0	282.3	106.3	245.7
Grassland	729.7	634.3	352.0	124.7	304.3

Table 2.8: SOC accumulation (to 30 cm) from a change in land use on cultivated land

New land use	t CO_2e ha^{-1} year^{-1}
Temporary grassland	1.28
Fertilised permanent grassland	4.40
Sown unfertilised grassland	3.67
Sown unfertilised grass margins	3.67
Natural reversion	1.65
Hedgerow	3.48
Scrub	3.48
Broadleaved woodland/tree strips	3.30
Conifer woodland	3.30

al., 2008). The highest SOC soils are often forestland and/or organic soils. A word of caution. The large SOC capacity associated with organic soils tends to result from management, or lack of, associated with pristine semi-natural habitat such as peat bogs and upland moorland. The presence of forestry on organic soils may actually result in a loss of SOC to the atmosphere as CO_2, due to the soil drying out. The large capacity of forestland may, therefore, be better employed on mineral soils. The majority of change in SOC on cultivated land occurs within the top 30 cm (the zone of disturbance; Smith *et al.*, 2000a, b) and so potential increases in SOC (Table 2.8) are generally limited to the top soil layers. The time taken to reach equilibrium varies depending on the potential equilibrium that may be reached, and the rate of C accumulation. It can be calculated using the following equation:

$$T = \frac{(SOCeqb_{(new)} - SOCeqb_{(baseline)})}{R_{(soc)}}$$

Where:

T = Time to establish new SOC equilibrium

$SOCeqb_{(new)}$ = potential SOC at equilibrium (t CO_2e/ha) of the new land use

$SOCeqb_{(baseline)}$ = SOC at equilibrium (t CO_2e/ha) of the baseline scenario (current land use)

$R_{(SOC)}$ = SOC accumulation rate (t CO_2e/ha/year) for a given change in land management

As highlighted previously, cultivated land contains less SOC at equilibrium than grassland or woodland for a given soil type (Bradley *et al.*, 2005; Dyson *et al.*, 2009). The frequent cultivation and disturbance of the soil exposes the SOM to oxygen causing increased oxidation of SOC to CO_2. This is not replaced due to smaller returns of plant residues to the soil (Smith *et al.*, 2000a, b; Falloon *et al.*, 2004). As a result, the quantity of SOC in the soil is typically smaller than other land uses. Methods to increase the SOC equilibrium of cultivated land without a change in land use include reduced frequency or depth of soil tillage through the inclusion of a temporary grass ley, minimum or zero tillage, stubble incorporation, prevention of soil erosion, avoidance of fallow, organic matter (crop residues, farmyard manure and straw) incorporation and avoidance of cultivation on peat soils (Janzen, 1987; Smith *et al.*, 2000a, b, c; Conant *et al.*, 2001; Falloon *et al.*, 2004; King *et al.*, 2004; Bradley *et al.*, 2005; Schils *et al.*, 2008; Dyson *et al.*, 2009; Ostle *et al.*, 2009).

Grassland has greater SOC compared to cultivated land due to the lower tillage frequency, permanent vegetation cover that returns greater plant residues. The rate of SOC accumulation in grassland may be increased by management that improves the rate of grass growth (IPCC, 2006). The SOC of grassland is reported to increase through the application of fertiliser, lime and the presence of clover in the sward (Conant *et al.*, 2001; Follett *et al.*, 2001; Ogle *et al.*, 2003; Soussana *et al.*, 2004). Broadly speaking, there are two main types of grassland, productive and unproductive. On productive grassland there is a high level of foliage off-take (i.e. removal from the system) which requires replenishment via new growth. This is achieved by the addition of supplementary nutrients. Non-productive grassland tends to receive nominal inputs. Grassland that is not grazed heavily has potential to increase in grass species diversity; this increased diversity permits 'resource partitioning', that is, a requirement for different nutrients from different layers within the soil profile at different times of the year. Competition between plant species is not as intense but biomass growth and return of SOM to the soil may potentially be high. This is subject to the underlying parent material and thus, for example, it would not be applicable to naturally species poor acid grassland. The timing (for example, removal during winter) and intensity (low to moderate stocking rates) of grazing is considered key by Conant *et al.* (2001; 2005) and Freibauer *et al.* (2004) in enhancing SOC in grassland. The natural susceptibility of a soil to compaction is also a key consideration for grassland management. Measures that reduce soil compaction risk include excessive stocking rates, grazing wet soils and the use of heavy machinery on wet soils (Louwagie *et al.*, 2009). Reduced stocking rates or, in areas of high precipitation, the removal of livestock from grassland during the winter, reduce the risk of topsoil compaction for which productivity may be decreased by up to 13% (Louwagie *et al.*, 2009). Compacted areas of grassland are often indicated by poor grass growth and vegetation cover. Where growth is poor, the return of residues tends to be more restricted and levels of SOM and SOC lower.

In both cultivated land and grassland, the SOC may be increased (subject to production displacement) where a change in land use to one of a potentially higher SOC equilibrium (e.g. woodland or forestry) results. Peat bogs sequester potentially huge quantities of C due to the low soil pH and low microbial activity coupled with anaerobic soil conditions that prevent oxidation of the C in SOM to CO_2. Carbon accumulates over thousands of years and peat depth may exceed 5 m. These habitats are identified as a priority for GHG mitigation and their maintenance is deemed essential (Schils *et al.*, 2008; Smith *et al.*, 2008; Regina and Alakukku, 2010). Emission factors used for national inventories that illustrate the importance of these habitats include 'drained shallow lowland peat soils' (4.0 t CO_2e/ha/year), 'drained deep lowland peat soils' (10.9 t CO_2e/ha/year), 'drained upland peat soils' (7.3 t CO_2e/ha/year) and 'cultivated peat soils' (15.0 t CO_2e/ha/year; Choudrie *et al.*, 2008; Freibauer, 2003). Where degradation occurs due to the creation of aerobic soil conditions by, for example, drainage, their restoration is critical as a means to prevent soil CO_2 emission (Jackson *et al.*, 2009; Schils *et al.*, 2008). Measures that remove drainage and restore these habitats potentially reverse the CO_2 release (Freeman *et al.*, 2001; Moorby *et al.*, 2007) although the impact may not be immediate depending on how long the land has been subject to drainage (Freeman *et al.*, 2001).

It can be seen that enhancement of SOM, and the C that it contains, is desirable for a number of reasons within the agricultural context, but an understanding of the reasons behind the decline is first necessary. Soils with low SOC tend to be subject to frequent cultivation, a general requisite of crop production. Reviews of opportunities for C sequestration in European agriculture have been undertaken by Conant *et al.* (2001),

Schils *et al.* (2008) and Ostle *et al.* (2009). More country-specific studies include Bradley *et al.* (2005), Dyson *et al.* (2009), Falloon *et al.* (2004), King *et al.* (2004) and Smith *et al.* (2000a, b, c).

Methane can also be emitted from soils. The emission of CH_4 from non-waterlogged soils, that is, arable crops, productive grassland and most woodland are negligible (Smith *et al.*, 2000a, b, c; Falloon *et al.*, 2004; Freibauer, 2003). The oxidation of CH_4 on aerobic soils results in the net removal of C from the atmosphere (Freibauer *et al.*, 2004). This process may be increased by a reduction of N fertiliser application, and is a potential benefit where land management intensity is reduced (for example, where supplementary nutrient application is decreased), although the restoration of the oxidation process may be subject to a time lag (Paustian *et al.*, 2004). Woodlands typically provide optimal moisture levels for abundant methanotrophic bacterial communities to thrive allowing them to act as methane sinks (Maxfield *et al.*, 2011).

Methanogenic microorganisms within anaerobic environments release CH_4 during the process of fermentation. The most significant natural environment in which this occurs is wetland soils, where the high water table means that methanogenesis occurs close to the soil surface, allowing methane to escape to the atmosphere in gaseous form. As a result, wetlands can act as both a source of CH_4 and a sink for CO_2, meaning that a complex balance may exist between the two properties. As a general rule, the emission of CH_4 from wet peat soils is correlated with depth, from which an estimated 0.5 to 3.8 t CO_2e ha^{-1} yr^{-1} may be emitted (Worrall *et al.*, 2003). Much of this CH_4 production arises from the presence of sedges including cotton grass (*Eriophorum vaginatum*), that function as a 'methane shunt', whereby CH_4 is extracted from the lower soil layers by the roots and is emitted rapidly to the atmosphere via the vegetative parts of the plant above the soil and water level.

Management practices that preserve or restore the water table though the blocking of drainage ditches have the potential to prevent the further release of CO_2 (Freeman *et al.*, 2001; Moorby *et al.*, 2007). Such options are restricted by soil type and location and are restricted to the northerly regions of the EU. Thus, paludiculture offers the potential for biomass production on peat soils while the C within peat does not oxidise to CO_2 and accumulation of C may continue. Increasing the capacity for water retention may also serve to provide a buffer and help attenuate flood peaks, thus potentially addressing another climate change objective. In the case of acid peat/blanket bogs in the uplands, the main function of paludiculture here will be capturing and storing precipitation and sequestering more carbon (rather than biomass production).

The exclusion of livestock from grassland vulnerable to flooding during the winter, or from areas adjacent to watercourses prevents damage to the soil structure and compaction. It maintains soil drainage, preventing the formation of anaerobic soil conditions, conducive with the formation of methane. This provides additional benefit in the mitigation of GHGs in the nitrogen cycle (Chapter 1).

In contrast, measures that increase the waterlogging of soils (for example, peat bog or fen restoration) have the potential to increase CH_4 emission to the atmosphere. This, however, is also dependent on the dominant vegetation structure. The restoration of plant communities dominated by *Sphagnum* spp. will not, according to Lindsay (2010), increase CH_4 significantly. Where habitats are dominated by species such as common cotton grass, there is potential for the methane shunt to proceed resulting in the rapid release of CH_4 from the lower soil layers. From a GHG mitigation perspective, the restoration process therefore benefits from ensuring that areas where *E. vaginatum* proliferates, namely those subject to periodic submergence (Rodwell, 2008), are not extensive. Permanent soil submergence, as typical of a pristine peat bog, is the restoration objective and would not therefore represent an issue when achieved, with *E. vaginatum*-dominated communities restricted to more peripheral areas.

Where high precipitation coincides with cool air temperatures, or where low-lying ground is inundated by water for the majority of the year the soils tend to be waterlogged and devoid of oxygen (anaerobic). These wet anaerobic conditions allow the formation over several hundreds of years of peat soils, a high C containing soil rich in humic material (Natural England, 2012). Maintenance of these conditions prevents the release of the significant quantities of C contained within the peat due to oxidation via decomposition, and the release of C as CO_2. The loss of anaerobic conditions through land drainage results in aerobic soil conditions, decomposition of the peat and the release of CO_2 (IPCC, 2006; Schils *et al.*, 2008). In consequence, the emission of CO_2 from drained peat may be substantial. Drained lowland and upland peat releases an estimated mean of 10.9 and 7.3 t CO_2e ha^{-1}year^{-1} in the UK, respectively (IPCC, 2006).

Operations conducive with increasing CO_2 emission include tillage, while the application of nitrogen risks an increase in N_2O. Where soil management inverts the soil on an annual basis, the loss of CO_2 from cultivated peat soils may be greater, estimated as 15.0 t CO_2e ha^{-1}year^{-1} by Freibauer (2003) although this is dependent on the climatic zone, with an estimated 5–10 t CO_2e ha^{-1}year^{-1} in cooler climates (IPCC, 2006). Agricultural peat soils may therefore constitute a significant potential source of CO_2

emissions (Schils *et al.*, 2008) rendering their restoration and maintenance a high priority, irrespective of the loss of production and displacement risk. The preservation of peat soils may be achieved by prevention of drainage/maintenance of the water table. The impact of restoration on CO_2 loss from peat soils is more uncertain. Freeman *et al.* (2001) report that the phenols that prevent peat decomposition are destroyed when the soil is drained. Decomposition and emission of CO_2 continues after restoration of the water table in response to the 'enzyme-latch effect'. The desired impact of re-flooding peat soils may not therefore be immediate.

Rice cultivation requires the creation of artificial wetlands and is a significant producer of CH_4 and the main cause of CH_4 emissions from agriculture that does not involve livestock. In Europe, however, rice cultivation is minor and the contribution of this sector to EU agricultural GHG emissions overall is small (COGEA, 2009; Leip and Bocchi, 2007). Strategies focus typically on the creation of aerobic soil conditions post harvest and pre-sowing through intermittent drainage coupled with the incorporation of composted residues under dry soil conditions (Smith and Conen, 2004; Xu *et al.*, 2000; 2003; Yan *et al.*, 2003).

The prevention of soil compaction and the maintenance of soil structure are key measures. Soil compaction may result from poaching by livestock due to grazing on wet soils. The removal of livestock from vulnerable grazing land during the winter is one mitigation strategy. A further strategy is reducing the compaction risk of the soil itself. The presence of organic matter helps prevent the compaction of soils, clay soils in particular, and maintain drainage. This helps prevent the creation of anaerobic soil conditions, and reduce the risk of denitrification. Measures that potentially enhance SOC, such as the inclusion of cover crops where

appropriate, or short-term leys are conducive with rendering soils less vulnerable.

The erosion of soil, either by water or
wind, results in soil loss (Kirkby *et al.*, 2004)
and consequently the emission of the SOC
as CO_2. Soil erosion causes a decline in soil
organic matter content (Mudgal and Turbé,
2010), and the loss of C contained within
the SOC component. Most SOM and SOC
is removed from the top 30 cm of the soil
profile, hence any removal of soil from this
layer exposes the soil beneath to the actions
of tillage. The risk of soil loss is site-specific
and dependent on soil type, local topography
(including gradient) and climatic variables
such as rainfall (see Chapter 5).

2.3.9. *Biomass*
Sequestration in biomass is, in a similar
manner to sequestration in soils, time limited, that is, it does not continue indefinitely.
The enhancement of plant biomass (Table
2.9) may be achieved through, for example,
a reduction in grazing intensity (Smith *et al.*,
2000a, b; Falloon *et al.*, 2004) in addition to
the more obvious approaches, for example
planting woodland (Smith *et al.*, 2000a, b;
Falloon *et al.*, 2004; Ostle *et al.*, 2009). The

rate of carbon sequestration and the equilibrium reached further depends on the tree
species, determined by the ecological zone.
According to the IPCC (2006) and summarised in Table 2.9, carbon sequestration in
land cover classed as forest varies from 36 t
$CO_2e/ha^{-1} yr^{-1}$ in boreal tundra woodland to
404 t $CO_2e/ha^{-1} yr^{-1}$ temperate oceanic forest.
Generally speaking, planting trees increases
carbon seque stration and the equilibrium.
Where the biomass is harvested and used
to substitute fossil energy, further emissions
reductions are possible.

The amount of C within boundary features may be enhanced by increased hedge
height, the planting of hedges on grass
boundaries or gaps in existing hedges with
new hedge plants and the addition of 'standard' trees at intervals along the hedge length
(Warner *et al.*, 2016). Plant species with
the greatest height potential (mature trees)
contain the largest quantities of C at equilibrium (Milne and Brown, 1997; Dawson
and Smith, 2007) albeit over a period of up
to 100 years. The annual biomass C accumulation rate assumes a linear annual rate of
accumulation although this is subject to the
age of the tree and the tree species (Milne

Table 2.9: Example land cover and biomass C at equilibrium

FAO ecological zone	RVC (CORINE land cover)	t CO_2e / ha^{-1} at equilibrium
All	Non-irrigated arable land	8.1
	Pastures (heavily grazed)	5.9
	Natural grasslands	8.8
	Moors and heathland	6.2
Temperate oceanic forest	Forests – broad-leaved	392.8
	Forests – coniferous	403.9
	Forests – mixed forest	398.4
Temperate continental forest	Forests – broad-leaved	259.8
	Forests – coniferous	289.5
	Forests – mixed forest	274.6
Temperate mountain systems	Forests	228.9
Boreal coniferous forest	Forests	130.0
Boreal tundra woodland	Forests	35.9
Boreal mountain systems	Forests	71.9
Sub-tropical dry forest	Forests	286.8

and Brown, 1997). The C sequestered within forest or woodland biomass when at equilibrium may be further influenced by FAO ecological zone (IPCC, 2006). Full biomass potential (maturity) is attained after different periods of time for each land use cover. Further, a change in land use may remove the biomass and C of the baseline land use during year 1, which is not representative of biomass accumulated during the years following establishment of the new land use.

Although a change in land use may increase the C sequestered, reversion back to a land use with lower equilibrium, irrespective of the time elapsed, will nullify the bulk of the C sequestered, with the loss of the carbon contained within emitted as CO_2 to the atmosphere. This might be through, for example, deforestation (IPCC, 2006; King *et al.*, 2004; Milne and Mobbs, 2006; Smith, 2005). Further, the loss proceeds typically at a rate faster than the original gain (Smith, 2004). These emissions savings are termed as being temporary, they can be lost by a reversion of the process. The CO_2 savings from, for example, decreased fuel production are termed permanent. If the original management is restored, the emissions are the same as previously but the emissions savings during the years of alternative management remain.

The composting of biomass is not without air pollution-related issues. The major concern is related to the release of bioaerosols which can form during material handling processes such as shredding for size reduction, screening and windrow turning (Wéry, 2014). This is mainly a concern for large-scale composting facilities rather than the small scale on individual farms but bioaerosols can also form in farm buildings where composting organic material is present, including livestock buildings (Rautiala *et al.*, 2003). Bioaerosols can impact on the health and welfare of farm workers

and livestock as they can contain potentially hazardous mycotoxins, endotoxins and glucans and due to the small particle size of bioaerosols (typical less than 10 µm) they are not filtered out by the nose and so can enter respiratory systems. For example, airborne *Aspergillus fumigatus* spores are often found in bioaerosols and this has been implicated in occupational allergic lung diseases such as farmer's lung disease and mushroom worker's lung disease as well as other lung infections (HSE, 2010).

There are various control options but as the very nature of composting relies on microbial activity, many control measures may actually reduce the efficiency of the composting process. For example, as bioaerosols are generally airborne, reducing dust should help reduce dispersal and so exposure. However, damping down may not always be appropriate as if the moisture content of the compost becomes too high this will facilitate the production of noxious odours and leachate, and can also encourage other toxins to form. Ensuring good ventilation can help reduce fungi and mould growth but too much will encourage bioaerosol dispersion. In enclosed spaces, dust extraction may be appropriate as may the use of personal protective equipment including respiratory protection, overalls, gloves and eye protection.

In large-scale composting facilities, monitoring programmes may be required that include air sampling and temperature management as high temperatures can facilitate bioaerosol dispersion.

2.3.10. Other activities

On-farm equipment and machinery that uses fuel or energy will emit air emissions either directly or indirectly. Farms often rely heavily on electricity for heating, lighting and ventilation. Dairy farms will also use electricity for collecting and cooling milk, heating water, and so on. Other farm buildings may also use electricity for managing air quality, air conditioning, grain dryers, and so on.

Whilst electricity use on-farm does not cause air pollution on-site, a wide range of air pollutants are caused during its production. These indirect emissions vary considerably depending on the way in which the electricity is generated and there are a number of different ways this can be done. Each approach will produce GHGs (particularly CO and nitrogen oxides) and other air pollutants in varying quantities through construction, operation (including fuel supply activities) and decommissioning. Some generation methods, such as wind power and nuclear power, release the majority of GHG and other air polluting emissions during construction and decommissioning. Others such as coal-fired power stations release the majority of pollutants during operation.

Coal-fired power stations are particularly problematic when it comes to air pollution and are considered by many to be seriously undermining Europe's efforts to mitigate climate change (CAN Europe, 2014). However, climate change is not the only issue. Burning coal releases nitrogen oxides, sulphur dioxide, particulate matter, dusts and heavy metals such as mercury and arsenic. These pollutants are a major cause of acid rain and ground-level ozone (smog), and are associated with a range of human health problems including respiratory diseases and cancer. These air pollutants are highly mobile and can be carried long distances in the atmosphere before they are converted, for example, into acids and deposited, potentially damaging plants, aquatic life and infrastructure. The extraction of coal is also problematic as it can cause local environmental destruction, contamination and depletion of water supplies. Modern coal-powered plants across Europe are regulated such that they must carefully control the emission of air pollutants using mitigation

technologies such as flue gas desulphurisation (gas scrubbers) which will remove a large proportion of the sulphur from the flue gas prior to its release to the atmosphere. Nevertheless, coal-fired power stations are still a major source of air pollution, particularly CO_2. Germany uses more coal to generate electricity than any other EU country, whilst the UK comes third in absolute coal consumption for power after Poland (CAN Europe, 2014). There are plans that coal-fired plants will be phased out across Europe but it will be some years, perhaps decades, before this process is complete. For example, The UK government currently plans to have phased out all coal-powered plants, replacing them with gas and/or nuclear, by 2025 but this date is potentially vulnerable to delay due to, for example, public debate regarding the use of nuclear energy, economics and the inevitable extensive planning process.

Farm equipment, machinery and vehicles that use fossil fuels such as petrol and diesel will also be responsible for generating air pollution both on-farm at the point of combustion and indirectly during the fuels production processes. During combustion, a wide range of pollutants are emitted including sulphur dioxide, nitrogen oxides, ground-level ozone, particulate matter, CO, CO_2, volatile organic compounds including benzene, some heavy metals and a number of other pollutants. Indirect air emissions from fossil fuels include a similar range of pollutants as when the fuel is burnt and these are emitted from the processes used to extract, refine and transport the fuel to the point of use.

Ozone-depleting substances (ODSs), that is, chlorofluorocarbons (CFCs) and hydrochlorofluorocarbons (HCFCs), are tightly regulated greenhouse gases which have been widely used in a range of equipment, mainly in refrigerators, freezers, air conditioners and fire extinguishers; all of which may be found on-farm. The use of ODSs is being phased out to protect the Earth's ozone layer from further damage and refrigerant gases are no longer permitted for maintenance and servicing of equipment. Nevertheless, estimates suggest that over 80 per cent of cooling equipment in current use still contains ODSs. Whilst these substances are in a closed system within the equipment, they can leak if there is not effective maintenance, if they are damaged or if they are not disposed of according to regulations.

References

Abberton, M.T., Marshall, A.H., Humphreys, M.W., Macduff, J.H., Collins, R.P. and Marley, C.L. (2008) Genetic improvement of forage species to reduce the environmental impact of temperate livestock grazing systems. *Advances in Agronomy*, 98, 311–355. DOI: 10.1016/S0065-2113(08)00206-X

ADAS (2007) *Rationale for the proposed NVZ Action Programme measures*. ADAS report to Defra – supporting paper D4 for the consultation on implementation of the Nitrates Directive in England. July 2007.

AHDB (2009) *Factsheet 5. Efficient milk production – Climate change. What you can do about ammonia emissions*. Agriculture and Horticulture Development Board (AHDB), Warwickshire.

Alcock, D. and Hegarty, R.S. (2005) *Effects of pasture improvement on productivity, gross margin and methane emissions of grazing sheep enterprises*. In Second Int. Conf. on Greenhouse Gases and Animal Agriculture, Working Papers (eds. C. R. Soliva, J. Takahashi and M. Kreuzer), pp. 127–130. ETH: Zurich, Switzerland.

Anderson, N., Strader, R. and Davidson, C. (2003) Airborne reduced nitrogen: ammonia emissions from agriculture and other sources. *Environment International*, 29(2–3), 277–286. DOI: 10.1016/S0160-4120(02)00186-1

Asseng, S., Ewert, F., Martre, P., Rötter, R.P., Lobell, D.B., Cammarano, D., Kimball, B.A., Ottman, M.J., Wall, G.W., White, J.W., Reynolds, M.P., Alderman, P.D., Prasad, P.V.V., Aggarwal, P.K., Anothai, J., Basso, B., Biernath, C., Challinor, A.J., De Sanctis, G., Doltra, J., Fereres, E., Garcia-Vila, M., Gayler, S., Hoogenboom, G., Hunt, L.A., Izaurralde, R.C., Jabloun, M., Jones, C.D., Kersebaum, K.C., Koehler, A-K., Müller, C., Naresh Kumar, S., Nendel, C., O'Leary, G., Olesen, J.E., Palosuo, T., Priesack, E., Eyshi Rezaei, E., Ruane, A.C., Semenov, M.A., Shcherbak, I., Stöckle, C., Stratonovitch, P., Streck, T., Supit, I., Tao, F., Thorburn, P.J., Waha, K., Wang, E., Wallach, D., Wolf, J., Zhao, Z. and Zhu Y. (2015) Rising temperatures reduce global wheat production. *Nature Climate Change*, 5, 143–147. DOI: 10.1038/nclimate2470

Audsley, E., Stacey, K.F., Parsons, D.J. and Williams, A.G. (2009) *Estimation of the Greenhouse Gas Emissions from Agricultural Pesticide Manufacture and Use*. Cranfield University, UK.

Beauchemin, K.A., Kreuzer, M., O'Mara, F. and McAllister, T.A. (2008) Nutritional management for enteric methane abatement: a review. *Australian Journal of Experimental Agriculture*, 48, 21–27. DOI: 10.1071/EA07199

Bending, G.D. and Turner, M.K. (2009) Incorporation of nitrogen from crop residues into light fraction organic matter in soils with contrasting management histories. *Biology and Fertility of Soils*, 45(3), 281–287. DOI: 10.1007/s00374-008-0326-y

Bouwman, A.F., Lee, D.S., Asman, W.A.H., Dentener, F.J., van der Hoek, K.W. and Olivier, J.G.J. (1997) A global high-resolution emission inventory for ammonia. *Global Biogeochemical Cycles*, 11, 561–588. DOI: 10.1029/97GB02266

Bradley, R.I., Milne, R., Bell, J., Lilly, A., Jordan, C. and Higgins, A. (2005) A soil carbon and land use database for the United Kingdom. *Soil Use and Management*, 21, 363–369.

Brentrup, F. and Pallière, C. (2008) Greenhouse gas emissions and energy efficiency in European nitrogen fertiliser production and use. *The International Fertiliser Society Proceedings*, 639.

Burnett, R.T., Arden Pope III, C., Ezzati, M., Olives, C., Lim, S.S., Mehta, S., Shin, H.H.,

Singh, G., Hubbell, B., Brauer, M., Anderson, H.R., Smith, K.R., Balmes, J.R., Bruce, N.G., Kan, H., Laden, F., Prüss-Ustün, A., Turner, M.C., Gapstur, S.M., Diver, W.R. and Cohen, A. (2014) An integrated risk function for estimating the global burden of disease attributable to ambient fine particulate matter exposure. *Environmental Health Perspectives*, 122(4), 397–403. DOI: 10.1289/ehp.1307049

Bussink, D.W. and Oenema, O. (1998) Ammonia volatilization from dairy farming systems in temperate areas: a review. *Nutrient Cycling in Agroecosystems*, 51(1), 19–33. DOI: 10.1023/A:1009747109538

Calverd, A.M. (2005) A radical approach to Kyoto. *Physics World*, July, 56.

Cambra-López, M., Aarnink, A. J., Zhao, Y., Calvet, S. and Torres, A. G. (2010) Airborne particulate matter from livestock production systems: a review of an air pollution problem. *Environmental Pollution*, 158(1), 1–17. DOI: 10.1016/j.envpol.2009.07.011

CAN Europe (2014) *Europe's dirty 30. How the EU's coal-fired power plants are undermining climate efforts.* Climate Action Network (CAN) Europe, HEAL, WWF, EEB and Klima Allianz, Germany. Available at: http://awsassets.panda.org/downloads/dirty_30_report_finale.pdf

Carson, R. (1962) *Silent Spring.* Houghton Mifflin Harcourt, USA.

Chadwick, D.R. (2005) Emissions of ammonia, nitrous oxide and methane from cattle manure heaps: Effect of compaction and covering. *Atmospheric Environment*, 39(4), 787–799. DOI: 10.1016/j.atmosenv.2004.10.012

Choudrie, S.L., Jackson, J., Watterson, J.D., Murrells, T., Passant, N., Thompson, A., Cardenas, L., Leech, A., Mobbs, D.C., Thistlethwaite, G., Abbott, J., Dore, C., Goodwin, J., Hobson, M., Li, Y., Manning, A., Ruddock, K. and Walker, C. (2008) *UK Greenhouse Gas Inventory, 1990 to 2006.* Annual Report for submission under the Framework Convention on Climate Change. ISBN 0-9554823-4-2.

Clemens, J., Trimborn, M., Weiland, P. and Amon, B. (2006) Mitigation of greenhouse gas emissions by anaerobic digestion of cattle slurry. Agriculture, Ecosystems and Environment, 112(2-3), 171–177. DOI: 10.1016/j.agee.2005.08.016

Cocks, A.T. (ed.) (1993) *The Chemistry and Deposition of Nitrogen Species in the Troposphere.* The Royal Society of Chemists, Cambridge, pp. 133.

COGEA (for the EC), evaluation of Common Agricultural Policy measures in the rice sector, November 2009. Available at: http://ec.europa.eu/agriculture/eval/reports/rice/fulltext_en.pdf

Conant, R.T., Paustian, K. and Elliott, E.T. (2001) Grassland management and conversion into grassland: effects on soil carbon. *Ecological Applications*, 11(2), 343–355. DOI: 10.1890/1051-0761(2001)011[0343:GMACIG]2.0.CO;2

Conant, R.T., Paustian, K., Del Grosso, S.J. and Parton, W.J. (2005) Nitrogen pools and fluxes in grassland soils sequestering carbon. *Nutrient Cycling in Agroecosystems*, 71(3), 239–248. DOI: 10.1007/s10705-004-5085-z

Crush, R. (2006) *Back to basics: a review of pesticide formulation types.* Research, Syngenta Crop Protection Inc., Greensboro, N.C. Available at: http://www.hort.cornell.edu/turf/shortcourse/BacktoBasics.pdf

Cuttle, S., Shepherd, M. and Goodlass, G. (2003) *A review of leguminous fertility-building crops, with particular reference to nitrogen fixation and utilisation.* Written as a part of Defra Project OF0316 'The development of improved guidance on the use of fertility-building crops in organic farming'.

Dabrowski, J.M. and Schulz, R. (2003) Predicted and measured levels of azinphos methyl in the Lourens River, South Africa: comparison of runoff and spray drift. *Environmental Toxicology and Chemistry*, 22(3), 494–500. DOI: 10.1002/etc.5620220305

Dawson, J.J. and Smith, P. (2007) Carbon losses from soil and its consequences for land-use management. *Science of The Total Environment*, 382(2-3), 165–190. DOI: 10.1016/j.scitotenv.2007.03.023

Defra (2007) *Direct energy use in agriculture: opportunities for reducing fossil fuel inputs.* Final report for Project AC0401. Undertaken by Warwick HRI and FEC Services Ltd. May 2007.

Defra (2010) *Fertiliser Manual (RB209)*, 8th Edition. Department for Environment, Food and Rural Affairs (Defra), The Stationery Office, London.

Defra (2012) *Agriculture in the United Kingdom 2011*. Department for Environment, Food and Rural Affairs.

Dentener, F.J. and Crutzen, P.J. (1994) A three-dimensional model of the global ammonia cycle. *Journal of Atmospheric Chemistry*, 19(4), 331–369. DOI: 10.1007/BF00694492

De Schampheleire, M., Nuyttens, D., Baetens, K., Cornelis, W., Gabriels, D. and Spanoghe, P. (2009). Effects on pesticide spray drift of the physicochemical properties of the spray liquid. *Precision Agriculture*, 10(5), 409–420. DOI: 10.1007/s11119-008-9089-6

De Vries, W., Kros, J., Oenema, O. and de Klein, J. (2003) Uncertainties in the fate of nitrogen II: a quantitative assessment of the uncertainties in major nitrogen fluxes in the Netherlands. *Nutrient Cycling in Agroecosystems*, 66(1), 71–102. DOI: 10.1023/A:1023354109910

Dobbie, K.E. and Smith, K.A. (2003) Nitrous oxide emission factors for agricultural soils in Great Britain: the impact of soil water-filled pore space and other controlling variables. *Global Change Biology*, 9(2), 204–218. DOI: 10.1046/j.1365-2486.2003.00563.x

Dobbie, K.E., McTaggart, I.P. and Smith, K.E. (1999) Nitrous oxide emissions from intensive agricultural systems: variations between crops and seasons, key driving variables and mean emission factors. *Journal of Geophysical Research*, 104(D21), 26891–26899. DOI: 10.1029/1999JD900378

Duncan, K. (2008) Agricultural practices that reduce greenhouse gases (GHGs) and generate co-benefits. *Environmental Toxicology*, II, 61–69.

Dyson, K.E., Thomson, A.M., Mobbs, D.C., Milne, R., Skiba, U., Clark, A., Levy, P.E., Jones, S.K., Billett, M.F., Dinsmore, K.J., van Oijen, M., Ostle, N., Foereid, B., Smith, P., Matthews, R.W., Mackie, E., Bellamy, P., Rivas-Casado, M., Jordan, C., Higgins, A., Tomlinson, R.W., Grace, J., Parrish, P., Williams, M.,

Clement, R., Moncrieff, J. and Manning, A. (2009) *Inventory and projections of UK emissions by sources and removals by sinks due to land use, land use change and forestry*. Ed: K.E. Dyson. Annual Report, July 2009. Department for the Environment, Food and Rural Affairs Climate, Energy and Ozone: Science and Analysis Division Contract GA01088.

EEA (2014) *Resource-efficient green economy and EU policies*. EEA Report No. 2/2014, European Environment Agency (EEA), Luxembourg: Publications Office of the European Union, ISBN 978-92-9213-465-5.

EEA (2015) *Air quality in Europe – 2015 Report*. European Environment Agency (EEA), no. 5/2015, Luxembourg: Publications Office of the European Union, ISSN 1977-8449.

EEA (2016) *Emissions of the main air pollutants in Europe*. European Environment Agency (EEA). Available at: http://www.eea.europa.eu/data-and-maps/indicators/main-anthropogenic-air-pollutant-emissions/assessment-3

Environment Agency (2008) *Greenhouse Gas Emissions of Water Supply and Demand Management Options*. SC070010/SR. Environment Agency, Bristol, UK.

Eriksen, J., Adamsen, A.P.S., Norgaard, J.V., Poulsen, H.D., Jensen, B.B. and Peterson, S.O. (2010) Emissions of sulfur-containing odorants, ammonia, and methane from pig slurry: effects of dietary methionine and benzoic acid. *Journal of Environmental Quality*, 39(3), 1097–1107. DOI: 10.2134/jeq2009.0400

Erisman, J.W. (2004) The Nanjing Declaration on management of reactive nitrogen. *Bioscience*, 54(4), 286–287. DOI: 10.1641/0006-3568(2004)054[0286:TNDOMO]2.0.CO;2

Erisman, J.W. and Wyers, G.P. (1993) Continuous measurements of surface exchange of SO_2 and NH_3: implications for their possible interaction in the deposition process. *Atmospheric Environment. Part A. General Topics*, 27(13), 1937–1949. DOI: 10.1016/0960-1686(93)90266-2

Erisman, J.W., Bleeker, A., Galloway, J. and Sutton, M.A. (2007) Reduced nitrogen in ecology and the environment. *Environmental Pollution*, 150(1), 140–149. DOI: 10.1016/j. envpol.2007.06.033

European Commission (EC) (2006) Commission staff working document accompanying the communication from the Commission to the Council, the European Parliament, the European Economic and Social Committee and the Committee of the Regions Thematic Strategy for Soil Protection impact assessment of the Thematic Strategy on Soil Protection {COM(2006)231 final}{SEC(2006)1165}. Brussels, 22.9.2006 SEC(2006)620.

European Union (2010) Rural development and climate change: implications from the Copenhagen Summit. *EU Rural Review: The Magazine from the European Network for Rural Development*, N°4, May 2010. European Union, Brussels, pp. 6–13.

Eurostat (2015) *Agriculture – Ammonia Emission Statistics*. Eurostat, Luxembourg City.

Falloon, P., Powlson D. and Smith, P. (2004) Managing field margins for biodiversity and carbon sequestration: a Great Britain case study. *Soil Use and Management*, 20(2), 240–247. DOI: 10.1111/j.1475-2743.2004.tb00364.x

FAO (2006) *Livestock's Long Shadow. Environmental Issues and Options*. Food and Agriculture, Organization of the United Nations, Rome, Italy.

Follett, R.F., Kimble, J.M. and Lal, R. (2001) The potential of U.S. grazing lands to sequester soil carbon. pp 401–430. In: R. F. Follett, J. M. Kimble and R. Lal (eds.) *The Potential of U.S. Grazing Lands to Sequester Carbon and Mitigate the Greenhouse Effect*. Lewis Publishers, Boca Raton, FL.

Freeman, C., Ostle, N. and Kang, H. (2001) An enzymic 'latch' on a global carbon store. *Nature*, 409, 149. DOI: 10.1038/35051650

Freibauer, A. (2003) Regionalised inventory of biogenic greenhouse gas emissions from European agriculture. *European Journal of Agronomy*, 19(2), 135–160. DOI: 10.1016/ S1161-0301(02)00020-5

Freibauer, A., Rounsevell, M.D.A., Smith, P., Verhagen, J. (2004) Carbon sequestration in the agricultural soils of Europe.

Geoderma, 122(1), 1–23. DOI: 10.1016/j. geoderma.2004.01.021

Ganuza, A. and Almendros, G. (2003) Organic carbon storage in soils of the Basque Country (Spain): the effect of climate, vegetation type and edaphic variables. Biology and Fertility of Soils, 37(3), 154–162. DOI: 10.1007/ s00374-003-0579-4

Hairsine, K. (2013) *Germany – controversially – still bombards forests with limestone to combat acid rain. Deutsche Welle*, 19/11/2013 Edition. Available at: http://dw.com/p/1AKi7

Hales, K.E., Parker, D.B. and Cole, N.A. (2012) Potential odorous volatile organic compound emissions from feces and urine from cattle fed corn-based diets with wet distillers grains and solubles. Atmospheric Environment, 60, 292–297. DOI: 10.1016/j.atmosenv.2012.06.080

Hamilton, S.W., DePeters E.J. and Milloehner, F.M. (2007) *Effects of dietary Rumensin on greenhouse gas and volatile organic compound emissions from lactating dairy cows*. Confidential report to ELANCO by Animal Science, UC Davis, USA.

Health and Safety Executive (HSE) (2010) *Bioaerosol Emissions from Waste Composting and the Potential for Workers' Exposure*. UK Health and Safety Executive, Derbyshire, UK.

Hesthagen, T., Sevaldrud, I.H. and Berger, H.M. (1999) Assessment of damage to fish population in Norwegian lakes due to acidification. *Ambio*, 28, 112–117.

Hilz, E. and Vermeer A.W.P. (2013) Spray drift review: the extent to which a formulation can contribute to spray drift reduction. *Crop Protection*, 44, 75–83. DOI: 10.1016/j. cropro.2012.10.020

Hobson, P.A., Miller, P.C.H., Walklate, P.J., Tuck, C.R. and Western, N.M. (1993) Spray drift from hydraulic spray nozzles: the use of a computer simulation model to examine factors influencing drift. *Journal of Agricultural Engineering Research*, 54(4), 293–305. DOI: 10.1006/jaer.1993.1022

Hodgkinson, R.A, Chambers, B. J., Withers, P.J.A and Cross, R. (2002). Phosphorus losses to surface waters following organic manure applications to a drained clay soil. *Agricultural Water Management*, 57, 155–173.

Holm-Nielsen, J.B., Al Seadi, T. and Oleskowicz-Popiel, P. (2009) The future of anaerobic digestion and biogas utilization. *Bioresource Technology*, 100(22), 5478–5484. DOI: 10.1016/j.biortech.2008.12.046

Hülsbergen, K.J. and Kalk, W.D. (2001) Energy balances in different agricultural systems – can they be improved? *The International Fertiliser Society Proceedings*, 476.

IPCC (2006) *2006 Intergovernmental Panel on Climate Change (IPCC) Guidelines for National Greenhouse Gas Inventories. Vol. 4, Agriculture, Forestry and Other Land Use*. Edited by S. Eggleston *et al.*, Institute for Global Environmental Strategies, Hayama, Japan.

IPCC (2007) *Contribution of Working Group I to the Fourth Assessment Report of the Intergovernmental Panel on Climate Change, 2007*. Solomon, S., D. Qin, M. Manning, Z. Chen, M. Marquis, K.B. Averyt, M. Tignor and H.L. Miller (eds.) Cambridge University Press, Cambridge, United Kingdom and New York, NY, USA.

IPPC (2001a) *The scientific basis for climate change*. Contribution of Working Group 1 to the Third Assessment Report to the IPPC. Cambridge University Press, Cambridge, UK.

IPPC (2001b) *Good practice guidance and uncertainty management in national greenhouse gas inventories*. IPCC National Greenhouse Gas Inventories Program Technical Support Unit, Kanagawa, Japan.

ISUS (2004) *Practices to Reduce Ammonia Emissions from Livestock Operations*. Iowa State University of Science (ISUS).

Jackson, J., Li, Y., Passant, N., Thomas, J., Thistlethwaite, G., Thomson, A., and Cardenas, L. (2009) *Greenhouse Gas Inventories for England, Scotland, Wales and Northern Ireland: 1990–2006*. Tech. rep., AEA Energy and Environment, Didcot, England.

Jaga, K. and Dharmani, C. (2003) Sources of exposure to and public health implications of organophosphate pesticides. *Revista Panamericana de Salud Pública*, 14(3), 171–185.

Jansen, K.L., Larson, T.V., Koenig, J.Q., Mar, T.F., Fields, C., Stewart, J. and Lippmann, M. (2005) Associations between health effects and particulate matter and black carbon in subjects with respiratory disease. *Environmental Health Perspectives*, 113(12), 1741–1746. DOI: 10.1289/ehp.8153

Janzen, H.H. (1987) Soil organic matter characteristics after long-term cropping to various spring wheat rotations. *Canadian Journal of Soil Science*, 67(4), 845–856. DOI: 10.4141/cjss87-081

Jarvis, S.C. and Bussink, D.W. (1990) Nitrogen losses from grazed swards by ammonia volatilization. In: *Proceedings of the 13th General Meeting of the European Grassland Federation*, pp. 13–17.

Johnson, K.A. and Johnson, D.E. (1995) Methane emissions from cattle. *Journal of Animal Science*, 73(8), 2483–2492. DOI: 10.2527/1995.7382483x

Johnston, A.E. (2008) Resource or waste: the reality of nutrient cycling to land. *The International Fertiliser Society Proceedings*, 630.

Junior, C.C., Cerri, C.E., Pires, A.V. and Cerri, C.C. (2015) Net greenhouse gas emissions from manure management using anaerobic digestion technology in a beef cattle feedlot in Brazil. *Science of The Total Environment*, 505, 1018–1025. DOI: 10.1016/j.scitotenv.2014.10.069

Kalk, W.D. and Hülsbergen, K.J. (1999) Dieselkraftstoffeinsatz in der Pflanzenproduktion. *Landtechnik*, 54, 332–333.

King, J.A., Bradley, R.I., Harrison, R. and Carter, A.D. (2004) Carbon sequestration and saving potential associated with changes to the management of agricultural soils in England. *Soil Use and Management*, 20(4), 394–402. DOI: 10.1111/j.1475-2743.2004.tb00388.x

Kirkby, M. J., Jones, R.J.A., Irvine, B., Gobin, A, Govers, G., Cerdan, O., van Rompaey, A.J.J., Le Bissonnais, Y., Daroussin, J., King, D., Montanarella, L., Grimm, M., Vieillefont, V., Puigdefabregas, J., Boer, M., Kosmas, C., Yassoglou, N., Tsara, M., Mantel, S., van Lynden, G.J. and Huting, J. (2004) *Pan-European Soil Erosion Risk Assessment: The PESERA Map*, Version 1 October 2003. Explanation of Special Publication Ispra 2004

No.73 (S.P.I.04.73). European Soil Bureau Research Report No.16, EUR 21176, 18pp. and 1 map in ISO B1 format. Office for Official Publications of the European Communities, Luxembourg.

Koerkamp, P.W.G.G. (1994) Review on emissions of ammonia from housing systems for laying hens in relation to sources, processes, building design and manure handling. *Journal of Agricultural Engineering Research*, 59(2), 73–87. DOI: 10.1006/jaer.1994.1065

Koerkamp, P.W.G.G., Metz, J.H.M., Uenk, G.H., Phillips, V.R., Holden, M.R., Sneath, R.W., Short, J.L., White, R.P., Hartung, J., Seedorf, J., Schroder, M., Linkert, K.H., Pedersen, S., Takai, H., Johnsen, J.O. and Wathes, C.M. (1998) Concentrations and emissions of ammonia in livestock buildings in northern Europe. *Journal of Agricultural Engineering Research*, 70(1), 79–95. DOI: 10.1006/jaer.1998.0275

Leip, A. and Bocchi, S. (2007) Contribution of rice production to greenhouse gas emissions in Europe. In: Bocchi, S., Ferrero, A. and Porro, A. (eds.) *Proceedings of the Fourth Temperate Rice Conference*, 25–28 June 2007, Novara, Italy, pp. 30–31.

Lewis, K.A., Green, A., Warner, D.J. and Tzilivakis, J. (2012) Carbon accounting tools: are they fit for purpose in the context of arable cropping. *International Journal of Agricultural Sustainability*, 11(2), 159–175. DOI: 10.1080/14735903.2012.719105

Lewis, K.A., Tzilivakis, J., Green, A., Warner, D.J., Stedman, A. and Naseby, D. (2013) Review of substances/agents that have direct beneficial effect on the environment: mode of action and assessment of efficacy. Final report for EFSA project CFT/EFSA/FEED/2012/02. 178 pp.

Lindsay, R. (2010) *Peatbogs and Carbon*. Royal Society for the Protection of Birds. Sandy, Bedfordshire, UK.

Louwagie, G., Gay, S.H. and Burrell, A. (eds.) (2009) *Final report on the project 'Sustainable Agriculture and Soil Conservation (SoCo)', JRC Scientific and Technical report*. Publications Office of the European Union, Luxembourg.

Lovett, D.K., Shalloo, L., Dillon, P. and O'Mara, F.P. (2006) A systems approach to quantify greenhouse gas fluxes from pastoral dairy production as affected by management regime. *Agricultural Systems*, 88(2–3), 156–179. DOI: 10.1016/j.agsy.2005.03.006

Machefert, S.E., Dise, N.B., Goulding, K.W.T. and Whitehead, P.G. (2002) Nitrous oxide emission from a range of land uses across Europe. *Hydrology and Earth System Sciences*, 6, 325–337. DOI: 10.5194/hess-6-325-2002

Martinelli, N., Olivieri, O. and Girelli, D. (2013) Air particulate matter and cardiovascular disease: a narrative review. *European Journal of Internal Medicine*, 24(4), 295–302. DOI: 10.1016/j.ejim.2013.04.001

Maxfield, P.J., Brennand, E.L., Powlson, D.S. and Evershed, R.P. (2011) Impact of land management practices on high-affinity methanotrophic bacterial populations: evidence from long-term sites at Rothamsted. *European Journal of Soil Science*, 62(1), 56–68. DOI: 10.1111/j.1365-2389.2010.01339.x

McInerney, M.J., Bryant, M.P., Hespell, R.B. and Costerton, J.W. (1981) Syntrophomonas wolfei gen. nov. sp. nov., an anaerobic, syntrophic, fatty acid-oxidizing bacterium. *Applied and Environmental Microbiology*, 41(4), 1029–1039.

Menzi, H., Katz, P.E., Fahrni, M., Neftel, A. and Frick, R. (1998) A simple empirical model based on regression analysis to estimate ammonia emissions after manure application. *Atmospheric Environment*, 32(3), 301–307. DOI: 10.1016/S1352-2310(97)00239-2

Mills, J.A.N., Kebreab, E., Yates, C.M., Crompton, L.A., Cammell, S.B., Dhanoa, M.S., Agnew, R.E. and France, J. (2003) Alternative approaches to predicting methane emissions from dairy cows. *Journal of Animal Science*, 81(12), 3141–3150.

Milne, R. and Brown, T.A. (1997) Carbon in the vegetation and soils of Great Britain. *Journal of Environmental Management*, 49(4), 413–433. DOI: 10.1006/jema.1995.0118

Milne, R. and Mobbs, D.C. (eds.) (2006) *UK emissions by sources and removals by sinks due to land use, land use change and forestry activities*. DEFRA Contract EPG 1/1/160.

Misselbrook, T.H., Van Der Weerden, T.J., Pain, B.F., Jarvis, S.C. and Chambers, B.J. (2000) Ammonia emission factors for UK agriculture. *Atmospheric Environment*, 34(6), 871–880. DOI: 10.1016/S1352-2310(99)00350-7

Monteny, G., Bannink, A. and Chadwick, D. (2006) Greenhouse gas abatement strategies for animal husbandry. *Agriculture, Ecosystems and Environment*, 112(2–3), 163–170. DOI: 10.1016/j.agee.2005.08.015

Moorby, J.M., Chadwick, D.R., Scholefield, D., Chambers, B.J. and Williams, J.R. (2007) *A review of best practice for reducing greenhouse gases*. Defra project report AC0206.

Mosier, A.R., Duxbury, J.M., Freney, J.R., Heinemeyer, O., Minami, K. and Johnson, D.E. (1998) Mitigating agricultural emissions of methane. *Climate Change*, 40(1), 39–80. DOI: 10.1023/A:1005338731269

Mosier, A.R., Halvorson, A.D., Reule, C.A. and Liu, X.J. (2006) Net global warming potential and greenhouse gas intensity in irrigated cropping systems in northeastern Colorado. *Journal of Environmental Quality*, 35(4), 1584–1598. DOI: 10.2134/jeq2005.0232

Mostafalou, S. and Abdollahi, M. (2013) Pesticides and human chronic diseases: evidences, mechanisms, and perspectives. *Toxicology and Applied Pharmacology*, 268(2), 157–177. DOI: 10.1016/j.taap.2013.01.025

Mudgal, S. and Turbé, A. (2010) *Soil biodiversity: functions, threats and tools for policy makers*. [Contract 07.0307/2008/517444/ETU/B1]. European Commission DG ENV. Final report February 2010.

Muniz, I.P. (1990) Freshwater acidification: its effects on species and communities of freshwater microbes, plants and animals. *Proceedings of the Royal Society of Edinburgh. Section B. Biological Sciences*, 97, 227–254.

Natural England (2012) *Environmental stewardship and climate change mitigation*. Natural England Technical Information Note TIN107, 2012.

Nuyttens, D., Devarrewaere, W., Verboven, P. and Foqué, D. (2013) Pesticide-laden dust emission and drift from treated seeds during seed drilling: a review. *Pest Management Science*, 69(5), 564–575. DOI: 10.1002/ps.3485

Odén, S. (1968) *Nederbördens och luftens försurning: dess orsaker, förlopp och verkan i olika miljöer*. Ekologikommittén, Statens naturvetenskapliga forskningsråd.

Oehme, M. (1991) Further evidence for long-range air transport of polychlorinated aromates and pesticides: North America and Eurasia to the Arctic. *Ambio*, 20(7), 293–297.

Ogle, S.M., Breidt, F.J., Eve, M.D. and Paustian, K. (2003) Uncertainty in estimating land-use and management impacts on soil organic carbon storage for U.S. agricultural lands between 1982 and 1997. *Global Change Biology*, 9(11), 1521–1542. DOI: 10.1046/j.1365-2486.2003.00683.x

Oliveira, R.B. de, Antuniassi, U.R., Mota, A.A.B. and Chechetto, R.G. (2013) Potential of adjuvants to reduce drift in agricultural spraying. *Engenharia Agrícola*, 33(5), 986–992.

Osman, K.A. (2011) Pesticides and human health. In: Stoytcheva, M. (ed.) *Pesticides in the Modern World – Effects of Pesticides Exposure*. InTech, Rijeka, pp. 205–230.

Ostle, N.J., Levy, P.E., Evans, C.D. and Smith, P. (2009) UK land use and soil carbon sequestration. *Land Use Policy*, 26(S1), S274-S283. DOI: 10.1016/j.landusepol.2009.08.006

Pan, M., Eiguren-Fernandez, A., Hsieh, H., Afshar-Mohajer, N., Hering, S.V., Lednicky, J., Hugh Fan, Z. and Wu, C-Y (2016) Efficient collection of viable virus aerosol through laminar-flow, water-based condensational particle growth. *Journal of Applied Microbiology*, 120(3), 805–815. DOI: 10.1111/jam.13051

Paustian, K., Babcock, B., Hatfield, J., Lal, R., McCarl, B., McLaughlin, S., Post, W., Mosier, A., Rice, C., Robertson, G., Rosenberg, J., Rosenzweig, C., Schlesinger, W. and Zilberman, D. (2004) *Agricultural mitigation of greenhouse gases: science and policy options*. Council on Agricultural Science and Technology (CAST) report, R141 2004, ISBN 1-887383-26-3, 120, May, 2004.

Poole, J.A. and Romberger, D.J. (2012) Immunological and inflammatory responses to organic dust in agriculture. *Current Opinion in Allergy and Clinical Immunology*, 12(2), 126. DOI: 10.1097/ACI.0b013e3283511d0e

Rautiala, S., Kangas, J., Louhelainen, K. and Reiman, M. (2003) Farmers' exposure to airborne microorganisms in composting swine confinement buildings. *AIHA Journal*, 64(5), 673–677. DOI: 10.1080/15428110308984862

RCEP (2005) *Crop Spraying and the Health of Residents and Bystanders*. Royal Commission on Environmental Pollution (RCEP), UK.

Regina, K. and Alakukku, L. (2010) Greenhouse gas fluxes in varying soils types under conventional and no-tillage practices. *Soil and Tillage Research*, 109(2), 144–152. DOI: 10.1016/j.still.2010.05.009

Rochette, P. and Janzen, H.H. (2005) Towards a revised coefficient for estimating N_2O emissions from legumes. *Nutrient Cycling in Agroecosystems*, 73(2–3), 171–179. DOI: 10.1007/s10705-005-0357-9

Rodwell, J.S. (ed.) (2008) *British Plant Communities. Volume 2. Mires and Heath*. Cambridge University Press, UK.

Sakellariou-Makrantonaki, M., Papalexisa, D., Nakosa, N. and Kalavrouziotis, I.K. (2007) Effect of modern irrigation methods on growth and energy production of sweet sorghum (var. Keller) on a dry year in Central Greece. *Agricultural Water Management*, 90(3), 181–189. DOI: 10.1016/j.agwat.2007.03.004

Sattar, S.A., Bhardwaj, N. and Ijaz, M.K. (2016) Airborne viruses. In: Yates, M.V., Nakatsu, C.H., Miller, R.V., and Pillai, S.D. (eds.) *Manual of Environmental Microbiology*, 4th Edition. American Society of Microbiology, USA. ISBN: 9781555816025. DOI: 10.1128/9781555818821

Sattler, B., Puxbaum, H. and Psenner, R. (2001) Bacterial growth in supercooled cloud droplets. *Geophysical Research Letters*, 28(2), 239–242. DOI: 10.1029/2000GL011684

Schils, R., Kuikman, P. and Hiederer, R. (2008) *Review of existing information on the interrelations between soil and climate change (CLIMSOIL)*. European Commission, Brussels.

Seedorf, J. (2004) An emission inventory of livestock-related bioaerosols for Lower Saxony, Germany. *Atmospheric Environment*, 38(38), 6565–6581. DOI: 10.1016/j.atmosenv.2004.08.023

Seinfeld, J.H. and Pandis, S.N. (1998) *Atmospheric Chemistry and Physics: From Air Pollution to Climate Change*. John Wiley and Sons, Inc., USA. ISBN 978-0-471-17816-3.

Shen, L., Wania, F., Lei, Y.D., Teixeira, C., Muir, D.C. and Bidleman, T.F. (2005) Atmospheric distribution and long-range transport behavior of organochlorine pesticides in North America. *Environmental Science and Technology*, 39(2), 409–420. DOI: 10.1021/es049489c

Silgram, M. and Harrison, R. (1998) Mineralisation of cover crop residues over the short and medium term. *Proceedings of the 3rd Workshop of EU Concerted Action 2108. Long-term reduction of nitrate leaching by cover crops*, 30 September–3 October 1997, Southwell, UK. AB-DLO, Netherlands.

Smith, J.U., Bradbury, N.J. and Addiscott, T.M. (1996) SUNDIAL: a PC-based system for simulating nitrogen dynamics in arable land. *Agronomy Journal*, 88(1), 38–43. DOI: 10.2134/agronj1996.00021962008800010008x

Smith, K.A. and Chambers, B.J. (1995) Muck: from waste to resource – utilization: the impacts and implications. *The Agricultural Engineer*, 50, 33–38.

Smith, K. A. and Conen, F. (2004) Impacts of land management on fluxes of trace greenhouse gases. *Soil Use Management*, 20(2), 255–263. DOI: 10.1111/j.1475-2743.2004.tb00366.x

Smith, P. (2004) Carbon sequestration in croplands: the potential in Europe and the global context. *European Journal of Agronomy*, 20(3), 229–236. DOI: 10.1016/j.eja.2003.08.002

Smith, P. (2005) An overview of the permanence of soil organic carbon stocks: influence of direct human-induced, indirect and natural effects. *European Journal of Soil Science*, 56(5), 673–680. DOI: 10.1111/j.1365-2389.2005.00708.x

Smith, P., Powlson, D.S., Smith, J.U., Falloon, P. and Coleman, K. (2000a) Meeting the UK's climate change commitments: options for carbon mitigation on agricultural land. *Soil Use and Management*, 16(1), 1–11. DOI: 10.1111/j.1475-2743.2000.tb00162.x

Smith, P., Powlson, D.S., Smith, J.U., Falloon, P. and Coleman, K. (2000b) Meeting Europe's climate change commitments: quantitative estimates of the potential for carbon mitigation by agriculture. *Global Change Biology*, 6(5), 525–539. DOI: 10.1046/j.1365-2486.2000.00331.x

Smith, P., Milne, R., Powlson, D.S., Smith, J.U., Falloon, P. and Coleman, K. (2000c) Revised estimates of the carbon mitigation potential of UK agricultural land. *Soil Use and Management*, 16(4), 293–295. DOI: 10.1111/j.1475-2743.2000.tb00214.x

Smith, P., Martino, D., Cai, Z., Gwary, D., Janzen, H., Kumar, P., McCarl, B., Ogle, S., O'Mara, F., Rice, C., Scholes, B., Sirotenko, O., Howden, M., McAllister, T., Pan, G., Romanenkov, V., Schneider, U., Towprayoon, S., Wattenbach, M. and Smith, J. (2008) Greenhouse gas mitigation in agriculture. *Philosophical Transactions Royal Society of London Biological Sciences*, 363, 789–813. DOI: 10.1098/rstb.2007.2184

Sommer, S.G. and Hutchings, N.J. (2001) Ammonia emission from field applied manure and its reduction—invited paper. *European Journal of Agronomy*, 15(1), 1–15. DOI: 10.1016/S1161-0301(01)00112-5

Sommer, S.G., Petersen, S.O., Sørensen, P., Poulsen, H.D. and Møller, H.D. (2007) Methane and carbon dioxide emissions and nitrogen turnover during liquid manure storage. *Nutrient Cycling in Agroecosystems*, 78(1), 27–36. DOI: 10.1007/s10705-006-9072-4

Soto-García, M., Martin-Gorriz, B., García-Bastida, P.A., Alcon, F. and Martínez-Alvarez, V. (2013) Energy consumption for crop irrigation in a semiarid climate (south-eastern Spain). *Energy*, 55, 1084–1093. DOI: 10.1016/j.energy.2013.03.034

Soussana, J.F., Loiseau, P., Viuchard, N., Ceschia, E., Balesdent, J., Chevallier, T., Arrouays, D. (2004) Carbon cycling and sequestration opportunities in temperate grasslands. *Soil Use Management*, 20(2), 219–230. DOI: 10.1111/j.1475-2743.2004.tb00362.x

Sutton, M.A. and Fowler, D. (2002) Introduction: fluxes and impacts of atmospheric ammonia on national, landscape and farm scales. *Environmental Pollution*, 119, 7–8.

Sutton, M.A., Pitcairn, C.E.R. and Fowler, D. (1993) The exchange of ammonia between the atmosphere and plant communities. *Advances in Ecological Research*, 24, 301–393. DOI: 10.1016/S0065-2504(08)60045-8

Sutton, M.A., Schjørring, J.K. and Wyers, G.P. (1995) Plant–atmosphere exchange of ammonia. *Philosophical Transactions: Physical Sciences and Engineering*, 351(1696), 261–278.

Sutton, M.A., Milford, C., Nemitz, E., Theobald, M.R., Hill, P.W., Fowler, D., Schjoerring, J.K., Mattsson, M.E., Nielsen, K.H., Husted, S., Erisman, J.W., Otjes, R., Hensen, A., Mosquera, J., Cellier, P., Loubet, B., David, M., Genermont, S.,Neftel, A., Blatter, A., Herrmann, B., Jones, S.K., Horvath, L., Führer, E., Mantzanas, K., Koukoura, Z., Gallagher, M., Williams, P., Flynn, M. amd Riedo, M. (2001) Biosphere–atmosphere interactions of ammonia with grasslands: experimental strategy and results from a new European initiative. *Plant and Soil*, 228(1), 131–145. DOI: 10.1023/A:1004822100016

Sutton, M.A., Nemitz, E., Erisman, J.W., Beier, C., Butterbach Bahl, K., Cellier, P., de Vries, W., Cotrufo, F., Skiba, U., Di Marco, C., Jones, S., Laville, P., Soussana, J.F., Loubet, B., Twigg, M., Famulari, D., Whitehead, J., Gallagher, M.W., Neftel, A., Flechard, C.R., Herrmann, B., Calanca, P.L., Schjoerring, J.K., Daemmgen, U., Horvath, L., Tang, Y.S., Emmett, B.A., Tietema, A., Peñuelas, J., Kesik, M.,Brueggemann, N., Pilegaard, K., Vesala, T., Campbell, C.L., Olesen, J.E., Dragosits, U., Theobald, M.R., Levy, P., Mobbs, D.C., Milne, R., Viovy, N., Vuichard, N., Smith, J.U., Smith, P., Bergamaschi, P., Fowler, D. and Reis, S. (2007) Challenges in quantifying biosphere–atmosphere exchange of nitrogen species. *Environmental Pollution*, 150(1), 125–139. DOI: 10.1016/j.envpol.2007.04.014

Sutton, M.A., Erisman, J.W., Dentener, F. and Möller, D. (2008) Ammonia in the environment: from ancient times to the present. *Environmental Pollution*, 156(3), 583–604. DOI: 10.1016/j.envpol.2008.03.013

Sylvester-Bradley, R. and Withers, P.J.A. (2012) Scope for innovation in crop nutrition to support potential crop yields. *International Fertiliser Proceedings*, 700.

Thauer, R.K., Jungermann, K. and Decker, K. (1977) Energy conservation in chemotrophic anaerobic bacteria. *Bacteriological Reviews*, 41(1), 100.

Thomas, C. (2004) *Feed into Milk (FiM): A New Applied Feeding System for Dairy Cows*. Nottingham University Press, UK.

Tzilivakis, J., Jaggard, K., Lewis, K.A., May, M. and Warner, D.J. (2005a) An assessment of the energy inputs and greenhouse gas emissions in sugar beet (*Beta vulgaris*) production in the UK. *Agricultural Systems*, 85(2), 101–119. DOI: 10.1016/j.agsy.2004.07.015

Tzilivakis, J., Jaggard, K., Lewis, K.A., May, M. and Warner, D.J. (2005b) Environmental impact and economic assessment for UK sugar beet production systems. *Agriculture, Ecosystems and Environment*, 107(4), 341–358. DOI: 10.1016/j.agee.2004.12.016

Tzilivakis, J., Warner, D.J., Green, A. and Lewis, K.A. (2015) *Guidance and tool to support farmers in taking aware decisions on Ecological Focus Areas*. Final report for Project JRC/IPR/2014/H.4/0022/NC, Joint Research Centre (JRC), European Commission.

UNECE (1999) *Control techniques for preventing and abating emissions of ammonia*. EB.AIR/WG.5/1999/8/Rev.1. United Nations Economic Commission for Europe (UNECE), Geneva, Switzerland, 37 pp.

Unsworth, J.B., Wauchope, R.D., Klein, A.W., Dorn, E., Zeeh, B., Yeh, S.M., Akerblom, M., Racke, K.D. and Rubin. B. (1999) Significance of the long range transport of pesticides in the atmosphere. *Pure and Applied Chemistry*, 71(7), 1359–1383.

van der Weerden, T.J. and Jarvis, S.C. (1997) Ammonia emission factors for N fertilisers applied to two contrasting grassland soils. *Environmental Pollution*, 95(2), 205–211. DOI: 10.1016/S0269-7491(96)00099-1

van Nevel, C.J. and Demeyer, D.I. (1996) Influence of antibiotics and a deaminase inhibitor on volatile fatty acids and methane production from detergent washed hay and soluble starch by rumen microbes in vitro. *Animal Feed Science Technology*, 37(1–2), 21–31 DOI: 10.1016/0377-8401(92)90117-O.

Velthof, G.L., Van Bruggen, C., Groenestein, C.M., De Haan, B.J., Hoogeveen, M.W. and Huijsmans, J.F.M. (2012) A model for inventory of ammonia emissions from agriculture in the Netherlands. *Atmospheric Environment*, 46, 248–255. DOI: 10.1016/j.atmosenv.2011.09.075

Venkataraman, S. (2016) *How is climate change affecting crop pests and diseases?* DownToEarth Blog 03 June 2016. Available at: http://www.downtoearth.org.in/blog/how-is-climate-change-affecting-crop-pest-and-diseases--54199

Verheijen, F.G.A., Bellamy, P.H., Kibblewhite, M.G. and Gaunt, J.L. (2005) Organic carbon ranges in arable soils of England and Wales. *Soil Use and Management*, 21(1), 2–9. DOI: 10.1111/j.1475-2743.2005.tb00099.x

Waghorn, G.C., Tavendale, M.H. and Woodfield, D.R. (2002). Methanogenesis from forages fed to sheep. *Proceedings of New Zealand Society of Animal Production*, 64, 161–171.

Warner, D.J., Tzilivakis, J. and Lewis K.A. (2008a) *Research into the current and potential climate change mitigation impacts of environmental stewardship*. Final report for Department for Environment, Food and Rural Affairs (Defra) project BD2302.

Warner, D.J., Tzilivakis, J. and Lewis K.A. (2008b) *The impact on greenhouse gas emissions of the revised Action Programme for Nitrate Vulnerable Zones*. Final report for Department for Environment, Food and Rural Affairs (Defra) project WT0757NVZ.

Warner, D., Tzilivakis, J., Green, A. and Lewis, K. (2016). Rural Development Programme measures on cultivated land in Europe to mitigate greenhouse gas emissions – regional 'hotspots' and priority measures. *Carbon Management*, 7(3–4), 205–219. DOI: 10.1080/17583004.2016.1214516

Watts, S.F. (2000) The mass budgets of carbonyl sulfide, dimethyl sulfide, carbon disulfide and hydrogen sulfide. *Atmospheric*

Environment, 34(5), 761–779. DOI: 10.1016/S1352-2310(99)00342-8

Webb, J., Menzi, H., Pain, B.F., Misselbrook, T.H., Dämmgen, U., Hendriks, H. and Döhler, H. (2005) Managing ammonia emissions from livestock production in Europe. *Environmental Pollution*, 135(3), 399–406. DOI: 10.1016/j.envpol.2004.11.013

Webb, J., Hutchings, N. and Amon, B. (2016) *EMEP/EEA air pollutant emission inventory guidebook 2016. 3.D.f, 3.I Agriculture other including use of pesticides.* European Environment Agency (EEA).

Wenneker, M., Heijne, B. and Van de Zande, J.C. (2004) Effect of natural windbreaks on drift reduction in orchard spraying. *Communications in Agricultural and Applied Biological Sciences*, 70(4), 961–969.

Wéry, N. (2014) Bioaerosols from composting facilities—a review. *Frontiers in Cellular and Infection Microbiology*, 4, 42. DOI: 10.3389/fcimb.2014.00042

Western, N. M., Hislop, E. C., Bieswal, M., Holloway, P. J. and Coupland, D. (1999) Drift reduction and droplet-size in sprays containing adjuvant oil emulsions. *Pest Management Science*, 55(6), 640–642. DOI: 10.1002/(SICI)1096-9063(199906)55:6<640::AID-PS985>3.0.CO;2-U

Williams, A.G., Audsley, E. and Sandars, D.L. (2006) *Determining the environmental burdens and resource use in the production of agricultural and horticultural commodities.* Main Report. Defra Research Project IS0205. Cranfield University and Defra, Bedford.

Williams, A.G., Audsley, E. and Sandars, D.L. (2009) *Environmental burdens of agricultural and horticultural commodity production – LCA (IS0205) version 3.* Cranfield University, UK.

Wolin, E.A., Wolf, R.S. and Wolin, M.J. (1964) Microbial formation of methane. *Journal of Bacteriology*, 87, 993–998.

World Bank (2016) *Agricultural nitrous oxide emissions (% of total).* World Bank, Washington, DC. Available at: http://data.worldbank.org/indicator/EN.ATM.NOXE.AG.ZS?year_high_desc=false

Worldwatch (2009) Livestock and climate change. *World Watch Magazine*, 22(6), 10–19.

Worrall, F., Reed, M., Warburton, J. and Burt, T. (2003) Carbon budget for a British upland peat catchment. *Science of the Total Environment*, 312(1–3), 133–146. DOI: 10.1016/S0048-9697(03)00226-2

WRI (2011) *Climate analysis indicators tool (CAIT).* Version 8.0. World Resources Institute (WRI), Washington, DC.

Xu, H., Cai, Z.C., Jia, Z.J. and Tsuruta, H. (2000) Effect of land management in winter crop season on CH4 emission during the following flooded and rice-growing period. Nutrient Cycling in Agroecosystems, 58(1–3), 327–332. DOI: 10.1023/A:1009823425806

Xu, H., Cai, Z.C. and Tsuruta, H. (2003) Soil moisture between rice-growing seasons affects methane emission, production, and oxidation. *Soil Science Society of America Journal*, 67(4), 1147–1157. DOI: 10.2136/sssaj2003.1147

Yan, T., Agnew, R.E., Gordon, F.J. and Porter, M.G. (2000) Prediction of methane energy output in dairy and beef cattle offered grass silage-based diets. *Livestock Production Science*, 64, 253–263. DOI: 10.1016/S0301-6226(99)00145-1

Yan, X., Ohara, T. and Akimoto, H. (2003) Development of region-specific emission factors and estimation of methane emission from rice fields in the East, Southeast and South Asian countries. *Global Change Biology*, 9(2), 237–254. DOI: 10.1046/j.1365-2486.2003.00564.x

Yin, J., Sun, Y., Wu, G. and Ge, F. (2010) Effects of elevated CO_2 associated with maize on multiple generations of the cotton bollworm, Helicoverpa armigera. *Entomologia Experimentalis et Applicata*, 136(1), 12–20. DOI: 10.1111/j.1570-7458.2010.00998.x

Zavala, J.A., Nabity, P.D. and DeLucia, E.H. (2013) An emerging understanding of mechanisms governing insect herbivory under elevated CO_2. *Annual Review of Entomology*, 58, 79–97. DOI: 10.1146/annurev-ento-120811-153544

Biodiversity

3.1. Setting the scene

3.1.1. Introduction

The term biodiversity is somewhat ambiguous. It can have a broad variety of definitions (Kaennel, 1998) ranging from some very strict metrics of biological diversity, such as the Shannon index (Shannon and Weaver, 1949), through to a generic term used for any aspect of biological or wildlife conservation. It is the latter of these that is used in this chapter, albeit what it actually constitutes is explored and discussed below. The first part of this chapter will explore the philosophical context of biological conservation, particularly with respect to why we should conserve and protect any species; what should be conserved and protected; what outcomes we are seeking; how we measure those outcomes; and what are some of the key issues and concerns, and current threats and future challenges for biological conservation.

It is also important to acknowledge that biological conservation is not undertaken in isolation from other activities. It is part of the process of sustainable development, not separate from it, and as such it is another objective among many, and a sustainable balance needs to be struck between competing and complementary objectives.

Consequently, there are often choices to be made, some of which will be compromises (trade-offs), and in other instances there are win-wins (synergies).

3.1.2. Why conserve and protect any species?

This may be a somewhat stark question to ask, but it is an important issue to explore. It has been determined that 99.9 per cent of all evolutionary lines that once existed on Earth have become extinct (Jablonski, 2004), thus the loss of species is a natural phenomenon. There is, however, concern that the current rate of extinction has accelerated to 100 to 1000 times the background level due to human activity, to the extent that it can be regarded as a mass extinction (Barnosky et al., 2011; Ceballos et al., 2015; De Vos et al., 2015). However, mass extinctions (when the Earth loses more than three quarters of its species in a geologically short interval) are not new. There have been a least five previous mass extinctions: Ordovician-Silurian (439 million years ago); Late Devonian (364 million years ago); Permian-Triassic (251 million years ago); End Triassic (199 million to 214 million years ago); and Cretaceous-Tertiary (65 million years ago). So, if species extinction, including mass extinctions, is within the realm of the normal evolutionary process, why is there concern over the loss of current species? Why don't we, humans, just let it happen? It's a question of what is being lost and the value that has to us, humans.

The value of biodiversity and the mechanisms to assess that value are topics that have been much debated over decades, to the extent that a whole book could be written on the subject (e.g. Pearce and Moran, 1994). There are many complex and differing arguments and these are discussed in detail by Cardinale *et al.* (2012), Collar (2003), Edwards and Abivardi (1998); and Ehrenfeld (1988). However, fundamentally the arguments can be classified into two themes that centre on (i) economic value and (ii) intrinsic value.

In many respects, the economic argument is the simplest. The Convention on Biodiversity (CBD, 2015) stated that at least 40 per cent of the world's economy and 80 per cent of the needs of the poor are derived from biological resources. This is not a great surprise considering that society depends on much of the world around us to survive, and no doubt a significant part of the economy is dependent on geological and mineral resources. This Convention recognised that biodiversity provides us with greater opportunities for new discoveries that could benefit us and also help us to adapt in a changing world, such as the consequences of climate change, and become more resilient. It clearly recognises the value of biodiversity for economic growth and prosperity.

There are numerous studies that place economic values on biological resources. For example, Cesar and van Beukering (2004) estimated that the net benefits of Hawaii's coral reef ecosystems (fisheries, tourism and natural protection against wave erosion) are worth US$ 360 million per year. Policy makers 'like' economic values as it makes decision-making easier, especially when

there are trade-offs. For example, if it were to cost US$ 100 million to protect Hawaii's coral reef ecosystems, then if they are worth US$ 360 million per year it makes the decision to spend US$ 100 million relatively easy. However, economic valuation of the environment does have some major flaws in some instances. Assessing direct economic values of fisheries (for example) is relatively straightforward as there are clear and direct economic processes and values. However, placing economic values on other aspects, such as habitats, landscapes or species that are not harvested for economic benefit is more difficult. To overcome such difficulties, techniques such as contingent valuation have been developed, which is usually based on a survey of people to determine how much they are 'willing to pay' to protect a species, habitat or landscape – thus deriving an economic value. However, these techniques can be flawed. For example, Figure 3.1 shows a scenic landscape in Somerset in the UK. Visitors to this location could be asked 'how much they would be willing to pay to protect and conserve this landscape and its wildlife?', and they may give a response of £10–20 per year perhaps. This provides an economic value which can then be taken into consideration in any decisions, and associated costs, of conserving the landscape and wildlife. However, if a slightly different but related question is asked, the response may be very different. For example, if visitors to the same location were asked 'how much they would need to be compensated for the loss of the landscape and wildlife?', say if a new high-speed railway were to be routed through this landscape, then a common response is 'they cannot put a value on it', 'it's priceless' (Ackerman and Heinzerling, 2002; Hausman, 2012; Ludwig, 2000). There are some aspects that transcend conventional economic frameworks and it is difficult, or impossible, to put monetary values

Figure 3.1: Rural landscape in Somerset in the UK
(Photo courtesy of: Dr John Tzilivakis)

on them, yet they are still valued. This is intrinsic value and it can often be connected to ethical, moral or spiritual values.

Clearly, there are big differences between economic and intrinsic values, however, they are not totally unconnected as they both relate to services that the biosphere provides us with. A framework that encompasses both economic and intrinsic values is the 'ecosystem services' approach. In some literature, the term 'ecosystem services' is used interchangeably with 'economic valuation' of the environment, but here it is discussed in the context of actual services and not their economic value.

Ecosystem services are 'the benefits provided by ecosystems that contribute to making human life both possible and worth living' (NEA, 2015). The term 'services' is used to encompass the tangible and intangible benefits obtained from ecosystems (sometimes separated into 'goods' and 'services'). There are many different definitions and classifications, but the one used by the NEA (2015) and Millennium Ecosystem Assessment (MEA, 2015) is one of the most holistic and commonly used for the top-level services, with the Common International Classification of Ecosystem Services (CICES) system (Haines-Young and Potschin, 2013) being commonly used for a more detailed classification. These classifications cover three main areas:

- Provisioning services: These include, for example, fresh water, food (e.g. crops, fruit, fish, etc.), fibre and fuel (e.g. timber, wool, etc.), genetic resources (used for crop/stock breeding and biotechnology), biochemicals, natural medicines, pharmaceuticals, ornamental resources (e.g. shells, flowers, etc.).

- Regulatory services: For example, air quality regulation, climate regulation (local temperature/precipitation, greenhouse gas sequestration, etc.), water regulation (timing and scale of run-off, flooding, etc.), natural hazard regulation (i.e. storm protection), pest regulation, disease regulation, erosion regulation, water purification and waste treatment, pollination.
- Cultural services: These are also broad and include, for example, cultural heritage, recreation and tourism, aesthetic value, spiritual and religious value, inspiration of art, folklore, architecture, social relations (e.g. fishing, grazing or cropping communities).

The provisioning services tend to have a more direct correlation to economic values, whilst the cultural services start to encompass more intrinsic values. The important aspect of ecosystem services is that the approach is holistic and all-encompassing. It is anthropocentric, recognising that all ecosystem services need to be presented in the context of benefits (and burdens) to humans. For example, even an 'altruistic' desire to conserve a species so that it has the opportunity to coexist alongside humans, is still driven by anthropocentric desires. What might seem like a selfish act to ensure the survival of another species is actually driven by the need of humans to achieve this outcome, even at an experiential or spiritual level.

The above provides some ideas on why we should protect and conserve biological resources, be it purely for economic and/or intrinsic/ethical reasons. However, we live in a world that needs to be managed, choices need to be made and actions implemented.

This raises a number of questions such as what issues and causes need to be tackled with priority, what we should protect and how this should be done.

3.1.3. *What to protect and conserve?*

When thinking about biological conservation the species that come to mind are often iconic animals, for example tigers, rhinos, pandas, elephants, red squirrels and polar bears. These species have often been at the forefront of campaigns for wildlife conservation. However, it is important to ask the question: 'why conserve and protect these species?' For example, it would be considered a tragedy if the polar bear was to become extinct and although there would be consequences for the polar ecosystem, the world would not come to a shuddering halt socially, economically and environmentally, it would adapt and carry on. However, there are other species, which are not so iconic, which could have very significant impacts if they were to become extinct, such as some plants, insects, fungi and even microbes and bacteria. For example, it has been estimated that insect pollination in the EU alone has an economic value of €15 billion per year (Gallai *et al.*, 2009). Thus, perhaps our efforts may be better placed trying to conserve and protect these species? However, the arguments are not that simple, and this is explored further with the example of the giant panda (*Ailuropoda melanoleuca*).

The giant panda is an interesting example that is often in the news, and is also the very symbol of the World Wide Fund for Nature (WWF) for nature. In 2009, the BBC TV naturalist Chris Packham argued in the *Guardian* newspaper (Benedictus, 2009) that perhaps we should let the panda go extinct. Packham made the following statements:

> We spend millions and millions of pounds on pretty much this one species, and a few others,

when we know that the best thing we could do would be to look after the world's biodiversity hotspots with greater care. Without habitat, you've got nothing. So maybe if we took all the cash we spend on pandas and just bought rainforest with it, we might be doing a better job.

We have to accept that some species are stronger than others. The panda is a species of bear that has gone herbivorous and eats a type of food that isn't all that nutritious, and that dies out sporadically. It is susceptible to various diseases, and, up until recently, it has been almost impossible to breed in captivity. They've also got a very restricted range, which is ever decreasing, due to encroachment on their habitat by the Chinese population. Perhaps the panda was already destined to run out of time.

Extinction is very much a part of life on earth. And we are going to have to get used to it in the next few years because climate change is going to result in all sorts of disappearances. The last large mammal extinction was another animal in China – the Yangtze river dolphin, which looked like a worn-out piece of pink soap with piggy eyes and was never going to make it on to anyone's T-shirt. If that had appeared beautiful to us, then I doubt very much that it would be extinct. But it vanished, because it was pig-ugly and swam around in a river where no one saw it. And now, sadly, it has gone forever.

Of course, it's easier to raise money for something fluffy. Charismatic megafauna like the panda do appeal to people's emotional side, and attract a lot of public attention. They are emblematic of what I would call single-species conservation: i.e. a focus on one animal. This approach began in the 1970s with Save the Tiger, Save the Panda, Save the Whale, and so on, and it is now out of date.

These statements by Packham touch upon some of the arguments presented above,

such as the charismatic megafauna appealing to human emotions, and thus having intrinsic value, and that perhaps the focus of conservation efforts should not be individual iconic species, but on whole habitats and ecosystems. However, there is a flip side to this argument which is presented by Mark Wright, chief scientist at the WWF. Wright does not agree with Packham's conclusions regarding the giant panda, but perhaps more importantly he highlights:

> Charismatic megafauna can be extremely useful. Smaller creatures often don't need a big habitat to live in, so in conservation terms it's better to go for something further up the food chain, because then by definition you are protecting a much larger area, which in turn encompasses the smaller animals.

Efforts to save mega-fauna, like the giant panda, require larger areas of habitat to be conserved in order to protect them and consequently many other organisms benefit, for example the giant panda shares its habitat with the red panda (*Ailurus fulgens*), the golden monkey (*Cercopithecus kandti*) and various birds that are found nowhere else in the world. So even though the target of conservation efforts is a single species, there are wider benefits for whole habitats and ecosystems. It is an issue of understanding objectives and outcomes.

The mix of focusing on single species and whole habitats and ecosystems is not uncommon, and this is also embedded into policies, laws and regulations (see Chapter 3.2) and is also reflected in how we monitor and measure biodiversity. Monitoring and measuring biodiversity is essential in order to understand the status of species, habitats and ecosystems and thus the impact humans may be having on them. However, we do not have the resources to monitor or measure everything. Consequently, we rely

Figure 3.2: Indices for farmland bird populations in England
(Source: Defra, 2015)

on what are commonly referred to as indicator species, working on the assumption that if the indicator species is doing well, then everything else in that habitat (including the habitat itself) is in good condition and doing well. For example, in the UK, the population of the farmland birds is a well-known headline indicator (see Figure 3.2).

Figure 3.2 shows there has been a steady decline in all farmland birds since the 1960s, thus raising concerns for these bird species. Clearly, this trend is something that we would like to reverse. So, a desirable outcome from conservation activities would be to see this indicator go up, and hopefully that would also mean other benefits to the wider ecosystem. For example, the key to improving farmland bird numbers may be to provide more food sources, such as seeds and insects, which in itself may benefit many other organisms. This may be true, but it must not be forgotten that it is

an assumption in absence of any monitoring data for other species.

It is important to also highlight that the choice of indicator species may result in a different picture being presented. Figure 3.3 shows while farmland birds have been declining, during the same period species such as jackdaw (*Corvus monedula*), rook (*Corvus frugilegus*) and wood pigeon (*Columba palumbus*) have increased, thus contradicting the picture presented when using farmland birds as an indicator for the overall health of the environment and biodiversity. Clearly, for some species of bird, the changes that have caused the decline of farmland birds have favoured them.

It is important to understand the causal mechanisms and processes at work that are leading to these population changes, and use this understanding to devise solutions and tools that will help achieve our desired outcomes.

Figure 3.3: Indices for generalist corvids, wood pigeon and kestrel in England
(Source: Defra, 2015)

3.1.4. Issues, threats and causes for concern

The causes of declining populations, species extinction and loss of biodiversity are inherently complex and can also be species-specific. Clearly, it is not possible to explore all issues for all species herein, however, it is possible to draw out some of the key concerns and illustrate these with examples. There are four main areas of concern: habitat loss; harvesting; exotic species; and climate change (Samways, 2005). Each of these is an issue in itself, but they are also interconnected and can exacerbate each other in some instances.

HABITAT LOSS

According to the WWF, habitat loss is the greatest threat that biodiversity faces. Habitats across the world continue to disappear as they are harvested for human consumption, degraded to a point they are no longer productive and cleared to make way for agriculture, housing, roads, pipelines and the other hallmarks of industrial development. There are many examples across Europe where important habitats are being lost.

For example, Britain has around 58,000 hectares of lowland heathland which is about 20 per cent of the total world resource. It is currently scattered across much of the British countryside including Cornwall, Devon, Pembrokeshire, west Glamorgan and west Gwynedd. However, over 80 per cent of Britain's lowland heathland has been lost since 1800 – 17 per cent in the last 50 years (Forestry Commission, 2017). Lowland heath is characterised by plants such as gorse and heather, and trees such as Scots pine (*Pinus sylvestris*) and birch (*Betula* spp.). It supports a range of breeding populations of rare birds, such as the Dartford warbler (*Sylvia undata*) and nightjar (*Caprimulgus europaeus*). It is habitat for the six types of reptile native to Britain, the adder (*Vipera berus*), grass snake (*Natrix natrix*), common lizard (*Zootoca vivipara*), slow worm (*Anguis fragilis*), sand lizard (*Lacerta agilis*) and smooth snake (*Coronella austriaca*). The sand lizard and smooth snake are rare and endangered species and are confined almost exclusively to this type of habitat. In addition, several rare arachnids (e.g. the ladybird spider, *Eresus sandaliatus*) and insects (e.g. the southern damselfly, *Coenagrion mercuriale* and black bog ant, *Formica picea*) are also found on lowland heathland. Thus, these species are all potentially lost should the habitat diminish further.

There are three aspects to habitat loss. The first is generally known as 'conversion' and refers to the straightforward loss of habitat due to conversion from a natural or semi-natural state into cultivated land, pasture or urbanisation. This might include, for example, land clearing, replacement of natural vegetation by cropping or orchards, establishment of reservoirs by permanent flooding, drainage of wetlands or surface mining.

The second is 'degradation' and this refers to a change in status – chemical or physical – that means the habitat can no longer support the species or ecosystem that used to depend upon it. This might include, for example, contamination of soils or waterbodies by pesticides, nitrates or other pollutants, acidification or eutrophication (see Chapter 2), loss of soil fertility due to intensive farming, damage to soil structure due to compaction or livestock poaching or physical disturbance such as excessive light or noise. Many of Europe's aquatic habitats are at threat of chemical contamination not only from agricultural practices but also from urban areas, polluted run-off from roads and municipal sewage treatment plants. As shown in Figure 3.4, more than half of the freshwater surface waterbodies in Europe are reported to hold less than good ecological status or potential.

The third aspect is 'fragmentation' and refers to the loss of continuous habitat, broken up by human development such as roads, buildings and urban sprawl, resulting in difficulties for some species to migrate or occupy the territory required to sustain them. Habitat fragmentation is often a limiting factor in the population growth of many species. A single species may require access to a range of different habitats to feed, breed and complete their life cycles. Small areas of habitat also have finite resources to support growing populations and their isolation can often make it

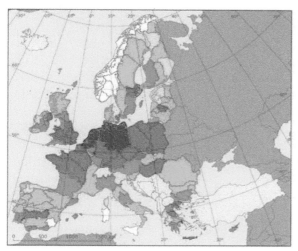
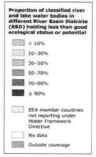

Figure 3.4: European rivers with water quality below desired ecological standards (Source: EEA, 2015a)

difficult for expansion into new areas due to the lack of established ecological connections between adjoining areas of the landscape (Haber, 1993; Jaeger and Holderegger, 2005). The isolation also makes it statistically more unlikely that two individuals of a particular species will find each other to reproduce. In addition, isolated populations are more vulnerable to stress factors such as extreme weather, natural hazards (fire, flood and disease) and disturbance by traffic noise, humans, dogs, and so on. Figure 3.5 shows the results of an analysis of landscape fragmentation across Europe undertaken by the European Environment Agency (EEA) and the Swiss Federal Office for the Environment (FOEN; EEA, 2011). It illustrates that the extent of landscape fragmentation in many parts of Europe is already considerable.

Whilst much of the cause of fragmentation, such as urban sprawl, is largely beyond the control of many farmers, historically farming practices have significantly contributed to the problem on their own land by, for example, removing hedges that acted as corridors between habitats. For example, in the UK farmers removed a quarter of the nation's hedgerows between 1946 and 1974,

amounting to 120,000 miles and it continued, albeit at a slower rate until the 1997 England and Wales Hedgerow Regulations were introduced making it illegal to remove hedgerows without prior consent. The main driver for this was agricultural intensification. Hedge removal enabled larger fields to be produced which were easier to cultivate and ensured that the larger machinery used was more easily manipulated.

However, there are some positive developments. For example, the EU has a total area of forestry and other wooded land area of 136 million ha accounting for about 36 per cent of its surface area (EC, 2003). Whilst in many parts of the world forest cover is declining, within the EU it is actually slowly but steadily increasing at a rate of approximately 0.3 per cent per year and forests are present in a huge variety of climatic, geographic, ecological as well as socio-economic conditions. This is not just beneficial for biodiversity but also will help mitigate climate change.

HARVESTING

Harvesting is an issue that relates directly to provisioning ecosystem services, whereby harvested resources, such as timber and

Figure 3.5: European landscape fragmentation (Source: EEA, 2011)

fish, have direct economic benefits. The main issue here is one of overexploitation or unsustainable use of renewable resources. Where it refers to a particular species, the negative impact is not only on the target species itself, for example a collapse in the population, but also any species that directly depends on the target species for food or shelter. A classic example here is overexploitation of fish in the seas and oceans. Across the pan-European region, 30 per cent of Community fish stocks are overfished outside safe biological limits that may not allow their recovery (EEA, 2015b).

As discussed in Chapters 4 and 5, there is also significant pressure on Europe's freshwater resources. Agriculture, energy production, industry, domestic households and tourism all pose a threat to water resources, with demand often exceeding availability. The increase in artificial storage facilities used to ensure stability of water supplies reduces the share of water allocated to natural systems. Overabstraction and prolonged periods of low rainfall or drought have frequently reduced river flows, lowered lake and groundwater levels and dried up wetlands. A recent report (WWF, 2016) has revealed that the Doñana World Heritage Site in Andalusia, Spain, Europe's leading Wetland, is drying up at an alarming rate and suffering from nitrate pollution, threatening hundreds of thousands of birds and damaging the wider environment. Already many species, assessed as threatened and present in the IUCN 'Red List' of endangered species have disappeared, including various dragonflies and damselflies, and there is evidence that duck populations have significantly declined. The report suggests that the underlying aquifer could take up to 60 years to recover even if the unsustainable water abstraction was controlled.

The Earth's natural resources have always been of fundamental importance to society and their economic development. However, as populations and economies grow, the rate of consumption has risen exponentially. Today, inhabitants of industrialised countries use up to eight times more resources than people living in societies that are less industrialised and more rural and agricultural-based. It is clear that our rate of consumption is currently unsustainable.

EXOTIC SPECIES

There are many well-documented cases of the impact of exotic species, and whether introduced deliberately or accidentally, they can have a devastating effect on indigenous organisms. The movement and introduction of species from one region to another has been greatly accelerated in recent centuries either by deliberate introduction of species with perceived agricultural or aesthetic benefits, or by accidental introductions, such as 'hitchhikers' on trading activities. Not all exotic species persist in a new environment, but those that do are often very successful, and when they are successful they can dramatically change an ecological community by being dominant competitors or effective predators and/or driving out many native species.

One of the best-known examples is the case of Japanese knotweed (*Fallopia japonica*) shown in Figure 3.6. This species first arrived in the UK in 1954 as a botanical specimen at the Royal Botanical Gardens in Edinburgh. Seen by many as an attractive addition to parks and gardens the plant was soon sold by many commercial garden centres and plant nurseries around the country. As the plant is easy to propagate by cuttings and rhizomes, distribution was rapidly exacerbated by: gifting between gardeners; vegetative spread naturally along watercourses; and artificial spread where soil containing rhizomes was moved during road building and construction. It is now widespread throughout the UK

Figure 3.6: Japanese knotweed (*Fallopia japonica*) (Photo courtesy of: Pixabay.com)

and the cause of considerable environmental damage. Japanese knotweed affects ecosystems by crowding out native vegetation and limiting plant and animal species diversity; scientific evidence of allelopathy (the release of chemicals that suppress the growth of other plants) has also been identified. It is a determined species and its rhizomes can grow through building foundations, asphalt and even concrete adding huge costs to development programmes and devaluing housing. Indeed, some mortgage providers will not support the purchase of property where knotweed is found and, similarly, some insurance companies exclude cover for the risk of knotweed damage. It is also a general nuisance as it can impede access to footpaths, and riverways. It has been estimated that it costs Great Britain around £165 million every year and the cost of eradication, if it was to be attempted UK-wide, could be more than £1.56 billion (Williams *et al.,* 2010).

There are many other examples where exotic or non-native species have caused problems. The grey squirrel (*Sciurus carolinensis*) was introduced into the UK in the 1870s as exotic animals in country-house parks and gardens. The species is now established across the UK and considered by many to be a major pest having driven the UK's native red squirrel (*Sciurus vulgaris*) into just a few isolated enclaves in the south and north into Scotland. In addition, grey squirrels can cause serious damage to woodland by stripping bark. Not only does this reduce the economic value of the timber but also makes the tree more susceptible to pests and disease. In Germany, the raccoon (*Procyon lotor*) was introduced for hunting during the 1930s, and in Russia and Belorussia in the 1950s, and has spread outwards from these locations. The raccoon is currently the only representative of its family in Europe; therefore, it occupies an entirely new ecological niche. It can be a vector for a number of zoonotic diseases and also impacts on numerous habitats and associated species. The Asian hornet (*Vespa mandarinia*) arrived in France in 2005, most probably in a container of pottery from China, and predates on native wasps and honeybees, causing important economic losses.

Within the aquatic environment, the North American signal crayfish (*Pacifastacus leniusculus*) was introduced into Europe in the 1960s as a farmed species for food. It has escaped from crayfish farms and is now out-competing native crayfish (*Austropotamobius pallipes* and *Astacus astacus*) in many places. Similarly, the floating pennywort (*Hydrocotyle ranunculoides*), a non-native invasive aquatic plant, was introduced as a result of discarded plants from garden ponds in the 1990s. It forms dense rafts, outcompeting other plants and deoxygenating the water, and can also block some waterbodies causing floods.

Problems such as those described above have led to interventions at both the national and EU level to try and control the spread of non-native species.

Climate change
Climate change (covered in detail in Chapter 2) can include changes in temperature (decreases and increases) and precipitation (amount, frequency and intensity). At the simplest level, rapid climate change may result in some species not being able to adapt quickly enough to survive in a new climate. Additionally, it can exacerbate the problems described above as some habitats may become further degraded due to, for example, drought or flooding. Changes in climate may result in certain species needing to migrate (e.g. northwards and/or in altitude) but unable to do so due to habitat fragmentation, whilst other species may not be able to recover well after harvesting due to additional stresses of climate change. In addition, the change in climate may suit exotic species and so lead them to thrive and spread further.

European impact
It is often the case that a particular species is simultaneously faced with a combination of the threats described above. The classic example is the current, well-debated, decline of many pollinator species across Europe and indeed the world. The buff-tailed bumblebee (*Bombus terrestris*) shown in Figure 3.7, is

Figure 3.7: The buff-tailed bumblebee (*Bombus terrestris*) collecting nectar (Photo courtesy of: Dr Doug Warner)

relatively common throughout Europe and appears to be able to survive in a wide range of habitats occupying many of Europe's temperate climate areas. However, other bumble species have not proved so adaptable and in the UK two species of bumblebee, Cullens bumblebee (*Bombus cullumanus*) and the short-haired bumblebee (*Bombus subterraneus*), have become extinct since the start of the 20th century and many other pollinator species are threatened. It is thought that there are several factors contributing to this loss. One cited reason is that changes in agricultural practices have reduced the abundance of wild flowers in the countryside, reducing pollinator food sources. Habitat fragmentation exacerbates the issue as it reduces the ability for the insects to move to other areas to look for pollen and nectar. The use of pesticides, particularly the neonicotinoids that affect the nervous system of insects, has also been blamed. In addition, there are also threats from pests and disease such as the Varroa mite, chalkbrood fungus and *Nosema apis*, a small unicellular parasite. The effects of climate change (see Chapter 2) are also having an impact by disrupting the synchronised timing of flower blooming and the timing at which bees pollinate as well as providing, at times, unfavourable weather conditions.

There are numerous other examples across Europe of threatened species and lost habitats. For example, in 2014 a report suggested that almost half of all the 160 habitats in Switzerland are threatened (Swiss Federal Office for the Environment, 2014). The island of Malta is home to a considerable number of species that are threatened and on the IUCN Red List, indeed a recent report stated that 6 per cent of the species assessed for inclusion on the Red List exist in Malta. Freshwater fish are a major concern due to agriculture pollution and forestry effluents (IUCN, 2013). In Portugal, nearly half of

its vertebrate species are threatened, especially the Iberian lynx (*Lynx pardinus*), monk seal (*Monachus monachus*) and black vulture (*Coragyps atratus*) due to habitat fragmentation and the destruction of rural habitats due to the construction of roads, dams and urban developments. Biological diversity in Iceland is small due to its isolation from other landmasses, meaning that it has few endemic species of fauna and flora compared with mainland Europe, as well as its harsh climate. As a result, many would consider its wildlife even more precious and it is a signatory to most biodiversity-related conventions and treaties. As a result, the country has seen increased emphasis on biodiversity protection over the last two decades, the focus of which has been to establish protected areas and to ensure that major developments such as roads, power plants and reservoirs have been undertaken in an ecologically sensitive manner (IINH, 2001).

3.2. Policy and interventions

Although the importance of biodiversity to a healthy and wealthy society has long been recognised, it has also been long recognised that human activity is a major contributor to declining populations the globe over. In different parts of Europe, species face different levels and types of threats as described above. Our understanding of biological species and biodiversity has developed and evolved over the last few centuries whilst at the same time the impact of humans on biodiversity has also grown in significance. As much as there are still areas of 'wilderness' on the planet, most species have to coexist alongside humans and human activity. If we are to conserve and sustain valued species and biological resources, then they need to be managed. This management can be in terms of practical and physical implementation of actions on the ground (covered in

Chapter 3.3), but also more strategic management in the form of conventions, agreements, strategies, policies and interventions. In consequence, all European governments and many international/global bodies such as the United Nations have extensive policies, regulations, schemes and initiatives in place that aim to protect biodiversity and their habitats.

During the last century, decreases in biodiversity were increasingly observed such that by 1960 it was abundantly clear that species were being lost in Europe and, indeed, globally, at an alarming rate. However, compared to some environmental concerns such as air pollution, the development history of biodiversity policy is relatively short and it only emerged onto political agendas in the early 1970s. Since that time, it has been influenced by a multitude of actors at the international and global level including the UN, World Wide Fund for Nature, World Heritage, EU and national level governments. During the period from 1970 to the early 2000s there were a significant number of international conventions and treaties that placed various conservation obligations on signatories such that national policies began to emerge along with the legislation, interventions and other initiatives required to slow and, hopefully, reverse the decline. These are summarised in Table 3.1.

Three of these conventions, in particular, proved to be particularly influential, namely the Bern Convention, Bonne Convention and the Convention on Biological Diversity. Most European countries including those within and outside of the EU and the EEA are signatories. These conventions, in particular, led to a number of policy initiatives in the EU, for example the Habitats Directive (Council Directive 92/43/EEC; EC, 1992) was the EU response to the Bern Convention (The Convention on the Conservation of European Wildlife and Natural Habitats).

The Bern Convention had the principal aim of conserving and protecting wild plant and animal species and their habitats as well as increasing cooperation between signatories of the convention. In order to ensure that EU Member States (MSs) delivered on their convention obligations, the EU Habitats Directive was established (Bern Convention, 2015).

The directive consists of 24 articles and 6 annexes, that aim to deliver its main objective of maintaining or restoring European protected habitats and species at a 'favourable conservation status'. What is meant by 'favourable conservation status' is also defined in the directive as (i) when the natural range and areas the site covers is stable or increasing and (ii) when the specific structure and functions necessary for its long-term maintenance exist and are likely to continue to exist for the foreseeable future. Protected species are listed in annexes which are revised on an ongoing process. The directive also contributed and helped establish a coherent European ecological network of protected sites by designating special areas of conservation (SACs) for certain habitats and bird species. In order for an area to be designated as a SAC, certain selection criteria must be met. The site must include habitats and/or species identified by the European Commission, in consultation with MSs, to be of 'Community importance' based upon, amongst other things, how representative the area is of the habitat type, the site's size and the degree of conservation of habitat structure and function. The concept of SACs linked to Special Protection Areas (SPAs) defined under the Birds Directive (see below). Together SACs and SPAs now make up the Natura 2000 Network (see below). In addition, the directive lays down requirements for MSs to ensure that appropriate conservation measures are in place to manage SACs, that there is encouragement at national level for

Table 3.1: International conventions and treaties

Convention/treaty name	Description
UN Convention on Wetlands of International Importance (Ramsar Convention). Introduced in 1971.	This is an intergovernmental treaty that provides the framework for the conservation and wise use of wetlands and their resources. It covers all aspects of wetland conservation and recognises that these ecosystems are extremely important for biodiversity conservation and for the well-being of human communities.
1972 UN Convention Concerning the protection of the World Cultural and Natural Heritage (WHC). Introduced in 1972.	This convention links together the concepts of nature conservation and the preservation of cultural properties, and recognises the way in which people interact with nature as well as the fundamental need to preserve the balance between the two. By signing the convention, each signatory pledges to conserve not only the World Heritage Sites situated on its territory, but also to protect its national heritage.
UN Convention for the Protection of the Marine Environment of the North Atlantic (OSPAR). Introduced 1972 with various amendments and revisions in 1974, 1992 and 1998.	This is the mechanism by which 15 countries of the western coasts and catchments of Europe, together with the EU, cooperate to protect the marine environment of the North-East Atlantic. The Convention was a consequence of the unification (in 1992) of the 1972 Oslo Convention against dumping at sea and the 1974 Paris Convention to cover land-based sources and the offshore industry (hence the name OSPAR: Oslo–Paris). In 1998, a new annex on biodiversity and ecosystems was adopted in to cover non-polluting human activities that can adversely affect the sea.
UN Convention on International Trade in Endangered Species of Wild Fauna and Flora (CITES). Introduced in 1975.	This convention aims to ensure that international trade in specimens of wild animals and plants does not threaten their survival. There are varying degrees of protection to more than 5600 species of animals and 30000 species of plants (CITES, 2015). CITES is legally binding on the parties, but it does not take the place of national laws. It provides a framework to be respected by each party, which has to adopt its own domestic legislation to ensure that CITES is implemented at the national level.
UN Convention for the Protection of the Marine Environment of the Mediterranean Sea (Barcelona Convention). Introduced in 1976 and subsequently extended.	This is a regional convention similar to OSPAR which aims to prevent and abate pollution from ships, aircraft and land-based sources. Signatories agree to cooperate and assist in dealing with pollution emergencies, monitoring and scientific research. It includes an Action Plan for the Protection of the Marine Environment and the Sustainable Development of the Coastal Areas of the Mediterranean. The convention has also given rise to seven protocols covering dumping (1975 and 1995); prevention and emergencies (2002); land-based sources of pollution (1980); special protected areas and biodiversity (1995); offshore activities (1994); hazardous wastes (1996); and integrated coastal zone management (2008; UNEP, 2015).
Council of Europe Convention on the Conservation of European Wildlife and Natural Habitats (Bern Convention). Introduced in 1979.	A binding international legal instrument, which covers most of the natural heritage of the European continent and extends to some States of Africa. It aims to conserve biodiversity and natural habitats, and to promote European cooperation. Signatories undertake to: • Promote national policies for the conservation of wild flora, wild fauna and natural habitats • Integrate the conservation of wild flora and fauna into national planning, development and environmental policies • Promote education and disseminate information on the need to conserve species of wild flora and fauna and their habitats. In 1998, the convention led to the creation of the Emerald network of Areas of Special Conservation Interest (ASCIs) which operate alongside the EU Natura 2000 programme.

Convention/treaty name	Description
UN Convention on the Conservation of Migratory Species of Wild Animals (Bonn Convention or CMS). Introduced in 1983.	This convention aims to conserve terrestrial, marine and avian migratory species by providing a global platform for conservation and sustainable use. It is the only global convention specialising in the conservation of migratory species, their habitats and migration routes. The signatories work together for conservation purposes by providing strict protection for the most endangered species. Those threatened with extinction are listed on Appendix I of the convention and those species that need or would significantly benefit from international cooperation are listed in Appendix II. The convention encourages the signatories to conclude global or regional agreements and so acts as a framework convention. The collection of agreements and memoranda of understanding (MoU) are often referred to as the CMS family. Currently, there are 19 international MoU and 7 agreements, all of which are based on sound management and conservation plans.
UN Convention on Biological Diversity (CBD). Introduced in 1993 and revised in 2010.	Currently commits signatories to five main objectives to • Address the underlying causes of biodiversity loss by mainstreaming biodiversity across government and society • Reduce the direct pressures on biodiversity and promote sustainable use • Improve the status of biodiversity by safeguarding ecosystems, species and genetic diversity • Enhance the benefits to all from biodiversity and ecosystem services • Enhance the implementation through participatory planning, knowledge management and capacity building.
UN Convention to Combat Desertification (UNCCD). Introduced in 1994.	This convention stems from a direct recommendation of the 1992 Earth Summit, where desertification, along with climate change and biodiversity loss, were identified as the greatest challenges to sustainable development. The convention addresses specifically the arid, semi-arid and dry sub-humid areas, known as the drylands, where some of the most vulnerable ecosystems can be found. Signatories work together to improve the living conditions for people in drylands, to maintain and restore land and soil productivity, and to mitigate the effects of drought. National action programmes are the key instruments to implement the convention. They are often supported by action programmes at sub-regional and regional levels. As the dynamics of land, climate and biodiversity are intimately connected, the UNCCD collaborates closely with the Convention on Biological Diversity (CBD) and the UN Framework Convention on Climate Change (UNFCCC) to meet these complex challenges with an integrated approach and the best possible use of natural resources.
The Food and Agriculture Organisation (FAO) of the United Nations International Treaty on Plant Genetic Resources for Food and Agriculture (ITPGRFA or International Seed Treaty). Introduced in 2001.	This is a comprehensive international agreement, in harmony with the Convention on Biological Diversity. Its main objectives are: • The conservation and sustainable use of plant genetic resources for food and agriculture, and • The fair and equitable sharing of the benefits arising out of their use, in harmony with the Convention on Biological Diversity, for sustainable agriculture and food security. It covers all plant genetic resources for food and agriculture, and its multilateral system of access and benefit-sharing covers a specific list of 64 crops and forages, which account for 80% of the food we derive from plants (ITPGRFA, 2015). It also includes provisions for farmers' rights and the sustainable use of crop genetic resources.

the management of landscape features that support the Natura 2000 network, and also to undertake monitoring of protected habitats and species populations, as well as reporting back to the EC on a six-year cycle.

The Birds Directive (Council Directive 79/409/EC and 2009/147/EC; EC, 2009) was an EU response to the Bern and Bonn Conventions. The latter (correctly entitled the 'Convention on the Conservation of Migratory Species of Wild Animals') seeks to provide protection to migrating wild species, particularly birds but also bats, small cetaceans (whales, dolphins and porpoises) and other species. The Birds Directive provides a framework for the conservation and management of, and human interactions with, wild birds in Europe. The main provisions include:

- Maintenance of populations of all wild bird species across their natural range via encouragement of various activities
- The identification and classification of Special Protection Areas (SPAs) for rare or vulnerable species (Annex I) and for all regularly occurring migratory species (particularly the protection of wetlands of international importance)
- Establishment of a general scheme to protect all wild birds
- Restrictions on the sale and keeping of wild birds
- Specification of conditions for hunting and falconry (Annex II lists huntable species)
- Prohibition of large-scale non-selective means of bird killing
- Encouragement of certain forms of relevant research
- Requirements to ensure that introduction of non-native birds do not threaten other biodiversity.

The establishment of SPAs are a key part of the directive and are sites of international importance for the breeding, feeding, wintering or the migration of rare and vulnerable species of birds. There are no formal criteria for selecting SPAs, but they can include, for example: if the area is used regularly by 1 per cent or more of the national population of a species listed in Annex 1 of the directive; if the area is used regularly by 1 per cent or more of the biogeographical population of a regularly occurring migratory species (other than those listed in Annex 1) in any season; if the area is used regularly by over 20,000 waterfowl or seabirds in any season; various combinations of criteria involving considerations such as population size and density, species range, breeding success, history of occupancy, multispecies areas, naturalness of site, severe weather refuges. SPAs form part of the Natura 2000 network.

Natura 2000 is the centrepiece of EU nature and biodiversity policy. It fulfils an EU obligation under the UN Convention on Biological Diversity and provides an EU-wide network of nature protection areas. The network is comprised of Special Areas of Conservation (SACs) and Special Protection Areas (SPAs) designated under the Habitats and Birds Directives. The sites include nature reserves, but most of the land is privately owned and unlike many nature reserves, Natura 2000 sites do not necessarily exclude human activities. Indeed, around 40 per cent of the total Natura 2000 area is farmed (EC, 2014) and many of the habitats and species that are protected are highly dependent on or associated with farming practices. However, these areas tend to be marginal farming land such as alpine meadows, open heathland and wet grasslands (EC, 2014) which offer difficult farming conditions, are often highly labour-intensive, economically vulnerable and so potentially at risk of abandonment. Consequently, the support offered to farmers within Natura 2000 areas can be vital for the continuing protection of these

The Natura 2000 and the Emerald networks

Natura 2000 sites

Candidate Emerald sites

Figure 3.8: Natura 2000 network (Source: EEA, 2012)

areas. Thus, the overall emphasis of Natura 2000 is on ensuring that management is sustainable (ecologically and economically). Surveillance of habitats and species is carried out on a six-year cycle. There are currently over 26,410 terrestrial Natura 2000 sites across the EU – 28 (EC, 2015) covering an area of 787,767 km^2 (see Figure 3.8).

In 2013, around 18 per cent of land in EU MSs was protected as part of the Natura 2000 network. The map shown in Figure 3.8 illustrates the distribution of Natura 2000 sites, as of 2012, and clearly illustrate the richness of Europe's natural heritage. For example, in Cyprus the Troodos National Park is a Natura 2000 site due to its unique beauty and its invaluable plant and animal habitats. These include 11 habitats listed on Annex 1 of the Habitats Directive, including four that are a priority for conservation (*Pinus nigra* subsp. pallasiana, *Quercus alnifolia*, and serpentinophilous and peat grasslands). It also hosts ten Annex II-listed species, three of which are priority plant species (*Arabis kennedyae, Chionodoxa lochiae* and *Pinguicula crystallina*), as well as a large number of endemic plants, the threatened Cyprus whip

snake (*Columber cypriensis*) and the critically endangered Cyprus beetle (*Propomacrus cypriacus*).

European MSs are responsible for ensuring that all Natura 2000 sites within their country are appropriately managed by designated conservation authorities. These authorities often work closely with other authorities, voluntary bodies, charities, local conservation groups and private landowners. The cost of the conservation activities is borne by each MS but there are central EU funds available for urgent and innovative activities as well as CAP structural funds and support under agri-environment measures.

In May 2011, the EC presented a new biodiversity strategy: 'Our life insurance, our natural capital: an EU biodiversity strategy to 2020' (EC, 2011). This built on a previous strategy – the 2006 Biodiversity Action Plan (EC, 2008) and was formerly adopted in April 2012 by the European Parliament. The new strategy has 6 targets:

- Continue to fully implement the Birds and Habitats Directives

- Maintain and restore ecosystems and their services
- Increase the contribution of agriculture and forestry to maintaining and enhancing biodiversity
- Ensure the sustainable use of fisheries resources
- Combat invasive alien species
- Help avert global biodiversity loss.

The strategy addresses two commitments made by EU leaders in March 2010. The first is the 2020 headline target: 'Halting the loss of biodiversity and the degradation of ecosystem services in the EU by 2020, and restoring them in so far as feasible, while strengthening the EU contribution to averting global biodiversity loss'; the second is the 2050 vision:

> By 2050, European Union biodiversity and the ecosystem services it provides – its natural capital – are protected, valued and appropriately restored for biodiversity's intrinsic value and for their essential contribution to human wellbeing and economic prosperity, and so that catastrophic changes caused by the loss of biodiversity are avoided.

It also addresses global commitments made (in Nagoya, Japan) under the Convention on Biological Diversity, where world leaders adopted a package of measures to address global biodiversity loss over the coming decade (EC, 2011).

Aquatic pollution issues have also seen a number of specific policy and regulatory initiatives. Rivers, lakes and coastal waters are vital natural resources. As well as providing drinking water and being an important resource for industry and recreation, they are crucial habitats for many different types of wildlife. Prior to 2000, EU water protection legislation was somewhat piecemeal and fragmented. European water legislation

began in the 1970s with standards for rivers and lakes used for drinking water abstraction in 1975, and in the setting of quality targets for drinking water in 1980. At an EC ministerial seminar held in Frankfurt in 1988, a number of improvements were identified that subsequently resulted in two important EU directives being adopted. The first was the Urban Waste Water Treatment Directive (91/271/EEC) which provided for secondary (biological) waste water treatment, and even more stringent treatment where necessary. The second was the 1991 Nitrates Directive (91/676/EEC) and this from an agri-environment perspective has the greater significance as the agricultural use of organic and chemical fertilisers was a major cause of water pollution in Europe. The photograph in Figure 3.9 shows livestock manure spilling into a water course potentially polluting the water with nitrates and pathogens. As well as causing environmental issues such as eutrophication, nitrate in drinking water has been linked with a number of health issues (such as methemoglobinaema in unborn infants and young children, and thyroid dysfunction and cancer in adults), hence nitrate pollution was one of the first aquatic agricultural pollution issues to be tackled.

The Nitrates Directive (see also Chapter 4) seeks to improve the quality of European waterbodies by introducing a number of measures to reduce nitrate pollution from agricultural sources. In particular, it concentrates on promoting better management of animal manures, chemical nitrogen fertilisers and other nitrogen-rich soil conditioners. Under the directive all EU MSs are required to:

- Identify all waterbodies (freshwater, groundwater, estuaries, coastal and marine waters) polluted or at risk of being polluted.

Figure 3.9: Livestock manure spilling into a watercourse
(Photo courtesy of: Dr Doug Warner)

- Designate 'Nitrate Vulnerable Zones' (NVZs) – areas of land which drain into polluted waters or those at risk of becoming polluted. Farmers and land-owners within these designated areas are required to comply with an action programme of stringent measures to reduce nitrate pollution. These measures vary across EU MSs but include the amount of total nitrogen applied to the land by organic and chemical fertilisers, mandatory times of the year when organic manures cannot be spread, specifications for the minimum amount of on-farm manure storage capacity and mandatory compliance with codes of good agricultural practice. These codes of practice are voluntary for farmers and land owners not in NVZs.
- Introduce a comprehensive monitoring programme and a four-year cycle

of reporting on nitrate concentrations in groundwaters and surface waters, eutrophication incidences and an assessment of the impacts of the NVZ action programmes on water quality and agricultural practices. NVZ designated areas and the action programme must be amended according to the findings of the monitoring programme.

Complying with the stringent regulatory demands of this directive has been a significant component of the EU agri-environmental landscape for a number of years (for farmers operating in affected areas), and it continues to be an integral part of the Water Framework Directive (WFD – also see Chapter 4). This latter legislation is now the central driver for environmental water policy throughout the EU.

The need for the WFD became apparent following a fundamental rethink of EU water policy beginning in the mid 1990s. Around that time, the EC embarked on a consultation process, the outcome of which was a consensus that whilst considerable progress had been made in tackling individual issues, current water policy was fragmented, in terms of objectives and means. There was also agreement on the need for a single piece of framework legislation to resolve these problems. In response to this, the EC presented a proposal for a Water Framework Directive.

The Water Framework Directive (WFD; Council Directive 2000/60/EC; EC, 2000) committed EU MSs to achieve good qualitative and quantitative status of all ground and surface waterbodies (including marine waters up to one nautical mile from shore) by 2015. A second phase from 2015–2021 updating the original objectives and introducing new measures for assessing water status is now in place. The WFD also requires EU MSs to establish river basin districts (RBDs) and for each of these a river basin management plan, which is prepared, implemented and reviewed every six years. There are four distinct elements to the river basin planning cycle: characterisation and assessment of impacts on river basin districts; environmental monitoring; the setting of environmental objectives; and the design and implementation of the programme of measures needed to achieve them. The Water Framework Directive is also discussed in terms of its implications for water resource and quality management in Chapter 4.2.

There are a number of objectives in respect of which the quality of water is protected. The key ones at European level are general protection of the aquatic ecology, specific protection of unique and valuable habitats, protection of drinking water resources and protection of bathing water. All these objectives must be integrated for each RBD. What constitutes 'good status' in relation to these objectives differs between surface and ground waters. Surface water is assessed for its 'ecological status' and 'chemical status'; and groundwater for its 'quantitative status' and 'chemical status'. With respect to biodiversity, clearly the ecological and chemical status of surface waters is of most importance; thus, these will be focused on herein. However, the quantitative and chemical status of groundwater can also be of relevance in instances where the groundwater is a source of water for surface water habitats.

In regard to ecological status, as no absolute standards for biological quality can be set which apply across the EU (because of ecological variability), the controls are specified as allowing only a slight departure from the biological community which would be expected in conditions of minimal anthropogenic impact. A set of procedures for identifying that point for a given body of water, and establishing particular chemical

or hydromorphological standards to achieve it, are provided in the WFD, together with a system for ensuring that each MS interprets the procedure in a consistent way to ensure comparability (Priestley, 2015).

Good chemical status is defined in terms of compliance with all the quality standards established for chemical substances at European level. The WFD also provides a mechanism for renewing these standards and establishing new ones by means of a prioritisation mechanism for hazardous chemicals. This will ensure at least a minimum chemical quality, particularly in relation to very toxic substances, everywhere in the Community. The European Commission is required to review the adopted list of priority substances every six years (EC, 2000). The latest review was in 2012 and led to a new directive amending the original list of priority substances. The new Priority Substances Directive (2013/39/EU) sets out a list of 45 'priority' substances for surface waters which must stay below specified levels that are safe for water bodies and human health (EC, 2000). The list includes many agricultural pollutants, particularly pesticides, and several possible or known carcinogens, such as benzene, lead and naphthalene. There is also a sub-set of 'priority' substances called the 'priority hazardous substances' list which is identified as such in the legislative list. Priority hazardous substances include cadmium and its compounds and mercury and its compounds. All MSs must stop any discharge of priority hazardous substances by 2020.

Much of the legal framework regarding biodiversity is focused on protecting and enhancing populations and their habitats. Whilst population expansions are often welcome, in the case of invasive non-native species this is not usually the case. There have been numerous examples when a species has been introduced deliberately or accidently outside of its normal distribution range which have subsequently led to impacts on local ecosystems, for example, by predation of native species or competition for light, water and nutrients/food. Invasive species are also responsible for billions of euros of damage to the European economy annually. Policy for managing these risks is directed by the EU Strategy on Invasive Alien Species adopted under the CBD and the Bern Convention, and enacted by the 2015 EU Regulation on Invasive Alien Species (1143/2014). This regulation establishes an international framework that includes a three-point management plan:

1. **Prevention:** This includes the identification of those species of greatest concern. This list is regularly updated and classifies species into three groups. Black-list species are prohibited or strictly regulated. The list includes, for example, the Siberian chipmunk (*Tamias sibiricus* Laxman) and the water hyacinth (*Eichhornia crassipes*). White-list species are low risk or benign and so have a low priority; best practice should be enough to manage them effectively. Grey-list species are those where the risk is, as yet, unknown. The black-list currently only covers a few of the species that are of concern at national level but these may be added in the future. These species include, for example, Japanese knotweed (*Fallopia japonica*), Himalayan balsam (*Impatiens glandulifera*; see Figure 3.10), giant hogweed (*Heracleum mantegazzianum*) and common ragweed (*Ambrosia artemisiifolia*), all of which are highly problematic in the UK, and the Pumpkinseed sunfish (*Lepomis gibbosus*) and Asian lady beetle (*Harmonia axyridis*), both of which are causing ecosystem damage in Italy. Preventative measures concentrate at the point of origin and include border and import controls and MS-level legal instruments.

2. **Early detection:** The main measure for this is the establishment of an alarm system and recognised communication channels for the prompt exchange of information between neighbouring MSs that may be affected.

3. **Rapid eradication:** Measures include a cooperation agreement between neighbours to address cross-boundary issues and establish appropriate control measures.

Regardless of what strategies, policies and other initiatives are in place, the day-to-day activities of farmers, landowners and rural communities play a major role in delivering the biodiversity outcomes sought. To this end, the CAP (see Chapter 1) and specifically

Figure 3.10: Himalayan balsam (*Impatiens glandulifera*): an invasive non-native species now common in the UK

(Photo courtesy of: Dr Doug Warner)

the link between farm payments, and provision of countryside management, as seen by many European countries, is critical. As described in detail in Chapter 1, there are two parts to this. First, direct payments to farmers and landowners within the EU and in some non-EU European countries depend upon compulsory compliance with a number of rules relating to environmental, biodiversity and climate objectives and provide a standard of farm management that seeks to ensure that the land is kept in good agricultural and environmental condition (cross compliance). In other non-EU countries, the objectives can vary although they all seek to ensure environmentally sound practices are adopted. Second, there are additional voluntary schemes that provide payments to farmers and landowners if they commit via contracts to environmentally-friendly farming techniques that go beyond legal obligations and the compulsory measures of cross compliance. In return, farmers receive payments that provide compensation for additional costs and income foregone as a result of applying those environmentally friendly farming practices in line with the stipulations of their agri-environment contracts. The stipulations within agri-environment contracts can be designed at national, regional and even local level so that they can take into account specific needs and environmental issues and so can provide a highly targeted means of addressing biodiversity concerns.

3.3. Farm level management and biodiversity protection

3.3.1. Introduction

Farmers rely heavily on ecosystem services provided by their land to survive. They need, for example, fertile productive soils, nutrient cycling, water cycling, pollination and biological pest control. So, even in the absence of farm payments and incentive schemes

there is every encouragement for them to look after their land, its environmental quality and biodiversity. How successful they are invariably depends on their personal knowledge and the mechanisms available to them for advice and support. Their choice of management practices that are adopted can, sometimes to extremes, affect the local environmental quality and the biodiversity on their land. Agriculture can maintain and enhance habitats, encouraging biodiversity or it can be the source of numerous problems including the loss of habitat, nutrient run-off, sedimentation of waterways, greenhouse gas emissions and the exposure of pesticide residues to non-target species, to mention just a few.

3.3.2. Understanding fragmentation

As discussed in Chapter 3.1, habitat loss, and particularly fragmentation, is a significant problem and, whilst farmers and landowners cannot control issues such as urban sprawl, they can take measures on their own farm to reduce habitat fragmentation so as to improve biodiversity. Habitat fragmentation can take many forms and the key to reducing it on the farm is to understand the different forms it takes and the various options available to farmers to minimise isolation. The options are highly dependent on the spatial location of the different patches, habitat type and the amount of land available. The main fragmentation scenarios are summarised in Figure 3.11.

The worst-case scenario is a number of small, isolated habitat patches surrounded by land unsuitable as habitat as shown in (a) of Figure 3.11. For a species to survive, a minimum number of individuals within a given habitat patch are required. This is known as the minimum viable population (MVP). The smaller the patch, the fewer individuals it will support and consequently its species richness will be lower, a concept

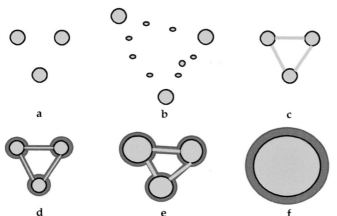

Figure 3.11: Habitat fragmentation and mitigation
In ascending order of biodiversity importance (low to high) (a) isolated habitat fragments or patches, (b) fragments or patches joined by 'stepping stones', (c) fragments joined by wildlife corridors, (d) fragments joined by wildlife corridors with buffer zones, (e) larger fragments joined by wildlife corridors with buffer zones, (f) mainland habitat with buffer zone.

described by the species–area relationship (Werner and Buszko, 2005). In general, a larger habitat patch will support greater species richness compared to a smaller patch of similar habitat quality (Hilty *et al.*, 2006). The species richness of the large patch shown in (f) (Figure 3.11) will typically possess and support a greater number of species. It also represents the desired state of a conservation area.

In the agricultural landscape, the highest priority will undoubtedly be food production and so allocating the entire land area to habitat creation is not possible, as a high proportion of the land area must be given to cropping and/or livestock and so a matrix of different land uses is created which must all coexist on the farm. This is represented by the intermediate stages (b–e) of Figure 3.11. An increase in connectivity between patches is more likely to support a greater number of individuals as fauna and flora interchange occurs between them, effectively increasing the area of habitat available to those individuals. The species richness then increases accordingly. At the lower end of the scale scenario, (b) utilises 'stepping stones', which are small patches of habitat that permit dispersal but that are not large enough in their own right for a species to complete its life cycle. If circumstances permit and the

stepping stones can be joined to form a continuous feature (c–e), then dispersal through the farm landscape becomes potentially easier for wildlife, with no requirement to enter vulnerable areas and open spaces at all.

The continuous features could be a hedgerow, ditch or a line of trees but it needs to allow movement between two different habitat patches, as shown in Figure 3.12 (blue triangle and red square), so that they provide contact between the two, equivalent to the 'conduit' or bridge (Hilty *et al.* 2006; Vogt, 2016) shown in (c) (Figure 3.12). The loop (d) (Figure 3.12) connects the same habitat patch (blue triangle) in two or more different locations. Such a situation may arise where the patch is elongated and irregular in shape and a length of linear habitat connects two sections of that habitat patch.

Linear habitats that dissect the cropped areas may perform a number of functions. If they also function as habitat or as a population source, their benefit is enhanced further. An example of this is beneficial insects such as predatory carabid beetles that use field margins as winter refugia, before dispersing into the crop during the spring (Holland *et al.*, 1999; Warner *et al.*, 2000). Filters (e) or barriers (f) (Figure 3.12) are either selective to certain species or prevent movement through them completely. Weak flying

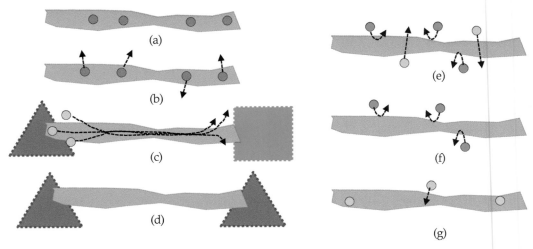

Figure 3.12: Form and function of wildlife corridors
(a) source, (b) habitat, (c) conduit or bridge, (d) loop, (e) filter, (f) barrier, (g) sink.
(Based on Hilty *et al.*, 2006 and Vogt, 2016)

insects such as midges (e.g. *Dasineura brassicae*) or the scarce copper butterfly (*Hoedes virgaurea*) may be prevented from direct flight through densely vegetated hedgerows or over embankments. However, barriers and filters can offer pest control benefits to the farmer by making it difficult for damaging pests to reach the crop. A worst-case scenario sees a linear feature function as a sink (g) (Figure 3.12) in which individuals colonise and remain within the linear feature, but do not successfully reproduce within it. For a conduit to function effectively, two different habitat fragments must be joined directly.

To ensure that a farm can provide the goods (e.g. food, fuel and fibre) it sets out to whilst still maintaining and enhancing habitats and biodiversity, activities must be taken to adopt farming practices and activities that contribute to the creation and maintenance of habitat quality, enhancing connectivity and so reducing habitat fragmentation and maintaining the ecological resilience (Defra, 2010).

3.3.3. Management of on-farm habitats

Generally speaking, structurally diverse habitats such as rough grassland, scrub, hedgerow or woodland in close proximity to each other provide a favourable habitat to a diverse array of species. Therefore, the creation of these types of habitats are invaluable on agricultural land. Farmed landscapes that have a mix of cultivated land and grassland surrounded by boundary features of tussocky grasses, hedgerows and farm woodlands, such as that shown in Figure 3.13, offer the possibility of dividing the area into distinct zones: habitat, linear boundary features, crop edge and the cropped area. The cropped area is further influenced by the individual management practices undertaken within that zone. Each of these zones may also function as a habitat fragment, stepping stone, corridor or buffer zone, depending on its size and location.

3.3.3.1. Heathland and moorland

Large areas of specific habitats are rare on-farm due to the very nature of agricultural

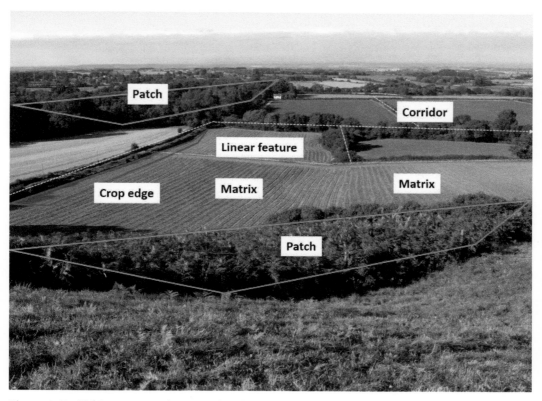

Figure 3.13: Habitat zones within agricultural land
(Photo courtesy of: Dr John Tzilivakis)

land management; most will be smaller habitat patches. An exception may be where neighbouring land managed extensively, for example upland moorland, is present adjacent to the farm boundary and is also present within it. Moorland is land with predominately semi-natural upland vegetation with or without rock outcrops and is primarily used as rough grazing. It includes wet and dry heathland, and mire communities. Heathlands are wide open landscapes dominated by low-growing shrubs such as ling heather (*Calluna vulgaris*), bilberry (*Vaccium myrtillus*) and gorse (*Ulex europaeus*). These areas are important for wildlife including rare birds such as black grouse (*Tetrao tetrix*) and invertebrates. The UK has around 75 per cent of the global resource of this type

of habitat with many notable sites including Bodmin Moor in Cornwall and Dartmoor in Devon. There are also sites in the Lake District, North York Moors and Scotland. There are also small but equally important sites in Austria (Tanner Moor) and Germany (GroßesTorfmoor). These sites are both peat moors and Natura 2000 sites.

Creation and restoration of these types of habitat tends to require specialist support as their vegetation tends to be highly complex. With respect to maintenance, one key issue is ensuring that soil fertility is managed appropriately. Heathland and moorland are typically nutrient-poor but deposition of nitrogen by grazing livestock will elevate the nutrient status and this can lead to a change in plant communities and so change the nature of the habitat and the biodiversity it supports. Options open to farmers and landowners to prevent this situation are to reduce stocking rates and remove livestock during the winter months. This will contribute to the preservation of moorland habitat quality through decreasing nitrogen inputs and maintaining nitrogen levels below their 'critical load' – the maximum nitrogen input that will not instigate a change in the plant community. In some areas, foraging animals such as deer can be responsible for seriously overgrazing these types of habitats and this situation may need specific management and professional guidance.

Where wetlands lie within the farm boundary, they may offer opportunities for extension or restoration as well as ongoing maintenance and support may be available for farmers in the EU under the CAP agri-environment schemes and initiatives (Chapters 1 and 3.2). For example, in England, the 2016 Environmental Land Management Scheme supported the creation, restoration and maintenance of moorland. An example of partially degraded upland heathland is given in Figure 3.14a.

The decline of heather and its replacement with grass due to overgrazing is evident in the foreground. An example of degraded wet heathland now dominated by mat-grass (*Nardus stricta*) is given in Figure 3.14b.

The area, formerly a combination of wet and dry heathland where species such as heather, the insectivorous sundew (*Drosera* spp.), sphagnum moss, heath spotted orchid (*Dactylorhiza maculata ssp. ericetorum*) and the bright yellow flowers of the bog asphodel (*Narthecium ossifragum*) (Figure 3.14c) would have once been abundant, has been subject to drainage and overgrazing before being abandoned. It has since been entered into a moorland restoration option to enable its transformation back to the original dwarf shrub composition.

3.3.3.2. Woodland and woodland edges
Farm woodlands often exist as fragments within intensive agricultural landscapes. The key to providing biodiversity benefits is to improve connectivity between those fragments by planting additional small woodlands (also known as trees in groups or field copses) which, when strategically located,

Figure 3.14a:
Heathland degradation with decline of heather cover and increase in grass-dominated areas
(Photo courtesy: Dr Doug Warner)

Figure 3.14b: Degraded wet heathland and mire communities now dominated by mat-grass (*Nardus strictae*) (Photo courtesy: Dr Doug Warner)

Figure 3.14c: Heath spotted orchid (*Dactylorhiza maculata*) and bog asphodel (*Narthecium ossifragum*) present in wet heathland (Photo courtesy: Dr Doug Warner)

function as stepping stones. Alternatively, the size of the individual woodland fragments can also be increased so that they can support larger populations. An increase in woodland size decreases the impact of 'edge effects', that is, a change in fauna and flora communities caused by changes in the microclimate (a modification to temperature, humidity, wind speed) and light levels compared to the woodland itself and the potential influx of generalist species that may be suited to the surrounding crops (Hilty *et al.*, 2006). Larger woodlands tend to support a greater number of woodland specialist species but that is not to say that woodland edges or small woodland fragments are not species diverse. Such habitats tend to contain a greater abundance of ecotone species, that is, species that move between both the woodland and the crop, a phenomenon noted by Usher *et al.* (1993) and discussed in greater detail below. Although in most cases a positive relationship exists between species richness and increased woodland area and connectivity, even small woodland fragments may support a relatively high species diversity due to the ecotone effect. Further, the creation of small farm woodlands provides habitat for several bird species currently in decline, for example the bullfinch (*Pyrrhula pyrrhula*) and mistle thrush (*Turdus viscivorus*).

Increasing the area of woodland or forest in a largely agricultural area will increase the habitat diversity of the agricultural landscape (Zanchi *et al.*, 2007). However, it is not solely an increase in area, further measures can be taken by the farmer or landowner to ensure that woodlands are managed to maximise their value to as broad a range of species as possible, a factor of great importance when size is a potentially limiting factor. For example, the wood mouse (*Apodemus sylvaticus*) displays a preference for larger woodland areas, or in their absence, smaller woodlands in close proximity (i.e. with a high degree of connectivity) to other woodland fragments (Fitzgibbon, 1993; Lazdinis *et al.*, 2005). Birds also show this type of behaviour; for example, a study that investigated bird populations during the winter and breeding seasons in woodlands created under CAP agri-environment schemes in southern and central England recorded over 50 different bird species on 65 farms (Vanhinsbergh *et al.*, 2002). Key variables that determine diversity in addition to size, include geographical location, area, age, tree species diversity, conifer content and structural diversity both within and around the woodland edge (Brockerhoff *et al.*, 2008; Sarlöv Herlin, 2001; Sarlöv Herlin and Fry, 2000). The characteristics of the surrounding farmland and the diversity and structure of the adjoining field boundaries are also of influence, as bird species richness is positively correlated with the extent of connections to hedgerows containing trees.

Increasing the number of tree species and the structural diversity within a given woodland will enhance the value of woodlands to biodiversity further. In particular,

the presence of fruit-bearing species, such as wild cherry (*Prunus avium*) or yew (*Taxus baccata*) provide an important food source for birds, mammals (Sarlöv Herlin and Fry, 2000) and invertebrates (Fry and Sarlöv Herlin, 1997). Species that flower for longer periods, for example lime (*Tilia x europea*) or willow (*Salix* spp.), are of greater value to invertebrates (Sarlöv Herlin and Fry, 2000), particularly pollinators such as bumblebees and solitary bees (Kirk and Howes, 2012).

However, it is not just the higher biodiversity taxa that benefit from diversity within a woodland. Fungal diversity is also potentially a function of host tree species diversity (Humphrey *et al.*, 2002). Maintaining structural diversity within the woodland understorey through either light grazing or the periodic coppicing of small areas is conducive to expanding woodland flora. It allows species with different habitat requirements, that is, those with preference for either shaded or warmer microclimates to coexist within close proximity to one another. One of the rarest UK butterflies, the heath fritillary (*Melitaea athalia*) has declined due to the absence of coppicing in woodlands and the continued maintenance of an open understorey. The presence of open spaces allows early successional habitat to coexist with the climax vegetation of the mature woodland.

Woodland management techniques will vary significantly with the location and the indigenous species supported. In many woodland management systems, such as those in common use in the UK, it is seen as important for farmers and landowners to prevent excessive grazing as this can potentially damage the understorey due to excessive trampling and the prevention of natural regeneration. Grazing is, therefore, carefully controlled.

In other European countries, different approaches for woodland management are often adopted that have been developed

Figure 3.15: English woodland and a 'hard' woodland edge
(Photo courtesy of: Dr John Tzilivakis)

to suit the landscapes and climates. Spain, for example, has considerably more woodland than the UK and these tend to have a very different structure and composition. Common trees are the Holm oak (*Quercus ilex*) and the Cork oak (*Quercus suber*). A popular management approach in the Iberian Peninsula is one that seeks to ensure the woodland works for biodiversity, the farm and the local community, and that a productive use of resources is attained to boost rural economies. The approach, known as 'Dehesa' in Spain or 'Montado' in Portugal, manages the woodland, its edges and surround pasture land as a single entity. A long-term view of tree management, typically 250 years, is adopted that allows sustainable harvesting. Grazing by goats, pigs, cattle and sheep is common and used to manage the grass and local herbaceous species such as the rock rose (*Jara pringosa*, a member of the Cistaceae family). Every ten years or so the woodland understorey is cleared to ensure the woodland composition is maintained and dominated by rampant vegetation. Fallen acorns, roots and wild herbs are eaten by the grazing livestock. Acorns are often collected to feed Black Iberian pigs (*Sus scrofa domesticus*), the ham (*jamón ibérico*) from which is seen as high quality and marketed as local produce, protected under the EU's Denomination of Origin Scheme which protects products from specific geographical areas (see Chapter 6). Other valued produce is also harvested from these areas, including wild game, mushrooms, honey and cork, and these help ensure that Dehesa areas have significant economic importance despite the land often having limited agricultural potential. These areas, also highly beneficial for biodiversity, are species-rich and support several endangered species, for example the Iberian lynx (*Lynx pardinus*) and the Spanish Imperial eagle (*Aquila adaberti*; Regato-Pajares *et al.*, 2004).

The woodland edge (Figure 3.15) is not deemed to be woodland proper but its value to biodiversity should not be underestimated. As mentioned previously, ecotones (including woodland edges) potentially contain a greater number of species than either of the habitat types they separate individually, since they contain species from both plus ecotone specialists. The woodland edge itself typically has a different microclimate to both the inner woodland and the adjacent agricultural land. These areas typically see an increase in light and nutrient levels with an associated increase in tree density and microhabitats (patches of bark loss, cracks, sap runs and epiphytes; Ouin *et al.*, 2015) and certain species utilise this to their advantage.

A woodland or forest edge may be termed either 'hard' or 'soft'. A hard edge denotes one that has no change in structure between the edge and the centre, that is, the interface consists of the extremes of the two habitats (Hilty *et al.*, 2006). A soft edge merges gradually with the second land use by the creation of a scrub sub-climax zone between the two habitats. The product is three distinct habitats in close proximity to one another. High-quality, structurally diverse, woodland edges, fulfil a number of roles which benefit biodiversity, many of which are similar to those performed by well-maintained hedgerows, including the provision of sites for breeding and foraging, shelter and dispersal corridors for a range of wildlife (Fry and Sarlöv Herlin, 1997). A number of studies have concluded that for farmland birds, for example, the more structurally diverse a woodland edge is,

the greater the diversity, richness and abundance of species present is likely to be (Dyda *et al.*, 2009; Fry and Sarlöv Herlin, 1997). This is particularly true for breeding birds, despite the increased likelihood of predation (Sarlöv Herlin, 2001). Similarly, several invertebrate groups, including butterflies, bumblebees and beetles, utilise the tall herbaceous vegetation on the field-ward side of the edge zone, as a source of either nectar or overwintering habitat, whilst a complex architecture favours web-producing spiders (Fry and Sarlöv Herlin, 1997). The unique microclimate and its similarity with hedgerows means that a structurally diverse and well-managed woodland edge may function as a corridor through which species suited to both habitats and the ecotone specialists may disperse, especially where woodland and hedgerow connectivity is present. It has been shown that such connectedness was important in the patterns of dispersal of woody plants from forest woodland edges, if their seeds were animal distributed (Sarlöv Herlin and Fry, 2000), whilst other authors have pointed to the fact that hedgerow species are more similar to those found in woodland edges than those in the woodland centre (McCollin *et al.*, 2000), meaning that a high-quality woodland edge zone is important in terms of the overall connectedness of habitats, with woodlands without such edges being less beneficial.

Through appropriate management it is possible to maximise the ecological benefits afforded by both new and existing woodland edge. In general, it is considered important in woodland edge management to ensure a structurally diverse habitat, with multiple layers of dense vegetation moving gradually, over some metres, from woodland to open farmland, including zones of shrub and herbaceous vegetation (Dyda *et al.*, 2009). Similarly, structural diversity at the scale of the local landscape will also serve to maximise overall biodiversity value (Fry and Sarlöv Herlin, 1997). However, guidelines for woodland edges contain few hard and fast management rules, although there are some clear rules of thumb. These include having sloped edges, that is, where vegetation becomes shorter as you move from woodland to farmland; indeed, woodland edges can be combined with a field margin/buffer strip, effectively extending the woodland edge ecotone further out into the field, and providing even greater diversity of species and habitats (Clarke *et al.*, 2011). It is also recommended that a woodland edge should have a curvilinear form, although this isn't always compatible with agricultural practices, and all edge forms should be appropriate for the region in which the feature is being developed. Indeed, there is no consensus as to how wide the edge zone should be, in part because this needs to be regionally appropriate, with most sources suggesting it should be as wide as possible (Fry and Sarlöv Herlin, 1997).

As far as the vegetation itself is concerned, as well as ideally being quite dense and multi-layered, tree and shrub species can be selected with biodiversity in mind. For example, in the case of UK woodland it may be appropriate to ensure the presence of pollen-bearing species such as the goat willow (*Salix caprea*) and blackthorn (*Prunus spinose*), as well as wild cherries, crab apples, hawthorn and rowan, although the choice should reflect local conditions and native species. It may also be necessary to control some plant types including brackens and bramble, although before doing so it has to be borne in mind that these species too may have considerable conservation value, despite often being viewed as a problem (Blakesley and Buckley, 2010).

There may, however, be circumstances in which the maintenance of dense vegetation within the woodland edge may not be appropriate, due to the need to deliver a regionally appropriate form, since open woodland is common in some areas, and may be a requirement of some valuable species of, for example, birds (Clarke *et al.*, 2011; Blakesley and Buckley, 2010). One way of achieving this is through the use of grazing or browsing livestock to clear some vegetation, but this must be done with great care, since livestock can suppress the development of saplings, and remove excessive ground vegetation. Through preferential grazing, they may also remove some palatable species more than other, less palatable such as gorse, changing the make-up of the plant community (Clarke *et al.*, 2011).

An important part of the woodland structure is dead, decaying and fallen trees (see Figure 3.16). These are a fundamental part of woodland and forest ecosystems. In the UK, for example, up to a fifth of all species found in a woodland environment rely on dead and decaying trees for all or part of their life cycle (Forestry Commission, 2012). The quantity of dead and decaying wood on the woodland floor is used as a key international indicator of the value of a woodland for biodiversity.

Historically, deadwood was removed from woodlands as it was considered to make the woodland look untidy and neglected, and because there was a common misconception that its presence endangered the forest health by harbouring pests and diseases. Therefore, it was often collected for firewood. However, the value of deadwood is now fully appreciated. Dead and decaying wood not only provides habitat and a source of food for a wide variety of species including moss, algae, fungi (see Figure 3.17), insects and mammals but also plays a role in sustaining soil fertility, hydrological processes and helps mitigate climate change by acting as a medium-term sink for carbon.

Woodland owners should therefore

Figure 3.16: Fallen and decaying woodland trees (Photo courtesy of: Prof. Kathy Lewis)

manage the deadwood element of their woodlands in a manner that supports biodiversity. However, management is not an onerous task with the main advice being to leave it alone especially in areas of high ecological value. It is important to aim for a range of species of deadwood at different stages of decay. In general, it is the larger trunks that are the most valuable as they remain for significant periods of time. If deadwood can be used to link habitat fragments such as linking banks of ponds and streams with the water zone then even greater value for biodiversity will be seen (Forestry Commission, 2012).

There may be other factors to consider when managing deadwood if the public have access to the farm woodland, such as ensuring public safety and considering the aesthetic impact of deadwood. In addition, leaving enough deadwood for biodiversity might need to be balanced against the economic benefits of timber and wood fuel harvest.

At a more basic level, there are other activities that farmer and landowners can do to increase the value of woodlands. Woodland patches and newer plantations devoid of mature trees in particular, may be enhanced by the appropriate siting of bat and bird boxes (Meddings *et al.*, 2011). Such structures act as surrogates for older trees where holes, crevices and loose bark are more frequent and it has been shown, by much research, that they can increase the number of breeding birds. Research findings suggest that some species happily use artificial nesting sites, including Eurasian kestrel (*Falco tinnunculus*), tree sparrows (*Passer montanus*), common starling (*Sturnus vulgaris*) and the long-eared owl (*Asio otus*) whilst others, such as carrion crows (*Corvus corone*) will not (e.g. Yom-Tov, 1974; Svensson, 1991; Field and Anderson, 2004). Similarly, artificial bug refugia, often called 'insect hotels' can be valuable if sited between habitat fragments. These are an assemblage of various natural materials created to provide shelter to overwintering insects. These will often be used by solitary bees and wasps, butterflies and earwigs. Bug hotels can be a simple collection of bamboo canes tied together or an

Figure 3.17: Fungi on woodland trees
(a) Bracket fungi (*Ganoderma*) on a living oak tree, (b) jelly fungus (*Tremellaceae*) on deadwood, (c) bracket fungi (*Trametes*) and moss on deadwood
(Photo courtesy of: Prof. Kathy Lewis)

Figure 3.18: A large bug hotel (Photo courtesy of: Prof. Kathy Lewis)

extensive construction such as that shown in Figure 3.18.

3.3.3.3. Trees

Trees outside of the woodland or forestry environment also offer significant biodiversity benefits, with these varying depending on their landscape arrangements, be it single trees within hedgerows, small numbers in copses, lines of trees acting as windbreaks or boundaries, or single trees isolated in a landscape. Their value is not just to birds but they also support a range of mammals (including bats, squirrels and mice), insects (including butterflies and moths) and flora such as mosses, lichen, algae and fungi which live or feed in their branches, barks and within their root system. In addition, trees provide a number of vital ecosystem services that support agricultural production including water and nutrient cycling, carbon sequestration and climate regulation, timber production, erosion control, airborne dust filtration and recreation. Other potential benefits include preventing access by pest species (Prevost and West, 1990) to a crop by creating a barrier and by mitigating pesticide drift subject to appropriate foliage porosity (Ucar and Hall, 2001).

Tree lines are features not typically planted to promote biodiversity; rather, they function as shelterbelts, reducing the impact of wind, softening the intensity of sunlight and screening visual impact. They can, however, function in a similar capacity to hedgerows, a key difference being the height of the canopy and density of vegetation within 1–2 m of the ground. Where herbicides are not applied to the area at and between the base of the trunks, a layer of herbaceous vegetation can develop. Where this is species diverse, it is itself of value to biodiversity. The accumulation of leaf litter and dead wood benefits detritivores (e.g. collembola) and provides refuges for a multitude of species, from invertebrates such as ground beetles through to amphibians during the terrestrial phase of their life cycle to hibernating mammals such as hedgehogs (*Erinaceus europaeus*).

In common with hedgerows and ditches, tree lines provide a linear feature, extending semi-natural woodland habitat and providing dispersal routes where two isolated habitat patches are joined. However, unlike hedgerows, dispersal is more likely to be beneficial to birds and flying insects as flight will enable them to utilise the higher canopy more effectively. In a similar manner to hedgerows, species such as bats will use tree lines as flight corridors (Boughey *et al.*, 2011; TBP, 2002).

Tree age and foliage density are important contributors to the value of tree lines to biodiversity. Karg (2004), for example, found a significant correlation between increased age of the shelterbelt and the abundance, biomass and diversity of wintering insects. The value of tree lines for the successful overwintering and survival of beneficial invertebrate species has been demonstrated throughout Europe (Gange and Llewellyn, 1989; Paoletti and Lorenzoni, 1989; Paoletti, 2002; Ryszkowski *et al.*, 2002; Wratten, 1998).

Isolated trees are, by definition (EC, 2015), single trees with a crown diameter minimum of 4 m, although trees recognised as valuable landscape features with a crown diameter below 4 m may also be considered. In the context of habitat fragmentation mitigation, isolated and scattered trees function mainly as 'stepping stone' habitat. They are often mature trees and sometimes those classed as 'veteran' trees which may be several hundreds of years old. Tree maturity enables them to exist as ecologically important habitat islands for several species (Gibbons *et al.*, 2008; Manning *et al.*, 2006; Tews *et al.*, 2004). Features of older trees include hollows, cracks, crevices, loose bark and deadwood of benefit to roosting bats, nesting birds and small mammals (DeMars *et al.*, 2010; Forestry Commission, 2005; JNCC, 2001; Mazurek and Zielinski, 2004). Oliver *et al.* (2006) note significant differences in terrestrial invertebrate assemblages beneath isolated trees compared with surrounding grazed pastures. Increased soil fertility, shade, increased soil moisture and the accumulated leaf litter layer create a microclimate favourable to ground-active and soil invertebrate assemblages (Huhta, 1976; Uetz *et al.*, 1979).

As well as having high landscape value, veteran trees (i.e. those that are considered particularly valuable because of great age, size or condition) support a wide range of fauna and flora and are the specific habitat of some rare and endangered species, such as those that require habitat that offers both live wood and deadwood. For example, ancient oaks provide habitat for the Cardinal Click beetle (*Ampedus cardinalis*), and the Oak polydore fungus (*Piptoporus quercinus*) occurs on the limbs and trunks of living or dead veteran oak trees that are more than 250 years old. Similarly, the critically endangered Cyprus beetle (*Propomacrus cypriacus*) is entirely dependent on veteran trees as it inhabits decaying heartwood. This is a very specific habitat type which is already highly fragmented and subject to continuing significant decline (Figure 3.19). This is a prime example of the vulnerability and rarity of many species across Europe.

All veteran trees (see Figure 3.20) have cultural heritage value, each being a relic of a former landscape and each having significant aesthetic appeal (see Chapter 6). Often, the distribution of veteran trees in a landscape can indicate a former land use pattern. The majority of veteran trees tend to occur in parkland. For example, Kedleston Park, Derbyshire, England is an important site for veteran trees, having over 100 specimens. However, they are still found within the agricultural landscape where they may mark old boundaries and some, such as willows, may map old water courses.

Ancient trees need to be carefully managed to balance the longevity of the tree with the needs of the species that use it as habitat and food. However, it is often the case that 'management' actually means little more than periodic inspection. Generally, it is best to leave decaying, dead and dropped wood in place providing this is unlikely to cause a hazard. Ideally, such trees should be

Figure 3.19: The habitat of the critically endangered Cyprus beetle (*Propomacrus cypriacus*)
(Photo courtesy of: Thulborn-Chapman Photography)

Figure 3.20: A declining veteran tree on Wicken Fen, England
(Photo courtesy of: Benjamin Hall)

fenced off to reduce the impact of livestock and if the tree is close to public areas, such as footpaths, it may be appropriate to divert them away from the tree to reduce the impact of tourists and visitors and the risk of vandalism. There may be occasions, however, when work may need to be conducted on the tree as it is unstable or it is considered in decline and vulnerable, in which case a management plan will be required. Developing the plan is a complex task, often needing the skills of an ecologist and tree care professional as the plans will be unique to the tree and the landscape it is in. Management of the tree needs to consider the land use around it and the contribution it makes to the local landscape value also needs to be considered. It may be the case that there are several potential conflicts of interest that will need to be resolved.

Funding opportunities for the management of veteran trees, particularly if specialist tree surgery is required, are sometimes available under MS agri-environment schemes or as grants via rural development programmes.

3.3.3.4. Waterbodies

Waterbodies take many forms, for example, surface waters (such as rivers, ponds and lakes), groundwaters and ditches (see Chapter 3.3.3.4). With respect to on-farm habitats, ponds and ditches are the most commonly found aquatic habitats. Ponds are typically defined as permanent or seasonal waterbodies between 1 m^2 and 2 ha in surface area (Wood *et al.*, 2013) and may have a wide range of surface areas, depths and origins (natural and man-made; European Pond Conservation Network, 2008). In terms of biodiversity, ponds are of considerable importance (Biggs *et al.*, 2005, 2007).

Natural pond systems are characterised by providing a wide range of habitats, including a variety of surface areas, deep open waters, shallow shorelines, open and shaded banks, and submerged, floating and emergent vegetation. They are ecologically varied, both individually and within a given geographical area (Céréghino *et al.*, 2014; Davies *et al.*, 2008). They can, like woodlands, exist as habitat islands in the agricultural landscape. Enhancing connectivity by an increase in number per unit area is conducive with maintaining biodiversity. For species that exist as meta-populations (small localised populations within an overall larger population but that exchange individuals between those localised populations), such as the European protected great crested newt (*Triturus cristatus*), the capacity to move between ponds is critical. Connectivity for amphibians and other species may be improved by linear features such as hedgerows but also by maintaining the riparian zones – the transitional areas that occur between land and the waterbody,

characterised by distinctive soil, hydrology and biotic conditions strongly influenced by the water (Naiman *et al.*, 2005).

Although pond biodiversity is in part controlled by natural site-specific factors beyond the scope of anthropogenic management, including climatic and biogeographic location, and elevation (Céréghino *et al.*, 2014), there is management available to maximise their ecological value. A key factor that contributes to their overall conservation value is the varied nature of the habitats they provide (Davies *et al.*, 2008). Fundamental principles, albeit subject to geographic location and site-specific factors, for example altitude (Biggs *et al.*, 1994; Gee *et al.*, 1997; Hinden *et al.*, 2005), include the hydroperiod (the period a pond contains water), water quality, the quality and extent of the vegetated (riparian) area, connectivity with other habitats and its size (although this is not as critical according to Oertli *et al.* (2002).

The CAP agri-environment measures have, in the main, contributed to improvements to water quality by encouraging the establishment of buffer zones around ponds, and improving their connectivity through the creation or maintenance of linear features such as ditches, hedgerows and tree lines. Management options to ensure the waterbody has high value for biodiversity includes ensuring clear water devoid of excessive nutrients (nitrogen and phosphorous causes eutrophication – see Chapter 3.1), with abundant aquatic invertebrate communities and submerged aquatic plants (optimal coverage 70–80 per cent; ARG, 2010; Baker *et al.*, 2011; Oldham *et al.*, 2000). A decline in quality is generally indicated by increasing discolouration and a decrease in aquatic invertebrate diversity and aquatic plant presence. Fencing to prevent access by livestock has benefits for both water quality and plant assemblages where temporary drying out occurs.

Ponds that dry up on occasions can potentially be of greater benefit to many biodiversity groups. Drying may only be for short periods during a given year, or even occur in intermittent years but, principally, this prevents the development of large fish populations, a key predator for many aquatic species, for example the larvae of amphibians or Odonata in the pond ecosystem. It does, however, come at a cost. Ponds prone to intermittent drying out are vulnerable to damage or loss by grazing livestock which damages the natural plant assemblage, either through eating the vegetation, or through excessively trampling the ground (Dimitriou *et al.*, 2006). The provision of fencing to prevent access by livestock mitigates such risks. A further factor, although not as directly obvious, is the lowering of local water tables due to abstraction. A natural water table and degree of water level fluctuation is an essential part of maintaining pond habitats, preventing excessive drying while allowing some, and providing an adequate drawdown zone (Dimitriou *et al.*, 2006).

Rivers and riparian zones (see Figure 3.21) play a major role in shaping the landscape and creating habitat for flora and fauna. However, they are not normally managed by farmers but tend, especially within the EU, to be managed using a catchment approach. Nevertheless, where rivers, and indeed other waterbodies, run through farmland, management of the riparian areas will be important as they are valuable to a wide range of different species. The dynamic equilibrium between the water and the directly adjacent

Figure 3.21: River and riparian zones
(Photo courtesy of: Dr John Tzilivakis)

land creates a corresponding dynamic equilibrium of life within the aquatic ecosystem. As long as the waterbody is allowed to freely interact with adjacent vegetated riparian areas, a diversity of habitats in various stages of ecological succession will be maintained. These areas are vital to link up terrestrial farm areas such as trees and hedgerows with waterbodies, to provide a gradual habitat succession, reducing fragmentation and ensuring, through this connectivity, wildlife that relies on both these areas can thrive. Riparian zones also provide other ecosystem services such as filtering, reducing the amount of nitrate and pesticide pollution that reaches the water by plant uptake, physical processes and chemical transformation together with trapping sediment-bound pollutants. These areas need to be carefully managed in a manner sensitive to the local area and needs of the landscape, including maintaining a mix of native trees, plants and shrubs valuable to local fauna species. Natural regeneration is seen as the best option with frequent inspection of regenerating riparian zones on a regular basis. Issues to be addressed include ensuring the right levels of light and shade are attained, that thinning and coppicing is undertaken if needed to ensure the correct plant communities develop and grazing is carefully managed (Gregory *et al.*, 1991).

One major issue for rivers is sedimentation. This occurs where eroded soil (see Chapter 5) is carried from land surfaces to a waterbody. Some sediments will occur naturally as a product of water movement on the river banks but it is often enhanced by human practices. For example, unvegetated river banks, stock movements close to the river bank, lack of functioning riparian zones, deforested areas and, often, simply by run-off from agricultural fields. Once they enter surface waters, eroded sediments can significantly change aquatic habitats. Sediments deposited on river beds may clog valuable gravel bed habitats, which are key in the life cycles of a number of benthic macroinvertebrates, and salmonid fish species, changing them into fine-grained muddy habitats, suitable for a very different species assemblage (Collins and Anthony, 2008). Fine particles can also stay suspended in the water column for a considerable period of time, greatly increasing the water turbidity, reducing light penetration which may kill vegetation and riverbed-living species.

A second issue is the interception of nutrients (nitrate and phosphate) from inorganic or organic sources, which can result in the eutrophication of the waterbody. Nutrient enrichment by organic pollutants stimulates microbial activity within the water. The increase in respiration by the microbes increases the consumption of oxygen, depleting the levels of oxygen within the water, also known as the Biological Oxygen Demand (BOD). The BOD of a waterbody or the level of organic pollution may be indicated by the aquatic invertebrates present from which a biotic index may be derived. A biotic index is a scale that indicates the quality of an environment in response to the organisms present (Mason, 1997). They use a relative scoring system proportional to the impact of a particular environmental variable, but are not a direct quantified measurement of that environmental variable itself. A freshwater biotic index works on two key principles (Abel, 1996): polluted water contains fewer species (lower species diversity) and, certain species or taxonomic groups as a whole sensitive to a particular pollutant are selectively removed in response to the presence of that pollutant. These are referred to as indicator species. Species or taxonomic groups are given a score depending on how sensitive they are to a particular pollutant. The site sampled can then be awarded an overall biotic score in response to the species or taxonomic groups caught, and water quality interpreted for that score depending on the biotic index used. In the UK, the Biological Monitoring Working Party (BMWP) biotic index was designed principally to detect high BOD and the presence of organic pollution in rivers (Abel, 1996). It awards a high score to families or taxa that are sensitive to low levels of dissolved oxygen within the water. The overall BMWP score is the sum of all the scores of each family or taxon caught at that site, or the taxon richness (i.e. it does not take account of abundance or numbers of an individual within a particular family caught; e.g. five individuals of the family Chloroperlidae in a sample will be given a score of 10, the same as one or two individuals). The Average Score Per Taxon (ASPT) prevents the potential bias of a purely species richness approach by calculating and basing a water quality classification on the average score for each family present within the site. A similar approach is applied to static waterbodies (Chapter 3.3.3.6) such as ponds using the Predictive System for Multimetrics (PSYM; Biggs *et al.*, 1996). A critical consideration is, however, the type of pollutant for which the biotic index has been developed. Taxonomic groups with higher BMWP or ASPT scores indicate sensitivity or intolerance to high levels of organic material and decreased oxygen within the water. Other potential pollutants such as zinc, to which stoneflies are not as sensitive (they may be sampled in relative abundance when zinc is present; Abel, 1996) would not be suitable for assessment with these indices. Stoneflies are also tolerant of low pH (acidic water), but their presence indicates that organic pollution is absent. When stoneflies are present but taxonomic groups sensitive to low pH (e.g. Gammaridae) are absent, it is possible to rule out one particular pollutant (in this case organic pollution) and ascertain the

likely presence of another potential issue (e.g. acidified waters).

3.3.3.5. Ditches

A ditch is an open channel, often a long field boundary, used to divert water to or from the field for drainage or irrigation. They include where open watercourses meet and have a width of between 2 and 12 m. Ditches are important biodiversity corridors and are used by a wide variety of species including amphibians and birds. Ditches can replace lost natural aquatic features, and may be one of the few remaining biodiversity resources (Blomqvist *et al.*, 2003), which in some cases at least can support considerable biodiversity, including species of conservation value (Clarke, 2014). This, however, is easily threatened by pressures such as the use of herbicides and other pesticides too close to the top of the bank, damage by grazing livestock, nutrient enrichment through fertilizer applications in the local area and falling water levels as a result of continued land drainage (Armitage *et al.*, 2003).

In terms of aquatic biodiversity, the biological communities associated with ditches can be expected to vary from those of other waterbodies as a result of their being linear features, with high edge ratios, and therefore subject to an intensive exchange of vegetation matter and biological organisms between themselves and the surrounding terrestrial environment (Herzon and Helenius, 2008). However, what is equally important is that ditches often have different characteristics to other waterbodies in the same area, allowing for the development of different ecological communities (Armitage *et al.*, 2003). Consequently, where comparisons with the biodiversity of other aquatic features exist, ditches are often the richest, suggesting that ditches, despite often being man-made features, make a valuable contribution to the overall biodiversity of agricultural landscapes (Armitage *et al.*, 2003). In Finland, for example, ditches which take the form of canalised natural streams (i.e. generally wet), have been found to support spawning grounds for brown trout (*Salmo trutta*), whilst amphibians have been found to benefit from ditches which periodically dry out, since such systems tend to exclude the predatory fish which can take a heavy toll on juveniles, yet they are still of assistance in facilitating movement across the agricultural landscape (Herzon and Helenius, 2008). Ditches can also be useful habitats for threatened crayfish, particularly if they are associated with overhanging vegetation and there is clean water, although it has to be acknowledged that this is uncommon in many agricultural areas. As far as aquatic flora is concerned, ditches are characterised by regular management followed by recolonisation and gradual change, leading to spatial and temporal variability in plant community composition and structure. Indeed, this is an important property of such environments, since the different stages of hydroseral succession support different communities of invertebrates, and therefore have value in their own right (Herzon and Helenius, 2008), and it is human intervention that ensures the continued presence of such a diversity of habitats.

Ditch systems (which include the bank, etc.) support a considerable diversity of terrestrial flora and fauna. For mammals, such as mice, voles, water vole (*Arvicola amphibius*) and otter (*Lutra lutra*) for example, ditch areas provide cover from predators, corridors for movement and feeding habitat. Equally, farmland bird species richness has been shown to relate well to the heterogeneity of farmland habits, something to which ditches and their surrounding area contribute significantly. They are particularly useful in terms of the provision of damp soil for probing feeders, permanent water for aquatic invertebrate foods, sparsely vegetated areas to allow access to soil invertebrates, rank emergent vegetation for nesting, and bush/tree groups for nesting and/or serving as singing posts (Herzon and Helenius, 2008). On a smaller scale, ditch banks and margins support invertebrate communities, the structure of which is dependent on the vegetation present (e.g. the presence or absence of nectar-producing plants and successional stage), as well as general site characteristics such as aspect, protection from wind and availability of shade. Such properties, for example, make ditches hydrologically unsuitable for species such as dragonflies (Herzon and Helenius, 2008).

In terms of terrestrial flora, ditches support a greater diversity of plants than either cropped areas or sown field margins, although most species are relatively common and can also be found elsewhere on the farm. In addition, such environments generally favour species tolerant of high nutrient status and common herbicides, which can lead to a reduction in plant biodiversity in favour of those species capable of taking advantage of a nutrient-rich environment, and thus outcompeting species better suited to poor nutrient conditions, thereby colonising new areas (Blomqvist *et al.*, 2003). As a result, some of the species present may be considered weeds. However, work carried out in the Netherlands has shown that species previously common in wetland areas, grasslands and dry hayfields, but which now have high conservation value, may now only be found in ditch banks (Herzon and Helenius, 2008). In general, a greater diversity of plant species can be found in ditch banks associated with grasslands than croplands (Clarke, 2014), although there are some which are less tolerant of grazing and which may therefore be more commonly associated with cropped areas.

As many species (e.g. invertebrates) have

limited scope for dispersal, ensuring the ditch is located close to other ditches and aquatic habitats is important for controlling a water-bodies species composition, as are its water quality and dissolved oxygen status (Herzon and Helenius, 2008). Some studies have indicated that the lack of suitably local recolonisation sources can explain the fact that the management of ditch banks to enhance plant species diversity, a popular element in the CAP agri-environment schemes of some countries, has in some cases failed to deliver the expected biodiversity benefits (van Dijk *et al.*, 2014; Geertsema and Sprangers, 2002). However, many ditches are connected to larger (often natural) waterbodies, and as such may themselves serve as important refuges from the impacts associated with high flow and pollution events, and pathways for recolonisation dispersal, allowing the wider aquatic environment to recover from those disturbances (Clarke, 2014; van Dijk *et al.*, 2014).

Most ditches have been installed for the purposes of improving drainage, and are, therefore, managed to maintain that function. This includes keeping water levels relatively low and regularly clearing vegetation, something which can be to the detriment of biodiversity (Clarke, 2014). Some management activities will influence ditch biodiversity to a greater or lesser degree. Twisk *et al.* (2003), for example, found that as well as a reduced frequency of ditch management being associated with higher aquatic plant diversity, methods to remove vegetation affected how much biodiversity impact was seen. A mowing drum had a more damaging effect on emergent plants than other forms of mechanical removal, whilst dredging using a suction pipe as opposed to a pull-shovel also led to reduced diversity. Equally, clearing late in the year compared with summer was also found to be more damaging, possibly as early clearance allowed for a degree

of vegetative recovery prior to the winter period. However, in general, the quality of ditch environments is a function of the following (Clarke, 2014; Herzon and Helenius, 2008):

- Geographical context: for example, being in former wetland locations where there is an underlying biodiversity pool.
- Physical structure: including the diversity of microhabitats present, and the cross-sectional form.
- Composition and structure of vegetation: successional stages, submerged/floating/emergent plants.
- Water quality: good quality water improves ecological value.
- Water availability: hydroperiod, depth, flow rate.

These characteristics are, of course, themselves a function of ditch type and size, as well as the surrounding agricultural and natural environment. It is clear from this list, however, that there are a number of ways in which these can be managed to the benefit of ditch biodiversity, with sympathetic management aiming to produce a range of water depths and bank profiles (Clarke, 2014). For biodiversity value to be maximised, good-quality management is required, for example by halting the hydroseral succession at a point where there is sufficient open water to allow for a varied macroinvertebrate community structure (Armitage *et al.*, 2003). In peat areas, low ditch water levels favour common plant species, so it has been suggested by some authors that raising water levels may be an effective means of improving plant biodiversity status (Twisk *et al.*, 2003). Similarly, ditch banks may be managed in such a way as to benefit biodiversity by ensuring that cutting is timed to maximise the opportunity for seed setting, which in turn produces a greater available seed resource for dispersal, aiding in ditch bank restoration (Leng *et al.*, 2011).

3.3.3.6. *Wetland habitats*

Wetlands can best be described as areas of land covered with either fresh or salt water or a combination of both having an intermediate status. Wetlands are being lost at an alarming rate and as they are considered to be highly valuable areas for biodiversity, this is a major concern. They are often highly species-rich supporting a range of fauna throughout their life cycles including mammals, birds, fish and invertebrates, as well as a wide range of flora. Wetlands also support the cultivation of rice, a staple in the diet of half the world's population. The total rice-growing area within the EU-27 is around 450,000 ha with around 70 per cent of this occurring in Spain and Italy. Wetlands also provide a significant number of other ecosystem services including water filtration, storm protection, flood control, carbon sequestration and recreation.

With the exception of a few crops, such as the aforementioned rice, European wetlands are not productive cropping areas and, indeed, one of the main reasons for the loss of these types of habitats is that they have been drained to claim them for agricultural production; this process will alter the ecological character of the habitat. Although wetland protection is officially a priority for the signatories that have ratified the Ramsar Convention (see Table 3.1), wetlands continue to be under threat. Where this has occurred, there are often significant impacts on biodiversity and some areas may no longer even qualify as wetlands (Verhoeven and Setter, 2010). However, if low-intensity agriculture is undertaken on these reclaimed wetlands, especially where the use of agricultural chemicals such as fertilisers and pesticides is minimised, the biodiversity of the wetland landscape may be high, although the species composition and setting differs strongly from that in its pristine situation. This 'secondary biodiversity' is often worthy of protection because it includes many rare and characteristic species (Verhoeven and Setter, 2010). In consequence, there may not be the desire to revert wetlands previously claimed for agriculture production back to their original status and the focus is on protecting the remaining areas on farm including marshland and water meadows. An example is purple moor grass and rush pastures. These wetland grasslands are located on poorly draining soils in areas with high precipitation or low-lying areas adjacent to watercourses, or they may have been deliberately created as part of a flood plain and flood mitigation strategy. They are typically grazed by cattle or horses, livestock able to utilise the coarser vegetation present, such as the soft rush (*Juncus effusus*) that is in the main, unpalatable to sheep. Other wetland habitats that have been restored and maintained under agri-environment schemes include fens and lowland raised bogs. Both habitats are highly sensitive to any form of management, but are especially vulnerable to drainage and nutrient enrichment. They both have declined substantially as a result of agricultural improvement, and examples exist where they have been afforded protection as Special Areas of Conservation. Lowland raised bogs, while relatively species-poor, are unusual in the habitat they provide and as such, many rare species of plants and invertebrates are associated with them. Particularly fascinating are the insectivorous plants, the sundew (*Drosera* spp.), butterwort (*Pinguicula* spp.) and bladderwort (*Utricularia*), that adapt to the low nutrient status by sourcing supplementary nutrients from insect prey.

The process of managing these types of areas for on-farm biodiversity benefits will be highly site-specific and depend upon the condition of existing wetland habitats. An assessment of that condition will identify if they have been adversely affected by drainage, pollution and overgrazing such that restoration might be necessary. Simple changes to drainage networks such as the blocking of drains or diversion of ditches, or reduction in fertiliser usage may be all that is required to restore some areas, given time. Light grazing (0.05–0.1 livestock units per ha) by cattle, especially in the summer months, can be essential and be useful to the farm as forage might be short in dryer areas. Frequent checks for undesirable plants such as exotic species (e.g. Japanese knotweed) or those that are harmful (e.g. ragwort *Senecio jacobaea* is toxic to horses) are necessary. The benefits are worthwhile, with residents of this habitat including the marsh fritillary butterfly (*Euphydryas aurinia*) and red-listed curlew (*Numenius arquata*).

3.3.3.7. *Hedgerows and green lanes*

A hedgerow, often referred to as a 'hedge', is a line of closely-spaced shrubs, bushes and tree species, planted to form a barrier or to mark the boundary of an area. In many European countries hedgerows are a common sight in the countryside. Indeed, England is characterised by its extensive networks of hedgerows that surround agricultural fields and form a 'patchwork quilt' effect. Hedgerows have been lost in many European countries for a range of different reasons including through the lack of management which has, over time, seen a hedgerow gradually convert to 'relict' hedges which are woody, linear features more like a line of trees or shrubs, often with many gaps.

Hedgerows, when well maintained, have the potential to function as corridors connecting habitat patches and enhancing dispersal between them and can act as extensions of semi-natural habitats. Those adjacent to waterbodies optimise the potential for amphibian dispersal between them. A number of species are known to utilise corridors within agricultural landscapes, some of which, such as the great crested newt (*Triturus crestatus*) that is on the IUCN Red List of threatened species, exist as meta-populations, and require dispersal between multiple ponds and habitat fragments within a given area to ensure overall population survival (Oldham *et al.* 2000). Hedgerows represent sub-climax vegetation, typically consisting of the species hawthorn (*Crataegus monogyna*), blackthorn (*Prunus spinosa*), wild or dog rose (*Rosa canina*) and hazel (*Corylus avellana*) planted in a line of one or two individuals deep, creating a width in the region of 2 m. Height can vary from as little as 0.5 m up to anything in the region of 3–4 m, although both represent extremes and are not typical of a well-managed and maintained hedge.

Ideally, a hedge should be 1.5–2 m in height, contain a mixture of species and a species-diverse understorey. They are important to multiple species groups including plants, breeding birds, small mammals and invertebrates including several predatory insect species that hibernate in field margins (e.g. Mountford *et al.*, 1994; Morandin and Kremen, 2013; Parish *et al.*, 1995; Cranmer *et al.*, 2012; Pfiffner and Luka, 2000). Many species of bats demonstrate a preference to fly along linear features such as hedgerows where they function as 'commuting habitats' (JNCC, 2001). Comparable to woodland edge, except existing as an isolated feature as opposed to buffering a mainland habitat, a structurally diverse hedgerow will provide hibernation sites for small mammals and invertebrates, perches and sources of seed for birds, and resources for pollinating insects (e.g. Holland and Luff, 2000; Hannon and Sisk, 2009). They have the potential to contain a diverse flora that permits flowering at different periods throughout the year, providing a sustained supply of nectar and pollen. The importance of this is illustrated by Hanley *et al.* (2011) who concluded that while 'mass flowering crops' (e.g. oilseed rape, linseed and beans) may be of some value to pollinators, their benefit is short-lived and limited only to the crop-flowering period. Once the crop ceases flowering, the pollinators have to locate suitable forage elsewhere.

The main activities required to enhance the biodiversity value of hedgerows, in addition to the provision of buffer zones, is enhancing structural diversity, either through planting new hedges, increasing the height, laying or gapping up. A further feature is the provision of standard trees within the hedge – trees planted at fixed distances along the hedge line but that do not require that the hedge increases in width, therefore maintaining its status as a linear feature as opposed to a habitat patch. Standard trees increase structural diversity with documented benefits for moth abundance and diversity (Merckx *et al.*, 2009; Wolton *et al.*, 2013), larger species in particular, and the frequency of bat flight (Boughey *et al.*, 2011). Taller hedgerows also offer benefit for trapping pests, and are most effective in reducing pesticide drift and protecting non-target areas, including ponds. Hinsley and Bellamy (2000) conclude that increases in height, width and overall volume coupled with the presence and abundance of trees were all positive influences on biodiversity. A lack of hedge trees decreases bird nesting opportunities (Sparks *et al.*, 1996), while increased volume and fewer gaps reduces corvid predation on songbirds. Typically, large, wide, bushy hedges support about 19 different species of bird whilst tidy, frequently mechanically cut hedges support only about eight breeding species (TBP, 2002), findings later corroborated by Woodhouse *et al.* (2005). Small mammals that forage in the hedge understorey are generally unaffected by hedgerow tree density. Cutting frequency is also a management factor. Under CAP, cross-compliance rules are in place that prevent hedges being cut at times of the year when breeding birds may be affected, that is, between March and August. Generally speaking, hedge cutting is not an annual activity. Within England, for example, cutting is normally undertaken every two years. If, instead, these hedges were managed under a three-year cutting regime, Staley *et al.* (2012) estimated that the biomass of berries would increase by about 40 per cent, resulting in a significant benefit for pollinators, birds and small mammals.

The phrase 'green lanes' is a term given to vegetated tracks that have a hedgerow on either side (Figure 3.22). Research has demonstrated that this arrangement offers particularly favourable conditions for biodiversity. A study that examined bird occurrence on 20 green lanes (farmland tracks with unsealed surfaces, bordered on each side by hedgerows) and on 20 paired single hedgerows during the 2002 breeding season in Cheshire, UK found that bird abundance, territories and species richness were all significantly greater on green lanes than on single hedgerows. The study also found that bird abundance on single hedgerows was influenced by the number of trees and amount of hawthorn (*Crataegus monogyna*) in the hedge (Croxton *et al.*, 2002). Other studies have shown that as well as the plant and tree communities, the spatial layout is important for the biodiversity as it creates a microclimate within the enclosed space, provides areas of shade and will buffer noise from external sources.

Management of green lanes is much the same as for single hedges and it is usually the case that regulations do not distinguish between regular hedgerows and green lanes. An important consideration is to ensure that clearance remains between the two hedges

Figure 3.22: A typical green lane in the English countryside (Photo courtesy of: Dr John Tzilivakis)

and the space is not permitted to gradually close.

3.3.3.8. *Traditional stone walls and terracing*

Traditional stone walls and terracing are common sites across much of Europe and used for stock containment, boundary identification and to enable sloping terrains to be cultivated safely and productively. They are also a significant part of Europe's cultural heritage and this aspect is discussed in Chapter 6.

Traditional stone walls consist of stones only, there is no mortar to hold them together, hence they are often referred to as 'dry stone' walls (Figure 3.23). In consequence, there is an abundance of access routes via cracks and crevices, for invertebrates to enter and use as refuges. The maximum and minimum specified width and height varies with EU MS but ranges between 0.25–4 m wide and 0.3–5 m high. Their value to biodiversity is greatly enhanced by their tendency to be present in the more upland areas often devoid of vegetation or other linear features due to

the unsuitability of conditions for trees or hedgerow species. The provision of a unique microclimate is recognised in a review by Collier (2013), supported by Francis (2010), who describes them as 'novel ecosystems'. Of interest is the contrast in plant communities on opposite sides of the wall. The warmer, dryer south-facing aspect may be more restricted to the presence of lichens, while ferns and mosses tend to develop on the north aspect where moisture levels are greater. The pioneer species, and those that tend to persist longest are typically lichens. In the absence of vascular plants that have yet to colonise, they provide the main source of nutrition for insect herbivores and molluscs. The Dry Stone Walling Association of Great Britain (DSWA, 2007) notes their value as perches for birds, as 'plucking stations' for birds of prey and holes for the provision of roosting sites for bats (DSWA, 2007). In general, their value increases where they consist of limestone, and with age and maintaining good structural condition.

Ideally, dry stone walls should be a functioning part of the landscape for stock

Figure 3.23: Dry stone wall, Dartmoor, England
(Photo courtesy of: Dr John Tzilivakis)

containment and management, and as important habitats for biodiversity, and so should be regularly maintained. A stone wall in good condition would have all or the vast majority of its top layer of stones still in place. It would be stable, with no or very few gaps or fallen stones and the top to the wall should be level with no sign of bowing or sinking and all sides would be straight. Regular inspections should be conducted to look for signs of decline including wilful damage and instability (DSWA, 2007). Vigorous and potentially damaging vegetation such as ivy and tree roots should be removed and capping stones that have become loose should be replaced. Without regular maintenance it is not uncommon for these walls to rapidly fall into a state of disrepair and so require significant attention by a dry stone wall professional (DSWA, 2011). Maintenance is best undertaken during the autumn, avoiding bird nesting periods. The

key is to use original stones wherever possible or at least find a good match. When replacing fallen stones, care should be taken to replace stones such that any moss or lichens on the stone are facing the same way that they grew. Where flora has colonised gaps in the wall these should be carefully removed and replaced in a suitable position after repair. If large sections of the wall are damaged, it should be repaired section by section over time so that plants and animals can recolonise. Cement should never be used.

Terraces are a complex reshaping of a sloping terrain which uses a technique similar to stone walls in construction whereby the wall acts as a retaining structure for a mixture of soil, sediments and stone to level a sloping area. The system usually incorporates a drainage system to avoid soil saturation. There are three general types. The first and most common type is the 'stepped' or 'benched' formation whereby several levels are created at different heights which can be linear or follow the land contour. The second, zigzags upwards on the slope and is known as a 'braided' formation. The third is where level 'pockets' are created to support an individual tree or building. As well as creating cultivation areas, they conserve moisture by reducing the speed of surface water run-off and reduce erosion as they reduce both the slope gradient and its length. They also offer hazard mitigation from overturned machinery and reduced labour requirements.

Terracing is commonly used across Europe and these structures need continuous care due to issues relating to slope instability, excessive run-off, rainfall infiltration, possible landslides (which a terrace collapse would exacerbate) and associated damage caused by earth tremors. As these mountainous areas are often economically fragile, many have been abandoned and are now slowly deteriorating (Tarolli *et al.*,

2014). For example, in Greece cultivated terraces are a common landscape characteristic, especially in the south of the mainland and on the many Greek islands, however many are in a poor state (Van-Camp *et al.*, 2004). Few terraces across Europe benefit from any legal protection, although Malta is one of the exceptions whereby under their Rubble Walls and Rural Structures (Conservation and Maintenance) Regulations 1997 (amended 2004) stone terraces, and indeed stone walls, are legally protected and it is illegal to deliberately demolish, modify, endanger or undermine them. Financial support for maintenance and renovation is available in many EU MSs, especially those with certain altitude and terrain conditions, via the agri-environment measures of the CAP. These schemes offer support measures to ensure the economic viability of these traditional land management systems without encouraging intensification, and reward the conservation of existing terraces. As these terracing systems tend to have been established in order to make harsh terrains more manageable, many are within designated Less Favoured Areas (see Figure 3.24) and so farmers receive additional farm payments in many EU MSs.

To keep terraces functioning properly and to achieve their original objectives, their management is essential and this is not dissimilar in many respects to that of dry stone walls. Regular checks should be made to ensure the top level of the terrace ridge is level and stable and has not been disturbed by tillage or other farming operations. Low spots should be filled and levelled, and any loose stones replaced. There should be regular checks to ensure there is no excessive soil erosion and where this is seen, mitigating action is required. Any drainage pipes or channels should be cleared of vegetation or foreign matter. Ideally, the wall should also be kept clear of any large rooting or invasive vegetation but considerable care

is needed if vegetation is well established as it may well be having a significant stabilising effect and its removal may cause the terrace to collapse. In addition, vegetation can help ameliorate soil erosion.

3.3.3.9. Grassland, meadows and pastures
Grassland can best be described as extensive areas where vegetation is dominated by a range of different grass species (Poaceae) with small populations of sedge (Cyperaceae) and rush (Juncaceae). Typically, these areas will have few shrubs or trees. Meadows are areas of grassland, usually used for hay production. Pasture is usually referred to in the context of livestock grazing and whilst it may also be grassland, it will often also include small shrubs and trees. Pasture will be a fundamental aspect of the farm's livestock management system. Whereas, often the main role of grazing livestock on grassland will be to help maintain the sward structural heterogeneity and so the biodiversity it supports. Grazing animals do this via treading, nutrient deposition from dung and urine and propagule dispersion.

The type of grassland is determined in part, by the underlying geology, although the climate will exert a further influence on the species present. On well-drained, shallow soils above chalk or limestone, soils of pH 6.5–8.5 give rise to calcareous grassland, among the most species-rich in Europe (Crofts and Jefferson, 1999). Typically restricted to areas where more intensive agricultural intensification is not practical, they are grazed mainly by cattle or sheep. These grasslands contain many rare species of plants and invertebrates including pasque flower (*Pulsatilla vulgaris*), monkey orchid (*Orchis simia*), wart-biter cricket (*Decticus verrucivorus*) and the chalkhill (*Polyommatus coridon*) (Figure 3.25) and adonis (*Lysandra bellargus*) blue butterflies (Natural England and RSPB, 2014).

Figure 3.24: Terrace dry stone wall failure within the terraced landscape of Tramonti (Salerno, Italy) (Photo courtesy of Dr Paolo Tarolli, described in Tarolli *et al.* 2014)

Figure 3.25: Chalkhill blue butterfly (*Polyommatus coridon*) on calcareous grassland (Photo courtesy of: Dr Doug Warner)

Nesting sites are provided for the amber-listed stone curlew (*Burhinus oedicnemus*). In contrast, acid grasslands occur on lime-deficient soils of pH 5.0 or below with underlying sandstone or sand deposits. In upland areas, they are present as a component of moorland habitat (Chapter 3.3.3.1), potentially indicative of degraded heather moorland that has been subject to overgrazing by sheep. They often represent relatively species-poor plant communities. Hicks and Doick (2014) cite a figure of <5 species per 4m², dominated by the grasses common bent (*Agrostis capillaris*), sheep's fescue (*Festuca ovina*) and wavy hair grass (*Deschampsia flexuosa*). Where managed appropriately, a five-fold increase in species diversity may be achieved, extending the floral interest to include species such as mossy stonecrop (*Crassula tillaea*) and spring speedwell (*Veronica verna*). The low nutrient status of the soil provides a favourable environment for rarer fungi including the wax cap (*Hygrocybe calyptriformis*) and fairy club fungus (*Clavaria zollingeri*). Fauna of note include the declining small heath butterfly (*Coenonympha pamphilus*), whose larval food plants include species of the abundant *Agrostis* and *Festuca* grasses. Populations of the adder (*Vipera berus*) also thrive in this habitat, as well as the red-listed nightjar (*Caprimulgus europaeus*), a ground-nesting species of bird. In addition to calcareous and acid grassland, a third key grassland type, located on mineral soils of pH 5.0–6.5, is neutral grassland. This type of grassland has typically been subject to some form of agricultural improvement, but where this improvement is subtle, species-rich swards continue to flourish. It includes pastures and hay meadows and, although biomass is cut and removed annually in the latter, the land is not ploughed and reseeded. Hay meadows have declined in many areas due to their replacement with silage as a means to provide winter feed for livestock. The swards of silage are typically species-poor. Pastures also offer a potentially species-diverse form of grassland. According to Jefferson *et al.* (2014), a typical sward may contain meadow foxtail (*Alopecurus pratensis*), the nectar-prolific common knapweed (*Centaurea nigra*) and meadow buttercup (*Ranunculus acris*). Species of conservation importance include the snake's-head fritillary (*Fritillaria meleagris*) and the green-winged orchid (*Orchis morio*) and greater butterfly (*Platanthera chlorantha*). The impact of soil pH and nutrient input on the floral composition of grassland is illustrated to great effect on the Park Grass field plot experiment, Rothamsted Research, UK. Experimental plots have, since 1856, received varying amounts of liming material and acidifying nitrogen fertiliser, ammonium

sulphate ($(NH_4)_2SO_4$). Consequently, there is a contrast in both soil pH and soil nutrient status between plots. Those highest in species diversity received no supplementary nutrients and had a pH maintained at neutral or above (Silvertown *et al.*, 2006).

The grassland habitat is under significant threat as much of it has been 'improved' for agricultural production, either through the addition of supplementary nutrients, cutting, or ploughing and reseeding (every five to eight years). In the UK, for example, meadows, grassland and pasture account for a third of all agricultural land with the majority of it having been improved. It is 'unimproved' grassland that offers the most benefit for biodiversity including rare wildflowers. More than half of all plant species found in Germany occur on unimproved grassland (Rook *et al.*, 2006). Avian and invertebrate populations will depend significantly on the availability, quality and diversity of seeds and insects in the sward. Therefore, from a biodiversity perspective, the species-richness of these areas can be enhanced by ensuring there are many different forage species present in the sward. This is of particular relevance to grasslands where hay is produced. Unlike silage, hay is cut later in the season allowing the flora present to flower and set seed, perpetuating sward species diversity for the following year. Hay, once cut, is required to be left on the surface to dry and reduce its moisture content. If it rains during this period, the process is delayed. Lower yields coupled with potential issues associated with a suitable drying period has made silage production a more attractive option. Silage is cut earlier in the year, preventing the flowering and seed set of many species present. As such, agri-environment measures are available to support hay production and maintain this more ecologically sound method of winter feed production. A further benefit of hay compared to silage is that the lower moisture reduces the risk of leachate during storage, a potential impact associated with silage. The highly nutrient-rich liquid carries an associated eutrophication risk, and high biological oxygen demand if it enters watercourses (Chapter 3.3.3.4).

All types of grassland are negatively impacted by overgrazing. Overgrazing results in a loss of species diversity and may cause bare soil and erosion. Continued deposition of nitrogen alters the soil nutrient status. Where high stocking levels are present, supplementary nutrient application elevates the nutrient status and decreases sward diversity further. Species-poor grasslands tend to be dominated by plants able to utilise nitrogen efficiently and grow rapidly. These species include perennial rye-grass (*Lolium perenne*), Yorkshire fog (*Holcus lanatus*) and creeping buttercup (*Ranunculus repens*; Scottish Natural Heritage, 2010). Undergrazing results in succession to scrub; in general, therefore, some form of management, albeit subtle, is required to maximise the quality of a given grassland type and its value to biodiversity. Some management practices affect pasture biodiversity indirectly. For example, using rotational grazing and managing pastures to leave more stubble and forage residue can influence beneficial insect and soil microbe populations which in turn will benefit birds and small mammals.

FIELD MARGINS AND BUFFER ZONES

Field margins are strips of land between the cropped area of the field and the field boundary. They serve a number of purposes, in particular functioning as buffer zones, protecting a habitat fragment from management activities within the crop such as cultivations and the application of agrochemicals (Norris, 1993). They can also, however, depending on their management, be habitats in their own right, functioning in a similar context to a linear habitat feature.

They provide living and nesting habitats and a source of food, seed, insects, pollen and nectar for a wide range of species. The presence of buffer zones either surrounding habitat patches or running parallel to linear habitat features functions to soften the impact of the crop on the main population or those individuals dispersing through the corridors.

The value to biodiversity is strongly linked to their vegetative composition and whether a seed mix is deliberately sown or is left to natural regeneration. If sown, the type of seed mixture used (grass or wildflower) and the management (if mown, frequency, frequency of resowing) will all affect the species attracted. The type of mixture, if full benefit is to be realised, needs to contain plants that flower at different times of the year and that will provide a continuous flowering resource (Goulson *et al.* 2005; Stanley and Stout, 2013). Any mixture that flowers only for a period of a few weeks during a season will not permit the margin to attain its full biodiversity potential. Flower-rich margins enhance bumblebee abundance strongly in the first two years following establishment, and then stabilise from year 3 onwards, with plant species diversity following a similar pattern (Korpela *et al.*, 2013). Of importance here is the establishment of a diverse floral mixture, preferably with the inclusion of at least one legume in the mix, and allowing a minimum of three years for the benefits to be fully realised (Lagerlöf *et al.*, 1992). Legumes within wildflower mixtures sown in buffer zones have been found to attract a range of bumblebee species (Stanley and Stout, 2013). Hoverflies (Syrphids) are predatory at the larval stage and their presence in agricultural areas may be enhanced by the planting of species such as phacelia (*Phacelia tanacetifolia*), mustard (*Brassica juncea*) and fennel (*Foeniculum vulgare*; Colley and Luna, 2000; Lovei *et al.* 1993). Adult females are attracted into the crop by these species where they then lay their eggs, positioning the larvae within the target area. Annual re-establishment of the mixture would not be appropriate.

Meek *et al.* (2002) note similar patterns for invertebrate diversity in a comparison of five types of field margin: cropped; sown with a tussocky grass mix; sown with a grass and wildflower mix; split margin (tussocky grass adjacent to hedge and grass and wildflower next to crop); and natural regeneration. Carabid abundance, a family of ground beetles, was greatest in an undisturbed, five-year-old strip where succession was allowed to proceed uninterrupted. Beetle numbers were greater in this scenario than in a two-year-old deliberately sown wildflower strip, and in an annually cultivated area with natural regeneration. The presence of vegetation cover and the frequency of cultivation are key underlying factors, with weed root systems considered to act as potential refuges. Lowest numbers were typically found in the cropped margin, except for the carabid beetle *Nebria brevicollis*, that was more abundant in the autumn.

Cropped margins with reduced chemical inputs (conservation headlands) and wild bird cover crops can provide relatively high food resources compared with a conventionally managed crop. However, resources are only present until harvest, their plant communities are relatively poor and arthropod abundance is usually lower than in uncropped margins (Vickery *et al.*, 2009). The value of field margins to foraging birds such as breeding yellowhammers (*Emberiza citronella*) was observed by Douglas *et al.* (2009), although it was found that their value increased where the margin was cut to increase food accessibility. The timing of cutting is critical here, cutting too early results in the loss of flowers for pollinators and subsequent seed development.

Grass buffer strips that occupy the area between the crop and a habitat feature, such as a watercourse, pond or woodland, provide specific protection for that feature as well as mitigating habitat fragmentation and improving biodiversity (Lovell and Sullivan, 2006; Cole *et al.*, 2012; Blackwell *et al.*, 1999). Buffer strips are typically sown grass but from which no production is derived. Management is restricted at the most to low-intensity grazing or cutting to prevent scrub encroachment. The targeted application of herbicide to remove broad-leaved dock (*Rumex obtusifolius*) or creeping thistle (*Cirsium arvense*) may also be beneficial. The habitat provided by grass buffer strips may not, however, be as valuable to biodiversity as a mix sown specifically for the purpose (e.g. sown wildflower mixtures) as discussed above. The contribution to biodiversity made by buffer strips tends to be in the prevention of surface run-off and nutrients or pesticide drift from entering sensitive habitat features that themselves are of value to biodiversity. That said, they do offer refuge to surface-active invertebrates, with carabid beetles, rove beetles (Staphylinids) and spiders (Araneae) having all been subject to scrutiny due to their natural enemy status and contribution to pest control. Whilst many of the species present are not rare, they do increase species richness and biomass, and support the trophic levels above them. Although, grassy strips are considered by some to be more limited as a food resource for birds (Vickery *et al.*, 2009) they potentially supply grass seeds and, depending on the complexity of the sward structure, a range of arthropods. A selective management programme is recommended (e.g. cutting small patches at different times of the year) and the inclusion of perennial forbs within the grass mixture to enhance species and structural diversity. This approach would benefit insectivorous birds, such as the skylark (*Alauda arvensis*) and small mammals in particular (Maisonneuve and Rioux, 2001; Josefsson *et al.*, 2013). Small mammals such as the wood mouse (*Apodemus sylvaticus*), bank vole (*Clethrionomys glareolus*), field vole (*Microtus agrestis*) and common shrew (*Sorex araneus*) utilise grass areas in association with other habitats (e.g. hedgerows) and so the margin in isolation would be of significantly lower value. The strategic location of grass strips to other habitat features, therefore, is important (Lovell and Sullivan, 2006). The benefit provided to aquatic biodiversity includes the providing of shade, reducing water temperatures, increasing dissolved oxygen concentrations, limiting primary production, and supplementing available food sources through the vegetative matter which may fall in (Barling and Moore, 1994).

3.3.3.10. Beetle banks

Beetle banks are grass strips, usually around 2–3 m wide that dissect large arable fields. They are often slightly elevated (around 0.4 m) above the crop level such that they form a ridge. Their primary purpose is to encourage predatory insects into the crop to provide biological pest control. Often, large fields were created by removing hedgerows and, as many predatory insects and spiders will not venture into the middle of large fields, beetle banks are a means to help ensure complete field coverage, in part providing a similar function to a hedgerow understorey.

Beetle banks can be havens for biodiversity (Macleod *et al.*, 2004; Thomas *et al.*, 2001a). In the early years following establishment they may not be as species-rich as field margins but are often better than grass strips. Species richness develops gradually over time so that those over a decade old are often comparable to field margins (Thomas *et al.*, 2001b). They provide vital habitat and refuge for many beneficial insects and spiders provided they are constructed and managed appropriately. They are also valuable to birds such as the European skylark (*Alauda arvensis*) and corn bunting (*Emberiza calandra*), and to small mammals such as mice and voles.

During establishments these strips should be sown with a grass seed mix that is around 60 per cent tussocky grass such as cocksfoot (*Dactylis glomerata*) together with native species such as fescues and bents. There are few disbenefits to farmers or biodiversity from introducing such features. They are useful for connecting habitat patches on opposing sides of large fields. However, depending on the farm layout, it may be necessary to leave a gap at one or both ends to allow farm machinery to pass. They can also help reduce soil erosion and run-off. They are inexpensive to create and to manage, as once established, they require only very little management, typically just a light trim every three years. One of the main issues is that these areas are highly vulnerable to pesticide drift and so pesticide spraying within 6 m should be avoided. Similarly, nitrogen-rich environments may encourage undesirable plant compositions.

3.3.3.11. Cropped area and soils as habitats

The cropped area is essentially a disturbed area of land, frequently cultivated and subject to multiple management interventions. Although the primary aim is one of production, there are measures available to increase the suitability of the crop for biodiversity (see Figure 3.26). The habitat conditions of cropped land typically contrast those of the farm's habitat fragments such as stepping stones and linear features which may exist within the cropped area.

AGROFORESTRY

Agroforestry is a form of intercropping; it does not remove the land from production although it has been noted to reduce yields

Figure 3.26: Goldfinch (*Carduelis carduelis*) in oilseed rape (Photo courtesy of: Dr Doug Warner)

in the region of 10 per cent. It combines the production of arable land with tree species planted within the productive agricultural area. A key benefit to the presence of trees on biodiversity within areas dominated by cultivation is the increase in structural diversity provided at the field scale, and the creation of a habitat mosaic and improved connectivity of habitat fragments at the landscape scale (Langeveld *et al.*, 2012; Nair, 2011). Although they possess certain elements in common with isolated trees, they lack the age and maturity to represent habitat in isolation. The trees are more representative of 'stepping stone' habitat or, where closer together in a line as suggested by Dix *et al.* (1995), form a linear habitat feature comparable to tree lines, but in the cropped area itself. The implementation of agroforestry areas as linear features offers potential for them to act as wildlife corridors, allowing the movement of populations between what would otherwise be isolated habitats (Hilty *et al.*, 2006; Dix *et al.*, 1995).

Several tree species provide viable options for agroforestry, each with a particular benefit to biodiversity associated with either the production of pollen for longer periods, the suitability of the flower to be utilised by a wide range of insect species, or the seeds or berries providing a source of nutrition for birds. The flowers of deciduous trees, especially *Salix* spp. (willow) and *Tilia* spp. (lime) provide a source of nectar and pollen for flying insects (Kirk and Howes, 2012), that are a source of food for bats (JNCC, 2001; BCT, 2013) and insectivorous birds. The fruit of species such as holly (*Ilex* spp.), crab apple (*Malus* spp.), wild cherry (*Prunus* spp.) and yew (*Taxus* spp.) are a potential food source for seed-eating birds. A review by Burgess (1999) identifies that both silvoarable and silvopastural systems improve arthropod and bird diversity, and silvoarable systems also improve small mammal diversity. This has potential benefits for insectivorous and predatory bird species (Sage and Tucker, 1998).

Agroforestry species enhance the vertical structure of the cropped habitat and modify the microclimate within the immediate cropped area. This, coupled with a decrease in tillage frequency and the potential for the development of tussocky grasses in the understorey, enhances the suitability of the habitat to amphibians during the terrestrial phase of their life cycle (ARG, 2010; Baker *et al.*, 2011; Oldham *et al.*, 2000). The enhanced numbers of surface-active invertebrates (Holland and Luff, 2000; Pfiffner and Luka, 2000; Thomas and Marshall, 1999) increases insectivorous bird numbers and insectivorous and grass-eating small mammal species (Kells and Goulson, 2003; Lye *et al.*, 2009). Natural regeneration of wildflowers in the understorey of deciduous plantations may increase wildflower diversity (Langeveld *et al.*, 2012) and also favour pollinating invertebrates (Hannon and Sisk, 2009; Hunter, 2002). Where the development of tussocky grasses occurs, it will provide potential nest sites for some species of bumblebee, or potentially increase the presence of small mammal nests that may also be utilised by bumblebees (Fussell and Corbet, 1992; Kells and Goulson, 2003; Lye *et al.*, 2009). Winter cover and hibernation sites are provided for surface-active insect predators of crop pests (Holland and Luff, 2000; Pfiffner and Luka, 2000; Thomas and Marshall, 1999). The limited cultivation and re-establishment frequency also favours predatory insect populations, although this is reported by Holland *et al.* (1999; 2005) as being dependent on sufficient ground cover being established and maintained on the soil surface, for example by beetle banks. Dix *et al.* (1995) report that although beneficial insects are enhanced by the presence of trees within cultivated areas, suitable hibernation areas (tussocky grass and herbaceous vegetation) need to be present at the base of the tree lines. Further, because such species need to utilise both the crop and the tree planted area in order to exert an impact on crop pest populations, they are essentially ecotone species and benefit most from a high edge to area ratio, that is, thin tree lines.

The use of agroforestry areas as buffer strips adjacent to wetland areas reduces the risk of nutrient run-off (Dimitriou *et al.* 2012; Kahle *et al.*, 2007) and eutrophication (Johnston and Dawson, 2005; Smith *et al.*, 1999; Withers and Lord, 2002), and any associated loss of biodiversity within waterbodies (Dudgeon *et al.*, 2006).

While grass buffer strips intercept surface run-off arising from agricultural land, the presence of agroforestry areas may reduce it at source, reducing the loading that the grass buffer strip must intercept. Nitrate leaching is reduced compared to arable cropping (Langeveld *et al.*, 2012) as supplementary nitrogen is not applied and tillage frequency is reduced. This decreases the rate of mineralisation and the release of nitrogen from organic matter. A permanent and deeper rooting system that the agroforestry trees provide utilises the nitrogen more effectively than that of an arable crop, rendering it unavailable to leaching (Nair, 2011; Nair and Graetz, 2004). Both coniferous and deciduous woodland significantly reduce the risk of soil erosion and surface run-off of phosphate (Dimitriou *et al.* 2012; Kahle *et al.*, 2007; Langeveld *et al.*, 2012). This increases with increased maturity and ground cover.

SHORT ROTATION COPPICE

Short Rotation Coppice (SRC) includes areas used for the production of biomass. Within the EU, each MS has its own list of approved species in order to avoid planting those that are not native. Supplementary mineral fertiliser and plant protection products may be used but the precise products and quantity depends on the MS. The impact of SRC on biodiversity is not well understood. The habitat it provides is different to arable cropping, with cultivation frequency diminished and structural diversity enhanced. How valuable it is to biodiversity is less certain, since the area remains a monoculture and is intermittently harvested on a two–four-year cycle (Forestry Commission, 2003) preventing full maturity. Its value, therefore, is most likely to be comparable to 'stepping stone' habitat rather than being a fragment of semi-natural habitat in its own right. That is not to say that it does not confer value, and, as highlighted for agroforestry, different species offer different potential benefits, notably *Salix* spp. (willow), *Prunus* spp. (wild cherry) and *Tilia* spp. (lime; Kirk and Howes, 2012). Their full benefit is unlikely to be realised in isolation; it needs to be present in combination with other habitats.

A review of the impacts of SRC on biodiversity by Langeveld *et al.* (2012) notes a general overall positive effect on what the authors consider to be two main categories of assessment: plant species (phytodiversity) and breeding bird species (zoodiversity), though they acknowledge that there is limited information on the length of time taken for the benefits to be realised. An increase in the structural diversity of the landscape relative to arable crops is attributed to the creation of a habitat mosaic effect. The authors reported an increase in the diversity of all plant species from representative habitat types, especially grassland species, but no impact on the abundance of endangered species. This increase for arable species is contradictory to the conclusions of others who cite the reduction in light penetrating the canopy and shading the soil as having a negative impact on species of open, arable habitats. Others also provide evidence that an increase in rare species may be seen, citing the examples of Pheasant's eye (*Adonis annua*), cornflower (*Centaurea cyanus*), broad-leaved spurge (*Euphorbia platyphyllos*) and shepherd's needle (*Scandix pectenveneris*; e.g. Persson *et al.*, 1989; Kirby, 1993).

Compared to arable land in isolation, SRC supports a greater diversity of invertebrate species, particularly diptera and arachnids (Sage *et al.*, 2006). Specialist species, such as the waved carpet moth (*Hydrelia sylvata*) of alder (*Alnus* spp.), may also add to the overall species richness of an area where SRC is present. This increase is attributed to the greater vertical diversity, the longer cropping cycle, and increased habitat stability and continuity. At the early stages of the cropping cycle, a greater variety of herbivores and phytophage invertebrate species are supported, and this increases species diversity at the higher trophic levels. Further, SRC does not typically receive applications of insecticide, although it may be applied occasionally to treat willow beetle outbreaks.

However, compared to mature woodland, SRC supports fewer species due to the lack of older mature trees, the associated reduction in microhabitat diversity (dead wood, gaps in the canopy) and the duration of time for which they have been available for colonisation by different species. Species with limited dispersal capabilities are likely to be absent from newly established crops. There may also be a decline in diversity and the potential loss and subsequent absence of such species at the end of each cropping cycle (Oxbrough *et al.*, 2010). During the first cropping cycle, arachnid species may be well represented but these tend to be removed during harvest and then fail to recolonize during the second and following crop cycles. Another reason for greater biodiversity in woodlands is that where gaps exist in the woodland canopy and sunlight is able to penetrate (largely absent in SRC crops), thermophilous (warmth-liking) species can coexist within a relatively small area with species that prefer cooler, shaded conditions. The absence of dead wood in SRC means they are also unsuitable for saproxylic invertebrate species.

SRC cropping has been shown to be beneficial to avian breeding populations. Generally, the number of species present are greater than in arable cropping although the extent of the benefits will depend on feeding (i.e. seed or insectivore) and habitat (whether species of open areas, shrub, forest, tall ruderal or reed areas and ecotones) requirements (Langeveld *et al.*, 2012; Berg, 2002; Hanowski *et al*, 1997). A further factor is the age and structure of the SRC crop, since height may vary from below 1 m when recently harvested to up to 8 m when fully grown.

Generally, woodland and shrub bird species tend to benefit the most from the introduction of SRC into areas dominated by arable cropping, however this is subject to the feeding strategy (Langeveld *et al.*, 2012; Coates and Say, 1999; Anderson *et al.*, 2004). Granivorous birds may not benefit due to the decline of understorey plant species and their seed production capacity and woodland birds that require mature trees or dead wood (e.g. nuthatch) are unlikely to benefit due to the relative immaturity of the SRC when it is harvested (Sage and Tucker, 1998). The species that tend to benefit are the insectivores that forage within leaf litter such as blackbirds and thrushes. The vertical structure of the SRC crop has also been shown to benefit ecotone species and those associated with ruderal habitats and reeds (Langeveld *et al.*, 2012).

The SRC crop is unlikely to benefit vulnerable species such as those given in the IUCN Red List. Those that thrive are generally the common and widespread species such as wren (*Troglodytes troglodytes*), blackbird (*Turdus merula*), robin (*Erithacus rubecula*) and chaffinch (*Fringilla coelebs*). However, there may be benefits for species regarded as having 'higher conservation concern' including the reed bunting (*Emberiza schoeniclus*) and song thrush (*Turdus philomelos*) and for vulnerable ecotone species such as the yellowhammer (*Emberiza citronella*), cirl bunting (*Emberiza cirlus*) and corn bunting (*Emberiza calandra*). The linear edge effect provided by the SRC–arable transition provides ideal habitat for small mammals and

hunting areas for birds of prey, especially owls (Sage and Tucker, 1998). The positive impact on small mammal populations, will in turn be beneficial to birds of prey by increasing available food. Larger mammals such as deer may also have potential to benefit from the enhanced structural diversity of the area, although they may also cause damage by grazing.

Some cropping systems can be particularly beneficial to soil biodiversity. Püttsepp *et al.* (2004), for example, noted a beneficial impact on soil microorganisms and soil fauna due to the potential separation of the soil into distinct zones or horizons under SRC associated with the removal of tillage and deposition of leaf litter, compared to the more uniform upper soil layers of arable land. Species such as willow (*Salix* spp.), alder (*Alnus* spp.), poplar (*Populus* spp.) and eucalyptus (*Eucalyptus* spp.) are able to form symbiotic mycorrhizal associations of two types (ectomycorrhizas and endomycorrhizas) simultaneously, increasing the diversity of fungi within the soil beneath SRC crops (Püttsepp *et al.*, 2004). Other soil-dwelling groups, namely worms, are also reported to benefit from SRC, due to the decreased mechanical damage from tillage, the increased soil moisture associated with no tillage and shading of the canopy, and the increase in deposition of organic material as leaf litter (Hubbard *et al.*, 1999; Whalen *et al.*, 2004).

CATCH CROPS OR GREEN COVER

Catch or cover crops represent areas of crop grown that are sown typically before a spring crop to mitigate the environmental loss of soil and nutrients during the preceding fallow period. The mitigation of soil and nutrient loss are the key objectives, however, additional value to biodiversity may be provided depending on the species of cover crop selected but, in general, this may be limited. A further factor to consider is that the land is not removed from production; the measure is integrated within a continued cropping cycle.

Farmland birds, in particular, are thought to benefit from catch cropping and green cover with the benefits being related to the species sown and, in turn, what is sown will depend upon the site, local weather and soil type. For example, some varieties of sorghum (family Poaceae, *Sorghum bicolor*) produce seed heads valuable for game birds but provide no livestock feed value and are susceptible to cold, wet summers. Whereas millet (family Poaceae, *Setaria* spp.), produces seed suitable for game and small birds but is not suitable for heavy, wet soils. Kale (family Brassicaceae, *Brassica oleracea*) is also suitable for game and small birds and once sown can last for two years but it may be hard to establish and is susceptible to flea beetle attack. Triticale (family Poaceae, *Triticosecale*) is also used for catch cropping and has value for a wide range of bird species through to late winter. It grows well in poor soils, requiring limited supplementary nutrients and is good as part of mixture. Some other crops can also be used but they tend to provide limited green cover, for example, quinoa (family Amaranthaceae, *Chenopodium quinoa*) and sunflower (family Asteraceae, *Helianthus annuus*; Thomson, 2014).

Cover crops may provide a nectar source in the spring if flowering is permitted and some shelter during the winter to soil surface-active predatory beetles (Holland *et al.*, 1999; 2005) compared to bare soil, although the impact is likely to be small. An improvement to soil conditions and enhancement of soil macro- and microorganisms, through the return of additional biomass upon destruction of the cover crop, is cited by Dabney *et al.* (2001).

Nitrogen-fixing crops

Nitrogen-fixing crops including legumes such as clover, fix atmospheric nitrogen (N_2)

into the soil (see Nitrogen cycle, Chapter 1). Within the EU, individual MSs have their own approved list of species (under cross-compliance regulations) but typically include faba bean (*Vicia faba*), pea (*Pisum* spp.), alfalfa (*Medicago*), lupin (*Lupinus*) and clover (*Trifolium*). The value of a particular plant species to pollinators depends in part, on the morphology of the flower combined with size of the pollinating insect and morphological adaptations to, for example, the mouthparts (Goulson *et al.*, 2005; Kirk and Howes, 2012). Therefore, from a biodiversity perspective, the choice of species could be chosen depending on local populations. Red clover (*Trifolium pratense*) is especially suitable for long-tongued bumblebee species, although they are also suitable for short-tongued species. Whilst red clover will be utilised by honeybees it is not as valuable a resource as white clover (*Trifolium repens*). Both clover species are acknowledged as having potential value to solitary bees although there is greater uncertainty due to lack of data. Field or faba bean (*Vicia faba*) is valuable to a broader range of pollinators including both short- and long-tongued bumblebee species, honeybees and solitary bees. Alfalfa (*Medicago*), also known as lucerne (*Medicago sativa*), is also attractive to long-tongued bumblebees, especially the common carder bee (*Bombus pascuorum*) and solitary bees such as the blue mason bee (*Osmia caerulescens*), although it is visited less by honeybees and short-tongued bumblebee species that appear to favour the lupin (*Lupinus*). In contrast, pea (*Pisum* spp.) with specific reference to garden pea (*Pisum sativum*), is considered to be of negligible value to bees, with only occasional visits to flowers observed.

Cover crop species that may only require re-establishment and tillage intermittently, such as birds foot-trefoil (*Lotus* spp.) or kidney vetch (*Anthyllis vulneraria*), may provide a more attractive habitat to soil-dwelling macroinvertebrates, partiucularly carabid beetles. Those that possess flowers with a long corolla and light colour are potentially more favourable to night-flying insects such as moths, a significant component of the diet of bats (BCT, 2013).

FALLOW LAND

Fallow land may be maintained as bare soil devoid of vegetation through continued frequent cultivation, or natural regeneration may be permitted to establish ground cover or it may be sown with a non-crop species mixture, however, within the EU what is selected will depend on what is approved at MS level. The key underlying factor is that it is not used for crop production. Where appropriately managed, fallow land improves the suitability of the cropped area to biodiversity. Ground cover may be manipulated with different species mixtures within designated fallow areas to benefit various groups of naturally occurring predator (or pollinators) depending on the species composition and the frequency that it requires re-establishment. Where sown, non-crop mixtures may be specifically formulated for pollinators and so will potentially enhance pollinator abundance. However, the precise benefit may be subject to the surrounding landscape and habitat type. Combinations of sown nectar and pollen plants attract greater numbers of bumblebees than areas allowed to regenerate naturally or sown with grass mixtures alone (Carvell *et al.*, 2007; Heard *et al.,* 2007). The aim should be to provide the longest period of continuous flowering coupled with resource provision early in the season. These two factors are critical in the enhancement and maintenance of bumblebee populations (Goulson *et al.,* 2005) as they will reduce the risk of colony starvation. The inclusion of generalist hemi-parasitic plants, semi-parasitic plants that help prevent dominance

by any one plant species, is recommended to maintain plant species diversity and this then offers the opportunity to enhance the diversity of invertebrates, pollinators and predatory arthropods (Joshi *et al.* 2000; Pywell *et al.*, 2011).

Crop cover is a key variable in determining the presence or absence of surface-active macroinvertebrates (e.g. ground beetles) within a given area, with longer-term fallow with ground cover most beneficial to surface-active invertebrates such as carabid beetles (Luff, 1994). The decrease in soil humidity associated with bare, exposed soil tends to be detrimental to the survival of carabid larvae, increasing mortality during the early life phase and decreasing adult numbers. The impact of annual cultivation is also a critical factor. Tillage and pesticide applications reduce carabid species diversity and abundance, due to both direct toxicity, mechanical damage and the removal of hibernation sites associated with ground cover (Kromp, 1999). Unsown fallow areas tend to have a low but potentially specialised associated fauna and flora, which may be rare. The overall species richness of invertebrates tends to be lower where soils are bare (Luff, 1994), however, rarities with specialist habitat requirements for disturbed land have been recorded in fallow agricultural land. Notable rare (IUCN Red List) specialist invertebrate species of open disturbed environments are cited by Schnitter (1994) in Germany, including *Harpalus zabroides*, *Amara littorea* and *A. municipalis*. On a similar theme, rare arable flora such as the ground pine (*Ajuga chamaepitys*) and fine-leaved fumitory (*Fumaria parviflora*) require areas of bare but frequently cultivated soils (Plantlife, 2000; Still and Byfield, 2007). Bare or sparsely vegetated soils potentially provide nest sites for solitary bees (Roulston and Goodell, 2011; Wuellner, 1999). At the higher trophic levels, bird species such as the skylark (*Alauda arvensis*) exhibit a preference for bare soils and open areas (Wilson *et al.*, 1997).

SOILS

Probably the most important biodiversity present in soils in agricultural areas are those species that confer benefits to crop production and ecosystem service provision, rather than rarity. Such species include earthworms, collembola (springtails), mycorrhizal fungi and the predatory larvae of insects (for example, carabid beetles). If we think about the soil habitat as three distinct layers: the soil surface, the upper topsoil layer (0–30 cm) and the deeper subsoil layer (below 30 cm), the magnitude of activity within these areas renders a given species more or less under the influence of agricultural practices. The soil surface has the potential to intercept agro-chemicals while the microclimate is influenced by the type and growth stage of the crop (height and density of canopy). Allowing crop residues to remain on the surface is a further determining factor. Descending into the upper topsoil layer, both this and the soil surface are impacted by the type and frequency of tillage. Only the deeper soil layer (>40 cm is beyond most sub-soiling operations) is not typically subject to the mechanical and chemical operations associated with crop production. Species that are active in the deeper soil layers would, therefore, be expected to be less influenced by any potential negative impacts from agricultural operations, or indeed benefit from positive management.

Earthworms and collembola function as detrivores, recycling the nutrients within dead plant or animal material. Detrivores break down the larger and more complex proteins into their constituent parts, including the readily available plant nutrient, ammonium (NH_4^+). This process completes the cycle between the consumers (herbivores and predators) and the primary producer (plants) within the agro-ecological food chain. There are three main types of earthworm (mostly from the family Lumbricidae) each with different zones of activity within the soil (Bouché, 1977). Epigeic species (litter dwellers) live and feed in plant material on the surface and do not burrow. Endogeic species (shallow earth dwellers) are present in the topsoil and dig extensive lateral branching systems of temporary burrows. Finally, anecic species (deep earth dwellers) dig semi-permanent vertical burrows up to around 1 m deep down into the subsoil. They feed on organic material that they collect from the soil surface before moving it into the deeper soil profile, facilitating the mixing of organic matter between soil layers. Further, these burrows maintain drainage and enhance the plant root zone, providing space within which the roots may extend.

The impact of tillage on earthworm populations is a phenomenon that remains not entirely understood as it is strongly site- and species-specific (Chan, 2001; Crittenden *et al.*, 2014). Comparisons of different tillage systems have produced contrasting results. In general, mechanical damage may result in the short term (a few weeks) but numbers often recover within a few months, and may actually increase due to the increased availability of organic material that has been incorporated into the upper soil profile and the loosening of the soil. Crittenden *et al.* (2014) report a positive correlation between earthworm numbers, albeit species-specific, and the soil moisture and organic matter content, coupled with lower 'penetration resistance', that is, lower soil compaction. All three factors are interlinked and influenced by the tillage method, frequency and timing. Soil moisture may decline immediately post tillage, although in areas of high precipitation the effect may be reduced. Soil organic matter typically declines in response to increased tillage frequency, but this may be partially offset by incorporating crop residues. Tillage also decreases the risk of soil compaction in annual cropping systems, allowing greater movement of soil-dwelling invertebrates within the topsoil, of particular benefit to the endogeic earthworm species.

Other soil-dwelling invertebrates include the larvae of many crop-active predatory beetles that also forage mainly in the upper soil layers, where they function in a potential pest control capacity (Thomas *et al.*, 1992; Sotherton, 1985). Some species such as *Pterostichus madidus* are able to utilise the lower soil profile and are capable of activity below the zone of cultivation (Holland and Luff, 2000). For those species active in or around the upper soil profile, tillage presents a potential risk to their abundance, either due to mechanical impact or burying them at a depth from which they are unable to emerge. Collembola may act as a source of prey for carabid beetle adults and larvae, important in the absence of pest prey in order to sustain predator populations throughout the cropping year. Their response to tillage is also represented by a decline in overall abundance, although the number of species tends to remain stable (Brennan *et al.*, 2006). Collembola are detrivores and/ or microphages, consuming and recycling decaying material or, importantly according to Gange (2000), feeding preferentially on non-mycorrhizal fungi, promoting the growth of mycorrhizal fungi species. There is some dispute among authors regarding the effect that collembola have on mycorrhizae; high populations may be detrimental (Bakonyi *et al.*, 2002) but would be kept in check by a healthy carabid population. Mycorrhizal fungi form a beneficial symbiotic relationship with plant roots, effectively increasing the root surface area to enable an increase in crop water uptake efficiency and increasing the plant's resistance to drought

(Miller and Jastrow, 1992). The impact of tillage is reported as being negative, either by decreasing hypha or spore density, or the rate of plant root inoculation (Kabir *et al.*, 1998). This impact is, to a degree, likely to be species-specific, with spore-prolific mycorrhizal species being affected the least, or even favoured, while species that produce and depend on extensive mycelial networks within the soil decline (Kurle and Pfleger, 1994).

Agri-environment measures that reduce mechanical disturbance, maintain soil moisture and enhance soil organic matter content, while simultaneously preventing compaction in the cropped area are those that will likely benefit the abundance of biodiversity with an important ecosystem service function. A decrease in tillage frequency may enhance the soil moisture and organic matter content, and reduce mortality from direct mechanical damage, but increase compaction risk. The introduction of biennial or perennial crops within a rotation, an element that may be possible with certain species of nitrogen-fixing crops, such as birds foot-trefoil (*Lotus* spp.) or kidney vetch (*Anthyllis vulneraria*), removes the need to undertake tillage or sowing annually. The risk of soil compaction decreases since machinery is not entering the cropped area as frequently.

3.3.3.12. *Artificial structures as habitats*

Farm buildings may, at first sight, not figure prominently in terms of biodiversity, however, many farm buildings offer significant value to wildlife, in particular birds and bats (JNCC, 2001). Although it is not uncommon for many farmers and landowners to provide artificial nesting and roosting sites, in reality these are rarely enough to enhance populations. Buildings made of natural materials such as wood are often the most attractive, especially those where there is access and space for nesting and roosting. Many bird species will rapidly take advantage of features such as eaves (house martins – *Delichon urbicum*), beams and ledges (swallows – *Hirundo rustica*, barn owls – *Tyto alba* and kestrels – *Falco tinnunculus*), access to roof spaces and crevices in walls (bats, spotted flycatcher – *Muscicapa striata*, European starlings – *Sturnus vulgaris* and sparrows – *Passer domesticus*).

A recent study undertaken in Poland identified that old Polish farms and villages are biodiversity hotspots as they are rich in different types of buildings, many of which have a complex structure, from old roof tiles and thatch to chimneys and timber beams, which provide ample opportunity for nesting and roosting birds. These areas are also highly connected to other nearby habitats such as alleys of old deciduous trees, traditional orchards, ponds and paddocks. These all contribute to the multitude of breeding and foraging sites for birds (Rosin *et al.*, 2016).

Old traditional farm buildings may also be important for plants such as ferns, mosses and lichens, and numerous insects although, in some cases, their presence can be detrimental to the building itself and so our natural heritage (see Chapter 6).

Generally speaking, farm reservoirs are built and maintained to ensure the farm's water security (see Chapter 4) and not for nature conservation. Indeed, these structures can, if not built and located sensitively, be detrimental to biodiversity by displacing natural habitats. However, this is not always the case and indeed the construction of winter storage reservoirs may reduce the need for abstraction from natural waterbodies during the summer months, helping to maintain their water levels, although it may not eliminate it entirely (See Chapter 5). It will, however, reduce the likelihood of earlier and more prolonged drying periods.

Many farm reservoirs have properties akin to those of ponds, with a similar set of characteristics tending to increase their biodiversity value. At their simplest, reservoirs are merely holes in the ground containing water, and, as such, may have little to make them attractive to many species of plant or animal, often being used by little more than aquatic birds (e.g. ducks) and insects such as mosquitos. However, with careful design, they can become amongst the most valuable biodiversity habitats on a farm. This starts with a reservoir's shape, since an irregular bank form is likely to be of greater value than a simple rectangular or circular design, particularly if it includes a gently sloping, contoured bank, resulting in shallow areas at the edges. There is evidence, for example, that the seasonal drawdown zone (i.e. that which intermittently dries out) of ponds, can be of particular value (Williams *et al.*, 1999), and similar traits may be expected to be found in some reservoirs. The inclusion of vegetation is also beneficial, both in its own right and in providing habitat for a wide range of animal species. Consequently, appropriate planting with vegetation both towards the edges and in at least some of the deeper water, maintains a range of valuable habitats, and maximises the biodiversity value of an on-farm waterbody. This approach can stretch well beyond the margins of a reservoir, however, and should be integrated with the planning of semi-natural habitats across the farm, including vegetation in the immediate area of a reservoir, and the broader habitats in order to reduce fragmentation. It is important to consider the environment within which a reservoir is set, since being close to or linked to other waterbodies (streams, rivers, ditches, ponds, etc.) increases the likelihood that a constructed reservoir will become an integral part of a wider network of habitats, providing resources either for more species or for species at different stages of their life cycle. There is naturally sometimes concern that the inclusion of sensitively planned features such as shallow edges and vegetation may reduce the storage capacity of the reservoir a little from that which might be achieved in the same area using a simple design and/or increase losses through evapotranspiration or leakage. However, so long as this is taken into account at the design stage, it should be possible to achieve both environmental and production goals at the same time.

Farm reservoirs should not be built on sites of scientific or environmental value nor those that have archaeological interest. Indeed, it is unlikely that planning permission within many European countries would be granted in this situation and, in any case, under EU law an environmental impact assessment would be required (Environmental Impact Assessment [EIA] Directive 2011/92/EU). Ideally, farm reservoirs should be linked to other farm habitats such as hedgerows and woodland. The surrounding areas should be planted with trees and shrubs to provide wildlife with shelter and breeding areas. Local plants should be encouraged to colonise margins to provide seed as a food source for breeding and over-wintering wildfowl. Marginal planting will also help reduce bank erosion.

3.3.4. *Crop management practice and its influence on biodiversity*

Subtle modifications to the management of the crop itself, while maintaining production and the crop cycle, may be available to improve the suitability of the crop to biodiversity. Unlike in the previous section the crop type is not altered, neither are additional crops added. The only modification is the management practice itself that is substituted with one that performs a similar function but which may be more beneficial to biodiversity.

Soil may not always come to mind when we think about habitat, but ensuring it is fertile and able to support a variety of plants, animals and soil microorganisms is important for many different ecosystems as well as ensuring the productivity of the farm. Each species relying on soil as habitat requires a slightly different environment in terms of the soil structure and composition. At the microorganism level, diversity is beneficial as different organisms are required during decomposition and nutrient cycling (see Chapter 1). A complex set of soil microbes can compete with disease-causing organisms, and prevent a problem-causing species from becoming dominant. Most soil organisms cannot grow outside of soil, so it is necessary to preserve healthy and diverse soil ecosystems in order to preserve beneficial microorganisms and so maintenance of the soil fertility is a critical management factor. These activities include undertaking practices that influence soil volume, structure, biological and chemical characteristics, and whether soil exhibits adverse effects such as reduced fertility, soil acidification, salinisation or erosion.

Inversion tillage and ploughing turns over residue from the crop, loosens the soil so that the next crop can establish easier and disturbs the root system of existing weeds so that they die. However, it can also exacerbate soil erosion and exposes buried weed seeds that, once exposed to light, germinate. If inversion tillage and ploughing is substituted with non-inversion shallow tillage or zero tillage via direct drilling, this will reduce soil disturbance, allow organic matter to build up and be beneficial for soil-dwelling macroinvertebrates and epigeic earthworm populations typically active in the top 20 cm of the soil profile.

Some modern farming systems have moved away from traditional crop rotations and have adopted monocropping. This refers to the practice of growing only one crop in a large area of land, year after year and offers advantages associated with the economies of scale; with just one crop to consider, agrochemicals can be bought in bulk and management is simpler. However, this practice is associated with a number of undesirable environmental and biodiversity impacts. It tends to reduce the variability of habitat structures on the farm and so decreases biodiversity, which includes fewer predatory birds and insects to help protect the crop. It also encourages the build-up of particular pests and diseases associated with the monocrop and, consequently, more chemical pesticides are needed which is also undesirable from a biodiversity perspective as well as adding costs for the farmer. When land is used for intensive monocropping, it takes its toll on the soil at a much faster rate than traditional fallow field farming and crop rotation, which allow the earth to recover. Soil fertility is reduced, organic matter decreases and the reliance on chemical nitrogen fertilisers increases in order to maintain yields. Changing to a system using crop rotations will avoid many of these issues, increasing biological activity and improving the soil fertility and its overall quality.

Optimising the use of crop protection products and using them in a sustainable manner will help minimise the impact on non-target species of potentially harmful substances. This will include using just enough to do the job of protecting yields, choosing pesticides with the least harmful profile, making use of no-spray buffer zones to protect sensitive features such as waterbodies, avoiding spraying field margins and beetle banks, and taking steps to reduce pesticide drift. The pest management system referred to as 'integrated pest management' (IPM) includes a common-sense approach to this by taking early planning steps to reduce the need to use pesticides, such as choosing pest and disease resistant varieties, good housekeeping on-farm (reducing clutter, overgrown vegetation and areas where pests and diseases may harbour) and monitoring pests and disease infestations, only spraying when it becomes economically sound to do so, that is, at the point when not spraying would result in an economic loss to the farm from a reduced yield and quality greater than the cost of spraying (chemical, fuel, time, environmental impact).

On pasture and grassland, adopting a land management system that includes a grazing regime with stocking densities that allow recovery and sustainable management of habitats will be advantageous to biodiversity. This is discussed above in relation to high nitrogen levels reducing the diversity of plant communities and, in some cases, changing the nature of these communities completely. Livestock can impact on soil in two ways. First, they can physically damage the soil structure and cause compaction where there is excessive and repeated livestock movement. Known as 'poaching', this can be visually apparent around drinking water troughs, entrances to fields, along river banks and other parts of the land where the animals congregate. The main issue is that damaged vegetation may not recover naturally once the grazing animal is withdrawn. Compacted soil can make it very difficult for new shoots to penetrate and often these areas fail to drain efficiently. The nitrogen cycle within the local area may also fail to function effectively as anaerobic, compacted zones in waterlogged soils will encourage denitrification which implies a loss of nitrogen and pollution of the atmosphere with N_2O. Second, there is the impact of the faeces and urine that the animal deposits directly to the soil. The amount of urine delivered to soil by a single grazing cow is around 2 litres per 0.4 m^2 (Addiscott *et al.*, 1991) which is equivalent to an instantaneous application of 400–1200 kg N ha^{-1}. This level of application can be highly corrosive to vegetation and excess nitrogen from deposited urine and manure will normally be as ammonia, dinitrogen and nitrous oxide (during denitrification) or as nitrate leaching (see Chapter 1 – Nitrogen cycle).

3.3.5. *Management of vulnerable and endangered species*

National rural development programmes and the agri-environment schemes in particular are increasingly important for restoring and maintaining farmland wildlife and their habitats. However, the general scientific opinion is that whilst these schemes can offer biodiversity benefits, these benefits are mostly seen by the more common species. Threatened species on the IUCN Red List (IUCN, 2013; Rodrigues *et al.*, 2006) rarely benefit as these often need more elaborate and targeted conservation measures (Kleijn *et al.*, 2006). Kleijn *et al.* (2006) also report that these measures are not particularly effective as the management prescriptions are often not sufficient to enhance the population densities and are sometimes implemented where the species would not naturally occur. Some agri-environment schemes, however, do include measures designed to protect vulnerable and endangered species, often via conservation of their favoured habitats. Some measures are geared specifically towards a particular species, for example, skylark (*Aiauda arvensis*) populations in the UK halved during the 1990s and are still declining. Undrilled patches in winter cereal fields have been shown to boost nesting opportunities for skylarks in areas of predominantly autumn-sown crops. The English agri-environment scheme (Countryside Stewardship) offers payments to farmers for establishing these 'skylark plots'. Other measures target species groups that contain a high proportion of Red List species such as amphibians. For example, the yellow-bellied toad (*Bombina variegate*) is restricted to central and south-eastern Europe. Over the last century, many populations of this species have disappeared or are in serious decline, mainly due to the loss of habitat and habitat fragmentation. The yellow-bellied toad is protected via a range of agri-environmental measures in different MSs, for example in France relevant measures include the restoration and management of forest ponds and the protection of frog habitat with fences.

References

Abel, M.L.A. (1996) *Water Pollution Biology*, 2nd Edition. CRC Press, London.

Ackerman, F. and Heinzerling, L. (2002) Pricing the priceless: cost-benefit analysis of environmental protection. *University of Pennsylvania Law Review*, 150(5), 1553–1584. DOI: 10.2307/3312947

Addiscott, T.M., Whitmore, A.P. and Powlson, D.S. (1991) *Farming, Fertilizers and the Nitrate Problem*. C.A.B. International, Wallingford, UK.

Anderson, G.Q., Haskins, L.R. and Nelson, S.H. (2004) The effects of bioenergy crops on farmland birds in the United Kingdom: a review of current knowledge and future predictions. *Biomass and Agriculture: Sustainability, Markets and Policies*. OECD, Paris, pp. 199–218.

ARG (2010) *Advice Note 5: Great Crested Newt Habitat Suitability Index*. Amphibian and Reptile Groups (ARG) of the United Kingdom.

Armitage, P.D., Szoszkiewicz, K., Blackburn, J.H. and Nesbitt, I. (2003) Ditch communities: a major contributor to floodplain biodiversity. *Aquatic Conservation: Marine and Freshwater Ecosystems*, 13(2), 165–185. DOI: 10.1002/aqc.549

Baker, J., Beebee, T., Buckley, J., Gent, A. and Orchard, D. (2011) *Amphibian Habitat Management Handbook*. Amphibian and Reptile Conservation, Bournemouth.

Bakonyi, G., Posta, K., Kiss, I., Fabian, M., Nagy, P. and Nosek, J.N. (2002). Density-dependent regulation of arbuscular mycorrhiza by collembola. *Soil Biology and Biochemistry*, 34(5), 661–664. DOI: 10.1016/S0038-0717(01)00228-0

Barling, R.D. and Moore, I.D. (1994) Role of buffer strips in management of waterway pollution: a review. *Environmental Management*, 18(4), 543–558. DOI: 10.1007/ BF02400858

Barnosky, A.D., Matzke, N., Tomiya, S., Wongan, G.O.U., Quental, B., Marshall, T.B.C., McGuire, J.L., Lindsey, E., Maguire, K.C., Mersey, B. and Ferrer, E.A. (2011) Has the Earth's sixth mass extinction already arrived? *Nature*, 471, 51–57. DOI: 10.1038/nature09678

BCT (2013) *Encouraging Bats. A Guide for Bat-Friendly Gardening and Living*. Bat Conservation Trust (BCT).

Benedictus, L. (2009) Should pandas be left to face extinction? *The Guardian*, 23 September 2009. Available at: http://www. theguardian.com/environment/2009/sep/23/ panda-extinction-chris-packham

Berg, A. (2002) Breeding birds in short-rotation coppices on farmland in central Sweden – the importance of *Salix* height and adjacent habitats. *Agriculture, Ecosystems and Environment*, 90(3), 265–276. DOI: 10.1016/ S0167-8809(01)00212-2

Bern convention (2015) *Bern Convention*. Convention on the Conservation of European Wildlife and Natural Habitats. Available at: http://www.coe.int/en/web/bern-convention/

Biggs, J. (2007) *Small-Scale Solutions for Big Water Problems*. Pond Conservation: The Water Habitats Trust, Oxford.

Biggs, J., Williams, P., Whitfield, M., Nicolet, P. and Weatherby, A. (2005) 15 years of pond assessment in Britain: results and lessons learned from the work of pond conservation. *Aquatic Conservation: Marine and Freshwater Ecosystems*, 15(6), 693–714. DOI: 10.1002/aqc.745

Biggs, J., Corfield, A., Walker, D., Whitfield, M. and Williams, P. (1994) New approaches to the management of ponds. *British Wildlife*, 5(5), 273–287.

Biggs, J., Williams, P., Whitfield, M., Fox, G. and Nicolet, P. (1996) *Biological techniques of still water quality assessment. Phase 3. Method development*. Pond Action R&D Technical Report E110, UK.

Blackwell, M.S.A., Hogan, D.V. and Maltby,

E. (1999) The use of conventionally and alternatively located buffer zones for the removal of nitrate from diffuse agricultural run-off. *Water Science and Technology*, 39(12), 157–164. DOI: 10.1016/S0273-1223(99)00331-5

Blakesley, D. and Buckley, G.P. (2010) *Managing Your Woodland for Wildlife*. Pisces Publications, Newbury.

Blomqvist, M.M., Vos, P., Klinkhamer, P.G.L. and Ter Keurs, W.J. (2003) Declining plant species richness of grassland ditch banks – a problem of colonisation or extinction? *Biological Conservation*, 109(3), 391–406. DOI: 10.1016/S0006-3207(02)00165-9

Bouché, M. (1977) Stratégies lombriciennes. *Ecological Bulletins*, pp. 122–132.

Boughey, K.L., Lake, I.R., Haysom, K.A. and Dolman, P.M. (2011) Improving the biodiversity benefits of hedgerows: how physical characteristics and the proximity of foraging habitat affect the use of linear features by bats. *Biological Conservation*, 144(6), 1790–1798. DOI: 10.1016/j.biocon.2011.02.017

Brennan, A., Fortune, T. and Bolger, T. (2006) Collembola abundances and assemblage structures in conventionally tilled and conservation tillage arable systems. *Pedobiologia*, 50(2), 135–145. DOI: 10.1016/j. pedobi.2005.09.004

Brockerhoff, E.G., Jactel, H., Parrotta, J.A., Quine, C.P. and Sayer, J. (2008) Plantation forests and biodiversity: oxymoron or opportunity? *Biodiversity and Conservation*, 17(5), 925–951. DOI: 10.1007/ s10531-008-9380-x

Burgess, P.J. (1999) Effects of agroforestry on farm biodiversity in the UK. *Scottish Forestry*, 53(1), 24–27.

Cardinale, B.J., Emmett Duffy, J., Gonzalez, A., Hooper, D.U., Perrings, C., Venail, P., Narwani, A., Mace, G.M., Tilman, D., Wardle, D.A., Kinzig, A.P., Daily, G.C., Loreau, M., Grace, J.B., Larigauderie, A., Srivastava, D.S. and Naeem, S. (2012) Biodiversity loss and its impact on humanity. *Nature*, 486(7401), 59–67. DOI: 10.1038/nature11148

Carvell, C., Meek, W.R., Pywell, R.F., Goulson, D. and Nowakowski M. (2007) Comparing the efficacy of agri-environment schemes to enhance bumble bee abundance and diversity on arable field margins. *Journal of Applied Ecology*, 44(1), 29–40. DOI: 10.1111/j.1365-2664.2006.01249.x

CBD (2015) *The Convention about Life on Earth*. Convention on Biodiversity (CBD). Available at: www.cbd.int

Ceballos, G., Ehrlich, P.R., Barnosky, A.D., García, A., Pringle, R.M. and Palmer, T.M. (2015) Accelerated modern human-induced species losses: entering the sixth mass extinction. *Science Advances*, 1(5), 1–5. DOI: 10.1126/sciadv.1400253

Céréghino, R., Boix, D., Cauchie, H.-M., Martens, K. and Oertli, B. (2014) The ecological role of ponds in a changing world. *Hydrobiologia*, 723(1), 1–6. DOI: 10.1007/s10750-013-1719-y

Cesar, H.S.J. and van Beukering, P.J.H. (2004) Economic valuation of the coral reefs of Hawaii. *Pacific Science*, 58(2), 231–242. DOI: 10.1353/psc.2004.0014

Chan, K. Y. (2001) An overview of some tillage impacts on earthworm population abundance and diversity—implications for functioning in soils. *Soil and Tillage Research*, 57(4), 179–191. DOI: 10.1016/S0167-1987(00)00173-2

CITES (2015) *Convention on International Trade in Endangered Species of Wild Fauna and Flora*. Available at: https://www.cites.org/

Clarke, S.A., Green, D.G., Bourn, N.A. and Hoare, D.J. (2011) *Woodland Management for Butterflies and Moths: A Best Practice Guide*. Butterfly Conservation, Wareham, Dorset, UK.

Clarke, S.J. (2014) Conserving freshwater biodiversity: the value, status and management of high quality ditch systems. *Journal for Nature Conservation*. DOI: 10.1016/j.jnc.2014.10.003

Coates, A. and Say, A. (1999) *Ecological assessment of short rotation coppice*. Report ETSU B/W5/00216/00/REPORT/1-3. A report for the Department of Trade and Industry.

Cole, L.J., Brocklehurst, S., McCracken, D.I., Harrison, W. and Robertson, D. (2012) Riparian field margins: their potential to enhance biodiversity in intensively managed grasslands. *Insect Conservation and Diversity*, 5(1), 86–94. DOI: 10.1111/j.1752-4598.2011.00147.x

Collar, N.J. (2003) Beyond value: biodiversity and the freedom of the mind. *Global Ecology and Biogeography*, 12(4), 265–269. DOI: 10.1046/j.1466-822X.2003.00034.x

Colley, M.R. and Luna, J.M. (2000) Relative attractiveness of potential beneficial insectary plants to aphidophagous hoverflies (Diptera: Syrphidae). *Environmental Entomology*, 29(5), 1054–1059. DOI: 10.1603/0046-225X-29.5.1054

Collier, M.J. (2013) Field boundary stone walls as exemplars of 'novel' ecosystems. *Landscape Research*, 38(1), 141–150. DOI: 10.1080/01426397.2012.682567

Collins, A.L. and Anthony, S.G. (2008) Assessing the likelihood of catchments across England and Wales meeting 'good ecological status' due to sediment contributions from agricultural sources. *Environmental Science and Policy*, 11(2), 163–170. DOI: 10.1016/j.envsci.2007.07.008

Cranmer, L., McCollin, D. and Ollerton, J. (2012) Landscape structure influences pollinator movements and directly affects plant reproductive success. *Oikos*, 121(4), 562–568. DOI: 10.1111/j.1600-0706.2011.19704.x

Crittenden, S.J., Eswaramurthy, T., De Goede, R.G.M., Brussaard, L. and Pulleman, M.M. (2014) Effect of tillage on earthworms over short-and medium-term in conventional and organic farming. *Applied Soil Ecology*, 83, 140–148. DOI: 10.1016/j.apsoil.2014.03.001

Crofts, A. and Jefferson, R.G. (eds.) (1999) *The Lowland Grassland Management Handbook*, 2nd Edition. English Nature/The Wildlife Trusts.

Croxton, P.J., Carvell, C., Mountford, J.O. and Sparks, T.H. (2002) A comparison of green lanes and field margins as bumblebee habitat in an arable landscape. *Biological Conservation*, 107(3), 365–374. DOI: 10.1016/S0006-3207(02)00074-5

Dabney, S.M., Delgado, J.A. and Reeves, D. (2001) Using winter cover crops to improve soil and water quality. *Communications in Soil Science and Plant Analysis*, 32, 1221–1250. DOI: 10.1081/CSS-100104110

Davies, B.R., Biggs, J., Williams, P.J., Lee, J.T.

and Thompson, S. (2008) A comparison of the catchment sizes of rivers, streams, ponds, ditches and lakes: implications for protecting aquatic biodiversity in an agricultural landscape. *Hydrobiologia*, 597(1), 7–17. DOI: 10.1007/s10750-007-9227-6

Defra (2010) *Fertiliser Manual (RB209)*, 8th Edition. Department for Environment, Food and Rural Affairs (Defra), The Stationery Office, London.

Defra (2015) *Indicator DE5: Farmland Bird Populations*. Defra Observatory Programme Indicators.

DeMars, C.A., Rosenberg, D.K. and Fontaine, J.B. (2010) Multi-scale factors affecting bird use of isolated remnant oak trees in agro-ecosystems. *Biological Conservation*, 143(6), 1485–1492. DOI: 10.1016/j.biocon.2010.03.029

De Vos, J.M., Joppa, L.N., Gittleman, J.L., Stephens, P.R. and Pimm, S.L. (2015) Estimating the normal background rate of species extinction. *Conservation Biology*, 29(2), 452–462. DOI: 10.1111/cobi.12380

Dimitriou, E., Karaouzas, I., Skoulikidis, N. and Zacharias, I. (2006) Assessing the environmental status of Mediterranean temporary ponds in Greece. *Annales de Limnologie – International Journal of Limnology*, 42(1), 33–41. DOI: 10.1051/limn/2006004

Dimitriou, I., Mola-Yudego, B., Aronsson, P. and Eriksson, J. (2012) Changes in some soil parameters in short-rotation coppice plantations on agricultural land in Sweden. *Bioenergy Research*, 5(3), 563–572. DOI: 10.1016/j.biombioe.2011.09.006

Dix, M.E., Johnson, R.J., Harrell, M.O., Case, R.M. and Wright, R.J. (1995) Influences of trees on abundance of natural enemies of insect pests: a review. *Agroforestry Systems*, 29(3), 303–311. DOI: 10.1007/BF00704876

Douglas, D.J., Vickery, J.A. and Benton, T.G. (2009) Improving the value of field margins as foraging habitat for farmland birds. *Journal of Applied Ecology*, 46(2), 353–362. DOI: 10.1111/j.1365-2664.2009.01613.x

DSWA – Dry Stone Walling Association of Great Britain (2007) *Dry Stone Walls and Wildlife*. Dry Stone Walling Association, Cumbria.

DSWA – Dry Stone Walling Association of Great Britain (2011) *Dry Stone Walls and Wildlife*. Dry Stone Walling Association, Cumbria.

Dudgeon, D., Arthington, A.H., Gessner, M.O., Kawabata, Z., Knowler, D.J., Lévêque, C., Naiman, R.J., Prieur-Richard, A.H., Soto, D., Stiassny, M.L. and Sullivan, C.A. (2006) Freshwater biodiversity: importance, threats, status and conservation challenges. *Biological Reviews of the Cambridge Philosophical Society*, 81(2), 163–182. DOI: 10.1017/S1464793105006950

Dyda, J., Symes, N. and Lamacraft, D. (2009) *Woodland Management for Birds: A Guide to Managing Woodland for Priority Birds in Wales*. The RSPB, Sandy and Forestry Commission Wales, Aberystwyth.

EC (1992) Council Directive 92/43/EEC of 21 May 1992 on the conservation of natural habitats and of wild fauna and flora. *Official Journal of the European Union*, L 206/7, 22.7.92. European Commission (EC).

EC (2000) Directive 2000/60/EC of the European Parliament and of the Council of 23 October 2000 establishing a framework for the Community action in the field of water policy. *Official Journal of the European Union*, L 327, 22.12.2000, pp. 0001–0072. European Commission (EC).

EC (2008) *The European Union's Biodiversity Action Plan. Halting the Loss of Biodiversity By 2010 – and Beyond*. European Commission (EC). Office for Official Publications of the European Communities, Luxembourg. ISBN 978-92-79-08071-5.

EC (2009) Directive 2009/147/EC of the European Parliament and of the Council of 30 November 2009 on the conservation of wild birds. *Official Journal of the European Union*, L 20/7, 26.1.2010. European Commission (EC).

EC (2011) *Our life insurance, our natural capital: an EU biodiversity strategy to 2020*. Communication from the Commission to the European Parliament, the Council, the Economic and Social Committee and the committee of the regions. European Commission (EC), Brussels, 3.5.2011, COM(2011) 244 final.

EC (2014) Commission Delegated Regulation

(EU) No 639/2014 of 11 March 2014 Supplementing Regulation (EU) No 1307/2013 of the European Parliament and of the Council Establishing Rules for Direct Payments to Farmers Under Support Schemes Within the Framework of the Common Agricultural Policy and Amending Annex X to that Regulation. European Commission (EC). *Official Journal of the European Union*, L 181,1–47.

EC (2015) *Natura 2000 Barometer.* Available at: http://ec.europa.eu/environment/nature/natura2000/barometer/

Edwards, P.J. and Abivardi, C. (1998) The value of biodiversity: where ecology and economy blend. *Biological Conservation*, 83(3), 239–246. DOI: 10.1016/S0006-3207(97)00141-9

EEA (2011) *Landscape fragmentation in Europe.* Joint EEA-FOEN report. European Environment Agency (EEA) Report No 2/2011. ISSN 1725-9177.

EEA (2012) *The Natura 2000 and the Emerald networks.* European Environment Agency (EEA). Available at: http://www.eea.europa.eu/data-and-maps/figures/the-natura-2000-and-the

EEA (2015a) *Proportion of classified river and lake water bodies in different River Basin Districts (RBD) holding less than good ecological status or potential for rivers and lakes.* Available at: https://www.eea.europa.eu/data-and-maps/figures/proportion-of-classified-surface-water-4

EEA (2015b) *Status of marine fish stocks.* This report was generated automatically by the EEA Web content management system on 28 Nov 2016.

European Communities (2003) *Sustainable Forestry and the European Union: Initiatives of the European Commission.* Office for Official Publications of the European Communities, Luxembourg. 2003. ISBN 92-894-6092-X

European Pond Conservation Network (2008) *The Pond Manifesto.* European Pond Conservation Network, Seville.

Ehrenfeld, D. (1988) Why put a value on biodiversity? In: Wilson, E.O. (ed.) *Biodiversity*, pp. 212–216. National Academy, Washington, DC.

Field, R.H. and Anderson, G.Q.A. (2004) Habitat use by breeding tree sparrows *Passer montanus. Ibis*, 146(s2), 60–68. DOI: 10.1111/j.1474-919X.2004.00356.x

Fitzgibbon, C.D. (1993) The distribution of grey squirrel dreys in farm woodland: the influence of wood area, isolation and management. *Journal of Applied Ecology*, 30(4), 736–742. DOI: 10.2307/2404251

Forestry Commission (2003) *The Forests and Water Guidelines*, 4th Edition. Forestry Commission, Edinburgh, UK.

Forestry Commission (2005) *Woodland Management for Bats.* Forestry Commission, Edinburgh, UK.

Forestry Commission (2012) *Managing Deadwoods in Forests and Woodlands, Practice Guide.* Forestry Commission, ISBN 978-0-85538-857-7.

Forestry Commission (2017) *England's Woods and Forests: Lowland Heath.* Forestry Commission Website http://www.forestry.gov.uk/forestry/Lowlandheath

Francis, R.A. (2010) Wall ecology: a frontier for urban biodiversity and ecological engineering. *Progress in Physical Geography*, 35(1), 43–63. DOI: 10.1177/0309133310385166

Fry, G. and Sarlöv Herlin, I. (1997) The ecological and amenity functions of woodland edges in the agricultural landscape: a basis for design and management. *Landscape and Urban Planning*, 37(1–2), 45–55. DOI: 10.1016/S0169-2046(96)00369-6

Fussell, M. and Corbet, S.A. (1992) The nesting places of some British bumble bees. *Journal of Apicultural Research*, 31(1), 32–41. DOI: 10.1080/00218839.1992.11101258

Gallai, N., Salles, J.M., Settele, J. and Vaissière, B.E. (2009) Economic valuation of the vulnerability of world agriculture confronted with pollinator decline. *Ecological Economics*, 68(3), 810–821. DOI: 10.1016/j.ecolecon.2008.06.014

Gange, A. (2000) Arbuscular mycorrhizal fungi, Collembola and plant growth. *Trends in Ecology and Evolution*, 15(9), 369–372. DOI: 10.1016/S0169-5347(00)01940-6

Gange, A.C. and Llewellyn, M. (1989) Factors

affecting orchard colonisation by the black-kneed capsid (*Blepharidopterus angulatus* (Hemiptera: Miridae)) from alder windbreaks. *Annals of Applied Biology*, 114(2), 221–230. DOI: 10.1111/j.1744-7348.1989.tb02099.x

Gee, J.H.R., Smith, B.D., Lee, K.M. and Griffiths, S.W. (1997) The ecological basis of freshwater pond management for biodiversity. *Aquatic Conservation: Marine and Freshwater Ecosystems*, 7(2), 91–104. DOI: 10.1002/(SICI)1099-0755(199706)7:2<91::AID-AQC221>3.0.CO;2-O

Geertsema, W. and Sprangers, J.T.C.M. (2002) Plant distribution patterns related to species characteristics and spatial and temporal habitat heterogeneity in a network of ditch banks. *Plant Ecology*, 162(1), 91–108. DOI: 10.1023/A:1020336908907

Gibbons, P., Lindenmayer, D.B., Fischer, J., Manning, A.D., Weinberg, A., Seddon, J., Ryan, P. and Barrett, G. (2008) The future of scattered trees in agricultural landscapes. *Conservation Biology*, 22(5), 1309–1319. DOI: 10.1111/j.1523-1739.2008.00997.x

Goulson, D., Lye, G.C. and Darvill, B. (2005) Decline and conservation of bumble bees. *Annual Review of Entomology*, 53, 191–208. DOI: 10.1146/annurev.ento.53.103106.093454

Gregory, S.V., Swanson, F.J., McKee, W.A. and Cummins, K.W. (1991) An ecosystems perspective of riparian zones, *Bioscience*, 41(8), 540–551. DOI: 10.2307/1311607

Haber, W. (1993) *Ökologische Grundlagen des Umweltschutzes*. Economica, Bonn.

Haines-Young, R. and Potschin, M. (2013) *Common International Classification of Ecosystem Services (CICES): Consultation on Version 4, August–December 2012*. EEA Framework Contract No EEA/IEA/09/003.

Hanley, M.E., Franco, M., Dean, C.E., Franklin, E.L., Harris, H.R., Haynes, A.G., Rapson, S.R. and Knight, M.E. (2011) Increased bumblebee abundance along the margins of a mass flowering crop: evidence for pollinator spill-over. *Oikos*, 120(11), 1618–1624. DOI: 10.1111/j.1600-0706.2011.19233.x

Hannon, L.E. and Sisk, T.D. (2009) Hedgerows in an agri-natural landscape: potential habitat

value for native bees. *Biological Conservation*, 142(10), 2140–2154. DOI: 10.1016/j.biocon.2009.04.014

Hanowski, J.M., Niemi, G.J. and Christian, D.C. (1997) Influence of within plantation heterogeneity and surrounding landscape composition on avian communities in hybrid poplar plantations. *Conservation Biology*, 11(4), 936–944. DOI: 10.1046/j.1523-1739.1997.96173.x

Hausman, J. (2012) Contingent valuation: from dubious to hopeless. *The Journal of Economic Perspectives*, 26(4), 43–56. DOI: 10.1257/jep.26.4.43

Heard, M.S., Carvell, C., Carreck, N.L., Rothery, P., Osborne, J.L. and Bourke, A.F.G. (2007) Landscape context not patch size determines bumble-bee density on flower mixtures sown for agri-environment schemes. *Biology Letters*, 3(6), 638–641. DOI: 10.1098/rsbl.2007.0425

Herzon, I. and Helenius, J. (2008) Agricultural drainage ditches, their biological importance and functioning. *Biological Conservation*, 141(5), 1171–1183. DOI: 10.1016/j.biocon.2008.03.005

Hicks, B. and Doick, K.J. (2014) *Lowland Acid Grassland: Creation and Management in Land Recreation*. Forest Research, Best Practice Guide, The Land Regeneration and Urban Greenspace Research Group, Note 16.

Hilty, J.A., Lidicker Jr., W.Z. and Merenlender, A.M. (eds.) (2006) *Corridor Ecology. The Science and Practice of Linking Landscapes for Biodiversity Conservation*. Island Press, London, UK.

Hinden, H., Oertli, B., Menetrey, N., Sager, L. and Lachavanne, J-B (2005) Alpine pond biodiversity: what are the related environmental variables? *Aquatic Conservation: Marine and Freshwater Ecosystems*, 15(6), 613–624. DOI: 10.1002/aqc.751

Hinsley, S.A. and Bellamy, P.E. (2000) The influence of hedge structure, management and landscape context on the value of hedgerows to birds: a review. *Journal of Environmental Management*, 60(1), 33–49. DOI: 10.1006/jema.2000.0360

Holland, J.M. and Luff, M.L. (2000) The effects of agricultural practices on Carabidae in temperate agroecosystems. *Integrated Pest Management Reviews*, 5(2), 109–129. DOI: 10.1023/A:1009619309424

Holland, J.M., Perry, J.N. and Winder, L. (1999) The within-field spatial and temporal distribution of arthropods in winter wheat. *Bulletin of Entomological Research*, 89(6), 499–513. DOI: 10.1017/S0007485399000656

Holland, J.M., Thomas, C.F.G., Birkett, T., Southway, S. and Oaten, H. (2005) Farm-scale spatiotemporal dynamics of predatory beetles in arable crops. *Journal of Applied Ecology*, 42(6), 1140–1152. DOI: 10.1111/j.1365-2664.2005.01083.x

Hubbard, V.C., Jordan, D. and Stecker, J.A. (1999) Earthworm response to rotation and tillage in a Missouri claypan soil. *Biology and Fertility of Soils*, 29(4), 343–347. DOI: 10.1007/s003740050563

Huhta, V. (1976) Effects of clear-cutting on numbers, biomass and community respiration of soil invertebrates. In: *Annales Zoologici Fennici*, pp. 63–80. Societas Biologica Fennica Vanamo.

Humphrey, J. Ferris, R., Jukes, M. and Peace, A. (2002) Biodiversity in planted forests. In: *Forest Research* (ed.) Annual Report and Accounts 2000–2001. pp. 24–33. The Stationery Office, Edinburgh.

Hunter, M.D (2002). Landscape structure, habitat fragmentation, and the ecology of insects. *Agricultural and Forest Entomology*, 4(3), 159–166. DOI: 10.1046/j.1461-9563.2002.00152.x

IINH – Icelandic Institute of Natural History (2001) *Biological diversity in Iceland: National report to the Convention on Biological Diversity*. Icelandic Ministry of the Environment, Reykjavik.

ITPGRFA (2015) *International Treaty on Plant Genetic Resources for Food and Agriculture (ITPGRFA)*. Website: http://www.planttreaty.org

IUCN (2013) *Malta's Biodiversity at Risk: A Call for Action*. International Union for Conservation of Nature, Red List publication from the EU Representative Office, Brussels.

Jablonski, D. (2004) Extinction: past and present. *Nature*, 427, 589. DOI: 10.1038/427589a

Jaeger, J. and Holderegger, R. (2005) Thresholds of landscape fragmentation (in German: Schwellenwerte der Landschaftszerschneidung), *GAIA*, 14(2), 113–118.

Jefferson, R.G., Smith, S.L.N. and MacKintosh, E.J. (2014) *Guidelines for the Selection of Biological SSSIs. Part 2: Detailed Guidelines for Habitats and Species Groups*. Chapter 3 Lowland Grasslands. Joint Nature Conservation Committee, Peterborough.

JNCC (2001) *Habitat Management for Bats. A Guide for Land Managers, Land Owners and their Advisors*. Joint Nature Conservation Committee (JNCC), UK.

Johnston, A.E. and Dawson, C.J. (2005) *Phosphorus in Agriculture and in Relation to Water Quality*. Agricultural Industries Confederation, Peterborough, UK. 72 pp.

Josefsson, J., Berg, Å, Hiron, M., Pärta, T. and Eggers, S. (2013) Grass buffer strips benefit invertebrate and breeding skylark numbers in a heterogeneous agricultural landscape. *Agriculture, Ecosystems and Environment*, 181, 101–107. DOI: 10.1016/j.agee.2013.09.018

Joshi, J., Matthies, D. and Schmid, B. (2000) Root hemiparasites and plant diversity in experimental grassland communities. *Journal of Ecology*, 88, 634–644.

Kabir, Z., O'Halloran, I.P., Widden, P, and Hamel, C. (1998) Vertical distribution of arbuscular mycorrhizal fungi under corn (*Zea mays* L.) in no-till and conventional tillage systems. *Mycorrhiza*, 8(1), 53–55. DOI: 10.1007/s005720050211

Kaennel, M. (1998) Biodiversity: a diversity in definition. In: Bachmann, P., Köhl, M. and Päivinen, R. (eds.) *Assessment of Biodiversity for Improved Forest Planning*. Kluwer Academic Publishers, Dordrecht.

Kahle, P., Hildebrand, E., Baum, C. and Boelcke, B. (2007) Long-term effects of short rotation forestry with willows and poplar on soil properties. *Archives of Agronomy and Soil Science*, 53(6), 673–682. DOI: 10.1080/03650340701648484

Karg, J. (2004) Importance of midfield shelterbelts for over-wintering entomofauna (Turew area, West Poland). *Polish Journal of Ecology*, 52(4), 421–431.

Kells, A.R. and Goulson, D. (2003) Preferred nesting sites of bumblebee queens (*Hymenoptera: Apidae*) in agroecosystems in the UK. *Biological Conservation*, 109(2), 165–174. DOI: 10.1016/S0006-3207(02)00131-3

Kirby, K.J. (1993) The effects of plantation management on wildlife in Great Britain: lessons from ancient woodland for the development of afforestation sites. In: Watkins, C. (ed.) *Ecological Effects of Afforestation*. CAB International, Melksham.

Kirk, W.D.J. and Howes, F.N. (2012) *Plants for Bees*. International Bee Research Association. Cardiff, UK.

Kleijn, D., Baquero, R.A., Clough, Y., Diaz, M., Esteban, J.D., Fernández, F., Gabriel, D., Holzscuh, A., Jöhl, R., Knop, E, Kruess, A., Marshall, E.J., Steffan-Dewenter, J., Tscharntke, T., Verhulst, J., West, T.M. and Yela, J.L. (2006) Mixed biodiversity benefits of agri-environment schemes in five European countries. *Ecology Letters*, 9(3), 243–254. DOI: 10.1111/j.1461-0248.2005.00869.x

Korpela, E.L., Hyvönen, T., Lindgren, S. and Kuussaari, M. (2013) Can pollination services, species diversity and conservation be simultaneously promoted by sown wildflower strips on farmland? *Agriculture, Ecosystems and Environment*, 179(1), 18–24. DOI: 10.1016/j.agee.2013.07.001

Kromp, B. (1999) Carabid beetles in sustainable agriculture: a review on pest control efficacy, cultivation impacts and enhancement. *Agriculture, Ecosystems and Environment*, 74(1–3), 187–228. DOI: 10.1016/S0167-8809(99)00037-7

Kurle, J.E. and Pfleger, F.L. (1994) The effects of cultural practices and pesticides on VAM fungi. In: Pfleger F.L., Linderman R.G. (eds.) *Mycorrhizae and Plant Health*. APS Press, Minnesota, pp. 101–131.

Lagerlöf, J., Stark, J. and Svensson, B. (1992) Margins of agricultural fields as habitats for pollinating insects. *Agriculture, Ecosystems and Environment*, 40(1–4), 117–124. DOI: 10.1016/0167-8809(92)90087-R

Langeveld, H., Quist-Wessel, F., Dimitriou, I., Aronsson, P., Baum, C., Schulz, U., Bolte, A., Baum, S., Köhn, J., Weih, M., Gruss, H., Leinweber, P., Lamersdorf, N., Schmidt-Walter, P. and Berndes, G. (2012) Assessing environmental impacts of short rotation coppice (SRC) expansion: model definition and preliminary results. *Bioenergy Research*, 5(3), 621–635. DOI: 10.1007/s12155-012-9235-x

Lazdinis, M., Roberge, J.-M., Kurlavičius, P., Mozgeris, G. and Angelstam, P. (2005) Afforestation planning and biodiversity conservation: predicting effects on habitat functionality in Lithuania. *Journal of Environmental Planning and Management*, 48(3), 331–348. DOI: 10.1080/09640560500067418

Leng, W., Musters, C.J.M. and de Snoo, G.R. (2011) Effects of mowing date on the opportunities of seed dispersal of ditch bank plant species under different management regimes. *Journal for Nature Conservation*, 19(3), 166–174. DOI: 10.1016/j.jnc.2010.11.003

Lovei, G.L., Hodgeson, D.J., Macleod, A. and Wratten, S.D. (1993) Attractiveness of some novel crops for flower-visiting hoverflies (Diptera: Syrphidae): comparisons from two continents. In: S. Corey (ed.) *Pest Control and Sustainable Agriculture*. CSIRO, Canberra, Australia.

Lovell, S.T. and Sullivan, W.C. (2006) Environmental benefits of conservation buffers in the United States: evidence, promise, and open questions. *Agriculture, Ecosystems and Environment*, 112(4), 249–260. DOI: 10.1016/j.agee.2005.08.002

Ludwig, D. (2000) Limitations of economic valuation of ecosystems. *Ecosystems*, 3, 31–35.

Luff, M.L. (1994) Starvation capacities of some carabid larvae. In: Desender, K., Dufrêne, M., Loreau, M., Luff, M.L., Maelfait, J.-P. (eds.) *Carabid Beetles, Ecology and Evolution*. Series, 51. Kluwer Academic Publishers, Dordrecht. pp. 171–175.

Lye, G., Park, K., Osborne, J., Holland, J. and Goulson, D. (2009) Assessing the value of rural stewardship schemes for providing foraging resources and nesting habitat for bumblebee queens (*Hymenoptera: Apidae*). *Biological Conservation*, 142(10), 2023–2032. DOI: 10.1016/j.biocon.2009.03.032

Macleod, A., Wratten, S.D., Sotherton, N.W. and Thomas, M.B. (2004) 'Beetle banks' as refuges for beneficial arthropods in farmland: long-term changes in predator communities and habitats. *Agriculture and Forest Entomology*, 6(2), 147–154. DOI: 10.1111/j.1461-9563.2004.00215.x

Maisonneuve, C. and Rioux, S. (2001) Importance of riparian habitats for small mammal and herpetofaunal communities in agricultural landscapes of southern Québec. *Agriculture, Ecosystems and Environment*, 83(1–2), 165–175. DOI: 10.1016/S0167-8809(00)00259-0

McCollin, D., Jackson, J.I., Bunce, R.G.H., Barr, C.J. and Stuart, R. (2000) Hedgerows as habitat for woodland plants. *Journal of Environmental Management*, 60(1), 77–90. DOI: 10.1006/jema.2000.0363

Manning, A.D., Fischer, J. and Lindenmayer, D.B. (2006) Scattered trees are keystone structures–implications for conservation. *Biological Conservation*, 132(3), 311–321. DOI: 10.1016/j.biocon.2006.04.023

Mason, C.F. (1997) *Biology of Freshwater Pollution*, 3rd Edition. Longman Publishers, Harlow, UK, p. 236.

Mazurek, M.J. and Zielinski, W.J. (2004) Individual legacy trees influence vertebrate wildlife diversity in commercial forests. *Forest Ecology and Management*, 193(3), 321–334. DOI: 10.1016/j.foreco.2004.01.013

MEA (2015) *Millennium Ecosystem Assessment (MEA)*. Website: http://www.millenniumassessment.org

Meddings, A., Taylor, S., Batty, L., Green, R., Knowles, M. and Latham, D. (2011) Managing competition between birds and bats for roost boxes in small woodlands, north-east England. *Conservation Evidence*, 8, 74–80.

Meek, B., Loxton, D., Sparks, T., Pywell, R., Pickett, H. and Nowakowski, M. (2002) The effect of arable field margin composition on invertebrate biodiversity. *Biological Conservation*, 106(2), 259–271. DOI: 10.1016/S0006-3207(01)00252-X

Merckx, T., Feber, R.E., Riordan, P., Townsend, M.C., Bourn, N.A., Parsons, M.S. and

Macdonald, D.W. (2009) Optimizing the biodiversity gain from agri-environment schemes. *Agriculture, Ecosystems and Environment*, 130(3–4), 177–182. DOI: 10.1016/j.agee.2009.01.006

Miller, R.M. and Jastrow, J.D. (1992). *The application of VA mycorrhizae to ecosystem restoration and reclamation*. Mycorrhizal functioning: an integrative plant–fungal process, Chapter 13, pp. 438–467.

Morandin, L.A. and Kremen, C. (2013) Hedgerow restoration promotes pollinator populations and exports native bees to adjacent fields. *Ecological Applications*, 23(4), 829–839.

Mountford, J.O., Parish, T and Sparks, T.H. (1994) The flora of field margins in relation to land use and boundary features. In: Boatman N.D. (ed.) *Field Margins: Integrating Agriculture and Conservation*. Monograph No. 58. British Crop Protection Council, pp. 105–110.

Naiman, R., Decamps, H. and McClain, M. (2005) *Riparia – Ecology, Conservation, and Management of Streamside Communities*. Academic Press, San Diego, CA. 448 pp. ISBN 9780126633153.

Nair, P.K. (2011) Agroforestry systems and environmental quality: introduction. *Journal of Environmental Quality*, 40(3), 784–790. DOI: 10.2134/jeq2011.0076

Nair, V.D. and Graetz, D.A. (2004) Agroforestry as an approach to minimizing nutrient loss from heavily fertilized soils: The Florida experience. *Agroforestry Systems*, 61–62(1–3), 269–279. DOI: 10.1023/B:AGFO.0000029004.03475.1d

Natural England and RSPB (2014). Lowland calcareous grassland. *NE546: Climate Change Adaptation Manual – Evidence to support nature conservation in a changing climate*. Section 20. Natural England.

NEA (2015) *UK National Ecosystem Assessment (NEA)*. Available at: http://uknea.unep-wcmc.org

Norris, V.O.L. (1993) The use of buffer zones to protect water quality: a review. *Water Resources Management*, 7(4), 257–272. DOI: 10.1007/BF00872284

Oertli, B., Joye, D.A., Castella, E., Juge, R., Cambin, D. and Lachavanne, J.-B. (2002) Does size matter? The relationship between pond area and biodiversity. *Biological Conservation*, 104(1), 59–70. DOI: 10.1016/S0006-3207(01)00154-9

Oldham, R.S., Keeble, J., Swan, M.J.S. and Jeffcote, M. (2000) Evaluating the suitability of habitat for the great crested newt (*Triturus cristatus*). *Herpetological Journal*, 10, 143–155.

Oliver, I., Pearce, S., Greenslade, P.J. and Britton, D.R. (2006) Contribution of paddock trees to the conservation of terrestrial invertebrate biodiversity within grazed native pastures. *Austral Ecology*, 31(1), 1–12. DOI: 10.1111/j.1442-9993.2006.01537.x

Ouin, A., Cabanettes, A., Andrieu, E., Deconchat, M., Roume, A., Vigan, M. and Larrieu, L. (2015) Comparison of tree microhabitat abundance and diversity in the edges and interior of small temperate woodlands. *Forest Ecology and Management*, 340, 31–39. DOI: 10.1016/j.foreco.2014.12.009

Oxbrough, A., Irwin, S., Kelly, T.C. and O'Halloran, J. (2010) Ground-dwelling invertebrates in reforested conifer plantations. *Forest Ecology and Management*, 259(10), 2111–2121. DOI: 10.1016/j.foreco.2010.02.023

Paoletti, M.G. (2002) Biodiversity in agroecosystems and bio-indicators of environmental health. In: M. Shiyomi and H. Koizumi (eds.) *Structure and Function in Agroecosystems Designing and Management*. Advances in Agroecology. CRC Press, Boca Raton, Fl., pp. 11–44.

Paoletti, M.G. and Lorenzoni, G.G. (1989) Agroecology patterns in northeastern Italy. *Agriculture, Ecosystems and Environment*, 27(1–4), 139–154. DOI: 10.1016/0167-8809(89)90080-7

Parish, T., Lakhani, K.H. and Sparks, T.H. (1995) Modelling the relationship between bird population variables and hedgerow and other field margin attributes. I. Species richness of winter, summer and breeding birds. *Journal of Applied Ecology*, 31(4), 764–775. DOI: 10.2307/2404166

Pearce, D.W. and Moran, D. (1994) *The Economic Value of Biodiversity*. IUCN – The World Conservation Union, Earthscan Publications Limited, London. ISBN: 1 85383 195 6.

Persson, T., Svensson, R. and Ingelog, T. (1989) Floristic changes on farm land following afforestation. *Svensk Botanisk Tidskrift*, 83, 325–344.

Pfiffner, L. and Luka, H. (2000) Overwintering of arthropods in soils of arable fields and adjacent semi-natural habitats. *Agriculture, Ecosystems and Environment*, 78(3), 215–222. DOI: 10.1016/S0167-8809(99)00130-9

Plantlife (2000) *Managing your Land for Ground Pine*. Back from the Brink Management Series. Plantlife, The Wild-Plant Conservation Charity, UK.

Prevost, Y.H. and West, R.J. (1990) Environmental architecture – preventing loss of seed production to insects in black and white spruce seed orchards. *Proceedings: Cone and seed pest workshop*, 4 October 1989, St. John's, Newfoundland, Canada. Information Report of Newfoundland and Labrador Region Forestry, Canada, pp. 100–117.

Priestley, A. (2015) *Water Framework Directive: achieving good status of water bodies*. Briefing paper Number CBP 7246, 31 July 2015. House of Commons Library, UK.

Püttsepp, U.A., Rosling, A. and Taylor, A.F.S. (2004) Ectomycorrhizal fungal communities associated with *Salix viminalis* L. and *S. dasyclados* Wimm. clones in a short-rotation forestry plantation. *Forest Ecology and Management*, 196(2–3), 413–424. DOI: 10.1016/j.foreco.2004.04.003

Pywell, R.F., Meek, W.R., Loxton, R.G., Nowakowski, M., Carvell, C. and Woodcock, B.A. (2011) Ecological restoration on farmland can drive beneficial functional responses in plant and invertebrate communities. *Agriculture, Ecosystems and Environment*, 140(1–2), 62–67. DOI: 10.1016/j.agee.2010.11.012

Regato-Pajares, P., Jiménez-Caballero, S., Castejón, M. and Elena-Rosselló, R. (2004) Recent landscape evolution in Dehesa woodlands of western Spain. In: S. Mazzoleni, G. D. Pasquale, M. Mulligan, P. D. Martino and F. Rego (eds.) *Recent Dynamics of the*

Mediterranean Vegetation and Landscape. John Wiley & Sons, Ltd, Chichester, UK, pp. 57–72.

Rodrigues, A. S., Pilgrim, J. D., Lamoreux, J. F., Hoffmann, M., and Brooks, T. M. (2006) The value of the IUCN Red List for conservation. *Trends in Ecology and Evolution*, 21(2), 71–76. DOI: 10.1016/j.tree.2005.10.010

Rook, A.J., Tallowin, J.R.B., Rutter, S.M. and Gibb, M.J. (2006) *Biodiversity in pasture based systems.* IGER Innovations No. 10. Available at: https://www.aber.ac.uk/en/media/depar tmental/ibers/pdf/innovations/06/06ch4.pdf

Rosin, Z.M., Skórka, P., Pärt, T., Żmihorski, M., Ekner-Grzyb, A., Kwieciński, Z. and Tryjanowski, P. (2016) Villages and their old farmsteads are hot spots of bird diversity in agricultural landscapes. *Journal of Applied Ecology*, 53(5), 1363–1372. DOI: 10.1111/1365-2664.12715

Roulston, T.H. and Goodell, K. (2011) The role of resources and risks in regulating wild bee populations. *Annual Review of Entomology*, 56, 293–312. DOI: 10.1146/ annurev-ento-120709-144802

Ryszkowski L., Karg J., Kujawa K., Goldyn H., Arczyńska-Chudy E. (2002) Influence of landscape mosaic structure on diversity of wild plant and animal communities in agricultural landscape of Poland. In: Ryszkowski, L.C. (ed.) *Landscape Ecology in Agroecosystems Management.* CRC Press, Boca Raton, pp. 185–217.

Sage, R.B. and Tucker, K. (1998) *Integrated crop management of SRC plantations to maximise crop value, wildlife benefits and other added value opportunities.* Report B/W2/00400/00/REPORT: ETSU, Oxford.

Sage, R., Cunningham, M. and Boatman, N. (2006) Birds in willow short rotation coppice compared to other arable crops in central England and a review of bird census data from energy crops in the UK. *Ibis,* 148(s1), 184–197. DOI: 10.1111/j.1474-919X.2006.00522.x

Samways, M.J. (2005) *Insect Diversity Conservation.* Cambridge University Press, Cambridge. ISBN-10: 0521789478.

Sarlöv Herlin, I. (2001) Approaches to forest edges as dynamic structures and functional concepts. *Landscape Research*, 26(1), 27–43. DOI: 10.1080/01426390120024466

Sarlöv Herlin, I.J. and Fry, G.L.A. (2000) Dispersal of woody plants in forest edges and hedgerows in a Southern Swedish agricultural area: the role of site and landscape structure. *Landscape Ecology*, 15(3), 229–242. DOI: 10.1023/A:1008170220639

Schnitter, P.H. (1994) The development of carabid communities from uncultivated fields and meadows in the first years of a succession. In: Dufréne, M., Loreau, M., Luff, M.L., Maelfait, J.-P. (eds.) *Carabid Beetles: Ecology and Evolution.* Kluwer Academic Publishers, Dordrecht/Boston/London, pp. 361–366.

Scottish Natural Heritage (2010) *Guide to Types of Species-Rich Grassland.* Perth, UK.

Shannon, C.E. and Weaver, W. (1949) *The Mathematical Theory of Communication.* The University of Illinois Press, Urbana, 117pp.

Silvertown, J., Poulton, P., Johnston, E., Edwards, G., Heard, M. and Biss, P. M. (2006). The Park Grass Experiment 1856–2006: its contribution to ecology. *Journal of Ecology*, 94(4), 801–814. DOI: 10.1111/j.1365-2745.2006.01145.x

Smith, V.H., Tilman, G.D. and Nekola, J.C. (1999) Eutrophication: impacts of excess nutrient inputs on freshwater, marine, and terrestrial ecosystems. *Environmental Pollution*, 100(1–3), 179–196. DOI: 10.1016/ S0269-7491(99)00091-3

Sotherton, N.W. (1985) The distribution and abundance of predatory Coleoptera overwintering in field boundaries. *Annals of Applied Biology*, 106(1), 17–21. DOI: 10.1111/ j.1744-7348.1985.tb03089.x

Sparks, T.H., Parish, T. and Hinsley, S.A. (1996) Breeding birds in field boundaries in an agricultural landscape. *Agriculture, Ecosystems and Environment*, 60(1), 1–8. DOI: 10.1016/ S0167-8809(96)01067-5

Staley, J.T., Sparks, T.H., Croxton, P.J., Baldock, K.C., Heard, M.S., Hulmes, S., Hulmes, L., Peyton, J., Amy, S.R. and Pywell, R.F. (2012) Long-term effects of hedgerow management policies on resource provision for wildlife. *Biological Conservation*, 145(1), 24–29. DOI: 10.1016/j.biocon.2011.09.006

Stanley, D.A. and Stout, J.C. (2013) Quantifying the impacts of bioenergy crops on pollinating insect abundance and diversity: a field-scale evaluation reveals taxon-specific responses. *Journal of Applied Ecology*, 50(2), 335–344. DOI: 10.1111/1365-2664.12060

Still, K. and Byfield, A. (2007) *New Priorities for Arable Plant Conservation*. Plantlife International – The Wild Plant Conservation Charity, UK.

Svensson, S. (1991) Preferences for nest site height in the starling *Sturnus vulgaris* – an experiment with nest-boxes. *Ornis Svecica*, 1, 59–62.

Swiss Federal Office for the Environment (2014) *Biodiversity in Switzerland – summary of Switzerland's fifth national report under the Convention on Biological Diversity*. Federal Office for the Environment (FOEN), Bern, Switzerland.

Tarolli, P., Preti, F. and Romano, N. (2014) Terraced landscapes: from an old best practice to a potential hazard for soil degradation due to land abandonment, *Anthropocene*, 6, 10–25. DOI:10.1016/j.ancene.2014.03.002

Tayside Biodiversity Partnership – TBP (2002) *Hedgerows and Tree Lines*. Tayside Biodiversity Partnership.

Tews, J., Brose, U., Grimm, V., Tielbörger, K., Wichmann, M.C., Schwager, M. and Jeltsch, F. (2004) Animal species diversity driven by habitat heterogeneity/diversity: the importance of keystone structures. *Journal of Biogeography*, 31(1), 79–92. DOI: 10.1046/j.0305-0270.2003.00994.x

Thomas, C.F.G. and Marshall, E.J.P. (1999) Arthropod abundance and diversity in differently vegetated margins of arable fields. *Agriculture, Ecosystems and Environment*, 72(2), 131–144. DOI: 10.1016/S0167-8809(98)00169-8

Thomas, M.B., Sotherton, N.W., Coombes, D.S. and Wratten, S.D. (1992) Habitat factors influencing the distribution of polyphagous predatory insects between field boundaries. *Annals of Applied Biology*, 120, 197–202. DOI: 10.1111/j.1744-7348.1992.tb03417.x

Thomas, S. R., Goulson, D. and Holland, J. M. (2001a) Resource pro(2)vision for farmland gamebirds: the value of beetle banks. *Annals of Applied Biology*, 139(1), 111–118. DOI: 10.1111/j.1744-7348.2001.tb00135.x

Thomas, S.R., Noordhuis, R., Holland, J.M. and Goulson, D. (2001b) Botanical diversity of beetle banks. Effects of age and comparison with conventional arable field margins in southern UK. *Agriculture, Ecosystems and the Environment*, 93(1–3), 403–412. DOI: 10.1016/S0167-8809(01)00342-5

Thomson, P. (2014) *Game Cover Crops Guide*. Game and Wildlife Conservation Trust, UK.

Twisk, W., Noordervliet, M.A.W. and ter Keurs, W.J. (2003) The nature value of the ditch vegetation in peat areas in relation to farm management. *Aquatic Ecology*, 37(2), 191–209. DOI: 10.1023/A:1023944028022

Ucar, T. and Hall, F.R. (2001) Windbreaks as a pesticide drift mitigation strategy: a review. *Pest Management Science*, 57(8), 663–675. DOI: 10.1002/ps.341

Uetz, G.W., Van Der Laan, K.L., Summers, G.F., Gibson, P.A. and Getz, L.L. (1979) The effects of flooding on floodplain arthropod distribution, abundance and community structure. *American Midland Naturalist*, 101(2), 286–299. DOI: 10.2307/2424594

UNEP (2015) *The Mediterranean Action Plan – Barcelona Convention and its protocols. A Framework for Co-operation and Policy*. United Nations Environment Programme (UNEP), Athens, Greece. Available at: https://wedocs.unep.org/rest/bitstreams/1298/retrieve (Last accessed: 16/10/17).

Usher, M.B., Field, J.P. and Bedford, S.E. (1993) Biogeography and diversity of ground-dwelling arthropods in farm woodlands. *Biodiversity Letters*, 1(2), 54–62. DOI: 10.2307/2999650

Van-Camp, L., Bujarrabal, B., Gentile, A.R., Jones, R.J.A., Montanarella, L., Olazabal, C. and Selvaradjou, S.K. (2004) *Reports of the Technical Working Groups established under the Thematic Strategy for Soil Protection*, EUR 21319 EN/2. Office for Official Publications of the European Communities, Luxembourg.

van Dijk, W.F.A., van Ruijven, J., Berendse, F and de Snoo, G.R. (2014) The effectiveness of ditch banks as dispersal corridor for plants in agricultural landscapes depends on species' dispersal traits. *Biological Conservation*, 171, 91–98. DOI: 10.1016/j.biocon.2014.01.006

Vanhinsbergh, D., Gough, S., Fuller, R. J. and Brierley, E.D. (2002) Summer and winter bird communities in recently established farm woodlands in lowland England. *Agriculture, Ecosystems and Environment*, 92(2–3), 123–136. DOI: 10.1016/S0167-8809(01)00301-2

Verhoeven, J.T. and Setter, T.L. (2010) Agricultural use of wetlands: opportunities and limitations. *Annals of Botany*, 105(1), 155–163. DOI: 10.1093/aob/mcp172

Vickery, J.A., Feber, R.E. and Fuller, R. J. (2009) Arable field margins managed for biodiversity conservation: a review of food resource provision for farmland birds. *Agriculture, Ecosystems and Environment*, 133(1–2), 1–13. DOI: 10.1016/j.agee.2009.05.012

Vogt, P. (2016) GuidosToolbox (Graphical User Interface for the Description of image Objects and their Shapes): Digital image analysis software collection. JRC, Ispra, Italy.

Warner, D.J., Allen-Williams, L.J., Ferguson, A.W. and Williams, I.H. (2000) Pest–predator spatial relationships in winter rape: implications for integrated crop management. *Pest Management Science*, 56(11), 977–982. DOI: 10.1002/1526-4998(200011)56:11<977::AID-PS224>3.0.CO;2-U

Werner, U. and Buszko, J. (2005) Detecting biodiversity hotspots using species–area and endemics–area relationships: the case of butterflies. *Biodiversity and Conservation*, 14(8), 1977–1988. DOI: 10.1007/s10531-004-2526-6

Whalen, J.K., Sampedro, L. and Wakeed, T. (2004) Quantifying surface and subsurface cast production by earthworms under controlled laboratory conditions. *Biology and Fertility of Soils*, 39(4), 287–291. DOI: 10.1007/s00374-003-0715-1

Williams, P., Biggs, J., Whitfield, M., Thorne, A. Bryant, S., Fox, G. and Nicolet, P. (1999) *The Pond Book: A Guide to the Management and Creation of Ponds*. Ponds Conservation Trust, Oxford, UK.

Williams, F., Eschen, R., Harris, A., Djeddour, D., Pratt, C., Shaw, R. S., Varia, S., Lamontagne-Godwin, J., Thomas, S.E. and Murphy, S. T. (2010) *The Economic Cost Of Invasive Non-Native Species on Great Britain*. CABI, Wallingford. 198pp.

Wilson, J.D., Evans, J., Browne, S.J. and King, J.R. (1997) Territory distribution and breeding success of skylarks *Alauda arvensis* on organic and intensive farmland in southern England. *Journal of Applied Ecology*, 34(6), 1462–1478. DOI: 10.2307/2405262

Withers, P.J.A. and Lord, E.I. (2002) Agricultural nutrient inputs to rivers and groundwaters in the UK: policy, environmental management and research needs. *Science of the Total Environment*, 282–283, 9–24. DOI: 10.1016/S0048-9697(01)00935-4

Wolton, R., Morris, R., Pollard, K. and Dover, J. (2013) *Understanding the combined biodiversity benefits of the component features of hedges*. Report of Defra project BD5214.

Wood, P.J., Greenwood, M.T. and Agnew, M.D. (2013) Pond biodiversity and habitat loss in the UK. *Area*, 35(2), 206–216. DOI: 10.1111/1475-4762.00249

Woodhouse, S.P., Good, J.E.G., Lovett, A.A., Fuller, R.J. and Dolman, P.M. (2005) Effects of land-use and agricultural management on birds of marginal farmland: a case study in the Llŷn peninsula, Wales. *Agriculture, Ecosystems and Environment*, 107(4), 331–340. DOI: 10.1016/j.agee.2004.12.006

Wratten, S.D. (1998) The role of field boundaries as reservoirs of beneficial insects. In: Park, J.R. (ed.) *Environmental Management in Agriculture: European Perspectives*. Belhaven Press, London, pp. 144–150.

Wuellner, C.T. (1999) Nest site preference and success in a gregarious, ground-nesting bee *Dieunomia triangulifera*. *Ecological Entomology*, 24(4), 471–479. DOI: 10.1046/j.1365-2311.1999.00215.x

WWF – World Wide Fund for Nature (2016) *Saving Doñana From Danger to Prosperity*, WWF Switzerland, ISBN 978-2-940529-45-2.

Yom-Tov, Y. (1974) The effect of food and predation on breeding density and success, clutch size and laying date of the crow (*Corvus corone* L.). *Journal of Animal Ecology*, 43(2), 479–498. DOI: 10.2307/3378

Zanchi, G., Thiel, D., Green, T. and Lindner, M. (2007) *Afforestation in Europe*: Final Version 26/01/07. Institute for European Environmental Policy, London.

CHAPTER 4

Water

4.1. Setting the scene

4.1.1. Introduction

Water is the most abundant substance on the Earth's surface, existing in the liquid, solid and gaseous states. Water's existence as a liquid under ambient conditions is one of the main characteristics that seems to set our planet apart from our neighbours in the solar system, and as far as we know, elsewhere. It is often considered to be an essential precursor to the development of life, without which evolution as we know it, may never have occurred. Consequently, it is hardly surprising that it plays a vital role in our survival as a species. In part, this has been achieved through the production of plant and animal foodstuffs, either naturally, or more recently (in evolutionary terms at least) through managed arable and livestock agriculture. Water as a physical entity and the business of farming are, therefore, intimately linked, with it serving as both an essential resource in all forms of agricultural production and as a receptor for a wide range of polluting emissions arising from farming activities. In addition, agricultural land management can significantly impact on the hydrological characteristics of the wider catchment, resulting in problems and/or benefits which

can extend well beyond a farm's local area and which can have implications for the way in which regional and national resources are managed and utilised.

Planet earth is, in the main, a 'blue' planet, with around 75 per cent of its surface covered by water in either liquid or solid form, and, as a result, it could be assumed that there would be few problems associated with the availability of water for agriculture or anything else. Indeed, in total there are approximately 1.39 billion km^3 of water on the planet. At any one time, however, around 97.5 per cent of this is in a saline form (see Figure 4.1), stored within the planet's oceans and seas, and to a much lesser degree within saline groundwaters. As a result, most of the planet's water is effectively unavailable for human use (other than as a transport conduit), despite the contribution made by modern desalination plants in rendering some salt water accessible. That leaves only around 2.5 per cent of the planet's water in the fresh form, and of this more than two thirds (69.6%) is frozen away in the ice caps, glaciers, permanent snows and permafrosts, making it equally inaccessible. Freshwater groundwaters account for a further 30.1 per cent but due to the constraints of modern technology and economics, not all of this is available for our use. Only 0.3 per cent of the planet's freshwater is found in relatively easily accessible surface waters (rivers, lakes, etc.; Shiklomanov, 1993).

Nevertheless, in theory, from a global perspective there is still enough available to meet all our needs. We currently withdraw around 3,918 km³ of water per year for use and 96 per cent of this is from freshwater sources, which is equivalent to about 20 per cent of what are described as the planet's Total Renewable Water Resources (TRWR). In this context, TRWR refers to that part of our water resources generated from precipitation and therefore available for ongoing use (FAO, 2014). Regardless of the planet's overall TRWR however, the most important issue is whether or not water is available when and where it is needed, and since the availability of water is both spatially and temporally variable, whilst some countries (or regions) have adequate supplies, others do not.

The picture across Europe as a whole is highly variable but in some areas the balance between water demand and that available is often highly stressed, mainly due to over-abstraction, low levels of precipitation and/or high temperatures. This can be demonstrated using the water exploitation index (WEI) which is simply the annual total water demand of a country expressed as a percentage of the total available resources. Whilst this index is rather simplistic, it is useful for comparison purposes since if the WEI is above 20 per cent then the country or area is considered to have a water scarcity problem. Based on 2010 data from across Europe, Iceland has the lowest WEI of around 0.5 per cent, for England and Wales the value was approximately 12 per cent and that of Malta 22 per cent (EEA, 2012). Cyprus is the most water-stressed European country with a WEI of over 65 per cent and this often has a significant impact on the Cypriot population. In periods of prolonged drought, water

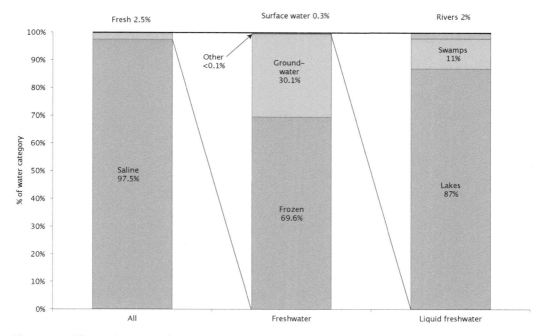

Figure 4.1: The environmental compartments occupied by the Earth's water (Created from: Shiklomanov, 1993)

supplies can reach critical levels forcing the government to take emergency action. For example, in 2008 there had been four consecutive years of low rainfall and the Cypriot government had to ship in 8 million cubic metres of water from Greece over a period of six months, as well as cutting domestic supplies by 30 per cent (Collins, 2009). The cost to the tax payer was estimated to be over €40 million.

As discussed in Chapter 1, water, in all its forms and in all its environmental stores, is not static and there is continuous movement and exchange between its stores and material states (liquid, solid, gas) governed by a number of processes that are collectively known as the water or hydrological cycle. This cycle influences many ecosystem services and undoubtedly the functioning efficiency and quality of the agri-environment in its widest sense. This chapter therefore, explores the complex issues that surround agriculture's relationship with water, and

considers ways in which local action can significantly improve that relationship, resulting in benefits for the farm and the environment. Water also plays a significant role with respect to wider society and sustainable resource uses and this is also touched upon in this Chapter, but is considered in greater depth, including its influence within the water–energy–food nexus, in Chapter 5.

4.1.2. *Water demand in agriculture and food*

The agriculture and horticulture sectors are amongst the world's largest and least efficient users of freshwater resources, accounting for around 70 per cent of all global abstractions, as shown in Figure 4.2 (FAO, 2014), equivalent to 2,722 km³ of water per annum. Some of this may be returned to the environment, and therefore be available for subsequent reuse, but nevertheless, this scale of demand inevitably places a significant strain

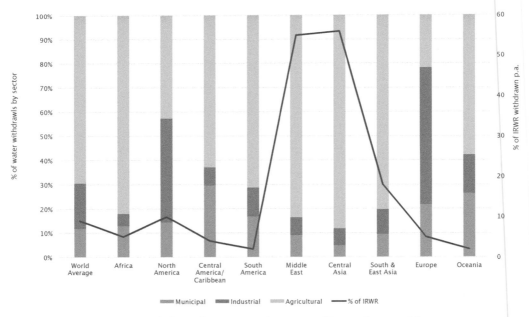

Figure 4.2: Per cent water withdrawn by sector and per cent of internal renewable water resources withdrawn per annum
(Created from: FAO, 2014)

on overall resources. The global statistics however, do not tell the whole story, as they mask important spatial and temporal differences in the way in which farming consumes water, the pressures that result from that consumption and consequently the manner in which water use is likely to develop in the future. At a global scale for example, parts of Asia and Africa use between 80 and 90 per cent of all the water they abstract for agricultural purposes, whereas in Europe the average is around 22 per cent. Whilst this may, in part, reflect greater water use efficiency on European farms it also reflects the significant amount of water used by the domestic and industrial sectors in what is a relatively densely populated and highly developed part of the world.

As would be expected, it is agriculture in southern Europe that has the heaviest demand. Spain for example, withdraws over 30 per cent of its TRWR and nearly 70 per cent of that withdrawn is to supply the needs of the country's farming industry (FAO, 2016a). Similar patterns can be found in many other Mediterranean countries, including Italy, for example, where in the late 1990s almost two thirds of all available resources were used for agriculture, primarily for irrigation (Massarutto, 1999). Since that time, there have been some improvements in efficiency but agriculture is still responsible for most of the demand. In contrast, agriculture in northern Europe appears to be a much less significant water consumer. In 2014, England utilised around 1.2 per cent of the water it abstracted from non-tidal sources for agricultural purposes (excluding fish farming), with almost 80 per cent being used for public water supplies and the electricity supply industry (Environment Agency, 2016). The UK used 5.5 per cent of its TRWR in 2012 (FAO, 2016b), which is also significantly lower than that of the Mediterranean countries. However, due to the form taken

by agricultural water use, this does not necessarily equate to a low level of pressure on resources. It is also worth remembering that all these figures relate only to the on-farm usage of water, and that off-farm industries associated with food and other agriculturally-derived products also use a considerable amount in processing. The majority of farm-used water is abstracted in order to supply the needs of irrigated field crops. Regions where irrigated agriculture is most common, such as in southern Europe, also tend to be those where the overall impact of agricultural water consumption is highest (EUREAU, 2009). It is also notable that many of the countries in southern Europe take advantage of their warm climate to produce high value horticultural crops (e.g. tomatoes, peppers) for export (primarily to other parts of Europe) on a year round basis. In such areas, these 'thirsty' crops are dependent on irrigation in order to allow economically-viable production, which has the result of increasing demand for water far beyond that which might be expected given the local population. Greece, Spain, Portugal and Italy for example, all abstract more than 400 m^3 per inhabitant each year for agriculture, whereas in the UK the equivalent rate of abstraction is less than 7 m^3 per inhabitant annually (EUREAU, 2009). Nevertheless, regional figures for England and Wales from the UK's Environment Agency (Environment Agency, 2016) reveal that during the period 2000 to 2015, between 65 and 85 per cent of all the water used in agriculture was used in spray irrigation, demand being highest in areas in which irrigated arable agriculture is a major element within the farming landscape, for example the East of England (equating to the Anglian Region shown in Figure 4.3) and parts of the Midlands and Yorkshire. The East of England region, in particular, is home to some of the country's most productive farmland, and therefore, to

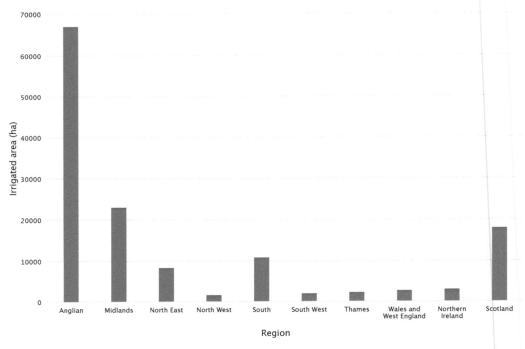

Figure 4.3: Areas of irrigated land in the United Kingdom
(Source: FAO, 2013)

the production of considerable areas of high value (but thirsty) field-grown potatoes and other vegetable crops. In England in 2010, these crops accounted for 49 per cent and 25 per cent of the irrigated area, and 54 per cent and 26 per cent of irrigation water use, respectively (Defra, 2011). However, it is also the driest part of the country, in which mean annual rainfall can be below 600 mm, and mean summer rainfall less than 150 mm. Summer, being the core growing season, is the time of year when UK-grown crops most need access to irrigation in order to maximise yields, reduce disease pressures and maintain produce quality (e.g. in potato cropping). As a result, there is considerable demand for abstracted water in this area at that time, which, despite the fact that water is generally perceived as being plentiful in the UK, can lead to considerable pressure on resources, albeit only during a restricted period of the year.

The livestock sector is also a major water user. The most obvious uses are for drinking, feed production, and health and welfare purposes. Like humans, farm animals are primarily composed of water and can rapidly lose it through a variety of biological processes (e.g. urination, perspiration, respiration, etc.), particularly in warm climates with a low humidity. As a result, they require a significant daily water intake to replace losses, maintain their health and welfare (LEAD and FAO, 2006) and to maximise yields. The requirement for drinking water varies with a number of environmental factors, most prominently air temperature but is also species and life stage dependent. For example, lactating dairy cows use considerable amounts of water producing milk and so require high water intake rates, as shown in Table 4.1 (University of Warwick, 2006). A restricted water supply will rapidly depress an animal's appetite leading to significant

Table 4.1: Drinking water requirements

Category	Water volume (l animal^{-1} day^{-1})
Cattle	
Dairy cow – lactating	104.5
Dairy cow – dry period	20.0
Dairy cow – overall	91.8
Beef cow	20.0
Dairy and beef bull	20.0
Calves	5.0
Pigs	
Gestating sows	18.0
Dry sows and gilts	6.0
Farrowing sows	30.0
Weaners	2.0
Growers	4.0
Finishers	5.5
Sheep	
Ewes	4.5
Rams	3.3
Lambs	1.7
Poultry	0.09 - 0.22

(University of Warwick, 2006)

weight loss, and if this situation becomes severe and maintained, death can occur in a matter of days (LEAD and FAO, 2006).

In some species and production systems, a significant proportion of the required water intake occurs through the animal's diet, with grass, for example, being composed of as much as 80 per cent water, although this can be as little as 10 per cent under hot, dry conditions (Pallas, 1986) As a result, between 10 per cent (in intensive production systems) and 25 per cent (in extensive systems) of an animal's water needs can be supplied in this form. Another 15 per cent can be produced as a result of metabolic processes (LEAD and FAO, 2006), but this still leaves a considerable amount needing to be provided in the form of fresh drinking water, either using totally natural sources (streams, rivers, etc.), or in more intensive systems, anthropogenically modified sources (wells, troughs, mains supplies, etc.).

In addition to drinking water, livestock systems use water to perform a number of other roles, such as the washing of animals, animal housing, dairy parlours and yards as well as equipment cleaning and machinery cooling. The washing of animals and their housing is usually undertaken relatively infrequently, perhaps between batches of animals in some production systems such as pigs and poultry. 'Mucking out', the process of removing soiled bedding and replacing it with fresh, is a more frequent practice. However, the strict hygiene standards imposed by regulations, standards and/ or manufacturers' guidelines required for milking parlours and the equipment associated with them, means that these are frequently cleaned and often with high-quality water (i.e. potable water). Regardless of the water source used, however, the hosepipes on which many systems are based may use considerable amounts of water, although the precise amount will depend on the equipment used and the skill and care of the person doing the work. Volume hoses, for example, which are quite efficient when dealing with loose material, may require 10 times the flow rate (reaching 80–150 litres min^{-1}) of pressure washers (using 8–12 litres min^{-1}), these being more effective for dried-on dirt (Milk Development Council, 2007).

Within dairy production systems (especially in the developed world), water is also an important component of the cooling systems used to ensure the quality and safety of the milk, with the UK's Milk Development Council, for example, recommending that a water to milk flow rate of 2:1 should be maintained in order to ensure optimum performance. This means that cooling alone can use twice as much water as the volume of milk being produced (Milk Development Council, 2007). Water is also used in disease control procedures, such as footbaths used to prevent and treat foot rot, a form of pododermatitis found in hooved animals such as cattle and sheep, and the dipping of sheep to prevent sheep scab, an allergic dermatitis caused by the mite *Psoroptes ovis* and its faeces.

As a consequence of the extensive range of uses water has within the livestock sector, it can be a major user of a nation's water resources, although as is the case in the arable sector, the figures vary considerably from country to country, and indeed regionally. Figures produced for England and Wales, for example, reveal that the livestock sector is responsible for just over a third of all agricultural abstractions (Environment Agency, 2016; University of Warwick, 2006) and studies have shown that livestock water consumption in parts of Northern Europe is almost as high as that for irrigation (Florke and Alcamo, 2004). Both Denmark and Ireland appear to have a particularly high percentage of water consumed by their livestock sectors with water consumption accounting for 28 per cent and 16 per cent, respectively (Mubareka *et al.*, 2013).

It is evident that water is vitally important to food production. However, it is rarely obvious just how much water is used to produce it and how this varies with location. The amount of water used to produce all goods and services is often described as its 'water footprint' which expresses the amount of water used on a per commodity basis. For example, whilst it varies depending on the production location, typically it takes about 27 litres of water to produce 1 kg of pasta in Italy, assuming it has some irrigation, which is between 4.5 and 5.5 litres per typical portion. Sixty litres of water are required to produce 1 kg of irrigated tomatoes in Spain, equivalent to around 10 to 13 litres per tomato and 10,000 litres of water to produce 1 kg of beef in Ireland, which increases to 16,000 litres in some other parts of Europe (Hess *et al.*, 2015; EC, 2010a). Figure 4.4 shows the amount of embedded water in some major agricultural products.

This raises questions as to why 'thirsty' products are produced in areas where water resources are stressed. The production and consumption of any food commodity is connected to a chain of impacts on the water resources and wider environment in the countries where it is grown and processed. Clearly, lower water use is desirable in any context but the amount of water 'embedded' in a product does not necessarily reflect the impact on water resources or the wider environment. If a product is produced in an area where water is plentiful, then its impact is likely to be low or negligible. Whereas if the same product, using the same quantity of water, is produced in a region where water is scarce or where demand outstrips supply, it may have a more significant impact on available water resources, and consequently impacts on aquatic habitats, biodiversity and human health as well as the quality of life will also be spatially variable. It could be argued

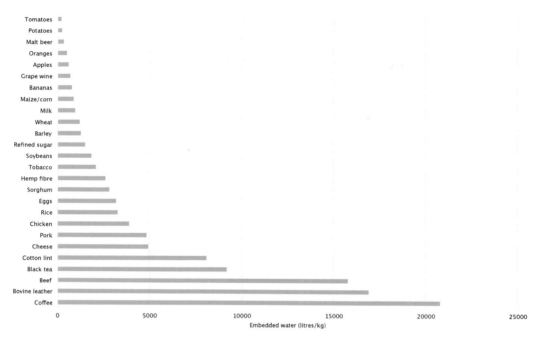

Figure 4.4: Global average embedded water in some agricultural products
(Created from: Chapagain and Hoekstra, 2004)

that just as different greenhouse gases have different impacts, in terms of their radiative forcing which is expressed using Global Warming Potential (GWP), the impact of water use in different locations should be reflected in a similar manner.

4.1.2.1. *Farm water use under pressure*
We live in a world in which there are increasingly limits to the extent to which we can, or should, be utilising natural resources of many sorts, and water is no exception (see Chapter 5). It has been estimated that it takes, on average, approximately 3,500 litres of water to produce the food a typical European consumes each day (Biointelligence Service, 2012) and, although a significant part of this is related to processing, it is inevitable that farmers around the world are either coming under pressure to control their use of this precious resource already, or are likely to do so in the very near future. This is for a variety

of reasons, the principal ones of which are detailed in Figure 4.5.

POPULATION INCREASE
The world's population currently stands at more than 7.2 billion people, and although growth rates have tended to slow in recent years, forecasts suggest that it could still be around 9.6 billion in 2050 and 11.2 billion in 2100 (UN, 2014). Forecasts of this sort inevitably come with a degree of uncertainty attached to them, but it is clear that even if all other factors remain constant, there will be considerable pressure for increased agricultural production in order to feed the increased population, a problem which is exacerbated by the way in which the structure of the world's population is expected to change over the same period. The majority of the 9.6 billion people likely to be around in 2050, for example, are expected to be living in the world's less developed countries, and

Figure 4.5: Key pressures on the availability of farm water supplies now and in the future

most of them in increasingly large urban areas, placing huge demands on the rural areas attempting to supply them. It is also likely (and indeed desirable) that although far too many people will still be living in impoverished conditions, considerably more people than at present will have seen their living conditions and wealth increase, although there is no suggestion that those people will have moved to a comfortable standard of living, and changing economic conditions could result in some of them falling back below the poverty line (World Bank Group, 2016). Nevertheless, such changes can be taken as an indication of the general way in which things are likely to change in the future, which is not only likely to empower them to be able to purchase more food, but also different types of food. To date, increased wealth has often been associated with a greater demand for animal-derived products (Foresight, 2011), which as well as being associated with increased direct water consumption by livestock, also leads

to an increase in the demand for feed crops (Rosegrant *et al.*, 2009a), which have to be produced on top of the arable crops needed for human consumption. In addition, the types of crops which people wish to buy may change, with increasing emphasis on higher value, but also more water-demanding crops, such as field-grown and/or covered vegetables (Rosegrant *et al.*, 2009a).

As a result of these changes, the FAO predicts that global food supply will need to increase by nearly 40 per cent by 2030 and up to 70 per cent by 2050 compared to the 2005/2007 period (in value terms). This equates to an increase in the production of cereals (including rice) of around 45 per cent and an increase in meat production of 75 per cent (by 2050); although rates will need to significantly exceed this in the developing world (Alexandratos and Bruinsma, 2012; FAO, 2009). To achieve this, it is probable that both a greater area of land will need to be turned over to agriculture, and that the intensity of production will increase markedly in many regions. Total harvested area for example, is expected to increase significantly over the first half of the 21st century (albeit contracting a little towards the end), and this is expected to be the result of only a 0.13 per cent per annum increase in the area of rain-fed crops (although given the area of such land in existence this is still considerable), whilst irrigated cropland has been forecast to expand by 0.24 per cent a year (Rosegrant *et al.*, 2009a). The magnitude of the resulting change in demand for irrigation water will be dependent on a number of factors, including the geographical location of the new land. Rosegrant *et al.* (2009a), for example, forecast a greater than average expansion of irrigated land in relatively wetter parts of the world, meaning that although more land may be irrigated, it could be used to meet a smaller proportion of the crop's water demand than is the case in drier areas. Nevertheless, the implication is that there will be an increase in the demand for water for this purpose, with the potential to significantly increase water stress in some locations. The consequences for resource management and agricultural productivity are discussed further in Chapter 5.

Competition from non-food crops

An additional concern in relation to the use of water by the agricultural sector, comes in the form of non-food crops, in particular biofuels (see also Chapter 5). Or, more precisely, the amount of water (and land) which might be needed to supply the biofuel industry with feedstocks (Bendes, 2002; FAO, 2008). The biofuels industry, when it does not use waste materials as a feedstock, impacts on overall water resources both directly, through the abstraction of water for irrigation (as well as off-farm in the biofuel conversion process itself), and indirectly by increasing rates of evapotranspiration, such that less water enters surface and

groundwaters, where it would be available for subsequent use (Rosegrant *et al.*, 2009b). In terms of irrigation water use, the impact of biofuels on a global scale has to date been limited, with only around 1 per cent of the water withdrawn for agriculture, being used to supply this type of crop. However, biofuels currently account for only a small part of the energy derived from liquid fuels, with the majority coming from fossil fuels (Dufey, 2006), so if this were to increase significantly, as some countries are intending, then the resulting pressure on water resources may impact both on agriculture aimed at food production, and water users more widely. This may be particularly important where the drive for biofuels is associated with water-stressed parts of the world, such as India and parts of China, both of which have significantly increased production in recent years (de Fraiture *et al.*, 2008).

Competition from non-agricultural sectors

Agriculture is far from the only user of water, with municipal (domestic) users accounting for 12 per cent of abstractions worldwide, and industrial users 19 per cent (FAO, 2014), and where resources are under pressure, these sectors compete for the available supply. This is particularly significant where non-agricultural users are a major component of the water market, such as in Western Europe, where industry uses almost 16 times as much water as agriculture, and municipal water supplies almost five times as much (FAO, 2014). It is also significant in areas where supplies are vulnerable. Industrial users, for example, will often take precedence when it comes to maintaining supplies, although in some cases, including those in the power industry, use is non-consumptive and as a result much of the water abstracted is then returned to the environment for further use (Defra, 2008). It is domestic consumers of water, however, who are (insofar as possible) unlikely to have their water supplies curtailed in preference to the agricultural industry, regardless of how essential water is considered to be in a farm's production system.

The consumption of water in people's homes is particularly high in the developed world. For example, households in the EU typically use between 294 (Romania) and 100 (Estonia) litres day^{-1}, with the UK using 150 litres day^{-1} (Waterwise, 2012) and 215 litres day^{-1} being typical in the US (Fry, 2006). However, only a small part (4%) of this water is used for drinking, with the rest being used in flushing toilets (30%), personal washing (33%), washing clothes (13%), and so on (Waterwise, 2012). Much of this is, after significant treatment, also returned to the environment, but there are considerable losses, and consequently, in densely populated parts of the developed world the pressure to maintain domestic water supplies can itself be a major issue, never mind having sufficient left over to meet agricultural needs. By way of contrast, domestic water use in the developing world is often much lower, typically being 52 litres day^{-1} in India and as little as 4 litres day^{-1} in Mali (Fry, 2006). However, if, as predicted, living standards continue to rise, this too will be an additional threat to the continued availability of water for agricultural purposes.

THE NEED FOR ENVIRONMENTAL PROTECTION

If agriculture is to be practiced in a sustainable manner, the needs of the natural environment cannot be ignored. Indeed, the environment can be viewed as a user of water resources in just the same way as those in the agricultural, industrial and municipal sectors. Water is essential for the delivery of thriving ecosystems and biodiversity (see Chapter 3), particularly aquatic biodiversity, although the importance of aquatic

systems in supporting riparian and terrestrial biodiversity should not be overlooked (Environment Agency, 2013a). The abstraction of water for agriculture may impact on natural flow regimes either directly, through the volume of water removed, or indirectly, for example as a result of groundwater abstraction lowering water tables and reducing recharge and/or causing drawdown in neighbouring watercourses (Environment Agency, 2013b; Nyholm *et al.*, 2002; Defra and Welsh Government, 2013); this latter case is particularly significant in the upper reaches of catchments where there is a strong hydraulic link between surface waters and groundwaters (Nyholm *et al.*, 2002), and where groundwaters are important in the maintenance of surface water flows during drier periods of the year (POST, 2012). In addition, abstraction may alter flow regimes (i.e. the variability in discharge in response to precipitation, temperature, evapotranspiration and basin drainage characteristics) through amendments made to support the needs of that abstraction, including changes to channel morphology and/or measures taken to regulate flow such as the use of dams, weirs and/or reservoirs, which change the way in which watercourses respond to natural climatic events within their catchments (Environment Agency, 2013b; Defra and Welsh Government, 2013; Jain, 2012). Therefore, although many abstractions may be considered sustainable, where this is not the case, a number of the features of the 'natural' flow regime may be amended, including water level, flow rate, peak discharge and flow timing (Poff *et al.*, 1997), resulting in ecological impacts (Environment Agency, 2013b; Defra and Welsh Government, 2013). However, there is now an increased emphasis around the world, for example in the EU's Water Framework Directive (see Chapters 3.2 and 4.2), on ensuring that, insofar as possible, surface waters maintain as natural an ecosystem as possible, which requires the maintenance of a suitable flow regime. This, in turn, suggests that in order to do so, non-environmental access to water, including that for irrigation, may need to be restricted.

SPATIAL AND TEMPORAL PATTERNS OF WATER DEMAND

As agricultural water demands increase under hotter drier conditions, it will come as no surprise that demands are greatest in those parts of the world or periods of the year, in which natural inputs of water in the form of precipitation are most limited. Consequently, the period in which the crops, demand for water is at its maximum (i.e. evapotranspiration is at its peak) corresponds precisely to the time of year when the natural availability of water is at a minimum (i.e. when precipitation is at its lowest), river levels may be restricted and reservoir resources are being depleted. This, combined with the demands of other users at the same time serves to put significant pressure on the farming industry to limit its use of water at just the time it most needs to expand it.

CLIMATE CHANGE

In addition to the above issues, the threats and uncertainties posed by climate change must be considered and they are likely to be significant in many parts of the world, including those not presently accustomed to pressure on water resources (Bozzola and Swanson, 2014). In the coming decades, global temperatures are expected to rise (IPCC, 2013) indeed the evidence is now clear that they already are, and have been doing so for a number of years (IPCC). This temperature increase is likely to be associated with an increase in precipitation at higher latitudes and a reduction at lower latitudes, whilst simultaneously climatic extremes will become more frequent (IPCC, 2013; Strzepek *et al.*, 2011), including drought and

flood conditions. At higher latitudes therefore, climate change could result in greater agricultural productivity, due to the presence of more favourable growing conditions and for longer periods of the year. The indications are, however, that negative impacts will be far more widespread than positive ones, primarily because many crops will struggle at very high temperatures (Porter *et al.*, 2014). The impact on crop water requirements and the availability of supplies will also be a significant factor. All other things being equal, higher temperatures tend to result in greater water losses through evapotranspiration, and therefore increased demand to replace those losses to maintain crop health and yields. However, limitations on our ability to do so may result in crops in many parts of the world where agriculture is currently highly productive being placed under increasing pressure. European studies consistently forecast that temperatures will increase across the continent, whilst precipitation will be higher in the north and lower in the south and east (Olesen and Bindi, 2002). Consequently, south-eastern Europe may be particularly badly hit, with the potential for increasingly regular and intense heatwaves and droughts in the summer but with limited scope for shifting cultivation to a less stressed time of year, due to what are often quite severe winter conditions (Olesen and Bindi, 2002). A similar pattern will be true for many livestock-producing areas as the water needs of animals have been shown to increase with ambient temperature (LEAD and FAO, 2006), such that in areas in which the livestock sector is a major consumer of available water, the demand for resources is also likely to increase significantly.

GENERAL WATER RESOURCE DEPLETION AND/OR CONTAMINATION

Many parts of the world are living beyond their means when it comes to the exploitation of water resources. This generally refers to countries which have relatively low levels of RWR, including those in North Africa and the Middle East, where intensive agriculture at least, relies heavily on the exploitation of groundwater resources which are only renewed very slowly, and consequently, in many parts of the world exploitation of water resources is unsustainable in the long term. However, this is not the general case in Europe since European countries rarely utilise more than a third of their internal RWR on an annual basis (Belgium ≈ 34%, Spain and Bulgaria ≈ 29%), although, as discussed previously, seasonal availability may still be restricted where high agricultural needs coincide with those of other users. The over-exploitation of coastal aquifers in a number of European countries, particularly those in the Mediterranean has, however, led to the intrusion of salt water into those aquifers, reducing their water quality and preventing application to land, as to do so would run the risk of damaging the land as far as crop production is concerned (Collins *et al.*, 2009) due to salinisation (see Chapter 5).

SOCIO-ECONOMIC PRESSURES

As well as the physical aspects of water resource management discussed above, socio-economic factors play a central role in driving both agricultural demand for water and its future supply. In the UK, for example, the main customers for field-grown vegetables, namely the major retailers and processors, impose exacting quality standards on their suppliers, which demand the use of irrigation in order to achieve the required benchmark (Knox *et al.*, 2009; Knox and Hess, 2014). As such, irrigation becomes an economic necessity for farmers in that sector, not only to maximise yields, but also to command the top prices available for their produce. In contrast, however, those self-same customers are increasingly scrutinising

their suppliers and placing them under pressure (e.g. through grower protocols) to understand and minimise the environmental impacts resulting from irrigation (Knox *et al.*, 2009). This is on top of the regulatory controls now often in place to protect the environment, and which threaten the continued availability of water to support farming. Equally, economic factors including global energy prices, can impact on the availability of water for irrigation on individual farms (de Fraiture *et al.*, 2008). An increase in oil price for example, can increase the demand for biofuel crops (see above) and therefore irrigation, whilst the cost of pumping water from groundwater sources may go up, limiting access to water (de Fraiture *et al.*, 2008).

4.1.2.2.   Impact on agricultural demand
It is clear that farms operate within an environment which can be influenced by a large number of natural and anthropogenic forces, which combine to determine both the demand for water, and our ability to satisfy that demand. Over the coming years this will undoubtedly continue to be the case, although the precise way in which this will develop over time is difficult to predict in detail. Whilst we can make predictions on how some factors, such as temperature, might change, any assessment of the impact that might have on irrigation demand assumes that the 'business as usual' model will continue into the future. This, however, may not be the case, as there are a number of areas in which either the balance of natural processes may change or producers could alter their behaviour. For example, although the patterns relating to the impact of climate change on crops may hold true in general, some plant types may respond somewhat differently. This is down to the fact that although the elevated levels of atmospheric carbon dioxide, which is associated with climate change, can lead to greater plant

growth and therefore higher evapotranspiration, it can also increase stomatal resistance thereby reducing evapotranspiration (Kundzewicz *et al.*, 2007). Therefore, since the balance between these two competing forces varies between different plant types, and is also affected by other environmental characteristics, determining the overall impact is far from straightforward (see also Chapter 2).

Equally, of course, producers have the option of amending the way in which they go about the business of farming, such that some of the negative impacts of climate change discussed above may be reduced or even taken advantage of so as to become benefits. Adaptation of this sort is one of the key themes of European climate change policy, as described in Chapter 2 (Olesen and Bindi, 2002), and may have considerable implications for water use globally, nationally and locally. Farmers across Europe are already adapting to climate change, in particular through changes to cultivation timings and the selection of alternative crop species and cultivars, which could reduce water demands. However, the fact that there may be an intensification in agricultural production in northern and western Europe (Olesen and Bindi, 2002), may increase pressure on resources in those areas far beyond that which might be expected from climate change alone. As a result, there are considerable uncertainties regarding the way in which demand for water for farming will change, but what is certain is that the more the industry can do to manage its use of this valuable resource in a sustainable way, the better for all concerned (the environment, farm businesses and other users).

4.1.3. *Impact of agriculture on water quality*

Despite the fact that in European countries between 50 and 98 per cent of the population

lives in urban areas (EU-27 mean ≈ 75%), large parts of the continent are given over to farmland of one sort or another. The proportion of agricultural land varies considerably from country to country, mainly due to variations in the amount of remaining natural habitat present (generally forest), but it can be very significant. In the UK, for example, over 71 per cent of the land area is given over to agricultural production, although the average across the EU-27 is nearer 44 per cent and only Finland and Sweden have less than 10 per cent as shown in Figure 4.6 (World Bank, 2013). As a result, the extent to which agriculture can impact on the quality of surface and groundwaters is much greater than any individual source may suggest, due to the vast number of pollutant sources spread over a very wide area. Consequently, it is perhaps unsurprising that agricultural sources of water pollution are high on the agenda for many environmental regulators, particularly since (in the developed world at least) the threat posed by other sources of aquatic pollution has diminished in recent decades. This is in part a direct result of the considerable work that has been done to tackle the major point sources of pollution, including discharges from the domestic (i.e. those from sewage treatment works) and industrial sectors. In England between 1990 and 2008, for example, the percentage of river length which (under the standards then in operation) achieved good chemical quality increased from only 55 to 79 per cent, while that achieving good biological quality went up from 55 to 72 per cent (HM Government, 2011). Similarly, point source emissions of phosphorus from industrial regions in Denmark and the Netherlands have decreased by up to 90 per cent since the mid 1980s (EEA and WHO, 2002). Therefore, although there are doubtless still improvements that can be made in relation to domestic and industrial pollution

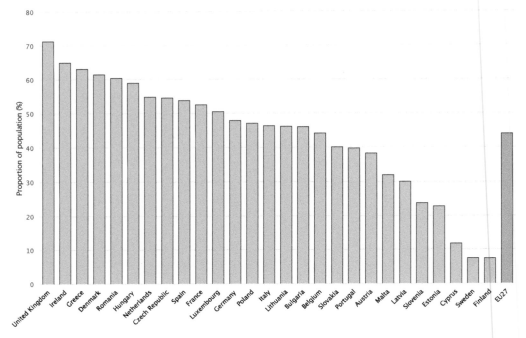

Figure 4.6: Percentage agricultural land cover in countries of the EU-27
(Source: World Bank, 2013)

in all European countries, the easy gains have already been achieved, focusing an increased level of attention on those sectors which were in the past considered less of a problem.

The post-war intensification in agricultural production led to an increase in the contaminant levels entering surface and groundwaters from that source, so that a considerably higher proportion of the observed pollution levels in European waters comes from agricultural sources than was previously the case. Indeed, work carried out in relation to the Water Framework Directive (see Chapters 3.2 and 4.2) has identified agriculture as a significant cause of waters failing to meet standards in many areas across Europe, including the UK and Germany (e.g. Holden *et al.*, 2015; Richter *et al.*, 2013). This has naturally enough led to the industry increasingly being criticised by environmentalists and regulators alike, with

increased levels of scrutiny on every aspect of the daily operation of farming businesses. Indeed, given the fact that so many of the activities which take place on a farm have the potential to be environmentally damaging in relation to water quality, it is perhaps understandable that this level of concern should exist. Due to the intimate relationship between farming and the environment within which it operates, a very wide range of substances could potentially find their way into waterbodies and therefore have a deleterious effect on other users, be they human-related or ecosystem services. Unlike in many other sectors, the impact that a farming business may have on the aquatic environment is, however, often determined by the natural context within which it operates, as well as the anthropogenic processes that form part of its operation. These natural factors include such things as geology and pedology which tend to be more or less fixed for a given location but spatially very variable, and patterns in climate and weather which are both temporally and spatially variable. The impact that farming practices have on the resulting pollutant levels of natural waters therefore varies spatially and temporally in ways which are dependent on the pollutant in question.

Farm-related sources of aquatic pollution can broadly be classified into those which are point source or non-point source (diffuse) in nature, albeit that for many contaminants a combination of both may result in the overall receiving water concentration. Point source pollution refers to that which enters a waterbody from a single, clearly identifiable source, and so in an agricultural context could refer to such things as leaking fuel tanks, slurry lagoons and chemical stores, and run-off from machinery cleaning areas. In some cases, such sources may result in acute pollution problems associated with a rapid influx of pollution at high concentrations. Non-point sources, in contrast, result in pollution being generated over a broad area, such as a catchment or sub-catchment. As such, they rarely result in acute pollution problems and instead are responsible for chronic issues linked to the long-term raising of pollutant levels. In an on-farm context, this is generally the pollution which results from within-field activities related to either arable or livestock production, such as the application of chemicals and the grazing of animals. Non-point sources are often those most heavily influenced by the site-specific characteristics of the farm and its location, as the mechanisms behind them may be climatically driven (through rainfall, run-off and infiltration, for example) and then influenced by properties such as soil, whereas point source are often driven by failings in equipment and/or management protocols. The apportionment of responsibility for point sources of pollution is (once identified)

a relatively simple matter, but for non-point sources it is far more difficult. Indeed, it has historically been an uphill battle to persuade farmers that what happens on their land can result in severe problems for the water industry and other users, as well as the environment. All too often, some in the industry have attempted to pass on all responsibility for water quality issues to their neighbours, or indeed other sectors altogether. The underlying reason for such doubts is that the contribution of an individual source to an environmental problem may be quite minor, it instead being the cumulative impact of actions across multiple farms that is a cause for concern. This, of course, presents a problem in persuading individuals to take action (at least unless regulatory steps are taken to make action compulsory), since in some cases, people have seen little motivation for changing their own practices when their neighbours may or may not be amending theirs. There are signs that this attitude has been changing for some years now, in no small part due to a number of well-publicised campaigns to persuade the industry of the need to take widespread voluntary action in order to avoid increased regulatory controls. For instance, in the UK, the agricultural industry-led Voluntary Initiative has championed the implementation of voluntary best practice so as to avoid taxes on pesticides or other mandatory measures (ECOTEC *et al.*, 1999) and the Metaldehyde Stewardship Group, established by a group of companies in the crop protection industry, through its 'Get Pelletwise' campaign, has done likewise to try and prevent restrictions being placed on the use of this molluscicide in order to meet water quality standards.

4.1.4. *Agricultural water pollutants*

Amongst the most commonly discussed aquatic pollutants resulting from agriculture, are the nutrients nitrogen (N) and

Figure 4.7: Global consumption of nitrogen and phosphorus fertilisers
(Source: FAO, 2016c)

phosphorus (P), both of which are essential plant nutrients. Nitrogen, in particular, is involved in almost all aspects of plant biochemistry (e.g. photosynthesis), so most growing crops require it in larger quantities than any other nutrient, whilst phosphorus is key in energy storage and transfer within plants, and is a component of DNA. Increasing the crop availability of these nutrients can, therefore, significantly increase yield and/or crop quality (although excess nitrogen can be detrimental to the quality of some crops – e.g. potatoes), and as a result the global consumption of fertilisers has increased considerably over the past half century or so, from around 11.5 million tonnes-N yr^{-1} and 10 million tonnes-P_2O_5 yr^{-1} in 1961 to nearly 109 million tonnes-N yr^{-1} and 47 million tonnes-P_2O_5 yr^{-1} in 2014 (see Figure 4.7).

In Europe, the pattern has been a little less dramatic in recent years, with use in much of the continent being relatively stable or even falling since the mid to late 1980s, albeit that consumption is still twice what it was in the early 1960s. The exception to this is in Eastern Europe, where agricultural production and therefore fertiliser use were both hit by the economic changes that took place in that part of the world as a result of the collapse of Communism, meaning that in just a few years, fertiliser use returned to levels not seen since the 1960s (Bumb and Baanante, 1996), a decline that was sufficient to be reflected in the global figures. This aside, however, it is clear that the increased availability of nutrients has played a significant part in increasing agricultural output over a similar period (Figure 4.8), something that has been made possible by access to readily available nutrients from inorganic sources.

Traditionally, nutrients would have been

Figure 4.8: Yields per hectare of key crops since 1961
(Source: FAO, 2016c)

obtained by adding organic manures to the land (based on either animal or human waste), and this is still done today, particularly in grassland systems. However, the specialisation which has occurred on farms in the developed world, means that the production of animal manures may no longer be conveniently located in terms of using them to fertilise arable farms, and although there are many benefits to adding organic matter to the land, the rapid crop response sought by many producers may not be achievable. In contrast, inorganic nutrient sources are often seen as easier to manage, and result in the rapid release of nutrients in a crop-useable form. Clearly, the benefits of fertiliser use have been considerable, particularly in relation to meeting the food needs of our growing population, however, this has not been without an environmental cost, particularly in relation to water quality. Agriculture is thought to contribute around 55 per cent of all the nitrogen entering European seas

(Bouraoui and Grizzetti, 2014) and is also a major problem for inland waters but with large regional differences. In the UK, for example, farming is estimated to account for around 70 per cent of all the nitrogen and 28 per cent of all the phosphorus delivered to UK watercourses (e.g. Edwards and Withers, 2008). In Denmark, the figure for nitrogen is around 80 per cent but less than 30 per cent in Finland (Bouraoui and Grizzetti, 2014). Point sources exist for both elements, particularly where fertilisers are spilt, leak from stores or are inadvertently applied directly to watercourses as a result of attempting to apply fertiliser too close to the edge of a surface waterbody. However, the main cause for concern is generally related to non-point source pollution resulting from the field application of inorganic fertilisers and/or organic manures and slurries, with the potential for harm being in essence a function of the two very different ways in which these nutrients tend to behave in the environment. The solubility of nitrogen means that once it enters the soil it has a tendency to be carried along by the movement of soil water (see Chapter 1, Nitrogen cycle). This process of 'leaching' is the main mechanism by which nitrogen is transported either to neighbouring surface waters (often with the aid of sub-surface drainage systems), or to deeper groundwater aquifers, and is a particular problem where there is a build-up of nitrogen in the soil due to its application at a rate higher than required by the crop or where it is applied at a time when crop growth may be minimal, but precipitation, and therefore sub-surface water flow, is high (i.e. too late in the growing season). These conditions occur most readily in soils with a high hydraulic conductivity (e.g. poorly structured sands) and when the potential for water movement is high (i.e. when inputs through rainfall and irrigation exceed the rate of evapotranspiration). Nitrogen levels in groundwaters tend to be lower than in surface waters but there is the potential for a considerable lag in the system, meaning both that pollution may not be immediately evident, and that once a problem has been identified, it may take quite some time for it to be resolved, even once the source of the problem is addressed (Wang *et al.,* 2013).

Phosphorus, in contrast, tends to be bound to fine particulates within the soil (e.g. clays), such that the amount present in solution, and therefore available for plant uptake, is a relatively small part of that present in the soil. For this reason, losses through leaching are generally minor (as is the threat posed to groundwaters), although they can occur, for example, where soils with low adsorption capacities have received high inputs. They can also be made worse where sub-surface drainage or macropores in the soil mean that added phosphorus bypasses potential adsorption sites. Instead, the major cause for concern is phosphorus transported through the erosion of sediments (see Chapter 1, Phosphorus cycle), and deposited within surface waters, particularly since it is the fine material to which phosphorus is bound that is most likely to be lost in this way. As a result, the phosphorus concentration in sediments delivered to a receiving waterbody can be considerably higher than in the field soil more generally. The factors likely to lead to phosphorus contamination of surface waters are, in essence, the same as those likely to result in contamination by eroded sediments (see below), with surface run-off being the key transport mechanism. In many parts of Europe, inputs of phosphorus to agricultural systems greatly exceed off-take (i.e. that taken up by growing crops, and subsequently removed), such that there has been a marked build-up of phosphorus in agricultural soils, making them a reservoir for environmental pollution. This problem is particularly bad in areas of very high-intensity livestock production, in which nutrients are brought onto the farm in the form of feed, and the resulting manures and slurries are then treated as a waste which must be disposed of, as under these conditions applications to land may well greatly exceed those required to 'feed' the crop (e.g. grass).

The enrichment of the aquatic environment with nutrients has the impact of disturbing the balance of natural ecosystems, favouring species best suited to take advantage of the new situation, and in some cases causing an explosion in the rate of biomass production (most notably by phytoplankton), with a resultant drop in the level of dissolved oxygen present, further damaging aquatic species as discussed extensively in Chapters 2 and 3. In some situations, if nitrogen pollution is present in drinking water, it has the potential for causing illness and disease. Most notably, methemoglobinaemia (also known as 'blue baby syndrome' as it is generally infants who are most vulnerable) can be caused by the conversion of leached nitrate to nitrite in the anaerobic conditions of the gut, which then binds to the haemoglobin in the blood, reducing its oxygen-carrying capacity (Camargo and Alonso, 2006). In addition, nitrate binds with amines to form carcinogenic nitrosamines, which have been linked to various forms of gastric cancer and there is also some evidence to suggest that they can contribute to the development of various other cancers, birth defects and a variety of other medical conditions (Camargo and Alonso, 2006), and consequently, they are more than an environmental concern.

The other major category of inputs used in agriculture is the various forms of crop protection chemicals (e.g herbicides, insecticides, fungicides and molluscicides) that are applied to fields (see Figure 4.9) and are central to the maximisation of crop yields and for ensuring the quality and/or safety of the final product. Traditionally, farmers relied on the use of appropriate crop rotations to help minimise pest and disease pressures, and thereby keep losses down to an acceptable level. In modern agriculture, however, the importance of rotations has been reduced, with many producers increasingly relying on an armoury of chemicals to control pests and diseases. Synthetic pesticides, therefore, have been of huge importance in the post-war development of agriculture; indeed, intensive conventional arable would have been extremely difficult without the use of such substances. By their very nature, however, the pesticides which have done so much to provide the quantity and quality of food we tend to take for granted in the developed world, can be harmful substances; after all, they are intended to kill something, meaning that their potential to cause environmental damage can be considerable. The impacts on biodiversity are covered in Chapter 3, but insofar as the water environment is concerned, they are a cause for concern not only due to their impact on aquatic species but also for the quality of water resources destined for our own consumption, although the actual impact of chemicals, especially 'cocktails' of such substances, is no easy matter to assess. European legislation (see Chapter 4.2) stipulates that the maximum concentration of any pesticide in drinking waters should be no more than $0.1 \, \mu g \, l^{-1}$, and that the maximum for the total of all pesticides should not exceed $0.5 \, \mu g \, l^{-1}$; however, these levels are not based on any clear idea of what the dangerous level for human consumption might be, instead they are based on what was considered (at the time) to be the minimum concentration that could be reliably assessed. It may well be that for many chemicals this is an overly cautious limit to set, whilst there is a possibility that for others it may be too high. For the molluscicide metaldehyde, for example, there is

Figure 4.9: Crop spraying
(Photo courtesy of: Dr John Tzilivakis)

good evidence that at a concentration of 0.1 µg l^{-1}, it would present no appreciable risk, although this is not reflected in the rules which must be adhered to. Consequently, the presence of pesticides in water resources (surface and ground) used for domestic supplies is a problem for water companies. The cost of removing pesticides from water supplies can be very high and, in the case of some chemicals, removal presents very significant practical and technological problems on top of the high cost. Anglian Water in the UK, for example, have estimated that the cost of setting up treatment facilities to remove metaldehyde in their area alone, would be £600 million, with an extra £17 million a year needed to run them (Anglian Water, 2016), something which is of interest in relation to the 'polluter pays' principle, as it is not the water industry that has caused the problem.

Consequently, within the EU a number of active substances have been lost from the farmer's arsenal precisely because of their tendency to be found in surface waters and the associated risk to biodiversity and/or human health. For example, the herbicides atrazine, simazine and isoproturon (IPU) have all been prohibited within the EU having regularly been found in waters at excessive levels, whilst others, including metaldehyde, continue to be under pressure. An added complication in determining the extent to which agricultural pesticides are found in waterbodies, is the mixture of potential sources which may be having an effect. Some substances, for example, are used extensively in the amenity and household sectors, in which controls on pesticide use and even more so training, can be significantly different to that which might be expected in the agricultural sector. This means that for some chemicals at least, conclusively tying the levels found in the environment to agricultural users, can be problematic.

Being a diverse group of substances, the mechanisms by, and the extent to which pesticides make their way into the aquatic environment, are many and varied, and can indeed be difficult to identify. The bulk of the agricultural pesticides appearing in surface and groundwaters are thought to do so as a result of non-point sources, which basically means the product applied at a field level. Older chemicals, in particular, can sometimes be quite persistent in the environment, with some continuing to appear in water samples long after the product itself has been removed from the market. This is something that can be even more of an issue when it comes to groundwater resources as these often exhibit a considerable delay in showing improvements. Many modern chemicals, in contrast, are designed to degrade relatively quickly after application, and although in some cases the products of degradation (metabolites) can themselves be problematic, this should mean that the risk to the environment is minimised. Nevertheless, those which are soluble can be leached in much the same way as might be the case for nitrogen, making them a threat to both surface and groundwaters, whilst others tend to bind to the soil in much the same way as the nutrient, phosphorus. This latter type of chemical tends to be less of a problem when it comes to groundwaters, but can be carried in eroded sediments or in water passing through larger pores and/or land drains, where there may be limited opportunities for degradation, and therefore be an issue when it comes to surface waters.

Point-sources of pesticide pollution may, for example, include occasions when a chemical spray or pellet is inadvertently applied directly to a waterbody, meaning that there is no chance for degradation to occur before the substance enters the environment, potentially at a very high concentration. Equally, locations at which chemicals are either spilt, or spray equipment is washed down, can be potent sources of contamination, particularly if those locations are directly linked to surface waters through the drainage system. Many areas of hardstanding in farmyards will have drains associated with them that are designed to carry rainwater to the nearest ditch or stream, but which will equally carry anything else which drains from the area. Chemical stores can also serve as point-sources of contamination if not properly managed and maintained, for example where buildings have no weatherproofing and chemicals are not stored in waterproof containers. Point sources of pollution may account for a small minority of the pollution entering the environment, however, they can result in high concentrations of pollution on a localised basis, with perhaps dramatic consequences for aquatic biodiversity and any downstream abstraction points.

Veterinary pesticides are used as treatments for a number of livestock conditions, most notably in the form of dips, footbaths and topical treatments for ectoparasites, for example. Where this is done, there is the potential for waste product, spillages, drippage from treated animals and so on, to pass through the drainage system into surface and/or groundwaters. In the past, for example, it was common practice in the UK (and indeed recommended), for waste sheep dip to be disposed of to land, although this is not current best practice. There is also a risk, albeit much reduced compared to topical treatments, of oral veterinary medications reaching surface waters via dung and urine. This is most likely to happen where livestock are able to access rivers and deposit wastes directly into the water, or where manures and slurries containing pharmaceutical residues are spread to the land and potentially hazardous residues move to the water with soil particles.

Livestock enterprises also present unique problems due to the potential threat posed by the inevitable levels of manures and slurries produced. To an extent this can be treated not as a waste but as a resource, being a valuable source of both soil organic matter and recycled nutrients; however, in some areas (particularly those with a lot of intensive livestock production) the amount of manure/slurry produced exceeds that which needs to be applied to land to maintain soil fertility and quality, meaning that it can rapidly become a waste disposal issue. The impact this can have on the risk of nutrient pollution has been discussed above, but there are also a number of other issues which need to be considered in relation to the impact of organic manures/slurries on environmental waters, particularly surface waters. Such pollutants can exert a very considerable biochemical oxygen demand (BOD) once they enter the aquatic environment. BOD is a measure of the amount of dissolved oxygen that would need to be present for aerobic organisms to be able to break down all the organic matter present. Slurries, for example, require a dissolved oxygen content of between 10,000 and 30,000 mg l^{-1} to break them down, whereas raw domestic sewage has a BOD of only 300 to 400 mg l^{-1} (Chambers *et al.*, 2001). This greatly reduces the level of dissolved oxygen within the water column, with potentially serious consequences for biodiversity. In addition, they may contain both chemical contaminants (e.g. heavy metals and veterinary products) and microbiological pollutants such as *E. coli* and *Cryptosporidium*, amongst others. Such contaminants clearly have the potential to be a threat to public water supplies, although water treatment will, in the majority of cases, mitigate against this. However, where people come into direct contact with untreated waters, perhaps in a recreational setting, the chance of infection is increased significantly.

Manures and slurries can enter the aquatic environment in a number of ways, including from both point and non-point sources. In the latter case, the issues revolve around both field-deposited urine and excreta (i.e. where livestock are grazing in the field) and, for example, those manures collected from animal housing and then spread on the land. These may subsequently either wash off the land in overland flow, or permit the leaching of contaminants to surface or groundwaters, particularly if deposits have been concentrated in particular areas (e.g. around feeding troughs, gates, trees) and/or the land has become poached (churned up by animal hooves). There is also the possibility that animals may deposit manure and urine directly into surface waters if they have direct access to it, for example where natural surface waters are used to provide drinking water for livestock. In this case, there is no time for biological degradation to occur or for microbial contaminants to die off, before they enter the aquatic system. Perhaps more of a concern, however, are the point sources of contamination, since these may result in considerable quantities of damaging material entering waters over a relatively short period. Manure heaps can seep polluting leachate and, especially in wet conditions, this may run off into surface waters. Slurry stores may at certain times of the year contain a great deal of potentially damaging liquid, which in the event of a leak or store failure could enter surface waters very quickly. If the slurry store is also used to store waste water from dairy parlours, potentially contaminated with milk, then the risk is even more significant as milk has a very high BOD: 140,000 mg l^{-1} (Bylund, 1995). Clearly, then, the management of manures and slurries is a key environmental issue faced by livestock enterprises, with considerable scope for things to go wrong.

Across Europe, agricultural activities are a major cause of sediments reaching surface waters (see also Chapter 5). In England, for example, it has been reported that around 75 per cent of this type of pollution can be traced back to farmland (Holden et al., 2015; Environment Agency, 2007). The risks posed by a number of sediment-bound pollutants, specifically phosphorus and some pesticides, have been discussed above, however, sediment material can also be both a source and a sink of heavy metals such as cadmium and zinc as well as other toxic or persistent substances due to its highly reactive interfaces and strong binding affinity. These substances can accumulate in drainage channels and waterbodies damaging their ecological value (Savic et al., 2015). There are also additional risks to aquatic biodiversity caused by the sediment itself, cited by some as the most detrimental form of aquatic habitat degradation (Cooke et al., 2015). It can significantly alter the character of the benthic environment by clogging gravel beds and it can reduce the clarity of the water due to suspended particulates, impacting on aquatic fauna and flora (Cooke et al., 2015). In addition, the risk of flooding can be increased through the subsequent deposition of eroded sediments.

The discussion above covers the major pollutant types relevant to the agricultural sector; however, there are a number of others, which may be important in some cases. For example, farms often store considerable volumes of diesel and fuel oil, to supply farm vehicles and heat buildings. If these stores fail, either catastrophically or through a slower leak, or if fuel is spilt when being transferred or transported, it may seep into surface drainage systems or groundwater resources, threatening both biodiversity and human resource exploitation. Equally, silage liquor can be extremely damaging should it find its way into the wider environment, due to its high BOD, and is one of the major sources of water pollution in some areas (SEPA, 1997).

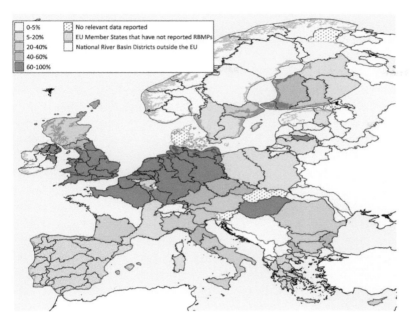

Figure 4.10: European surface waterbodies affected by pollution pressures associated with agriculture
(Adapted from: European Commission, 2012a)

Although point source pollution incidents have declined since the issue was publicised and legislation introduced in many European countries (Environment Agency, 1998), these and other types of pollution mean that a significant proportion of European surface waters are affected by pollution associated with agriculture to at least some extent, with northern European countries in particular reporting that between 60 and 100 per cent are affected in some river basin districts (see Figure 4.10), although data is in some cases difficult to apportion in this way, so there is much the industry needs to do to address the problem.

4.1.5. The relationship between agriculture and flood risk

When it comes to water management and food productivity, concerns are usually regarding scarcity and pollution, however, the problem of flooding cannot be ignored. Flooding is one of the most catastrophic acts of nature. Although it is often the case that the focus of attention is on urban areas, the impact on rural communities and the farming sector can be equally devastating. There are several types of floods including those along coastal regions, caused by storms resulting in extreme tidal conditions, known as coastal surges, and those that are independent of waterbodies altogether and caused by extreme levels of hillside run-off flowing, unchecked by infiltration, into urban areas; these are known as pluvial floods. However, those that most affect rural communities along river corridors are fluvial, or riverine, floods caused by overbank flooding. This occurs simply because the volume of water flowing through the river is greater than its bank capacity and this excessive water volume is caused by heavy rainfall, snow melt or, in some instances, a blockage downstream such as an ice jam or debris preventing or slowing water flow causing it to back up and eventually overflow.

Europe's aquatic waterbodies, particularly surface waters, have seen wide-ranging morphological changes caused by anthropogenic activities which have led to an increased risk of fluvial flooding, a risk that is exacerbated by climate change. These activities have included altering the shape and/or direction of rivers to better suit societal needs, for example straightening meandering rivers effectively shortening them and so reducing their capacity and increasing the rapidity with which flood waters arrive downstream. Other examples are draining seasonally flooded areas along river corridors for agricultural production, the use of land drains (surface and/or sub-soil) to either bring marginal land into production or to allow machinery access to land earlier in the year, the construction of embankments to protect buildings, urban areas and other assets as well as the building of roads, housing and industry, and dams and weirs to divert water to generate electricity. There are also other land management issues that can exacerbate fluvial flooding. For example, where farm practices do not effectively manage run-off and erosion, surface water flows can not only lead to pollution of watercourses and flooding but also, over longer periods of time, lead to a buildup of washed-off soils and sediments on the river bed (often referred to as 'silting up') which gradually reduce the volume of water that the river can hold and thus the risk of flooding is exacerbated further. As a result, the countryside is less able to store water.

Equally, deforestation has the effect of reducing the amount of water intercepted by vegetation, and therefore increasing the amount which reaches the ground (Wheater and Evans, 2009). Trees also transpire considerable quantities of water, thereby recycling it back to the atmosphere, such that tree removal can increase the likelihood of overland flow, which equates to rapid and increased run-off. In addition, intensive farming has often led to a reduction in the organic matter content of soils, increased compaction and degraded their structure (e.g. through the use of heavy machinery), all of which tend to reduce the extent to which

they can absorb and hold on to water, allowing more to be passed in to the wider catchment, whilst the move towards larger fields has increased average slope lengths, increasing the likelihood of overland flow (Wheater and Evans, 2009). Consequently, changes to land management such as these have meant that modern farming (along with a number of other changes in the way in which catchments are managed) can increase the flood risk problems faced further downstream, in both urban and rural areas, something which is almost certain to get worse as a result of climatic changes (see Chapter 2).

Whilst agriculture may be adding to the problem, flood events can have a significant impact on the sector. The year 2012 saw one of the wettest springs on record in the UK and tens of thousands of hectares of farmland were adversely affected by flood waters (Morris and Brewin, 2013). The following year, serious floods along the river Danube and other central European rivers cost the agricultural sector in excess of €1 billion. More than 400 farms in Germany reported significant crop losses and 20 per cent of vegetables in the Czech Republic were also lost (Euractiv, 2013). The type of damage incurred was extensive and not just related to field crop losses. Typically, these events will cause damage to farm buildings and infrastructure which can result in the loss of stored goods such as harvested crops, conserved forage and livestock feeds. Damp buildings may not be suitable for housing livestock for many months and equally cannot be used for storing feed and forage due to the risk of hazardous fungal pathogens developing. As a result, the cost of 'away grazing' will be incurred. Livestock can also suffer in terms of their health. Flood events can result in injury and, in addition, animals can often suffer loss of body condition due to stress and so are more vulnerable to ill health and disease. Farmers are unable to access parts of their land for long periods and also have significant costs associated with the clean-up and necessary repairs.

With respect to field crops, the seriousness of the damage caused by flooding can be crop-specific. Oilseed rape (canola), for example, has a long season and so statistically, is more at risk than crops with a shorter season. Farms are at their most vulnerable as they approach harvest periods as it is at this point when the highest economic impact can occur. Harvest can be delayed and if the crop is totally lost, not only has the farm lost a source of income but also the costs incurred in growing the crop are wasted. In addition, if flooding occurs late in the season, it is often too late to replant. Some crops, such as soya bean, are known to survive flooding quite well whereas others, such as maize do not. Consequently, it is important to ensure that crop choice takes the likely risk of flooding in to account, as is already happening in many areas. In the south and east of Germany, for example, the period for harvesting potato crops coincides with the high-risk period for summer floods and therefore, potato production tends to be restricted to the north of the country (Klaus *et al.*, 2016).

Even if the crop appears to survive the flood, it may exhibit problems as it develops. As a flooded field drains, the drainage waters can carry away nutrients and minerals leaving a poor growing environment for the crop that remains. Fields under water or at water-holding capacity for any length of time have reduced levels of oxygen, without which the plant cannot sustain respiration, water uptake or root growth. Larger plants can suffer more than small ones (assuming small plants are not just dislodged from the soil and washed away) as they need more soil oxygen. Waterlogged soils can also accumulate carbon dioxide, which is toxic to plants at high concentrations. Crops in these environments tend to be more prone to disease (e.g. root rot from pythium) and weak plants are more susceptible to insect attack. Thus, even if the plant survives the deluge of water, it may not perform well and yields are likely to be significantly lower than expected.

4.2. Policy and interventions

4.2.1. Introduction

Water-related issues have occupied an important place in European agricultural policies for a number of decades, such that there is now a considerable amount of legislation in place. Water policy is broad and deep in terms of the diverse areas it covers and the scientific approach used to protect resources. It not only seeks to ensure water is available and of good quality for human and industrial consumption but also that it can support biodiversity, ecosystem services and recreation. In addition, policy must address the converse issue of too much water in the wrong place (i.e. flood control). Water was one of the first areas to be covered by EU environmental policy and there are now over 25 different water directives and related mechanisms as well as many reciprocal and equivalent instruments in non-EU European nations. Indeed, within the Common Agriculture Policy itself, there are a number of areas in which water takes centre stage, with funding being potentially available for a range of actions that are intended either to reduce the flow of pollutants from agricultural holdings, or to reduce water demand either in total or at sensitive times of the year (see Chapter 1).

The development of modern EU water policy began with the first Environmental Action Programme in 1973 which established the main objectives and principles that would shape water and other environmental policies in the coming years. By the early 1990s, there were increasing concerns regarding the eutrophication of waters and the general quality of water resources. Consequently, two new legal instruments were adopted setting strict rules on the treatment of waste water (the Urban Waste Water Directive) and the use of nitrates in agriculture (the Nitrates Directive), which constituted a considerable tightening of EU water policy. However, despite concerted efforts and a wide range of regulatory interventions, water quality and its availability continued to be major concerns. The general concensus was that EU water policy was fragmented and, in some instances, conflicting. Following a period of consultation and discussion across EU Member States (MSs) it was concluded that a Water Framework

Directive (WFD) was required to focus, integrate, rationalise and standardise, as well as improve the efficiency of European water protection legislation (Nilsson *et al.*, 2004). The WFD subsequently came into force in December 2000.

This broad scope means that the WFD is a highly ambitious piece of legislation providing a holistic framework for the management of both surface and groundwaters across the EU that builds on and/or integrates with a number of other directives and regulations, many of which have implications for the agricultural sector. Table 4.2 provides a summary of the main EU regulatory measures.

Table 4.2: EU directives affecting water

Instrument	Description	Relevant areas			
		PC	PH	RA	BE
Common Agriculture Policy	Main objectives are to provide a stable, sustainably-produced supply of safe food at affordable prices and decent living standards for agricultural workers. 'Greening' of the CAP introduced more environmentally sound farming methods and greater care of natural resources. Discussed in Chapter 1.	✓		✓	✓
Water Framework Directive (2000/60/EC)	Overarching legislative instrument for all EU water policy, the purpose of the directive is to establish a framework for the protection of inland surface waters, estuaries, coastal waters and groundwater. Introduces a number of key concepts including catchment management of waterbodies and environmental quality standards (EQS) for polluting substances. Discussed in Chapters 3 and 4.	✓		✓	✓
Groundwater Daughter Directive (2006/118/EC)	Establishes a regime which sets groundwater quality standards and introduces measures to prevent or limit inputs of pollutants into groundwater to protect quality and guard against deterioration. It initiates a scientifically sound response to the requirements of the WFD.	✓	✓		
Drinking Water Directive (98/83/EC)	Relates to the quality of water intended for human consumption. Its objective is to protect human health from the adverse effects of any contamination of water by ensuring that it is wholesome and clean.	✓	✓		
Nitrates Directive (91/676/EEC)	Primary purpose is to improve water quality by protecting water against pollution caused by nitrates from agricultural sources. In particular, it is about promoting better management of animal manures, chemical nitrogen fertilisers and other nitrogen-containing materials spread onto land.	✓	✓		✓
Floods Directive (2007/60/EC)	Introduces a requirement for all MSs to undertake flood risk assessments of their river basins and associated coastal zones and develop flood risk management plans. Plans are reviewed every 6 years in a cycle coordinated with the WFD implementation cycle.		✓	✓	✓

Instrument	Description	Relevant areas			
		PC	PH	RA	BE
Environmental Quality Standards Directive (EQSD) (2008/105/EC)	Closely linked to the WFD and establishes EQSs for priority substances and certain other pollutants defined in the WFD with the aim of achieving good chemical status of surface waters.	✓	✓		✓
Habitats and Birds Directives (92/43/EEC and 2009/147/EC)	Provides a comprehensive approach to managing biodiversity including the protection of aquatic habitats. Discussed in Chapter 3.	✓			✓
Thematic Strategy on the Sustainable Use of Pesticides, 2002	Seeks to achieve a more sustainable use of pesticides by reducing the risks and impacts of pesticide use on human health and the environment in a way that is consistent with crop protection, including avoiding water pollution.	✓	✓		✓
Directive on the Sustainable use of Pesticides (2009/128/EC)	Establishes a framework for Community action to ensure pesticides are used more sustainably by reducing the risks and impacts of pesticide use on human health and the environment.	✓	✓		✓
Plant Protection Products Regulations (1107/2009)	Aims to ensure a high level of protection of both human and animal health and the environment and at the same time to safeguard the competitiveness of Community agriculture. It defines harmonised rules for the approval of active substances and the placing on the market of plant protection products.	✓	✓		✓
Biocides Regulation (528/2012)	Covers a diverse range of substances that are not directly crop protection-related including disinfectants, pest control products and preservatives. It defines harmonised rules for the approval of such substances.	✓	✓		✓
Directive on Industrial Emissions (2010/75/EU)	Replaces the Integrated Pollution Prevention and Control Directive along with several other related instruments. It establishes a framework for the control of the main industrial activities, giving priority to intervention at source, ensuring prudent management of natural resources and taking into account, when necessary, the economic situation and specific local characteristics of the place in which the industrial activity is taking place. Regulations apply particularly to intensive pig and poultry farms.	✓	✓	✓	✓
Priority Substances Daughter Decisions – Related to the WFD and EQSD	Identifies substances or groups of substances which are a major concern for European waters. Substances are classified depending on the 'level' of concern based on best available knowledge; this level then relates to Community Actions required to mitigate risks.	✓	✓	✓	✓
Bathing Water Directive (2006/7/EC)	It applies to all surface waters that can be used for bathing. Seeks to improve the quality of bathing waters to protect public health and environmental quality. This is achieved via sampling and monitoring, and awareness raising.	✓	✓		✓
Landfill Directive (99/31/EC)	Seeks to mitigate the effects on the environment, in particular the pollution of surface water, groundwater, soil and air including the greenhouse effect, as well as any resulting risk to human health from the landfilling of wastes.	✓	✓		✓

PC – pollution control, PH – public health, RA – water resource availability, BE – biodiversity and ecosystem services.

At the current time, the WFD remains a key instrument relating to all aspects of water management within the European Union, as well as influencing water policy in those countries that are associated with it, such as Norway and Switzerland and many of its neighbours further afield (Boeuf and Fritsch, 2016; ENPI, 2010). Water quality and water scarcity are a major problem in the majority of European countries. River basins and water networks often traverse international boundaries and pollution originating in one country can cause problems in a neighbouring one. Consequently, many of the EU's neighbours have adopted similar water policies and there are a number of collaboration agreements in place. For example, Norway and Iceland as part of the EEA, have been obliged to fully implement the WFD and have transposed it into national regulations. Similarly, whilst Switzerland has not gone as far as adopting the WFD it does actively cooperate on the protection of international river basins such as the Rhine (Nilsson *et al.*, 2004). The European Neighbourhood and Partnership Instrument (ENPI) has been operational since 2007 and is now well established as a principal vehicle for cooperation with the EU's eastern and southern neighbours including Belarus, Moldova and Ukraine, as well as countries further afield such as Jordan and Israel. Whilst water protection is one of its main focuses due to its importance to economic development as well as environmental protection, this initiative is not restricted to water or indeed the environment alone but covers a wide range of sectors including energy, transport, information society, research and innovation. With respect to water policy, the main focus has been to encourage partner countries to converge their own policies and regulatory instruments with that of the WFD and several other WFD-related directives. For example, Moldova is one of the most vulnerable countries in the region in terms of water resources, having the least water supply per capita in Europe. Many of its rivers and waterways are polluted and have been overabstracted; the River Bic, for example, has been exploited to the point that it is now on the verge of disappearance. Under the ENPI, the EU Water Governance Project is providing assistance to Moldova and its neighbours to establish systems for the effective management of transboundary water resources that will put systems in place to enable them to move towards more sustainable use (ENPI, 2010).

The WFD has been discussed in the context of biodiversity and aquatic habitats in Chapter 3 but the directive also has significant implications for water management more generally. The most important concept in this respect is that it takes an integrated catchment-based approach to water resource management and requires MSs to develop river basin management plans (RBMPs) for each of the 110 defined river basin districts (RBDs) across the EU, including those requiring cross-border cooperation. In some cases, where catchments are very large, sub-plans have also been produced. Consequently, RBMPs encompass the entire aquatic system associated with a river (or sometimes a group of smaller rivers), from its source to its mouth and some way beyond, including both surface and groundwaters. They are intended to ensure that all waters achieve and maintain a good status as shown in Table 4.3. Importantly, it should be recognised that there are a variety of different quantifiable methods used to define both the quality and quantity of water resources of different types. In principle, a surface waterbody is considered as being of good quality if its condition is close to that which might be found within an unaffected body of that sort and in that location, although where waterbodies have been heavily modified by man

Table 4.3: Classification of surface and groundwaters under the Water Framework Directive

Assessment class	Surface waters	Groundwaters
Ecological status (ecological potential if artificial or heavily modified)	Five classes, based on biological/physico-chemical parameters.	Not applicable.
Chemical status	Two classes: good or fail. Based on specific priority and other pollutants.	Two classes: good or poor. Based on core parameters and specific other pollutants.
Quantitative status	Not applicable, although assessed in relation to ecological status.	Two classes: good or poor.

there is an acknowledgement that this may not be ecologically possible and the concept of good ecological status is replaced by one of good ecological potential (i.e. as close to good ecological status as is possible given the constraints present).

The following sections within this chapter discuss water policies and interventions that have particular relevance to the agricultural sector, although given the multitude of ways in which farming and water interact with each other, it must be acknowledged that this cannot be an all-encompassing list, with others too having a role to play, either directly or indirectly.

4.2.2. *Water availability*

The EU's decade-long growth strategy, 'Europe 2020', sets out three mutually reinforcing priorities for development, namely the need for (EC, 2010b):

1. Smart growth: developing an economy based on knowledge and innovation.
2. Sustainable growth: promoting a more resource-efficient, greener and more competitive economy.
3. Inclusive growth: fostering a high-employment economy delivering social and territorial cohesion.

The 'sustainable growth' element within this is mainly reflected in the goal of moving towards a low-carbon economy; however, the principle can equally be applied to the way in which we exploit resources of other sorts, including water (see also Chapter 5). The subsequent 'Roadmap to a Resource Efficient Europe' (European Commission, 2011), therefore, explicitly recognised a much broader range of resources that are under pressure, and stated clearly that water abstraction should remain below a threshold of 20 per cent of the available renewable water resource (European Commission, 2011). This signposted the increasing importance being placed on the sustainability of water resource exploitation in major European policy instruments; something that was further emphasised in the 'Blueprint to Safeguard Europe's Water Resources' which outlined actions needed to improve the implementation of European water policy (European Commission, 2012b). Central to delivering this, is the Water Framework Directive, which now has a major influence on the way in which resource exploitation is controlled across Europe and as a result has major implications for agriculture, particularly with respect to the cost of supplies and water availability.

With respect to the cost of water at farm level, economic instruments of various sorts can be used to reduce consumption or at least reduce wastage and drive more sustainable use in any industry. The WFD enshrines the concept of recovering the costs of water services into European water policy, and, as a result, this affects the amount paid by agricultural producers for their water use. According to the WFD, in principle, this should be a reasonable reflection of the true costs involved, including social and environmental costs, something which in many parts of Europe has not been the case in the past. The ineffectiveness of water pricing in the EU was considered in the Communication 'Addressing the Challenge of Water Scarcity and Droughts in the European Union' (European Commission, 2007), which highlighted the fact that despite the WFD's requirements, market-based economic measures have not been a major feature of the water resource management policies of MSs. Some countries, for example, currently exempt irrigation from volumetric metering (Poláková, et al., 2013) for both technical and political reasons. First, although metering of farm buildings relying on mains supplies (e.g. some of those used in the livestock sector) is a relatively simple matter, directly abstracted water use is considerably more problematic to police, with private wells for example, being difficult to control. This is particularly the case in countries in which water has traditionally been viewed as being freely available to anyone who wants to abstract it (Poláková, et al., 2013). Second, many farmers have traditionally been reluctant to see charging based on metering imposed upon them, fearing the impact that it might have on their business's productive capabilities. In addition, governments have often shied away from the use of pricing mechanisms that accurately reflect costs in the agricultural sector, as they often have in the domestic sector, because they too are concerned about the impact that doing so may have on farmers' ability to access water for their businesses, and consequently the effect that could have on food security

(production and price). However, volumetric charging is done in some European countries including Malta, Cyprus and Luxembourg. Whilst this has been successful on some levels, there is some evidence that it can lead to illegal abstractions, which is a recognised problem across Europe, especially in Spain and Cyprus. Illegal abstractions can take a number of different forms, including the drilling of unlicensed groundwater wells, exceeding the licensed abstraction quota or by using portable pumping equipment to remove water from rivers. The problem for authorities is that enforcement is very difficult as it is time-consuming and highly expensive (EEA, 2012).

Despite these concerns, there is now an increasing emphasis being placed on the use of compulsory metering in all sectors including agriculture (European Commission, 2007), as a necessary precursor to being able to accurately reflect usage in cost-recovery systems (European Commission, 2012b). For those nations that reject the metering approach, there are alternative mechanisms available for introducing a form of pricing into the agricultural industry, including area-based schemes which charge on the basis of the area of land being irrigated and/ or on which particular crops are grown. Alternatively, quotas could be imposed more widely across Europe, such as those attached to abstraction licenses in the UK, France and Spain, and a fee charged commensurate to the true cost. This would allow the use of a given volume of water, which may or may not be dependent on the crop being grown (Poláková, et al., 2013). Whatever system is adopted, however, it is clear that pricing will be an increasing feature of the way in which European water resources are managed in the future.

As water is used extensively throughout the farm, ensuring an adequate supply of the right quality of water is a fundamental requirement of all farms and, indeed, food security. The WFD places a requirement on all MSs to ensure that surface waters (with some exceptions) achieve 'good ecological status' or 'good ecological potential', both of which have inherent within them a requirement for quantitative control. Overabstraction, either directly from surface waters or from groundwaters which are hydrologically linked to them, can put considerable pressure on supplies as well as on aquatic ecosystems (Environment Agency, 2013b; Klaar, *et al.,* 2014). England's Environment Agency, for example, estimates that flows are insufficient to maintain healthy ecosystems in around 13 per cent of English rivers, most notably in the Thames, Anglian, Severn and Humber basins (Environment Agency, 2013b; POST, 2012). The WFD requires a catchment-based approach to be taken to the management of abstraction across the EU in an attempt to ensure equitable access, as well as protecting aquatic ecosystems.

The situation in the UK provides a good case study illustrating these issues. In England and Wales Catchment Abstraction Management Strategies (CAMS) are used to manage water abstractions. These strategies consider the environmental requirements and the amount of abstraction already licensed in order to determine whether or not there is any scope for further exploitation, or indeed whether restrictions may be required (Environment Agency, 2013a). Historically, in the UK 'licences of right' were issued to water resource users, with little or no consideration of the environmental impact that such uses may be having and regardless of what they might have in the future (Environment Agency, 2013a). In addition, many licences had a fixed water allocation based on daily and annual abstraction limits (Defra and Welsh Government, 2013), with little scope for abstraction to be linked to the prevailing

climatic and/or environmental conditions. These licences were also granted more or less in perpetuity, such that they could only be amended through a slow legal process (Defra and Welsh Government, 2013; POST, 2012). This situation is further complicated by the fact that a third of such licences, amounting to 55 per cent by volume, are unused (Defra and Welsh Government, 2013). In managing abstraction through the CAMS process, it is necessary to take into account the abstraction which is licensed, rather than that which is presently being exploited (as where a licence exists it may be reactivated), meaning that the efficient allocation of available resources is severely hampered (POST, 2012; Defra and Welsh Government, 2013). Therefore, in recent years there has been considerable pressure on authorities to review the UK's abstraction licensing system, to provide greater protection for the environment (Environment Agency, 2013a). Whilst some progress has been made by amending legislation such that all new licences are time-limited, with periodic reviews to determine whether changes should be made and/ or licences revoked (POST, 2012), there are still many abstraction licences that are held indefinitely, and contain few if any restrictions aimed at reducing abstraction in times of particular stress. The UK Government does appear to be committed to reforming the system further however, by better linking abstraction to water availability and increasing the scope for reviewing and amending licences (Defra and Welsh Government, 2013). Current thinking is moving towards a risk-based approach to permit reviews, which, whilst ensuring that no licences are held in perpetuity, would allow a knowledge of catchment conditions to inform the frequency with which reviews take place (Defra, 2016).

The situation in Spain is a good comparison to that in England and Wales. Spain has

Figure 4.11: Parma River in the Emilia-Romagna region of North Italy, rapidly drying in June
(Photo courtesy of: Prof. Kathy Lewis)

one of the highest rates of water abstraction in Europe and this is managed by the allocation of 'water rights' at the river basin level. The management process seeks to actively encourage farmers to improve the sustainability of their water usage and does this by systematically reducing their volume quotas annually. For example, farmers within the Guadalquivir river basin, the Guadalquivir being the second longest river with its entire length in Spain, have had their allocated abstraction volumes gradually reduced from around 7,000 m^3 ha^{-1} in 1985 to 5000 m^3 ha^{-1} in 2005. The result of this has been significant improvements in farming practices, irrigation efficiency and cropping plans (Dworak *et al.*, 2007).

The need to ensure a safe level of abstraction whilst still maintaining sufficient flow rates for biodiversity is a primary objective for water security policies. This can be illustrated with a situation in the municipality of Parma, in Italy. Most of the water needs of the urban areas in the region are acquired from groundwater sources, some of which are frequently in deficit and there are also concerns regarding water quality, mainly due to high nitrate concentrations arising from agricultural activities. As a consequence, studies are ongoing regarding the feasibility of extracting water from the local rivers (e.g. Rivers Taro, Cerno and Parma which are tributaries of the River Po), which are of better quality. However, these rivers already suffer serious low flows (See Figure 4.11), mainly due to agricultural abstractions for irrigation, for long periods with seasonal drying not uncommon (Santato *et al.*, 2016). All rivers throughout the Po catchment are valued for their rich biodiversity, for example, the River Taro supports a wide range of aquatic species including the south European nase, barb and Italian chub as well as many different amphibians. In addition, a designated Special Protection Area follows the path of the River Taro and this nature reserve hosts more than 250 species of migratory birds, 17 habitats of European interest and over 800

Table 4.4: Summary of approaches used to estimate e-flows

Approaches	Description	References
Utilising hydrological data	Simplest and most widespread approach, accounting for around 30% of global systems. Uses historical hydrological data in order to establish an e-flow profile.	Tharme, 2003; AMEC, 2014
Hydraulic rating methods	Use physical hydraulic variables as surrogates for factors believed to impact on particular target species. E-flows are determined from the relationship between flow rate and the chosen parameter (e.g. between discharge and flow depth).	Tharme, 2003; Gippel and Stewardson, 1998; AMEC, 2014
Habitat simulation techniques	These approaches are an extension of the hydraulic rating methods and use a variety of mathematical models to establish the relationship between flow and ecological status on a site-specific basis.	Tharme, 2003; AMEC, 2014
Holistic approaches	Take a broad-brush approach, and focus on addressing the e-flow requirements of the whole riverine system and attempt to derive flow-ecology relationships which are transferable across wide geographic areas.	Tharme, 2003; Mathews and Richter, 2007

species of plants including many species protected by the Habitats Directive (Tourism Emilia Romagna, 2017). Therefore, water managers must consider how to implement the changes needed to supply local demand whilst minimising the impact on the aquatic environment and the ecosystems it supports (Bonzanigo and Sinnona, 2014).

Given the importance of ecological systems both intrinsically and within the water regulatory framework, the need to identify the flow properties of the waterbodies necessary to protect them has achieved ever greater significance in recent years (Klaar, *et al.,* 2014; Richter, *et al.,* 2006). Ecological-flows, often referred to as 'e-flows', can in part be defined as 'the quantity, timing, and quality of water flows required to sustain freshwater and estuarine ecosystems' (Klaar, *et al.,* 2014; Acreman and Dunbar, 2004; Gippel, 2001), and, as such, are far more complex than just the maintenance of minimum flow rates per se. Indeed, disturbances such as

drought and flooding are an integral part of many ecosystems and essential in maintaining natural assemblages. Consequently, as the EU's Blueprint makes clear (European Commission, 2012b), if the WFD is to be complied with, mechanisms are required for defining what the flow quantity requirements are likely to be. Although it is clear that flow regime is a significant factor controlling the make-up of aquatic ecosystems (Jain, 2012), predicting and quantifying precisely what impacts might result from perturbations in that regime is far from straightforward (Bunn and Arthington, 2002), albeit that it is now widely accepted that ensuring that the flow regime is as natural as possible is likely to be desirable. This is an active area of research and there are a number of potential approaches being used, as summarised in Table 4.4. These techniques are helping regulatory authorities understand e-flows within their catchments and so improve water management.

For example, in England, the Environment Agency is using a broadly hydrological approach supported by hydraulic rating and habitat simulation for determining e-flows (AMEC, 2014). However, as yet there is no preferred European approach.

The CAP (see Chapter 1) also has elements within it to encourage sustainable water use, although the focus is more on water quality than efficient use. As far as mandatory actions required under cross compliance is concerned, none of the statutory management requirements (SMRs) specifically target the conservation of water resources and only one of the standards for Good Agricultural and Environmental Condition (GAEC 2 – water abstraction) is relevant. GAEC 2 requires those wishing to abstract water for irrigation to do so in compliance with the authorisation procedures established in their country and so is closely related to the abstraction control issues discussed above. Other GAECs have indirect relevance through, for example, soil practices that conserve water and therefore reduce the need for irrigation, such as the maintenance of soil organic matter (GAEC 6). Within Pillar 2 funding streams, the options are a little more extensive, with the availability (if selected by a MS for implementation) of payments through agri-environmental schemes and capital grants, in particular, as well as funding for training in best practice, and so on (Poláková, et al., 2013). Capital schemes may include funding for more water-efficient technologies, farm reservoirs and other storage systems (Poláková, et al., 2013) and may assist in maintaining production in more or less its present form in what is a changing political and environmental context. Agri-environment scheme payments, on the other hand, generally require production systems to be amended in some way, and are often intended to compensate producers

for the reduced yields which may result from doing so. This might, for example, include converting from irrigated to rain fed production, or at least reducing a business's reliance on supplemental water supplies, or the implementation of management techniques which may help to conserve soil water such as cover crops (Poláková, et al., 2013). Consequently, although payments for activities designed to reduce water consumption have to date been less prominent than those discussed below in relation to water quality, with increasing pressure on resources, they are likely to become a more important part of the water management policies being applied to the agricultural sector.

The bottom line in any successful policy aimed at reducing agricultural water use is to ensure that it is implemented on-farm. To a large extent, this is done by legislation and the use of European and/or national funded initiatives such as the CAP agri-environment schemes. However, there are a number of other mechanisms often instigated by government–industry or other stakeholder partnerships that can help deliver the desired policy outcomes and provide a balance between regulation and voluntary actions. With respect to water efficiency, these can often be successful simply because water is an essential farm input that is becoming more expensive and farmers are likely to respond positively to anything that saves them money. Many of these types of schemes are knowledge transfer-based, designed to raise awareness of new and emerging technologies and best practice via, for example, workshops, demonstration farms, farm tools and free literature. For example, 'Waterwise', a joint initiative by the UK's Environment Agency, National Farming Union and LEAF (Linking Environment and Farming), which produced a step-by-step guide for farmers to help them identify areas where water could be saved (Environment Agency *et al.*, 2007). More recently, the Sustainable Agriculture Initiative Platform, a multi-industry research partnership with a global outlook, launched an easy-to-use farm water assessment tool intended to encourage communication and engagement between a farm and its supply chain customers (SAI, 2015). In terms of new technologies, a partnership between Monsanto, the University of Milan and drip irrigation experts from Israel has launched a solution for optimising irrigation efficiency that combines in-field sensors and drip irrigation which, together with specialised training, has, according to press reports, resulted in a 234 per cent increase in irrigation water productivity and 335 per cent increase in fuel cost efficiency for maize farmers using the system in Italy (Monsanto, 2016). Taking this a step further, initiatives such as awareness-raising campaigns can also be used to steer consumer behaviour by encouraging the choice of products that are more sustainable in terms of the water used. This can put pressure on food supply chains to improve water use and so gain a market advantage. To this end, the use of labelling, certification and farm/food assurance schemes have a role to play.

4.2.3. Flood risk

Whilst much of the focus of European water policy is on tackling the issue of water scarcity (as well as pollution), it also needs to address the converse problem of flooding. After the catastrophic flooding events in central Europe around the start of the millennium, addressing this issue climbed the European political agenda and in 2007 the Floods Directive was introduced. This directive seeks to mitigate, as far as possible, the adverse consequences of flooding for human health, homes and infrastructure, the environment and cultural heritage as well as the economic impacts. It introduced

the requirement for all MSs to undertake flood risk assessments of all watercourses and coast lines at risk from flooding within their borders. It also required MSs to map the flood extent as well as the assets and humans at risk in these areas, and to take adequate and coordinated measures to reduce that risk. This is done by the development of flood risk management plans that are reviewed on a six-year cycle. These plans detail what activities are in place and what will be developed during the management cycle for the prevention of flooding and the protection of all 'at risk' peoples and assets. The directive also requires MSs to coordinate their plans with those of their neighbours, especially where there are shared river basins.

Whilst the Floods Directive has been broadly welcomed, it is not without its critiques. One of the problems is linked to the flexibility available to MSs in terms of the approach taken to determining and reporting flood risk. For example, in the first phase of implementation, MSs were able to use existing flood risk assessments as long as they were suitable for identifying areas at significant risk of flooding from all potential sources and in all parts of their territory. Alternatively, if existing risk assessments did not meet these requirements, then a new preliminary assessment was required. This resulted in a rather fragmented and somewhat unharmonised picture across Europe. Some MSs, for example, quickly undertook preliminary flood risk assessments across their entire territories and for all potential sources of flooding whilst others were much slower to respond and introduced transitional approaches before fully complying with the directive. This meant that not all assessments were done with up-to-date data, nor were they all simultaneously available, with the same coverage or using the same methodology. Even today, mapping and reporting approaches differ. For example, 'flood hazard and flood risk' maps have been prepared by most MSs, but some just look at the potential for river flooding, whilst others also consider coastline events. Some MSs have combined flood risks from all sources into one map whilst others produce a separate map for each flood source.

Second, the assessment of flood risk should ideally include an assessment of the impact of climate change, changes in land use and other aspects that may have an impact on flooding consequences. However, this is not a requirement of the Floods Directive and so not many MSs have considered these aspects when preparing their flood maps and associated management plans (European Commission, 2015). Consequentially, plans may not be sufficiently forward-looking to enable future increases in flood risk to be anticipated.

Another criticism is that the Directive is very demanding regarding the highly accurate topographical and socio-economic data needed to undertake detailed assessments and these data are not available for every river basin. Undertaking assessments without all the necessary data can significantly affect the quality of the risk maps and management plans produced such that, in some instances, they may not be fit for purpose (Tsakiris *et al.*, 2009). It is also widely acknowledged that in many MSs the effectiveness of communications between water managers and crisis managers from the emergency services needs to be significantly improved. However, generally speaking, there is broad acceptance that the move towards a risk-based policy is a move in the right direction although there is still much to be done.

4.2.4. *Water quality*

As far as water quality issues are concerned, there are a number of key European directives which are either a direct result of the introduction of the WFD or linked to it, and which are intended to address one or more forms of pollution and/or types of waterbody (see Table 4.2). At its simplest, the protection of water quality can be seen as a requirement to enable the utilisation of surface and groundwaters and provide the water needed for drinking and in food products. In order to ensure this, the first Drinking Water Directive (80/778/EC – Council of the European Union, 1980) was introduced in 1980 and subsequently replaced in 1998 (98/83/EC as amended – Council of the European Union, 1998), the new Directive being introduced to further protect water consumers from the deleterious effects of contamination in various forms.

The directive itself and the national legislation introduced to enshrine it into the legal systems of MSs (e.g. Water Supply (Water Quality) Regulations 2000 in the UK and the Trinkwasserverordnung 2001, as amended 2016, in Germany) are of more direct relevance to the water industry than farmers per se. However, the requirements of the Drinking Water Directive do have a number of indirect implications for those in the agricultural sector. It sets out a minimum quality standard that is mandatory across all nations of the EU and EEA. This quality standard is comprised of almost 50 chemical, microbiological and so called 'indicator' standards, which must be monitored and tested for on a regular basis. Individual MSs may add to this basic list or, indeed, raise the minimum standards, should they wish. Many of these pollutants are of little relevance to the agricultural sector but others can arise from the farm-based activities described earlier in this chapter, including pathogenic bacteria (e.g. *E. coli*, for which a limit of zero mg l^{-1} is set), nitrate (50 mg l^{-1}) and pesticides (0.1 µg l^{-1} for individual chemicals and 0.5 µg l^{-1} for total pesticides), for example. Consequently, there is a demand for reduced agricultural pollution to contribute to meeting the standard and so comply with the Drinking Water Directive.

Perhaps of more direct importance to farmers is the Nitrates Directive (Directive 91/676/EEC), which has been in existence since 1991, and is intended to reduce the extent to which nitrogen from agricultural sources becomes a problem in surface and/or groundwaters (see Chapter 3 in terms of biodiversity). This directive requires MSs to monitor the nitrate content of waterbodies, identify those that are either polluted or at risk of becoming polluted (i.e. they could contain in excess of the 50 mg l^{-1} limit imposed by the Drinking Water Directive), and establish Nitrate Vulnerable Zones (NVZs) within which a mandatory code of practice must be implemented to manage the problem. As discussed in Chapter 3, NVZs have had a major impact on the way in which agricultural nitrogen is managed in many parts of the EU, and have, in particular, placed significant restrictions on the way in which organic manures and slurries are handled (e.g. Defra, 2009; Scottish Executive, 2005). There are, however, a number of variations in the way different MSs have implemented the overall requirements of the directive in their domestic legislation and guidance.

In England, for example, the latest iteration of the rules is laid down in the Nitrate Pollution Prevention Regulations 2015, with equivalent legislation having been implemented in the devolved administrations of Scotland, Northern Ireland and Wales. This stipulates (with some scope for derogations):

1. A crop-dependent N-max limit for the amount of crop available-N which can be applied to different crops.
2. A limit of 170 kg ha^{-1} (as stipulated in the directive itself) for the amount of nitrogen in organic manures/slurries that can be spread (or deposited) on a farm in any calendar year (averaged over the farm).
3. An additional limit of 250 kg ha^{-1} for total N applied in organic form (not including that deposited by in-field livestock) on any given hectare of land.
4. Specific limits for compost applications.
5. Closed periods during which organic manures with a high, readily available N content cannot be spread (dependent on soil type and land use).
6. Storage and management requirements for manures and slurries.

In contrast, in Denmark, the approach taken to managing nitrogen pollution arising from agriculture is based on three main strands, namely fertiliser accounting, harmonised rules and regional zoning. Under the fertiliser accounting rules, farmers must submit annual fertiliser accounts for their farms, which detail the nitrogen balance for the past year and provisional fertilisation plans for the coming year (Le Goffe, 2013). A nitrogen quota is then calculated for the farm using nitrogen requirement standards based on crop type, soil, irrigation and yields that are below those expected if a farm was following a policy of maximising output (although this can be challenged if a business can demonstrate that their yields are higher than those used as standard). Farmers can be fined if their nitrogen usage is greater than their quota, such that the economic benefits of exceeding the limit are significantly reduced or even wiped out. The Danes have also set general manure spreading (see Figure 4.12) limits that are even stricter than those set by Directive 91/676/EEC, and in ecological zones (that are sensitive to nitrogen, ammonia and odour) even more stringent limits are imposed, meaning that overall, the approach taken in Denmark goes beyond that required by the directive.

As part of the Water Framework Directive,

Figure 4.12: Manure spreading
(Photo courtesy of: Pixabay.com)

a requirement is imposed on the European Parliament and Council to propose measures specifically intended to prevent and/or control the pollution of groundwaters and as a result allow the required good chemical status to be achieved. This is currently executed through the Groundwater Daughter Directive 2006/118/EC (GDD). This directive is closely related to the WFD, hence the term 'daughter', and replaced earlier legislation (referred to simply as the Groundwater Directive) introduced in 1980. The current directive stipulates the requirements for both determining what constitutes good chemical status in groundwater terms, and for the identification and reversal of upward trends in contamination. In the former case, the standards which must be applied for many of the pollutants likely to be generated from agricultural sources, namely nitrate and pesticides, are defined in other legislation (i.e.

the Drinking Water Directive). For others, however, MSs have been given a good deal of flexibility to define (and publish in relevant river basin management plans) threshold values which are appropriate to their own national and/or river basin district priorities, although a minimum set of pollutants is stipulated for which they must 'consider' establishing them (see Table 4.5).

To a large extent, the GDD is of more direct relevance for catchment-scale management planning than farm-scale decision-making, but such requirements nevertheless have implications which filter down to those working at the farm level. For example, under Article 6 of the GDD there is a requirement to take 'all measures necessary to prevent inputs into groundwater of any hazardous substances', where hazardous substances are in essence the pollutants named in the WFD, and to limit contamination by other

Table 4.5: Minimum list of pollutants and their indicators for which MSs have to consider establishing threshold values in accordance with Article 3 of the GDD (as amended)

Substances or ions or indicators which may occur both naturally and/or as a result of human activities		
Arsenic	Mercury	Sulphate
Cadmium	Ammonium	Nitrites
Lead	Chloride	Phosphorus (total)/Phosphates
Man-made synthetic substances		
Trichloroethylene	Tetrachloroethylene	
Parameters indicative of saline or other intrusions		
Conductivity		

pollutants. As a result of national legislation, for example, the Groundwater (England and Wales) Regulations 2009 and the Portuguese Water Law no. 58/2005 (which both, in part, transpose the GDD into national legislation), authorisations may be needed in order to dispose of some chemicals to land (e.g. sheep dip), and in addition MSs must establish a framework by which codes of practice can be approved in order to provide land managers and others with the practical advice they need to comply with the top-level legislation. These are intended to translate esoteric top-level directives into something that will have tangible benefits on the ground.

It is perhaps unsurprising that, given their hazardous nature and the fact that they form such an important strand in legislation such as the GDD, plant protection chemicals have come in for specific attention within the legislative framework. Although the Pesticide Authorization Directive, 1991 (91/414/EEC) has now been superseded, it established rules regarding the placing of plant protection products on the market, and in particular that no product should be on the market unless 'in the light of current scientific and technical knowledge' it poses no harmful effect to human or animal health, directly or indirectly (e.g. through drinking water, food or feed), or to groundwater and has no

unacceptable influence on the environment, this latter point referring specifically (in as far as water quality is concerned) to the fate of pesticides in the environment, and, in particular, the likelihood that their use will result in the contamination of surface and/or groundwater (including drinking water sources). This, along with the biodiversity concerns discussed in Chapter 3, led to the removal of a number of products from the market following the introduction of this directive. Of around 1,000 products which were approved for sale in one or more MS prior to 1993 only around 26 per cent passed the review programme that Directive 91/414/EEC started, 67 per cent were never submitted for re-approval (either because they were already considered obsolete or economically unviable) and 7 per cent of products were put forward but failed to gain approval (European Commission, 2009). Consequently, the number of existing products fell by almost 75 per cent to approximately 250, with another 180 new ones being added to the market more recently (Chapman, 2014).

Following the release of the Thematic Strategy on the Sustainable Use of Pesticides in 2006, Directive 91/414/EEC was replaced by Regulation (EC) 1107/2009, which has even more stringent requirements for the registration of new products and the renewal of old ones, than the previous legislation. As such, where substances are deemed to be particularly hazardous (e.g. carcinogenic), a hazard-based approach is now taken which takes into account the likely impact in real-life usage situations (King, 2011). Many of the most restrictive requirements under the new legislation relate more to human health issues than the aquatic environment per se, although where chemicals are considered hazardous and banned as a result, there are implications for aquatic pollution management, as these particular substances will cease to be a threat. However, it does not necessarily follow that those falling into this category will be those most likely to be found in the aquatic environment. A parallel piece of legislation (the Directive on the Sustainable Use of Pesticides, 2009), on the other hand, is specifically intended to reduce the impact of pesticide use on the environment (as well as human health), with the aquatic environment and water supplies being specifically highlighted. This is done through the promotion of a more holistic approach to pest management (Integrated Pest Management – IPM), which makes maximum use of non-chemical approaches to pest and disease problems, and as such is of clear on-farm relevance. As part of this process, MSs are required to develop 'national action plans' with two main objectives. First, the plan must lay out the measures that are to be implemented within that MS to reduce the risks to the environment (and human health) of agriculturally-applied pesticides and what activities would be undertaken to encourage IPM and comply with EU legislation. Second, there is a requirement for MSs to set out quantitative objectives associated with the implemented measures, and the timetable for achieving them. In addition, under the Directive on the Sustainable Use of Pesticides, MSs are required to establish procedures for ensuring that professional users of pesticides, such

as those in the agricultural industry, have access to suitable certified training, including a form of continuing professional development and that application equipment is subject to regular (generally at least every five years until 2020 and at least every three years from then onwards) inspection.

The fact that IPM is specifically encouraged by the directive and so also by the domestic legislation it is transposed into, should in itself minimise water quality issues by maximising the extent to which alternative (cultural or biological) approaches are utilised, and therefore reducing the amount of chemical used. Application rates should be as low and as infrequent as practical by making use of pest warning systems and other technologies, taking due care not to encourage the development of pesticide resistance. The associated regulations also require that where pesticides are used, reasonable precautions are taken to protect the environment and this has a number of implications for farming practices. For example, steps must be taken to ensure that, as far as possible, chemical applications are contained within the crop/area being targeted and the risk of drift should be prevented or at least reduced to the minimum. In addition, where the use of pesticides poses a risk to the aquatic environment, a product which is less hazardous in this respect and does not contain any priority substances (as defined in the WFD), should be used as a matter of preference.

As discussed above, in as far as agriculture is concerned, diffuse pollution sources are often considered to be the major sources of aquatic pollutants, but point sources can sometimes be significant, and should problems occur they have the potential to cause considerable damage. To an extent these are addressed by some of the legislation discussed previously, including manure/slurry stores (the Nitrates Directive) and the disposal of sheep dip (the Groundwater Daughter Directive), for example. However, there is also legislation that specifically addresses point sources. Within England, for example, issues associated with the storing on-farm of silage, slurry and fuel oil, have been addressed through the Water Resources (Control of Pollution) (Silage, Slurry and Agricultural Fuel Oil) (England) Regulations 2010 (as amended), with similar regulations having been introduced in other parts of the UK. These regulations stipulate construction, management and placement rules which must be adhered to, when storing any of these substances on a farm. In the case of fuel oil stores, for example, there are comprehensive rules that govern the size and locations of storage areas, including that they must be fully surrounded by a bund capable of storing at least 110 per cent of the capacity of the largest enclosed tank (or 25 per cent of the total storage volume if that is greater, although in many agricultural situations this is unlikely to be the case). The bund and base must both be impermeable and the storage area must be at least 10 m from any surface waterbody into which any leakage could flow. In implementing such rules, it is hoped that the risks posed by the accidental, or even deliberate, release of the sort of damaging substances which are often stored on farms, should be minimised.

A farmer's major interactions with the above legislation are often through one or more elements of the CAP, more specifically through financial incentives given for taking environmentally beneficial measures. As far as water quality is concerned, the requirement to manage this problem is clearly reflected in the rules governing Pillar 1 funding (direct payments to farmers) through the cross compliance system, with water quality concerns being a key driver behind a number of the statutory management requirements (SMRs). For example,

farmers whose land falls within an NVZ must comply with the NVZ rules in Directive 91/676/EEC as transposed into the applicable national legislation system. Water quality legislation is also a key driver behind a number of the standards for good agricultural and environmental condition (GAECs), including the need to protect watercourses using buffer strips (GAEC 1) and the need to have prior authorisation before releasing any substance that may harm or pollute groundwater (GAEC 3). Under many of the agrienvironment and other schemes under Pillar 2 (rural development) of the CAP, farmers may also be encouraged to further protect surface waters using enhanced buffer strips or wetland creation, for example (Poláková, *et al.*, 2013). Point source pollution might be tackled through capital grants for such things as better storage facilities for manure, slurry or silage, improvements to livestock handling areas, or fencing to prevent direct animal access to surface waters (Poláková, *et al.*, 2013). Consequently, a significant proportion of the non-commercial income of agricultural businesses can depend on their being able to demonstrate that they have taken steps to mitigate against their impact on the quality of surface and groundwaters.

In addition to the legislative requirements mentioned above, and just as there are nonregulatory initiatives intended to ensure a secure water supply, in many European countries there are a range of voluntary programmes aimed at addressing water quality issues, as well as systems for disseminating best practice advice to those on the ground. At a governmental level, these include various codes of best practice to help growers comply with regulator requirements such as the Italian Code of Good Practice for the Protection of Groundwater from Nitrate (Codice di buona pratica per la Protezione delle acque sotterranee da Nitrati), the UK's Protecting our Water, Soil and Air: A

Code of Good Agricultural Practice and the Code of Practice for using Plant Protection Products used in England and Wales. It also includes targeted advice channels such as the Catchment Sensitive Farming initiative active in England, which raises awareness of the diffuse source of pollution problems associated with the industry, with a focus on giving free training and advice to those operating in catchments at particular risk (Defra, 2016). Another example is the various technical support agencies in Italy that promote the use of IPM and have developed targeted programmes for specific farming sectors such as that implemented in the grapes/wine sector in Friuli-Venezia Giulia (Massarutto, 1999). Whilst in the Netherlands, Environmental Impact Cards (a ranking system of pesticides based upon their environmental behaviour) have been implemented that provide guidance to farmers to help them select less polluting substances (PAN Europe, 2007). Farm assurance schemes and labelling/certification schemes also address water quality issues by including related standards in their farming protocols. For example, the 'Melinda' trademark used for apples in Trentino, Italy specifies a strict maximum content of pesticide residues among their quality criteria and the Belgian FRUITNET label also complies with IPM standards (PAN Europe, 2007). The motivation for industry to get involved is often a result of the commercial benefits which might be associated with ensuring that pollution is kept to a minimum, although this might be being a little simplistic, since many in the industry also have more altruistic motives. This is most clearly illustrated in the field of pesticides, in which the crop protection industry has a vested interest in ensuring that its products are not banned due to their impact on aquatic environments, such programmes including those instigated by the Voluntary Initiative which 'promotes the responsible use of pesticides'

and the Metaldehyde Stewardship Group, through their 'Get Pelletwise' initiative, both in the UK. Equally, the water industry has an interest in keeping substances such as crop protection chemicals out of waters they use to supply domestic consumers, either due to the difficulty in or the costs associated with treatment, such that there are a number of examples of water utilities providing producers in the relevant catchments with advice, or even paying them to adopt specific practices, for example by encouraging them to choose alternative products (Anglian Water, 2016). In Germany, for example, farmers and the water supply company WVZ Maifeld-Eifel have been collaborating since 2015 to reduce nitrate leaching and improve water quality. Under the scheme, farmers receive compensation payments for undertaking a less intensive form of cultivation funded by the WFD programme in Rhineland-Palatinate (WVZ Maifeld-Eifel, 2016). These types of collaborations are popular in the UK and in other parts of Europe as many governments would rather a voluntary approach is adopted to solving as many environmental problems as possible, despite the mounting evidence that these approaches have limited impact and at best ensure that just the minimum legal standard is implemented (McCarthy and Morling, 2015; Arimura *et al.*, 2008).

4.3. *Farm level management and protection*

4.3.1. *Introduction*

For European agriculture to make a valuable contribution to wider efforts to address both water resource management and water quality issues, it is essential that those in the farming community have access to the information and tools they need to do the job. Fortunately, the scope for making changes to on-farm practices and infrastructures in this respect is quite good. A number of amendments are possible, and these often amount to little more than the implementation of current best practice. Although some require a considerable input in terms of finance or other resources, others, in contrast, can be put in place at little or no cost, or even make sound financial sense. Consequently, although some measures may be restricted to those who are particularly environmentally enthusiastic, almost anyone in the industry should be able to find something which suits their business, and which will aid the situation in some way. In doing so, the contribution made by the agricultural industry as a whole could be very significant. What is suitable in a given situation will, however, depend on the site and business objectives of that farm, including their type of holding, local soils, geology, climate, and so on, so there is no 'one size fits all' solution to water management problems in an agricultural setting. Instead, farm businesses need to be free to select options that suit their business objectives and the environmental imperatives we are all faced with. However, it is obvious that, due to the environmental and resource management issues faced by the industry, progress has to be made, as the 'business as usual' model is not sustainable in the long run.

4.3.2. *Management of water resource use in agriculture*

Within farming, the need to ensure the 'efficient' use of this essential resource is the constant headline message, although in this context the term 'efficiency' tends to mean different things to different stakeholders. From the farm business perspective, 'efficient use' is generally gauged in terms of economic and/or production-based metrics (e.g. yield, income), possibly benchmarked against others in the sector. In recent years, benchmarking water use (i.e. comparing actual usage by one farm against that of others) has become quite popular, although given the uniqueness of any individual farm in terms of its location, crops grown and farming practices, achieving practical comparisons is far from easy. With respect to crop production, managing water for efficient use, therefore, often means ensuring that the maximum possible 'crop per drop' is obtained from the water available, or the maximum possible yield per unit of irrigation (see Figure 4.13). Producers, however, generally aim to achieve the maximum crop yield per hectare of land and around harvest time each year, articles appear in the agricultural press highlighting the achievements of producers in obtaining particularly high yields. From some perspectives, of course, this may indeed be considered a success, as it maximises farm incomes and contributes to food security. However, maximising farm incomes is not necessarily the same as maximising profits. The law of diminishing returns means that there comes a point for many inputs including water, when the increase in yield per additional unit of input starts to level off or decline. In Figure 4.13, increases in crop yield are plotted against water usage and the point at which the curve levels off is that at which the use of additional water resources ceases to be justified on purely economic terms, assuming that each unit of input has an equal financial cost. This is because any further inputs may result in greater yield and so income, but this will be more than offset by the increase in costs in supplying it, such that overall profits will fall and the enterprise will no longer be operating at maximum efficiency.

In many areas of Europe, however, including some parts of the UK, water has been a relatively cheap input to the business, with farmers often facing few, if any, restrictions on their access to it. In addition, water use is not

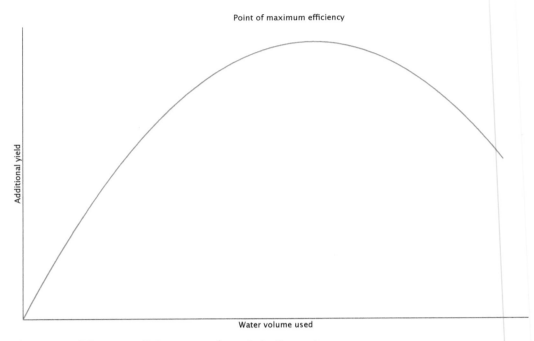

Figure 4.13: Water use efficiency curve for an irrigation system

always charged for on a 'per unit' basis. For example, regulatory control of water resources via abstraction licences is well established in Europe, although the fees charged vary considerably. An abstraction licence may entitle a user to access up to a maximum amount of water, but below that maximum the cost of use is a function of the fuel used in pumping, the time taken, and so on, and therefore not easily related to the volume of water used per se. However, access to abstraction licences is becoming much more controlled as the need to protect resources for other users and the environment becomes more significant. Indeed, as far as many regulators are concerned, increasing water use efficiency in agriculture is not a matter of getting the most 'crop per drop' but simply one of reducing consumption to protect natural resources. In addition, real business decisions need to take into account a much broader suite of objectives than just those relating to yield. Society must also consider how much water

should be given over to different consumers including industry and agricultural production. Value judgements need to be made in relation to how far it is desirable to go, and this could be influenced by such things as the influence of crop quality on sales price (e.g. in potato production, water may be added to improve quality and command a higher price), whether crops/livestock have a value that goes beyond the purely financial (e.g. in relation to food security) and whether other ecosystem services benefit from production. Nevertheless, it is clear that the conservation of water resources will be of benefit to the farm business, the environment and other users, and the following sections consider a number of ways this can be achieved in a farming context.

4.3.2.1.   *Water auditing*

The old adage that 'if you can't measure it, you can't manage it' was originally coined for use in business but it can equally be applied to water management, in that unless the land manager has a clear understanding of how water is being used on their holding, it is very difficult to make informed decisions about whether it can be improved, and if so, how. In this sense, the use of water audits as a tool for informing decisions, and thereby increasing efficiency, has been around for many years. A water audit is a means of reviewing current practice and performance with a view to identifying areas of concern, and options for improvement. A number of different approaches exist, but in essence, there are a number of steps which are common to most (see Figure 4.14).

The first step is to determine as precisely as possible what the baseline position is to provide a basis on which to improve. In order to do so, it is necessary to identify the source of all the water being used on the farm, how much is being used from each source and what the financial costs involved in doing so are. Most businesses use a variety of sources of water, including the mains supply, abstraction from surface and/or private groundwater wells, from reservoirs and water recycled from other sources, each of which has its own characteristics in terms of cost and suitability for use in different aspects of an enterprise's operation. Where possible, water metering should be used to determine exactly how much is being used, but in some cases estimates may need to be made, possibly based on the capacity of pumping equipment and the time for which it is running, or indeed from published standard figures. The logical next step is to apportion the water used to the various activities that the farm undertakes and the different pieces of equipment or locations used. The information gathered can then be used to determine whether simple reductions in water use can be made or whether the sources of water being used for the different activities being carried out on the farm are the most appropriate (Environment Agency *et al.*, 2007). For example, whether potable water from the mains supply, which tends to be the

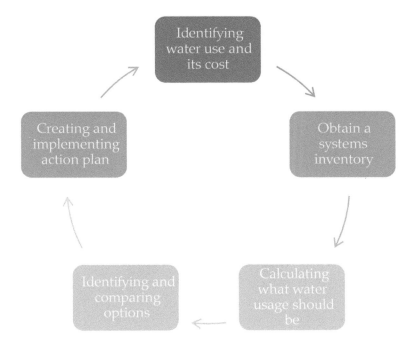

Figure 4.14: General steps in a water auditing process (Adapted from: Environment Agency *et al.*, 2007)

most expensive, is being used when directly abstracted or even recycled water would be perfectly adequate. It should also be possible to determine whether the amounts of water being used are as expected and reasonable. This might be achieved by comparison with previous usage rates (which might help identify leakages if amounts used are higher than expected) or where available, published data. Another approach is to use bench-marking either by simply working with other local farmers and sharing data, or using the various online schemes and systems available. Regardless of which technique is used, it is essential to review the data regularly to identify unexpected changes and react accordingly.

The data gathered and monitored as described above should provide a sound basis on which to identify suitable, cost-effective solutions to any problems identified and areas in which savings, in terms of water use, may be possible. Some of the options for implementation are discussed later in this chapter, each of which may be more or less appropriate for a given set of circumstances, so that businesses will need to make value judgements as to whether they are suitable for their farm. Whatever is selected, however, should then form part of a written water management plan, which clearly sets out what will be done, including both one-off capital projects and procedures for inclusion in everyday management processes, as well as the time frame over and frequency with which it should happen. It is also important to specify what the expected results of any implemented measures are, so that at some future date this can be compared with a subsequent set of monitoring data. This then sets a timetable for the implementation of improvements, against which progress can be assessed. However, the production of a water management plan should not be seen as a one-off process but rather part of a repeated cycle of improvement which itself should be reviewed and updated on a regular basis. What constitutes a regular basis will, however, vary from business to business, and is dependent on a combination of how much water is being used and the risks to overall catchment water resources.

4.3.2.2. Irrigation scheduling

As discussed previously, irrigation accounts for a considerable amount of on-farm water use in cropping systems and, to help reduce water demand, the amount of water used in this manner needs to be optimised. One way this can be done is by considering the way in which soil and water are managed together to meet crop needs most effectively and central to this is the process of irrigation scheduling. Irrigation scheduling refers to the use of water management strategies to determine the amount of water required by a growing crop to prevent overapplication (wastage) or underapplication, which may lead to yield loss from drought stress. The approach seeks to determine how much water needs to be applied through irrigation (as opposed to occurring naturally), and the most appropriate timing for those applications. This is achieved by using a set of predetermined criteria describing how much water to apply in a given situation, and when. Whilst this may sound an obvious approach, it is sometimes dismissed as being excessively time-consuming but if done correctly it can sometimes save both time and money. Its planning, however, does demand a degree of forethought and may require a reasonably detailed set of input data to operate effectively.

This data will include parameters that are specific to the type and condition of the irrigation equipment used, such as its specific application rate and how uniformly application can be achieved. Crop-related parameters are another consideration as water requirements vary depending on crop and sometimes the specific variety, as some are more vulnerable to drought stress than others. Irrigation scheduling may also require information on how much water stress can be tolerated and the likely rates of evapotranspiration, which is controlled by a combination of crop and climatic factors. Obviously, weather plays a huge part in whether or not irrigation is required with important parameters including both past and predicted temperature and precipitation, as it may be possible to maximise efficiency if projected rainfall can be taken advantage of. In addition, there may be a number of practical concerns that might need to be taken into consideration, including, for example, labour requirements, cost and possibly equipment availability (Martin *et al.*, 1990).

There are a number of direct ways of measuring or estimating the moisture content and/or capacity of a soil. At its simplest, this could rely on an assessment of the look and feel of the soil. An experienced land manager can often make an informed judgement as to whether irrigation is necessary or not, and if so, how much, based on an understanding of the soil's properties, and indeed the needs of the crop being grown in it. Although this approach is very quick and simple, it may not be particularly accurate, even when carried out by someone who thoroughly knows their land, and as a result there are now a number of pieces of equipment which allow the accurate quantification of soil moisture content or availability, including:

1. **Tensiometers**: These are relatively cheap instruments that measure soil water potential (Ψ), which is an indicator of the amount of energy required to move water within a soil, or in terms of crop management, for plant roots to extract water from their growth medium (soil).

2. **Capacitance probes**: This approach is particularly suitable for continuous monitoring, and uses electrodes in the soil to measure its ability to conduct electricity (or more accurately the dielectric constant of the soil), a property which is related to the soil's moisture content.

3. **Electrical resistance blocks:** These use the fact that water within a soil lowers its electrical resistivity, so if a block of material (e.g. gypsum) including probes for measuring that electrical resistivity, is placed in the soil so that it exhibits the same water content characteristics, an assessment of water content can be obtained.

4. **Neutron probes:** These are perhaps more suitable for specialist and/or research use, as in some European countries a licence is required to use them and they can be difficult to calibrate; however, these instruments can provide a reasonably good assessment of a soil's water content. They are based on a radioactive source and detector, with the source emitting neutrons out into the surrounding soil, and the detector picking up the slowdown of those neutrons on contact with hydrogen (i.e. that present in water molecules).

The alternative is the water balancing approach, in which mathematical models are used to estimate what the soil water content will be at any given time, based on a series of parameters describing the soil, crop and climate, for example. Central to this is an understanding of the available water content (AWC), which is the difference between the soil's water content at saturation and at permanent wilting point and is therefore the potential amount of plant available water the soil can contain. Farmers will need to manage the soil water content so that it stays somewhere between the two extremes to prevent both excessive plant stress (yield loss, death, etc.) and excessive irrigation (water use, cost etc.), unless deep percolation is desirable for the removal of salts. This type of modelling, therefore, allows the inputs from irrigation and precipitation and the outputs through evapotranspiration and percolation to be evaluated, and the water content of the soil to be balanced; but, as with most forms of modelling, the results will tend to become increasingly inaccurate over time, such that regular in-field calibration is also needed by using, for example, one of the measurement systems described above.

The evidence suggests that, at the present time, around a third of farmers rely on operator judgement based on an assessment of a soil's look and feel, to decide on how much irrigation water is added. A further third will use some sort of in-field soil moisture measurement, whilst around a quarter will utilise calculations of a soil's water balance (Hess *et al.*, 2008), although as discussed above, even these producers will need at least some in-field measurement. Regardless of which system is used to inform the process, however, accurate irrigation scheduling should allow the water status of a field's soils to be managed, and controlled more accurately than would otherwise be the case. Indeed, it becomes possible to take into account the needs of particular crops at specific points in the growing season. In addition, there are a number of benefits which a well-run system of irrigation scheduling may be able to provide:

1. **Water savings:** The main benefit of irrigation scheduling is that it optimises water use, which does not necessarily always equate to reduced usage, but where unnecessary irrigation is avoided, savings may accrue, freeing up resources for the irrigation of other crops

or alternative uses (public water supply, environmental protection, etc.).

2. **Increased management flexibility:** The ability to schedule irrigation according to specific crop needs, can aid in ensuring the more efficient rotation of application equipment among fields.

3. **Reduced cost of labour:** Where the maximum benefit is obtained from soil moisture storage, fewer irrigations may be required, reducing the man-hours required for the work, and potentially the labour costs involved.

4. **Reduced fertiliser costs:** Excess irrigation can result in the loss of nutrients through surface run-off and/or leaching.

5. **Increased income:** Optimising the way in which water is used in the production of crops, can increase returns by raising crop yields and/or the quality of produce (commanding a higher sales price).

6. **Reduced waterlogging:** Excess irrigation may result in the waterlogging of soils (particularly in vulnerable areas of a field), with all the implications for soil structure and water quality that entails. It may also impact negatively on crop yields.

7. **Better control of soil salinity:** As discussed above and in Chapter 5, in some locations (albeit minimally in the UK) soil salinisation can be a significant problem associated with the practice of irrigation. More accurately controlling the amount of water applied to the soil can minimise the build-up of salts, by ensuring sufficient leaching from the soil profile.

4.3.2.3.   Optimising the performance of irrigation networks

Irrigation has been a feature of agricultural systems around the world since ancient times, with the civilizations of Egypt and Mesopotamia first diverting the annual flood waters of the Nile, Tigris and Euphrates around 8,000 years ago, while in Europe, the Greeks and Romans too made considerable use of irrigation to boost the productivity and reliability of their farming systems and this is still the objective today. Historically, the practice of irrigation has been dominated by the use of surface waters, although groundwater sources were also sometimes tapped, with the main developments being to the way in which water was extracted in order to supply the system (e.g. the shaduf, water wheels, Archimedes screw, etc.). Regardless of the extraction system employed, however, gravity was then used to transport water from its source to the field, via manmade irrigation channels, such that it could be spread over the growing area through a flood- or furrow-based application system, the former being one in which the entire surface of the field is intermittently covered by a layer of anthropogenic flood water, whilst the latter was one in which water was flooded through a series of channels within the field to wet the surrounding soil, this method being more suited to crops grown in defined rows and where the soil has a reasonably high hydraulic conductivity. In more recent times, however, there have been significant developments in the way in which water is transported to and applied to the field, allowing considerably more scope for managing that water use in such a way as to reduce pressure on resources.

The first step needed to optimise the farm's water network is to determine the most suitable means of transporting water from its point of origin (surface or ground) to its point of use (the field or fields in which the crop is being grown). On a large scale the choice of conveyancing system is likely to be a matter for local and/or national planning authorities rather than individual farmers, but businesses may be faced with a number of options once water reaches their farm, such that a balance needs to be struck

between the capital costs involved in implementation, the cost of water, the efficiency with which a given system is likely to operate and the need to maximise efficiency (i.e. are there limits placed on water availability or is water freely available?).

The best way of achieving this will also depend on a number of geospatial factors, including the distance to be covered, the underlying soil, geology and climate. In the UK, for example, water is rarely carried far from its source, often being abstracted on the farm intending to use it. As a result, the most common form of transfer is through temporary or permanent pipework, which, if properly maintained, will minimise losses through leakage, although care needs to be taken to ensure that this occurs in reality. In many parts of Europe, however, open irrigation canals are commonplace, such that losses through leakage and evaporation can be significant. Estimates vary, but research suggests that losses from pipelines may be only 5 per cent, whilst concrete channels lose 15 per cent and earthen channels, 30 per cent of water during transit (Karamanos *et al.*, 2007). The significance of the impact will also depend on the length of the channel and the soil from which the channel is constructed (Table 4.6). Regardless of how water arrives at its point of use, land managers are then faced with a number of options for applying that water to the field, which vary considerably in complexity, cost, usability (in different situations) and the efficiency with which they allow water resources to be utilised, as discussed below.

Despite the fact that flood and furrow forms of irrigation have been around for many centuries, they are still frequently used for irrigation around the world, and Europe is no exception. This is particularly the case in parts of southern Europe, where it remains the dominant form of irrigation, with around 60 per cent of all Spanish

Table 4.6: Indicative values of conveyance efficiency for open channels

| Canal length | Earthen canals – conveyancing efficiency | | | Lined canals |
	Sandy soil	Loamy soil	Clay soil	
Long (> 2 km)	60%	70%	80%	95%
Medium (200 m–2 km)	70%	75%	85%	95%
Short (< 200 m)	80%	85%	90%	95%

(Brouwer *et al.*, 1989).

Table 4.7: Indicative values of the field application efficiency

| Irrigation method | Field application efficiency | |
	Brouwer *et al.* (1989)	Karamanos *et al.* (2007)
Surface irrigation (e.g. flood, furrow)	60%	75%
Sprinkler irrigation	75%	85%
Drip irrigation	90%	90%

irrigation for example, taking a gravity-fed flood or furrow form (Baldock *et al.*, 2000). The efficiency of such systems is relatively low, having an indicative field application efficiency (that part of the water applied which is used by the crop) of around 60 per cent (Wriedt *et al.*, 2008) and it can be significantly lower, although such systems can have their efficiency improved in a number of ways. One of the key limitations of any system in which the uniform application of water is dependent on gravity, particularly in flood irrigation systems, is that the distribution surface needs to be managed so as to ensure that it remains reasonably level, as even a slightly uneven surface will result in some parts of a field receiving more water than others. Consequently, the use of modern levelling equipment can aid the efficient running of the system. In addition, since water is fed into one end of the field and flows towards the other, the time taken for the water to cross the field can result in the inflow side receiving considerably more water than the outflow side, due to the longer period during which infiltration can occur. Therefore, if the outflow side of the field is to receive sufficient water, so much

water will have infiltrated into the soil at the inflow end that a significant amount may be lost through deep percolation (i.e. move beyond the depth at which the crop can utilise it). Therefore, surge irrigation systems, in which a furrow irrigation system is activated intermittently over a period of time, rather than constantly, are becoming more popular. This approach creates a cycle of in-furrow wetting and drying, causing fine particles of soil to deposit on the furrow bed. In turn, this reduces the rate of infiltration and allows water to distribute across the field more rapidly, increasing the uniformity of application, and reducing overall water use (Hess and Knox, 2013; Rogers and Sothers, 1995). Studies have demonstrated that surge irrigation of alternate furrows in a cotton crop, for example, could reduce water use by 44 per cent and result in an application efficiency rate of nearly 85 per cent (Horst *et al.*, 2007). Another way of increasing the overall efficiency of surface irrigation systems is to catch water flowing off the outflow end of a field. Some run-off is inevitable from surface irrigation systems and in many cases that run-off is lost (although it may return to surface waters), so if it can be intercepted and stored on-farm, it can then be recycled for further use, reducing the amount that needs to be abstracted from waterbodies (Hess and Knox, 2013).

Although all of the above techniques increase the costs involved in what is traditionally a low-cost form of irrigation, they are well suited for integration into the sort of systems which operate in relatively wealthy parts of the world, such as Europe. Nevertheless, there are a number of alternatives, with the potential to increase the efficiency of water use well beyond that generally obtainable from surface water irrigation systems. In many other parts of Europe, for example, sprinkler systems dominate and they are increasingly being used in those parts of the continent traditionally relying on gravity systems. In France, for example, 85 per cent of irrigation is through some form of sprinkler system (Baldock *et al.*, 2000). However, sprinklers vary considerably in their form and therefore their operational characteristics, and their potential to minimise water use and other environmental impacts.

The dominant form of irrigation in the UK, is through the use of pressurised rainguns (Figure 4.15a). These are viewed as providing a convenient, reliable system that can be used on uneven or sloping fields and which require only limited labour input to operate (Van der Molen *et al.*, 2007) as they can be set up and then left running unsupervised until it is time to relocate them. Spraying water over such distances, however, requires considerable energy input, and runs the risk of resulting in an uneven distribution of water, particularly in windy conditions, and as a result can be quite wasteful, although in well-run systems and under suitable operating conditions, it has been shown that they are able to limit waste to as little as 10 per cent (Knox *et al.*, 2008). As a result of these limitations, alternative forms of sprinkler irrigation are becoming increasingly popular,

although they too have both advantages and disadvantages.

In travelling-boom-based systems, a mechanised spray boom on wheels moves slowly across a field (Figure 4.15b) and sprays water in a downward direction onto the soil/crop; in doing so they avoid many of the problems associated with water being wasted through drift, the blowing of spray away from the intended target, and can provide very uniform water application (Smith *et al.*, 2002; Irmak *et al.*, 2011). In addition, the operating pressure of such systems is considerably lower than that of rainguns, meaning that the fuel cost involved in operating the necessary pumps and, therefore, the associated emission of greenhouse gasses, is also reduced. On the downside, however, there are considerable capital costs involved in obtaining the equipment, so they only tend to be suitable for high-value crops such as some field-grown vegetables. They also require a reasonably flat topography and regular field shapes to operate most effectively and as they generally apply all the water needed in a single pass, application rates can be high, running the risk that overland flow will be induced, which may result in water running off the field, causing erosion and being wasted (Van der Molen *et al.*, 2007; Smith *et al.*, 2002). An alternative is a centre pivot-type system, in which water is fed into the boom from a central point around which the boom rotates. On a global basis, these systems are probably more common than other boom systems, but, although there are a significant number of these systems in Europe, particularly in France and Spain, they are perhaps most usually associated with areas such as North America, North Africa and the Middle East. Centre pivot systems have many of the benefits of other boom systems and can often have lower labour requirements as there is no need to move the equipment. However, they are less compatible with pre-existing field shapes as they deliver water in a circular pattern, whereas European field systems are more usually rectangular or irregular in form.

Solid-set sprinkler systems (Figure 4.15c) often use a similar spraying technology to that of the raingun (Figure 4.15a), although rotary systems are also common, but on a much-reduced scale. The energy requirements for operation using these systems are significantly lower than for their larger cousins but require a high level of initial investment. However, they do permit considerable flexibility in the way in which water is applied (Knox and Kay, 2011). They are generally set up and left in place for the lifetime of a particular crop, and as they are always in place, they can be switched on at any time and with any frequency, often remotely or automatically by some of the soil moisture monitoring systems discussed above. They therefore avoid the need to apply extra water, which may be lost, in order to allow for the time it may take for irrigation to occur again, and so can ensure that water supply is optimised to the needs of the crop. In addition, and unlike booms and centre pivot systems, they can be applied to fields with irregular shapes and/or uneven surfaces making them potentially more flexible. However, the set-up costs involved mean that in the past they have mainly been restricted to use in permanent and semi-permanent crops (e.g. fruit crops), although they are now increasingly being seen in high-value horticultural crops, in part due to technical advances making systems quicker and easier to set up (Ayars *et al.*, 1999).

Potentially, the most efficient methodology of all, however, is the use of trickle (or drip) irrigation, albeit only for certain crops. In essence, trickle irrigation systems use a porous or perforated pipe laid alongside a crop in order to allow water to seep from the pipe to wet the soil in and around the immediate location of the hole. In doing this it is possible to ensure that water is only placed in the area of the soil in which plants are growing, so that none is wasted in wetting uncropped areas and/or the crop canopy. Such systems also operate under significantly lower pressures (and therefore use relatively little energy in operation), and once in place can be automated, limiting the demand for labour. However, as pores and holes can become clogged, a relatively clean source of water, free of excessive sediment or algae is required, if ongoing maintenance is not to be an issue. Savings in water can be significant, but it has to be remembered that crops still require the same amount of water to grow successfully, so it is only ever going to be possible to reduce water application to a certain degree. Nevertheless, if a crop has a clearly defined root zone, then it is a system that, despite the capital costs involved, can offer significant benefits. For this reason, it has most often been associated with semi-permanent crops grown in rows (e.g. fruit crops), although it is now being used alongside other row-grown crops including potatoes and salad crops (Ayars *et al.*, 1999).

4.3.2.4.   Optimising water availability

When it comes to on-farm water use, application is only one of the elements that needs to be considered, the other major one being related to the source of water used. In many areas, the abstraction of water, either from watercourses or boreholes, for immediate use has been the mainstay of agricultural water use, at least where mains water supplies are not used. However, for the reasons described above, obtaining and keeping abstraction licences is becoming increasingly difficult, and, as a result, farm reservoirs are becoming an increasingly popular option, particularly in areas in which there is a marked seasonal pressure on resources. Most

Figure 4.15a: Raingun irrigation
Photos courtesy of:
Pixabay.com

Figure 4.15b:
Travelling boom irrigator
Photos courtesy of:
Pixabay.com

Figure 4.15c: Solid-set sprinkler irrigation
Photos courtesy of:
Pixabay.com

irrigation is undertaken in the summer when crop demand is at a maximum and precipitation inputs at a minimum. At other times of year, in contrast, precipitation inputs and river levels are much higher, whilst demand, at least from arable producers, may be negligible. Consequently, there may be far fewer restrictions on the amount of water farm users may be permitted to abstract, allowing water to be taken relatively freely, and stored for use when demand increases, although a licence to do so will generally still be needed. This means that the potential impact of abstractions on surface waters in periods of low flow, particularly those on aquatic biodiversity, can be minimised, and as a result the use of on-farm storage systems in areas in which the pressure on surface waters is considerable, is often supported by environmental regulators. In addition, as discussed in more detail in Chapter 3, on-farm reservoirs offer significant potential for enhancing farm biodiversity, although to maximise this potential the design needs to mimic the form of natural waterbodies and, ideally, be linked to other habitats through wildlife corridors or stepping stones (see Chapter 3).

Equally, however, there are marked advantages for producers themselves, since reservoir storage provides them with much greater control and security of irrigation supplies, reducing the potentially negative yield and quality implications of being unable to supply the needs of growing crops at particularly sensitive times. However, this comes at a cost, specifically a financial one in terms of construction, although this is heavily dependent on whether an impermeable lining is required, or whether use can be made of the local pedology/geology (for example through the compaction of a clay soil). At 2012 prices in the UK, for example, it was estimated that the capital costs of a 100,000 m³ structure would cost more than £100,000 if unlined and double

that if lined. In addition, there may be significant costs involved in feasibility studies, design, obtaining planning permission and suitable abstraction licences, for example, which can add another 10 to 15 per cent to the cost (Weatherhead *et al.*, 2014), and take considerable time. The investment required is therefore considerable, which means not only that farm reservoirs are generally restricted to areas and crops in which the implications of not having sufficient water may also result in a heavy financial penalty, but also that only businesses with access to sufficient investment capital can afford the initial outlay. Although, in some cases there may be opportunities for producers to share the burden with their neighbours, by developing shared water storage schemes (Weatherhead *et al.*, 2014) or to obtain funding to improve water management including support for reservoirs under national and EU funding schemes. However, access to such schemes varies significantly between countries and is liable to frequent change in response to the political or financial priorities of the day.

Where reservoirs are being planned, there are a number of issues that need to be addressed, most notably whether a suitable area of flat land, ideally located centrally within the area that needs irrigation, can be identified for the location of the reservoir. Both the physical area and land quality will need to be considered, as the land used for the reservoir is likely to mean that land is taken out of production or lost from the natural environment. Ease of access to the site for construction equipment and the potential impact on local communities will also need careful consideration. The latter can act as a barrier to construction if the local community believes that the impact could be significant, for example in relation to the local environment, aesthetics or safety. Some of these concerns can be allayed by careful consideration at the design stage, for example, by careful siting of a reservoir away from areas which are too close to local properties or places to which people may have access (e.g. footpaths). The inclusion of appropriate fencing may limit safety concerns and may also help reduce a new reservoir's aesthetic impact. Care should also be taken to ensure that the skyline is not broken by the reservoir banks (Suffolk Coast and Heaths AONB, 2010).

Assuming that a suitable site can be found, then size requirements will need to identified and this is likely to be a compromise between the volume of water required, considering the level of access to other resources (e.g. summer abstraction, mains supplies, etc.) and the cost.

4.3.2.5. Water efficiency in livestock systems
The water audit process has an equally important role in providing a basis for increasing water use efficiency in the livestock sector, as it does in the arable sector. Indeed, in many ways the increased range of water uses that are associated with managing animals, means that it may be even more important as well as potentially more complex. As discussed previously, water has a multitude of applications in livestock systems and each of these uses needs to be identified and as far as possible quantified. However, it is also important to consider non-metered sources of water, such as animals helping themselves to drinking water from unfenced surface waters. Such sources make a valuable contribution to livestock businesses, and if removed must be replaced with water from another source, as all animals have a level of required water use which cannot be reduced without affecting animal health and productivity. Such sources can, however, be difficult to quantify, and may need to be obtained from published estimates of livestock requirements (e.g. Environment

Agency *et al.*, 2007). Once a water audit has been used to establish the baseline situation, a number of options then exist for reducing water use.

In terms of drinking water, there is only limited room for manoeuvre as far as overall water consumption is concerned, although there is scope for amending the sources of water used, and for ensuring that any water distribution systems are free of leaks and other sources of inefficiency. For example, it may be desirable to make the maximum possible usage of natural sources of water (e.g. springs, streams, etc.), so as to minimise the requirement for anthropogenic inputs. This, however, relies on a source of suitable quality being available in the necessary location or close by so that piping it in may be feasible. There are, however, environmental concerns associated with free access to surface waters, particularly in relation to the impact on water quality and erosion (see Chapters 3 and 5).

There are a greater number of options available when it comes to the cleaning of buildings and yards, since although hygiene requirements may necessitate the use of potable water in the cleaning of dairy parlours, this may not be the case for other areas. For example, some floors and other hardstanding areas can be scraped to remove most solid material (either as the sole form of cleaning or prior to washing), a process which can be aided by keeping surfaces damp so as to prevent dust. Indeed, the need for cleaning yards in the first place can be reduced if the time that animals spend there, for example before and/or after milking cattle, is minimised. Another option is the use of recycled water (for example, warm water from plate-cooler systems) for cleaning purposes (Milk Development Council, 2007). With the aid of suitable storage and capture facilities, it may also be possible to harvest rainwater from the roofs of farm buildings in

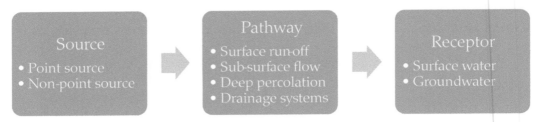

Figure 4.16: Source–Pathway–Receptor model of pollutant delivery

order to supplement other sources of water, as this water is of an adequate quality for many washing purposes. Finally, washing of yards and buildings is often carried out using either volume hoses or pressure hoses which, as discussed in Chapter 4.1.2, have different water usage characteristics and are suited to somewhat different forms of cleaning, such that choosing the correct tool for the job is essential in minimising water use (Milk Development Council, 2007).

4.3.3. On-farm control of water quality concerns

The management of water pollution problems within an agricultural context, in particular those from non-point sources, can be a complex issue due to the multiple factors at play, some of which may be highly dependent on site- and time-specific circumstances. Consequently, a holistic approach is needed that attempts to take into account all the relevant natural and management factors, so as to effectively and efficiently minimise the problem. In relation to diffuse sources, it is also important that controls are implemented broadly across the industry, since the contamination appearing in surface and groundwaters is rarely from a single farm business, tending to be generated by many farms across a catchment, and therefore, piecemeal adoption by a few of the more proactive farmers within an area, is unlikely to have the desired effect. This is an approach advocated by the CAP (see Chapter 1) and

the WFD (see Chapters 3 and 4.2), as well as being used by water companies to help ensure that raw water quality in their region is as good as possible and that the farmers and landowners are given all the support they need to deliver that. As with many pollution issues, the basic concept of controlling the source–pathway–receptor system of pollutant movement, as described in Chapter 1 and shown schematically in Figure 4.16 (Holman *et al.*, 2008), holds true, providing a number of opportunities for intervention. Ideally, the most desirable option is to stop the pollution at source as it is considerably less likely to fail than methods that attempt to prevent the movement of pollutants between the source and a receiving waterbody, or those that try to deal with pollution after it has entered the aquatic environment. However, this is not always possible and so other interventions to mitigate pollutant releases may be required.

4.3.3.1. Pollution control at source

The ultimate expression of source control is, of course, some sort of land use change, and although this may seem a little extreme, in situations particularly vulnerable to pollution, it may be the only option available. Nevertheless, it is unlikely that producers will be keen on making such wholesale changes to their activities, by, for example, converting intensive arable land to low-intensity grazing land, in all but those most extreme cases (Cuttle *et al.*, 2007).

However, partial changes can also be effective, for example decreasing the intensity of production by reducing inputs of inorganic fertiliser may help minimise the presence of excess soil nitrogen and therefore reduce nitrate leaching. In a grassland context, less intensively managed land also tends to have a greater soil organic matter content than that managed intensively, which can lead to the immobilisation of nitrogen, effectively locking it away in a less mobile form and preventing leaching. Such changes can, of course, be associated with a reduction in yield, and therefore income and potentially profitability. However, even where such significant changes to the production system are deemed to be inappropriate or undesirable, there are a number of ways in which the land management system can be amended in order to minimise the threat to waterbodies. To illustrate, arable land receiving organic manures tends to have a higher organic matter content (Van Beilen, 2016) than land that does not and so may also see increases in nitrogen immobilisation. Organic manures also tend to improve soil structure, which promotes efficient nutrient uptake by crops and so limits the amount available for loss to the environment. It will also aid soil water retention, reducing losses through surface run-off and/or sub-surface flow, as well as reducing erosion (see Chapter 5). Although clearly, if this is done to excess, the organic manures can themselves become a source of pollution.

In addition to the above changes in production systems, there are a number of other management options for controlling water pollution at source, many of which have implications for more than one pollutant, and involve little more than following the sort of best practice that many farmers will already be familiar with. In an arable enterprise or where grass is being produced as a crop, for example, a key in-field option for reducing losses of both phosphorus and nitrogen, is to ensure (as far as practical) that applications are attuned to the needs of the growing crop. In the main, this revolves around the determination of crop nutrient requirements, the soil nutrient content (e.g. through regular testing) and consequently the deficit, if there is one, which needs to be made up through the application of organic and/or inorganic fertilisers. A number of formalised systems for making this assessment exist across Europe, including look-up tables, spreadsheets calculators and software. These include the PLANET software system (Dampney and Sagoo, 2008) and the RB209 fertiliser recommendation look-up tables (Defra, 2010) both used in England and Wales; AZOFERT, a decision support tool developed by INRA in France (Machet *et al.*, 2007); the RISSAC-ARI environmentally-friendly fertilisation system designed for Hungary (Fodor *et al.*, 2011); and the mathematical model, EU-Rotate-N (Rahn *et al.*, 2007), developed for horticultural cropping across Europe. Some cover a range of crops (e.g. Defra, 2010; Heppell *et al.*, 2016) whilst others provide more detailed advice and are crop-specific (e.g. Rahn *et al.*, 2001; Gallardo *et al.*, 2014). The utilisation of this informed approach should, in principle at least, ensure that excess applications of nitrogen and phosphorus are minimised. Nevertheless, it is difficult to make precise recommendations, due to the temporal and spatial variations which are inherent in factors such as yield and climate. Consequently, some excess application is inevitable from time to time. In some senses, a similar approach can be taken in relation to pesticides and the risk of contaminating watercourses (and other environmentally-sensitive areas), in that it is essential to ensure that before any crop protection products are applied, there is a genuine need to do so. This can be achieved by adopting an integrated pest management

(IPM) approach to crop protection in which steps are taken to try and avoid problems occurring in the first place, thereby negating the necessity for chemical treatment (see Chapter 5.3 for full details).

If we accept that the excess application of crop protection chemicals and, particularly, nutrients, is on occasions inevitable, it becomes important to keep as much of the applied substance as possible on the field for as long as feasible. In doing so, it allows time for plant uptake or degradation, and so reduces the in-field reservoir of such substances, minimising the pollutant source. It also ensures that as much benefit as possible is obtained from the application. As far as phosphorus is concerned, the movement of this pollutant tends to be closely aligned with that of sediment, since as discussed above, it tends to be bound to fine soil particulates which may be mobilised by the processes associated with erosion. Consequently, the field management techniques used to minimise erosion for agronomic reasons, will have a commensurate benefit in terms of the levels of phosphorus and sediment in the surface waters of the surrounding catchment. The techniques available to achieve this are discussed in Chapter 5. Equally, some pesticides (i.e. those considered to be hydrophobic, usually with a high soil absorption coefficient [K_d]) will also bind to sediments and, like phosphorus, can be transported to waterbodies along with the movement of that sediment. Consequently, methods to control and minimise soil erosion will also go some way to reducing the impact of pesticides in the aquatic environment. Pesticides bound in this manner and which reach surface waters tend to be locked into the sediment, which becomes a sink for those pollutants, potentially mitigating environmental impacts. However, the pesticide can be unlocked by re-suspension during high discharge events

or by the burrowing and feeding, known as bioturbation, of some aquatic species, for example benthic invertebrates (Bundschuh *et al.*, 2016).

In relation to nitrogen, techniques such as ensuring that autumn crops are sown as early as possible, in northern Europe at least, have been shown to be beneficial, as has the use of winter cover crops (Cuttle *et al.*, 2007). Both practices encourage the late season uptake of nitrogen from the soil, leaving only the minimum available for subsequent mobilisation. Consequently, by the time the wet winter conditions are likely to result in increased sub-surface water flow and therefore nitrogen leaching, the size of the nitrogen reservoir has been diminished. Switching to spring cultivation, means that new nutrient additions do not need to be made until after sowing, so that nitrogen mineralisation is limited prior to the presence of a vigorously growing crop that is able to use any mobile nitrogen present. The application of nutrient too late in the autumn or early in the spring should be avoided if possible. Direct cultivation (e.g. a farming system in which soil tillage is avoided) may have a similar result, as this too tends to reduce the rate at which soil nitrogen mineralisation occurs (Cuttle *et al.*, 2007).

In livestock systems, animals can be managed so as to limit the deposition of manure and urine in the field and/or the amount collected during housed periods, which subsequently needs to be field-applied in either manure or slurry form. Reducing the length of the grazing season, for example, will reduce field deposition (Cuttle *et al.*, 2007), and, although problems may still occur as a result of the spreading of manures and slurries, this provides the opportunity for controlled application in which high-risk areas, such as those close to watercourses, can be avoided, as can the need to spread when conditions are unsuitable such as when the soil is wet. Although it will, inevitably, reduce farm income through reduced productivity, reducing the stocking density of an enterprise will reduce manure/slurry production altogether and therefore the quantity that needs to be applied to land (Cuttle *et al.*, 2007). Additionally, reducing stocking density will ease the pressures placed on farm storage systems. Ensuring the farm has sufficient manure and slurry storage capacity is important for several reasons. First, it allows for a greater degree of spreading flexibility as land managers may be forced to spread in unfavourable conditions when stores are full. It is also particularly important in areas designated as NVZs where there may be stringent controls on the amount of organic manures that can be spread, as well as closed periods during which it is not permitted at all.

Livestock waste production and in-field deposition can also be managed, to a degree, by the manipulation of livestock diets. Livestock in intensive systems receive considerably more phosphorus than they need, meaning that a significant proportion passes through the body. Similarly, the adsorption efficiency with which nitrogen enters the body is never 100 per cent such that some will always be present in deposited urine and excreta. If adsorption efficiencies for these nutrients can be increased, not only does it bode well for the environment but it also means lower costs and improved livestock productivity, both of which are good for the economic health of the farm. There are several feeding strategies that can be adopted to reduce nitrogen and phosphorus excretion and ensuring rations have a high degree of digestibility is important in this respect. Reducing the particle size of pig and poultry feed via pelleting, for example, will improve the digestibility of rations and reduce excretion by up to 15 per cent (Sutton and Beede, 2003). An increasingly popular

approach is to add enzymes such as phytase (Lewis *et al.*, 2015) to the diet. In many rations, phosphorus is bound to a substance called phytate and the phytase enzyme has been shown to increase phosphate availability in feed by releasing it from the phytate making it more available to the animal and so reducing the amount excreted, possibly by up to 33 per cent (Spiehs, 2005). Phase feeding is another option, whereby rations are formulated to better match the animal's growth stage and nutrient requirements. However, this will have practical limitations as it will mean animals need to be grouped together according to growth stage or weight during feeding.

Although the options discussed above may reduce the loss of potential pollutants from farmland, in the modern farming systems typical of many European countries, the need to apply substances to crops and/or the land is unlikely to diminish completely. Even in organic systems, manures and slurries are key to maintaining soil fertility, and a number of pest control products are approved for use, albeit a greatly reduced number (Soil Association, 2016; Weldon and McBride, 2009). Whenever any substance is spread or sprayed on agricultural land, therefore, it is vital to ensure that it is done in such a way as to minimise the potential for aquatic pollution. In particular, it is important to avoid high-risk locations and times, such as when rainfall is forecast, as this aids in the prevention of surface wash-off of pesticides, and organic and inorganic fertilisers, something which can be further helped if manures and slurries, in particular, are incorporated into the soil soon after application. Equally important is the use of appropriately calibrated application equipment, as this helps avoid uneven spreading, which may cause overapplication in some locations, whilst others may not receive enough. It also reduces the temptation to overapply in order to overcome the short-comings of an ineffective spreading/spraying pattern. There are also a number of options when it comes to product choice, which have the potential to reduce the risk of pollution. Slow release fertiliser pellets, in which nitrogen is in a less soluble form, for example, can reduce the risk that nitrogen will be released too fast for a crop to utilise it before it leaches away. Similarly, some pesticide formulations have been formulated to be more stable. For example, some metaldehyde slug pellets have been designed to ensure that they are eaten by pests rather than breaking down and releasing their contents into the environment. Conversely, some pesticide formulations have been designed to be less stable so that the active substance breaks down before it has time to reach a waterbody. Indeed, in the case of pesticides, most products are considerably less stable in the environment than was the case only a few years ago.

As far as point sources are concerned, the key issues which might need to be addressed revolve around the need to ensure that a site is prevented from becoming a pollutant source in the first place. For example, legislation establishes a number of rules which should be followed when storing potentially polluting substances on a farm, in particular silage, slurry, manure and fuel oil. In England, for example, this is enshrined in the Water Resources (Control of Pollution) (Silage, Slurry and Agricultural Fuel Oil) (England) Regulations 2010 (as amended) which seeks to ensure that stores of a number of potential pollutants are less likely to fail and cause a problem. In essence, the key requirements are to ensure that stores are not overfilled, that they of a sound construction, and in some cases capable of retaining effluent seepages (silage stores, manure heaps) and leakages. In following these rules, it is hoped that any potential pollution is retained at its

source such that no further measures are required. When it comes to the in-field point sources associated with livestock, such as the areas likely to become nutrient enriched as a result of the deposition of excreta and urine (e.g. around gates, trees and feeding/watering points), it is necessary to try and avoid such areas becoming a problem. Feed and water troughs should be moved regularly, for example, so that deposition occurs over a wider area and nutrients do not build up to unacceptable concentrations (Cuttle *et al.*, 2007). Similarly, field gates should, where possible, be located in areas less prone to poaching, and therefore erosion and leaching.

4.3.3.2. Pathway control

If the generation of pollutants cannot be avoided, and, in all probability, they will never be eliminated altogether, the next best thing is to prevent them from reaching either surface or groundwaters, by disconnecting the source from the receptor. This may be a simple matter of ensuring that pollutants are not deposited directly within surface waters (effectively reducing the pathway length to zero), with little or no chance for degradation. In an arable setting, for example, it is vital to ensure that neither fertilisers nor pesticides are spread or sprayed too close to the boundary of a waterbody, or in windy conditions liable to cause drift into non-targeted areas such as ditches, ponds or rivers. To this end, regulations in most European countries go a long way to preventing this type of poor practice. For example, the CAP cross compliance regulations restrict spreading of inorganic fertilisers and manures close to waterbodies and where an unacceptable risk exists due to the toxicity of a particular pesticide to aquatic organisms, pesticide product labels will specify a minimum width of no-spray zone to mitigate the potential for water pollution.

Figure 4.17: Beef cow drinking with direct access to surface water
(Photo courtesy of: Dr John Tzilivakis)

Within a grazing environment, livestock should, where possible, be prevented from gaining direct access to surface waters (see Figure 4.17). This may not be practical in some more extensive forms of agriculture, where livestock range over large areas, but in intensive grazing situations, it is advisable to fence off waterbodies, so that livestock cannot deposit manure and urine directly into them. This will significantly reduce the likelihood of nutrients and microbiological pollutants contaminating the aquatic environment. Given the considerable value to production of animals consuming naturally occurring water, it may be that naturally occurring sources can still be used albeit channelled in such a way as to ensure that contaminated flow is not returned to the main waterbody. For example, it could be used to supply drinking troughs within the field. Similarly, where animals need to cross watercourses during movement around the farm, it may, within planning constraints, be advisable to avoid the use of fords and replace them with bridges, since this again

prevents livestock from urinating and defecating in the water (Cuttle *et al.*, 2007).

In a similar vein, on-farm storage facilities for materials with the potential to cause pollution, such as silage, slurry, fuel oil, pesticides and fertilisers, need to be sited appropriately, to ensure that should leakages or spills occur, the likelihood that any release will find its way into the aquatic environment is minimised, and that there will be sufficient opportunity to deal with any problems that may occur. Although the exact regulations tend to vary with material stored and from country to country, typically this is at least 10 m from a surface waterbody with even greater distances required in relation to boreholes, wells and springs, typically 50 m or more. No store should be sited in a location prone to flooding. It is also essential to consider any man-made drainage systems that may be in place and therefore increase the ease with which substances can be passed to a waterbody. Many areas of hardstanding, for example, are associated with a rainwater drainage system that could result in leakages and spillages in that area getting into receiving waters much faster than would otherwise be the case, and with very little opportunity to do anything about it once it has gone down the drain. The washing of farm machinery contaminated with noxious substances, such as pesticides, presents a particular problem in this respect as in many cases this has traditionally been done in the farmyard, where run-off can get into drainage systems and subsequently surface waters. A number of solutions have been suggested to tackle this sort of contaminated run-off from hard surfaces, including soakaways, sedimentation boxes, infiltration trenches and biobeds, for example, each of which has its own merits. Soakaways as a method of dealing with run-off from various sorts of hard surface and allowing it slowly to infiltrate into the soil have been around for some time, and may be very good for assisting in lowering peak flows in watercourses, and in removing sediments and sediment-bound pollutants (Avery, 2012). They are, however, less useful in dealing with pollutants such as nitrogen, which are held in solution, and which may still leach into soil and groundwaters. Where run-off is likely to contain significant quantities of coarse sediments, some form of sedimentation box or trap in which material is given an opportunity to settle out before flow is passed onto the surface water network or other treatment system may be a good mitigation option. These systems will also offer benefits in relation to sediment-bound pollutants, although some very fine particles to which pollutants are preferentially bound, may pass through. On the other hand, biobeds and biofilters are retention systems/soakaways that are specifically designed to deal with pesticides in equipment washings and spillages during filling, and are generally associated with a dedicated equipment handling area. These systems consist of a pit, which is usually lined, filled with a mixture of straw, soil and compost which temporarily retains pesticides so that bacteria can break them down. Their effectiveness has been shown to be dependent on the specific chemicals involved but they can reduce levels of pesticides in waste waters very significantly (Spliid *et al.*, 2006).

When it comes to diffuse sources of pollution, anything which disrupts the passage of water, sediment and other pollutants from the field to the receiving waters, is likely to be of benefit in relation to at least some of the pollutants discussed above. Restricting or stopping flow through sub-surface field drainage networks will restrict the sub-surface transport of pollutants, and provide time for many (e.g. some pesticides) to degrade, whilst others (e.g. nutrients) may be taken up by vegetation, before they can be transported to receiving waters. However, this may have unwanted implications in terms of drainage, with land becoming significantly wetter, or it being later in the season before heavy farm machinery can go onto the land without damaging soil structure. This could have implications for productivity and erosion (see Chapter 5). Land drainage systems have usually been installed for very good reason, and not using them may, therefore, not be a viable option. In addition, it may result in negative environmental impacts, for example, nitrate leaching may increase and the risk of run-off and erosion will also be exacerbated.

Fortunately, there are a number of methods for reducing the movement of pollutants from fields that do not involve such a drastic change to its hydrological properties. These include methods designed to reduce erosion by preventing soil mobilisation (e.g. increasing soil organic matter, appropriate crop selection, contour ploughing, terraces, etc.) and those designed to promote the infiltration of overland flow and, therefore, any pollutants being carried in it. Within field grass, filter strips running along contours, for example, effectively reduce the slope length over which flow occurs and reduce its ability to carry sediment. They also filter out particulates and promote infiltration and so are useful for addressing sediments and sediment-bound pollution. They work best on relatively uniform, gentle slopes, where flow concentration is less of an issue, as this has a tendency to lead to strips being overwhelmed in certain areas, greatly reducing their efficacy. In some cases, areas of woodland can be planted to have a similar effect, with the added benefit of acting as a shelter belt (Avery, 2012).

Farm tracks, field tramlines and unvegetated spaces between row crops can become a key factor in pollutant delivery to waterbodies, since they have the potential to act as a conduit for the rapid movement of dissolved and sediment-adsorbed substances. Research has shown that up to 80 per cent of run-off from arable land is associated with unvegetated and compacted tramlines (Natural England, 2011). With respect to farm tracks, the problem can be minimised if the potential for overland flow is considered when tracks are being sited, although in most cases there is limited scope for moving them. That being the case, it is important to ensure that run-off is prevented from rapid waterbody entry by, for example, the use of cross drains in order to direct flow towards a soakaway, or other device for the control of pollution. Where tramlines are causing a problem, using low ground pressure tyres on equipment can help, as will disrupting the surface of tramlines with tines to allow the water to soak into the soil. Unvegetated strips between row crops, such as potatoes, can also act as conduits for excess water. This can be mitigated by growing catch crops in the row gaps, aligning crop rows at right angles to the normal direction of run-off and, especially where the problem is severe, considering changing the crop.

The final opportunity to disrupt the flow of water and its contained pollutants comes at the bank edge, where the use of riparian buffer strips has become a common site in European agricultural landscapes. These features have been promoted as a means of addressing a number of pollution problems over the years, as well as providing valuable biodiversity habitat (see Chapter 3). Many studies have shown that these types of buffers can significantly reduce diffuse pollution but their effectiveness will depend upon site conditions and buffer width. Such systems consist of a band of natural or semi-natural vegetation running alongside the bank of the protected waterbody, and although strictly speaking they do not alter the rate at which pollutants are lost from the productive area of the farm, they can be used

to intercept them before they reach a watercourse. Much of the early work was done in relation to their impact on nutrient pollution, with a range of processes interacting to reduce the nitrogen and phosphorus content of surface and sub-surface run-off. As in the case of in-field filter strips, sediments and the phosphorus attached to them enter the buffer strip in surface run-off, and are then filtered out by the vegetation, particularly grasses and other dense low-level vegetation, such that any nutrients present can then be taken up by that vegetation (Haycock and Muscutt, 1995). Dissolved nitrogen in sub-surface flow may also be taken up by growing vegetation and/or biologically transformed by bacteria into nitrogen gas through the process of denitrification, which generally occurs under wet anaerobic conditions. Indeed, some buffer strips are maintained, as much as possible, in a wet condition, so as to maximise the occurrence of this process (Avery, 2012). Subsequently, a number of studies have been undertaken to evaluate their usefulness in removing other pollutants with significant success being demonstrated in relation to microbial pathogens such as *Cryptosporidium parvum* oocysts, for example (e.g. Hussein *et al.*, 2008). There have also been good levels of success in relation to some pesticides; however, the diversity of their chemical properties means that efficacy is very variable, with sediment-adsorbed substances often being efficiently removed, whilst those found in solution are less so, but with removal nevertheless increasing with buffer strip width (Krutz *et al.*, 2005; Kay *et al.*, 2009). Indeed, a comprehensive review by Reichenberger *et al.* (2007) concluded that, as a 'rule of thumb', on average a 5 m buffer strip will have a pesticide reduction efficiency of 50 per cent rising to 97.5 per cent for a 20 m buffer width.

If buffer strips are to work to their maximum efficacy however, a number of conditions must be met. First, the presence of sub-soil drainage undercutting the buffer strip will significantly reduce the extent to which pollutants are removed, particularly those moving through the soil in solution, as they will tend to bypass the beneficial effects of the buffer. Equally, it is important that the vegetation is present at those times when the delivery of pollutants is expected to be at a maximum, which in some cases may be the autumn and winter months. Although for filtering purposes the vegetation may not necessarily be growing, for proper vegetative uptake, it must. Breaching can also be a problem, in that buffers work best when flow enters them fairly evenly across their length. If an uneven topography means that flow is focused in small areas, it can result in overland flow simply flooding vegetation, whilst sub-surface flow moves through too quickly for proper interception. Similarly, trampling and grazing by livestock can damage vegetation, reducing buffer strip effectiveness, so, if possible, large animals should be kept off. It should also be remembered that nutrients (and perhaps other pollutants) may build up within the buffer zone itself such that it becomes a potential pollutant reservoir in its own right. Contaminated and/or nutrient-enriched sediments may accumulate due to the deposition of fine particles from the field, which can have a significantly higher concentration of bound chemicals than the field soil in general. Regular management of the area, including cutting and removal of vegetation, will reduce the likelihood of this problem occurring for at least some pollutants.

4.3.3.3. Receptor management

Although it is always best to keep pollutants out of water resources wherever and whenever possible, it is highly likely that at least some agricultural pollutants will find their way to receiving waters. Where this happens

to a significant degree it may be necessary to adopt 'within-receptor' pollution management as the last resort. Most of the available methods for doing this concentrate on the provision of an environment within which sedimentation can occur and/or one in which chemical degradation or biological uptake can reduce contaminant levels, and all have the aim of ensuring that the pollutant levels present in the outflow to the system (i.e. that passed to the downstream drainage network) are lower than those at the inflow. Grassed waterways, for example, are broad, open channels within which dense grass or similar vegetation is grown. They are on the periphery of 'within-receptor' approaches as they are also a form of pathway control. These channels can either be established within fields in areas in which overland flow tends to concentrate or used as a means of controlling the movement of water around the farm. Being relatively broad, they tend to reduce the velocity of flow, encouraging the deposition of solids, whilst the vegetation acts as a filter for particulates, with trapped nutrients then being utilised by the growing vegetation, and other pollutants degrading before getting into the wider environment. Depending on the soil forming the bed, grassed waterways can also encourage infiltration, reducing overall run-off volumes, with benefits for both the volume of contaminated water being generated and for downstream flood alleviation (see Chapter 4.1.5). They tend to be established in areas of a field which might otherwise be prone to erosion and so the vegetative cover will help mitigate this, and therefore prevent additional pollutants becoming part of the problem. Nevertheless, where grassed waterways are created within agricultural fields (as opposed to amendments to existing intermittent waterways), they can result in significant areas of land being taken out of production, with all the economic implications that may have. For this reason, they have, to date, been less popular in parts of Europe than in areas of North America, where productive space is less of an issue; although, if land would otherwise be subject to erosion, they can still be a useful tool for controlling such damage.

Ditches and other watercourses can also be amended in ways which encourage the deposition of sediment and its associated contaminants, by installing seepage barriers, check dams and other structures, which are designed to retard flow, and lower its sediment-carrying capacity. Such systems may have limited effectiveness under high flow conditions, since they will generally be overwhelmed, with the bulk of the flow simply going over the top of them. However, they are much more effective under low flow conditions and will hold back at least some storm flow for slow release and some contamination will be removed from that portion of the flow. Deposited material must then be removed to prevent an excessive build-up of silt, although this can often be recycled to the land, so long as this is done with care so that the sediment is properly entrained within the soil. In this way, the nutrients contained in this material can be used as a contribution towards the crop's requirement. The same is true for systems such as those involving specifically designed detention/retention ponds and infiltration basins, which come in a variety of forms. Some are generally dry and are only intended to be used in times of overland flow generation, whilst others are wet most of the time and used on a permanent basis. However, regardless of the form they take, all are intended to store a portion of the flow generated and reduce flow velocities in order to encourage infiltration to groundwaters and the deposition of sediments. Consequently, they are particularly well suited to addressing the problems associated with sediments and sediment-bound

pollutants, but can also be effective against nitrogen if vegetative uptake can be encouraged and groundwater is not at risk

The use of constructed wetlands has been proposed for the treatment of water pollution of several types, in both urban and rural settings. These are specifically designed to take advantage of a number of biochemical processes in order to degrade and reduce the pollutant content of inflow waters. In doing so they emulate the sort of processes which have always formed part of the water purification system associated with natural aquatic ecosystems. Constructed wetlands designed for water treatment, as opposed to biodiversity management, for example, are formed of an integrated series of environments which allow a variety of processes to come into play, including those above and below ground, and related to plants and microorganisms. Typically, they involve a series of interconnected shallow pools with an impermeable bed, containing emergent vegetation (often referred to as reed beds) separated by areas in which surface or subsurface flow occurs (although this latter consideration is sometimes ignored in favour simply of a series of interconnected reed beds). They clean water through a number of different processes including the deposition and filtering out of particulates (sediment etc.), the uptake of nutrients by both plants and microorganisms, and microbial action in aerobic and anaerobic conditions (reduction of nitrogen and BOD). In addition, they encourage the precipitation of dissolved phosphorus with subsequent adsorption to bed sediments as well as the predation and die-off of pathogenic microorganisms (Carty *et al.*, 2008). These processes benefit if the retention time within the wetland system is maximised and/or the flow velocity minimised, providing time for processes to occur. Therefore, they are most efficient when the influent arrives at a moderate rate and with a relatively low level of contamination (either arable or livestock-based) and should therefore not be used for the treatment of highly contaminated run-off such as silage effluent or slurry, for example (Carty *et al.*, 2008). The advantage of such systems is that they are relatively cheap to install and operate, particularly if the local soil and geology mean that artificial liners are unnecessary, and come associated with a number of co-benefits, such as the provision of valuable habitat and the attenuation of floods. However, they do require a suitable topography, specifically gently sloping land, and a significant area of land to be set aside for them, as failure to do so may mean that retention times are inadequate for sufficient water cleaning to occur. It is also worth keeping in mind that in some climates and at times of the year (i.e. the winter in northern Europe) pollutant removal capabilities may be significantly reduced, due to a combination of shorter retention times and reduced vegetative growth, and this may be the point in time when pollutant run-off might peak. They can also become clogged by sediment if the content of the influent waters is too high, and can be overwhelmed by high flows during storm events, possibly becoming a source of pollution in their own right if retained sediments are then mobilised, something which is a concern in relation to any system designed to trap polluted sediments. These concerns can be minimised through appropriate design, and if necessary by connecting them to other systems (such as an initial sedimentation ponds), but nevertheless they should be considered at the outset, and it should not be assumed that no maintenance will be required, even though when appropriately sited and designed, ongoing maintenance needs should not be excessive.

Although the above can never hope to be an all-encompassing description of the techniques available for reducing the levels of pollution associated with European agricultural production, it nevertheless provides a broad introduction to the approaches available for on-farm implementation currently. In the future, however, a number of new options may become available, particularly in relation to dealing with specific problems, and be added to the toolbox of farmers, environmentalists, regulators and water companies alike.

4.3.4. Managing water at a catchment scale

In many cases, the waterbodies of a catchment, both above and below ground, are interconnected, and activities carried out in one part of the catchment can have a significant impact on the availability and quality of water in another. Physical changes to the river and channel morphology, changes in the management of the surrounding catchment and pollution events (e.g. run-off from farmland and roads, industry and sewage discharges, etc.), will all affect the ability of the catchment as a whole to support its various ecosystem services, including supporting biodiversity, flood protection and providing a clean water supply. The large-scale management of such properties is generally the responsibility of national and/or regional bodies (e.g. the Environment Agency in England), who are increasingly making use of a catchment-based approach to management, as promoted by the Water Framework Directive. Management based on a whole catchment perspective, essentially seeks to decentralise some of the management of the water environment to farmers, local communities and businesses (including water companies) and other interested groups, the aim being to prevent water pollution occurring, manage resources and deliver a range of other benefits for the economy and environment. The idea is to bring people together to enable a broader, more integrated approach to be taken in relation to identifying solutions to catchment-wide problems. As much of the land significant in this respect will be owned by farmers and private landowners, their cooperation and collaboration is vital if such an approach is to work. Essentially, the main focus of catchment management is to address diffuse nitrogen, phosphorus and pesticide pollution, minimise the impact on resource availability and ensure suitable flows for the environment, but the approach can also help with flood and biodiversity management more generally, as well as a range of other pollutants.

With respect to diffuse pollution from farms within a catchment, for example, the approach tends to revolve around raising awareness via the use of seminars, workshops, demonstration farms and targeted literature, websites, and so on. Often local water companies will employ an agronomist or other advisor, who can offer targeted advice and support for farmers. These companies may also provide financial or other incentives to try and induce farm-level changes in practices, so as to meet catchment objectives. For example, by subsidising the swap from one pesticide product to another less polluting, but perhaps more expensive, one. Similarly, many of the farm-scale measures, suggested earlier in this chapter as mechanisms for reducing water use, are promoted by catchment scale regulators and managers as a means to achieve the water resource conservation objectives necessary to maintain environmental flows and ensure supplies for other users.

There are also catchment-scale measures that can reduce the risk of flooding and, generally, the approach here will rely on a large number of small-scale interventions to reduce the magnitude and flashiness of the run-off generated within the catchment. Many of these interventions will need to be applied by individual farmers and land managers and indeed there is much that they can do at the farm scale to assist in delivering catchment-wide flood management benefits. Such activities might potentially include many of the techniques described in this book for reducing overland flow and minimising erosion, but also the recreation of wetlands, blocking drainage channels to increase water retention and the planting of trees within the catchment. It is also important that farmers undertake ditch management, including dredging and thinning of vegetation to ensure that the ditch has the capacity to hold excess water and that this water can freely flow through the farm's water network. Small-scale interventions such as these will not have much effect if just done on a few farms, however, but if implemented across a whole catchment the impact can be significant. Another option is to divert flood waters into woodlands or areas of open land where damage will be minimal. Indeed, in areas that are particularly prone to flooding, it has been suggested that agricultural land itself should be used to store flood waters. Initiatives such as 'Making Space for Water' in England and Wales (Defra, 2005) and 'Room for Rivers' in the Netherlands (Rijke *et al.*, 2012) have called for the reappraisal of land management on flood plains, such that they can be used to store flood waters and mitigate the risks posed to other parts of the catchment, particularly areas used for housing. Flooding the land in this manner can also have benefits for biodiversity, however, it is also highly controversial within the farming industry, as a result of the potential damage done to farming activities. As a result, it is probable that some form of incentive scheme will be required, if farmers and landowners are to agree to their land being used in this manner; that is unless it becomes a regulatory requirement.

References

Acreman, M. and Dunbar, M.J. (2004) Defining environmental river flow requirements – a review. *Hydrology and Earth System Sciences*, 8(5), 861–876. DOI: 10.5194/hess-8-861-2004

Alexandratos, N. and Bruinsma, J. (2012) *World Agriculture Towards 2030/2050: The 2012 Revision*. Food and Agriculture Organization of the United Nations. Rome.

AMEC (2014) *Environmental Flows: Learning Good Practice from Other Countries*. AMEC Environment and Infrastructure UK Limited, Newbury, UK.

Anglian Water (2016) *Love every drop: Got a spare £600million?* Anglian Water online. Available at: http://www.anglianwater.co.uk/news/got-a-spare-600million.aspx

Arimura, T.H., Hibiki, A and Katayama, H. (2008) Is a voluntary approach an effective environmental policy instrument? A case for environmental management systems. *Journal of Environmental Economics and Management*, 55(3): 281–295. DOI: 10.1016/j.jeem.2007.09.002

Avery, L.M. (2012) *Rural Sustainable Drainage Systems* (RSuDS). Environment Agency, Bristol, UK.

Ayars, J.E., Phene, C.J., Hutmacher, R.B., Davis, K.R., Schoneman, R.A., Vail, S.S. and Mead, R.M. (1999) Subsurface drip irrigation of row crops: a review of 15 years of research at the Water Management Research Laboratory. *Agricultural Water Management*, 42(1), 1–27. DOI: 10.1016/S0378-3774(99)00025-6

Baldock, D., Dwyer, J., Sumpsi, J., Varela-Ortega, C., Caraveli, H., Einschütz, S. and Petersen, J.E (2000) *The Environmental Impacts of Irrigation in the European Union*. Institute for European Environmental Policy, London.

Bendes, G. (2002) Bioenergy and water: the implications of large-scale bioenergy production for water use and supply. *Global Environmental Change*, 12(4), 253–271. DOI: 10.1016/S0959-3780(02)00040-7

Biointelligence Service (2012) *Water saving potential in agriculture in Europe: findings from the existing studies and application to case studies.* Final Report to the European Commission, DG Env.

Boeuf, B. and Fritsch, O. (2016) Studying the implementation of the Water Framework Directive in Europe: a meta-analysis of 89 journal articles. *Ecology and Society*, 21(2), Art 19. DOI: 10.5751/ES-08411-210219

Bonzanigo, L. and Sinnona, G. (2014) Present challenges for future water sustainable cities: a case study from Italy. *Drinking Water Engineering and Science*, 7(1), 35–40. DOI: 10.5194/dwes-7-35-2014

Bouraoui, F. and Grizzetti, B. (2014) Modelling mitigation options to reduce diffuse nitrogen water pollution from agriculture. *Science of the Total Environment*, 468, 1267–1277. DOI: 10.1016/j.scitotenv.2013.07.066

Bozzola, M. and Swanson, T. (2014) Policy implications of climate variability on agriculture: water management in the Po river basin, Italy. *Environmental Science and Policy*, 43, 26–38. DOI: 10.1016/j.envsci.2013.12.002

Brouwer, C., Prins, K. and Heibloem, M. (1989) *Irrigation Water Management: Irrigation Scheduling – Training Manual No. 4.* Food and Agriculture Organization of the United Nations, Rome.

Bumb, B.L. and Baanante, C.A. (1996) *World Trends in Fertilizer Use and Projections to 2020.* : International Food Policy Research Institute, Washington, DC.

Bundschuh, M., Schletz, M. and Goedkoop, W. (2016) The mode of bioturbation triggers pesticide remobilization from aquatic sediments. *Ecotoxicology and Environmental Safety*, 130, 171–176. DOI: 10.1016/j.ecoenv.2016.04.013

Bunn, S.E. and Arthington, A.H. (2002) Basic principles and ecological consequences of altered flow regimes for aquatic biodiversity. *Environmental Management*, 30(4), 492–507. DOI: 10.1007/s00267-002-2737-0

Bylund, G. (1995) *Dairy Processing Handbook*. Tetra Pak Processing Systems AB, Lund.

Camargo, J.A. and Alonso, A. (2006) Ecological and toxicological effects of inorganic nitrogen pollution in aquatic ecosystems: a global assessment. *Environment International*, 32(6), 831–849. DOI: 10.1016/j.envint.2006.05.002

Carty, A.H., Scholz, M., Heal, K., Dunne, E., Gouriveau, F. and Mustafa, A. (2008) *Constructed Farm Wetlands (CFW) – Design Manual for Scotland and Northern Ireland, Stirling, Scotland Belfast, Northern Ireland.* Scottish Environment Protection Agency and Northern Ireland Environment Agency.

Chambers, B., Nicholson, N., Smith, K., Pain, B., Cumby, T. and Scotford, I. (2001) *Managing Livestock Manures. Booklet 2. Making Better Use of Livestock Manures on Grassland.* MAFF, London, UK.

Chapagain, A.K. and Hoekstra, A.Y. (2004) *Water Footprints of Nations, vols. 1 and 2.* UNESCO–IHE Value of Water Research Report Series No. 16. Available at: http://waterfootprint.org

Chapman, P. (2014) Is the regulatory regime for the registration of plant protection products in the EU potentially compromising food security? *Food and Energy Security*, 3(1), 1–6. DOI: 10.2166/wp.2008.054

Collins, R. (2009) Water scarcity and drought in the Mediterranean. *Change Magazine.* Available at: http://www.changemagazine.nl/doc/jaargang_5_nummer_3/water-scarcity-and-drought-in-the-mediterranean.pdf

Collins, R., Kristensen, P. and Thyssen, N. (2009) *Water Resources Across Europe – Confronting Water Scarcity and Drought.* European Environment Agency, Copenhagen.

Cooke, S.J., Chapman, J.M. and Vermaire, J.C. (2015) On the apparent failure of silt fences to protect freshwater ecosystems from sedimentation: a call for improvements in science, technology, training and compliance monitoring. *Journal of Environmental Management*, 164, 67–73. DOI: 10.1016/j.jenvman.2015.08.033

Council of the European Union (1980) Council Directive 80/778/EEC of 15 July 1980 Relating to the Quality of Water Intended for Human Consumption. *Official Journal of the European Communities*, L 229: 11–29.

Council of the European Union (1998) Council Directive 98/83/EC of 3 November 1998 On the Quality of Water Intended for Human Consumption. *Official Journal of the European Communities*, L 330: 32–54.

Cuttle, S.P., Macleod, C.J.A., Chadwick, D.R., Scholefield, D., Haygarth, P.M., Newell-Price, P., Harris, D., Shepherd, M.A., Chambers, B.J. and Humphrey, R. (2007) *An inventory of methods to control diffuse water pollution from agriculture (DWPA).* User Manual (DEFRA Project ES0203), UK, 113 pp.

Dampney, P.M.R. and Sagoo, E. (2008) PLANET–the national standard decision support and record keeping system for nutrient management on farms in England. In: *Agriculture and Environment VII–Land Management in a Changing Environment. Proceedings of the SAC and SEPA Biennial Conference*, pp. 239–245.

de Fraiture, C., Giordano, M. and Liao, Y. (2008) Biofuels and implications for agricultural water use: blue impacts of green energy. *Water Policy*, 10(S1), 67–81. DOI: 10.2166/wp.2008.054

Defra (2005) *Making Space For Water: Taking Forward a New Government Strategy For Flood and Coastal Erosion Risk Management in England.* Product code PB10516, Defra publications, London.

Defra (2008) *Future Water: The Government's Water Strategy for England.* H.M. Government, London.

Defra (2009) *Protecting our Water, Soil and Air: A Code of Good Agricultural Practice for Farmers, Growers and Land Managers.* The Stationery Office, Norwich, UK.

Defra (2010) *The Fertiliser Manual (RB209)*, 8th Edition. TSO, Norwich.

Defra (2011) Water Usage in Agriculture and Horticulture Results from the Farm Business Survey 2009/10 and the Irrigation Survey 2010. Newport, Wales: Office for National Statistics.

Defra (2016) *Catchment sensitive farming: reduce agricultural water pollution.* [Online] Available at: https://www.gov.uk/guidance/catchment-sensitive-farming-reduce-agricultural-water-pollution [Accessed 11 January 2017].

Defra and Welsh Government (2013) *Managing

Abstraction and the Water Environment.* Defra, London.

Dufey, A. (2006) *Biofuels Production, Trade and Sustainable Development: Emerging Issues.*: International Institute for Environment and Development, London.

Dworak, T., Berlund, M., Laaswer, C., Strosser, P., Roussard, J., Grandmougin, B., Kossida, M., Kyriazopoulou, I., Berbel, J., Kolberg, S., Rodriguez- Diaz, J.A. and Montesinos, P. (2007) *EU water saving potential.* European Commission report, Brussels, Belgium.

EC – European Commission (2007) *Communication from the Commission to the European Parliament and the Council: Addressing the Challenge of Water Scarcity and Droughts in the European Union* – COM(2007) 414 Final. European Commission, Brussels.

EC – European Commission (2009) *EU Action on Pesticides: Our food Has Become Greener.* Directorate-General for Health and Food Safety, Brussels.

EC – European Commission (2010a) *Water Scarcity and Drought in the European Union.* European Union Publications Office, Brussels.

EC – European Commission (2010b) *Communication from the Commission: Europe 2020 – A Strategy for Smart, Sustainable and Inclusive Growth* – COM(2010) 2020 Final. European Commission, Brussels.

EC – European Commission (2011) *Communication from the Commission to the European Parliament, the Council, the European Economic and Social Committee and the Committee of the Regions: Roadmap to a Resource Efficient Europe* – COM(2011) 571 Final. European Commission, Brussels.

EC – European Commission (2012a) *European surface waterbodies affected by pollution pressures associated with agriculture.* Available at: http://ec.europa.eu/environment/water/water-framework/facts_figures/pdf/Agricultural_pressures2012.pdf

EC – European Commission (2012b) *Communication from the Commission to the European Parliament, the Council, the European Economic and Social Committee and the Committee of the Regions: A Blueprint to Safeguard Europe's Water Resources* – COM(2012) 673 Final. European Commission, Brussels.

EC – European Commission (2015) *Communication from the Commission to the European Parliament and the Council on the WFD and the Floods Directive: Actions towards the 'good status' of EU water and to reduce flood risks.* SWD (2015) 51 final. Commission Staff Working Document, Brussels.

ECOTEC Research and Consultancy, University of Hertfordshire, Central Science Laboratory and the University of Newcastle upon Tyne. (1999) *Design of a Tax or Charge Scheme for Pesticides.* ISBN 1 851121 60 9, DETR.

Edwards, A.C. and Withers, P.J.A. (2008) Transport and delivery of suspended solids, nitrogen and phosphorus from various sources to freshwaters in the UK. *Journal of Hydrology,* 350(3–4), 144–153. DOI: 10.1016/j.jhydrol.2007.10.053

EEA – European Environment Agency (2012) *Towards efficient use of water resources in Europe.* EEA Report 1/2012.

EEA – European Environment Agency and the WHO (2002) *Water and health in Europe: a joint report from the EEA and the WHO Regional Office for Europe.* World Health Organisation Regional Publications, European Series No. 93. ISBN 92 890 1360 5.

ENPI (2010) *Borders divide countries not rivers.* ENPI Info Centre, Feature no. 21, Available at: http://www.euneighbours.eu/files/features/FT21%20east%20Mold%20water%20%20EN.pdf

Environment Agency (1998) *Silage Pollution and How to Avoid It.* Environment Agency, Bristol, UK.

Environment Agency (2007) *The unseen threat to water quality: diffuse water pollution in England and Wales.* Report – May 2007. Environment Agency. Bristol, UK.

Environment Agency (2013a) *Managing Water Abstraction.* Environment Agency. Bristol, UK.

Environment Agency (2013b) *Significant Water Management Issues: Abstraction and Flow Problem.* Environment Agency. Bristol, UK.

Environment Agency (2016) *ENV15 – Water*

abstraction tables. Available at: https://www.gov.uk/government/statistical-data-sets/env15-water-abstraction-tables

Environment Agency, LEAF and NFU (2007) *WaterWise on the Farm – A Simple Guide to Implementing a Water Management Plan (Version 2.* Environment Agency. Bristol, UK.

Euractiv (2013) *Flooding costs farmers €1 billion as damage continues*. €urActiv.com website, Available online at: http://www.euractiv.com/section/agriculture-food/news/flooding-costs-farmers-1-billion-as-damage-tally-continues/

EUREAU (2009) EUREAU *Statistics Overview on Water and Wastewater in Europe 2008: Country Profiles and European Statistics*. European Federation of National Associations of Water and Wastewater Service, Brussels.

FAO – Food and Agriculture Organisation (2008) *The State of Food and Agriculture 2008 – BIOFUELS: prospects, risks and opportunities*. Food and Agriculture Organization of the United Nations, Rome.

FAO – Food and Agriculture Organisation (2009) *Global agriculture towards 2050*. High-level Expert Forum: How to feed the world 2050, Rome, 12–13 October 2009.

FAO – Food and Agriculture Organisation (2013) *Global map of irrigation areas in the United Kingdom*. Available at: http://www.fao.org/nr/water/aquastat/irrigationmap/gbr/GBR-gmia.pdf

FAO – Food and Agriculture Organisation (2014) *AQUASTAT: Water Withdrawal by Sector, Around 2007 – September 2014 Update*. Food and Agricultural Organisation of the United Nations, Rome.

FAO – Food and Agriculture Organisation (2016a) *Country factsheet: Spain*. Available at: http://www.fao.org/nr/water/aquastat/data/cf/readPdf.html?f=ESP-CF_eng.pdf

FAO – Food and Agriculture Organisation (2016b) *Country factsheet: United Kingdom*. Available at: http://www.fao.org/nr/water/aquastat/data/cf/readPdf.html?f=GBR-CF_eng.pdf

FAO – Food and Agriculture Organisation (2016c) FAOSTAT. Available at: http://www.fao.org/faostat/en/#home

Florke, M. and Alcamo, J. (2004) *European outlook on water use*. Final Report, EEA/RNC/03/007, 83. Centre for Environmental Systems Research, University of Kassel.

Fodor, N., Csathó, P., Árenás, T. and Németh, T. (2011) New environment-friendly and cost-saving fertiliser recommendation system for supporting sustainable agriculture in Hungary and beyond. *Journal of Central European Agriculture*, 12(1), 53–69. DOI: 10.5513/JCEA01/12.1.880

Foresight (2011) *The future of food and farming: final project report*. The Government Office for Science, London.

Fry, A. (2006) *Water: Facts and Trends*. World Business Council for Sustainable Development, Conches-Geneva, Switzerland.

Gallardo, M., Thompson, R.B., Giménez, C., Padilla, F.M. and Stöckle, C.O. (2014) Prototype decision support system based on the VegSyst simulation model to calculate crop N and water requirements for tomato under plastic cover. *Irrigation Science*, 32(3), 237–253. DOI: 10.1007/s00271-014-0427-3

Gippel, C.J. (2001) Hydrological analyses for environmental flow assessment. In: F. Ghassemi and P. Whetton (eds) *Proceedings MODSIM 2001*, International Congress on Modeling and Simulation, The Australian National University, Canberra. Canberra: Modeling and Simulation Society of Australia and New Zealand, pp. 873–880.

Gippel, C.J. and Stewardson, M.J. (1998) Use of wetted perimeter in defining minimum environmental flows. *Regulated Rivers: Research and Management*, 14(1), 53–67. DOI: 10.1002/(SICI)1099-1646(199801/02)14:1<53::AID-RRR476>3.0.CO;2-Z

Haycock, N.E. and Muscutt, A.D. (1995) Landscape management strategies for the control of diffuse pollution. *Landscape and Urban Planning*, 31(1–3), 313–321. DOI: 10.1016/0169-2046(94)01056-E

Heppell, J., Payvandi, S., Talboys, P., Zygalakis, K.C., Fliege, J., Langton, D., Sylvester-Bradley, R., Walker, R., Jones, D.L. and Roose, T. (2016) Modelling the optimal phosphate fertiliser and soil management strategy for crops. *Plant and Soil*, 401(1–2), 135–149. DOI: 10.1007/s11104-015-2543-0

Hess, T.M. and Knox, J.W. (2013) Water savings in irrigated agriculture: a framework for assessing technology and management options to reduce water losses. *Outlook on Agriculture*, 42(2), 85–91. DOI: 10.5367/oa.2013.0130

Hess, T.M., Knox, J.W. and Kay, M. (2008) *Managing Water Better: The Agronomic, Economic and Environmental Benefits of Good Irrigation Practice*. UK Irrigation Association, Cranfield, UK.

Hess, T., Lankford, B., Lillywhite, R., Cooper, R., Challinor, A., Sutton, P., Brown, C., Meacham, T., Benton, T. and Noble, A. (2015) *Water use in our food imports*. Farming and Water Report 3. Global Food Security Programme. Online, Available at: http://www.foodsecurity.ac.uk/assets/pdfs/water-used-in-imports-report.pdf

HM Government (2011) *Water for Life*. The Stationery Office, Norwich, UK.

Holden, J., Haygarth, P.M., MacDonald, J., Jenkins, A., Sapiets, A., Orr, H.G., Dunn, N., Harris, B., Pearson, P.L., McGonigle, D. and Humble, A. (2015) *Farming and Water: Sub-Report 1 – Agriculture's Impacts on Water Quality*. BBSRC, Swindon, UK.

Holman, I.P., Howden, N.J.K., Whelan, M.J., Bellamy, P.H., Rivas-Casado, M. and Willby, N.J. (2008) *An improved understanding of phosphorus origin, fate and transport within groundwater and the significance for associated receptors*. SNIFFER Project WFD85 Final Report.

Horst, M.G., Shamutalov, S.S., Gonçalves, J.M. and Pereira, L.S. (2007) Assessing impacts of surge-flow irrigation on water saving and productivity of cotton. *Agricultural Water Management*, 87(2), 115–127. DOI: 10.1016/j.agwat.2006.06.014

Hussein, J., Ghadiri, H., Lutton, M., Smolders, A. and Schneider, P. (2008) The effect of flow impedance on deposition of Cryptosporidium parvum oocysts with or without a vetiver buffer strip. *Soil Biology and Biochemistry*, 40(10), 2696–2698. DOI: 10.1016/j.soilbio.2008.06.022

IPCC (2013) Climate Change 2013: *The Physical Science Basis. Contribution of Working Group I to the Fifth Assessment Report of the Intergovernmental Panel on Climate Change*. Cambridge University Press, Cambridge, United Kingdom and New York, USA.

Irmak, S., Odhiambo, L.O., Kranz, W. L. and Eisenhauer, D.E. (2011) *Irrigation Efficiency and Uniformity, and Crop Water Use Efficiency*. University of Nebraska, Lincoln: EC732.

Jain, S.K. (2012) Assessment of environmental flow requirements. *Hydrological Processes*, 26(22), 3472–3476. DOI: 10.1002/hyp.9455

Karamanos, A., Aggelides, S. and Londra, P. (2007) Water use efficiency and water productivity in Greece. In: N. Lamaddalena, Shatanawi, M (eds.) *Water Use Efficiency and Water Productivity: WASAMED Project*. CIHEAM, Bari, pp. 91–99.

Kay, P., Edwards, A.C. and Foulger, M. (2009) A review of the efficacy of contemporary agricultural stewardship measures for ameliorating water pollution problems of key concern to the UK water industry. *Agricultural Systems*, 99(2–3), 67–75. DOI: 10.1016/j.agsy.2008.10.006

King, R. (2011) *Crop Production Technology: The Effect of the Loss of Plant Protection Products on UK Agriculture and Horticulture and the Wider Economy*. The Andersons Centre, Melton Mowbray, UK.

Klaar, M.J., Dunbar, M.J., Warren, M. and Soley, R. (2014) Developing hydroecological models to inform environmental flow standards: a case study from England. *Wiley Interdisciplinary Reviews: Water*, 1(2), 207–217. DOI: 10.1002/wat2.1012

Klaus, S., Kreibech, H., Merz, B., Kuhlmann, and Scroter, K. (2016) Large-scale, seasonal flood risk analysis for agricultural crops in Germany. *Environmental Earth Sciences*, 75, 1289. DOI: 10.1007/s12665-016-6096-1

Knox, J. and Hess, T. (2014) *A Water Strategy for UK Horticulture: Technical Report*. Agriculture and Horticulture Development Board, Kenilworth, UK.

Knox, J. and Kay, M, (2011) *Save Water and*

Money – Irrigate Efficiently. UK Irrigation Association, Cranfield, UK.

Knox, J., Kay, M. and Weatherhead, K. (2008) *Switching Irrigation Technologies: Does Switching from Rainguns to Booms, Sprinklers or Trickle Save Water, Energy and Money – Fact or Fiction?* UK Irrigation Association, Cranfield, UK.

Knox, J., Weatherhead, K., Rodríguez Díaz, J. and Kay, M. (2009) Developing a strategy to improve irrigation efficiency in a temperate climate: a case study in England. *Outlook on Agriculture*, 38(4), 303–309. DOI: 10.5367/000000009790422160

Krutz, L.J., Senseman, S.A., Zablotowicz, R.M. and Matocha, M.A. (2005) Reducing herbicide runoff from agricultural fields with vegetative filter strips: a review. *Weed Science*, 53(3), 353–367. DOI: 10.1614/WS-03-079R2

Kundzewicz, Z.W., Mata, L.J., Arnell, N.W., Doll, P., Kabat, P., Jimenez, B., Miller, K., Oki, T., Zekai, S. and Shiklomanov, I. (2007) Freshwater resources and their management. In: M. L. Parry (eds). *Climate Change 2007: Impacts, Adaptation and Vulnerability. Contribution of Working Group II to the Fourth Assessment Report of the Intergovernmental Panel on Climate Change*. Cambridge University Press, Cambridge, UK, pp. 173–210.

LEAD and FAO (2006) *Livestock's Long Shadow: Environmental Issues and Options*. Food and Agriculture Organization of the United Nations, Rome.

Le Goffe, P. (2013) *The Nitrates Directive, Incompatible with Livestock Farming? – The Case of France and Northern European Countries*. Jacques Delors Institute, Paris and Milan.

Lewis, K.A., Tzilivakis, J., Green, A. and Warner, D.J. (2015) The potential of feed additives to improve the environmental impact of European livestock farming: a multi-issue analysis. *International Journal of Agricultural Sustainability*, 13(1), 55–68. DOI: 10.1080/14735903.2014.936189

Machet, J.M., Dubrulle, P., Damay, N., Duval, R., Recous, S. and Mary, B. (2007) *Azofert: a new decision support tool for fertiliser N advice based on a dynamic version of the predictive balance sheet method*. In: 16th International Symposium of the International Scientific Centre of Fertilizers, Gand (BEL), pp. 16–19.

Martin, D.L., Stegman, E.C. and Fererese, E. (1990) Irrigation scheduling principals. In: *Management of Farm Irrigation Systems*. American Society of Agricultural Engineering, St Joseph, MT, pp. 155–203.

Massarutto, A. (1999) *Agriculture, water resources and water policies in Italy*. Available at: http://www.feem.it/userfiles/attach/Publication/NDL1999/NDL1999-033.pdf

Mathews, R. and Richter, B.D. (2007) Application of the indicators of hydrologic alteration software in environmental flow setting. *Journal of the American Water Resources Association*, 43(6), 1400–1413. DOI: 10.1111/j.1752-1688.2007.00099.x

McCarthy, C. and Morling, D. (2015) *Using regulation as a last resort? Assessing the performance of voluntary approaches*. Royal Society for the Prevention of Birds, Sandy, Bedfordshire.

Milk Development Council (2007) *Effective Use Of Water On Dairy Farms*. Milk Development Council, Cirencester, UK.

Monsanto (2016) *AQUATEK helps farmers save water, energy and time*. Available at: http://monsantoblog.eu/aquatek-helps-farmers-save-water-energy-time-enit/#.WK2pY02mk2c

Morris, J and Brewin P. (2013) Impact of seasonal flooding on agriculture: Spring 2012 floods in Somerset, England. *Journal of Flood Risk Management*, 7(2), 128–140. DOI: 10.1111/jfr3.12041

Mubareka, S., Maes, J., Lavalle, C. and de Roo, A. (2013) Estimation of water requirements by livestock in Europe. *Ecosystem Services*, 4, 139–145. DOI: 10.1016/j.ecoser.2013.03.001

Natural England (2011) *Protecting water from agricultural run-off: an introduction*. Natural England Technical Information Note TIN098, 1st Edition. Natural England Publishing, Sheffield.

Nilsson, S., Langaas, S. and Hannerz, F. (2004) International river basin districts under the EU Water Framework Directive: identification and planned cooperation. *European Water Management Online*, 2, 1–20.

Nyholm, T., Christensen, S. and Rasmussen, K.R.(2002) Flow depletion in a small stream caused by ground water abstraction from wells. *Groundwater*, 40(4), 425–437. DOI: 10.1111/j.1745-6584.2002.tb02521.x

Olesen, J. and Bindi, M. (2002) Consequences of climate change for European agricultural productivity, land use and policy. *European Journal of Agronomy*, 16(4), 239–262. DOI: 10.1016/S1161-0301(02)00004-7

Pallas, P.H. (1986) *Water for Animals AGL/ MISC/4/85*. Food and Agriculture Organization of the United Nations, Rome.

PAN Europe (2007) *Pesticide Use Reduction Strategies in Europe: Six Case Studies*. Pesticide Action Network Europe, London, UK.

Poff, N.L., Allan, J.D., Bain, M.B., Karr, J.R., Prestegaard, K.L., Richter, B.D., Sparks, R.E. and Stromberg, J.C. (1997) The natural flow regime. *BioScience*, 47(11), 769–784. DOI: 10.2307/1313099

Poláková, J., Farmer, A., Berman, S., Naumann, S., Frelih-Larsen, A. and von Toggenburg J. (2013) *Sustainable management of natural resources with a focus on water and agriculture: study – final report*. European Union, Brussels.

Porter, J.R., Xie, L., Howden, M., Iqbal, M.M., Lobell, D., Travasso, M.I., Garrett, K., Lipper, L., McGrath, J., Aggarwal, P. and Hakala, K. (2014) Food security and food production systems. In: C.B. Field, y otros edits. *Climate Change 2014: Impacts, Adaptation, and Vulnerability. Part A: Global and Sectoral Aspects*. Contribution of Working Group II to the Fifth Assessment Report of the Intergovernmental Panel on Climate Change. Cambridge University Press: Cambridge, UK and New York, USA, pp. 485–533.

POST (2012) *POSTNOTE No. 419: Water Resources Resilience*. The Parliamentary Office of Science and Technology, London.

Rahn, C., De Neve, S., Bath, B., Bianco, V., Dachler, M., Cordovil, C., Fink, M., Gysi, C., Hofman, G., Koivunen, M., Panagiotopoulos, L., Poulain, D., Ramos, C., Riley, H., Setatou, H., Sorensen, J., Titulaer, H., Weier, U. (2001) A comparison of fertiliser recommendation systems for cauliflowers in Europe. *Acta Horticulturae*, 563, 39–45.

Rahn, C.R., Zang K., Lillywhite, R.D., Ramos, C., De Paz, J.M., Doltra, J., Riley, H., Fink, M., Nendel., C., Thorup-Kristensen, K. and Pedersen, A. (2007) *Development of a model based decision support system to optimise nitrogen use in horticultural crop rotations across Europe – EU ROTATE N*. Final Scientific Report QLK5-2002-01100. Wellesbourne, UK. DOI: 10.17660/ActaHortic.2001.563.3

Reichenberger, S., Bach, M., Skitschak. A. and Frede, H-G. (2007) Mitigation strategies to reduce pesticide inputs into ground- and surface-waters and their effectiveness: a review. *Science of the Total Environment*, 384(1–3), 1–35. DOI: 10.1016/j. scitotenv.2007.04.046

Richter, B.D., Warner, A.T., Meyer, J.L. and Lutz, K. (2006) A collaborative and adaptive process for developing environmental flow recommendations. *River Research and Applications*, 22(3), 297–318. DOI: 10.1002/ rra.892

Richter, S., Völker, J., Borchardt, D. and Mohaupt, V. (2013) The water framework directive as an approach for integrated water resources management: results from the experiences in Germany on implementation, and future perspectives. *Environmental Earth Sciences*, 69(2), 719–728. DOI: 10.1007/ s12665-013-2399-7

Rijke, J., van Herk, S., Zevenbergen, C. and Ashley, R. (2012) Room for the river: delivering integrated river basin management in the Netherlands. *International Journal of River Basin Management*, 10(4), 369–382. DOI: 10.1080/15715124.2012.739173

Rogers, D.H. and Sothers, W.M. (1995) *Kansas State University Cooperative Extension, Irrigation Management Series: Surge Irrigation*. Available at: https://www.bookstore.ksre.ksu.edu/pubs/ L912.pdf

Rosegrant, M.W., Ringler, C. and Zhu, T. (2009a) Water for Agriculture: Maintaining Food Security under Growing Scarcity. *Annual Review of Environment and Resources,*

34, 205–222. DOI: 10.1146/annurev. environ.030308.090351

Rosegrant, M.W., Fernandez, M. and Sinha, A. (2009b) Looking into the future for agriculture and AKST (Agric. Knowledge, Sci. Technol.). In: *Agriculture at a Crossroads – Global Report: International Assessment of Agricultural Knowledge Science and Technology.* Island Press, Washington, D.C., pp. 307–376.

SAI – Sustainable Agriculture Initiative (2015) *Water Stewardship in Sustainable Agriculture: Farm and Catchment Level Assessments. A publication of the Water Committee of SAI Platform, Brussels EUREAU, 2009. EUREAU Statistics Overview on Water and Wastewater in Europe 2008: Country Profiles and European Statistics.* European Federation of National Associations of Water and Wastewater Services, Brussels.

Santato, S., Mysiak, J. and Pérez-Blanco, C. D. (2016) The Water Abstraction License Regime in Italy: a case for reform? *Water*, 8(3), 103–118. DOI: 10.3390/w8030103

Savic, R., Ondrasek, G. and Josimov-Dundjerski, J. (2015) Heavy metals in agricultural landscapes as hazards to human and ecosystem health: a case study on zinc and cadmium in drainage channel sediments. *Journal of the Science of Food and Agriculture*, 95(3), 466–470. DOI: 10.1002/jsfa.6515

Scottish Executive (2005) *Prevention of Environmental Pollution from Agricultural Activity: A Code of Good Practice.* Scottish Executive. Edinburgh.

SEPA (1997) *Silos and Silage Effluent.* SEPA, UK, Stirling.

Shiklomanov, I. (1993) World fresh water resources. In: P.H. Gleick (Ed.) *Water in Crisis: A Guide to the World's Fresh Water Resources.* Oxford University Press, New York, pp. 13–24.

Smith, R., Baillie, C. and Gordon, G. (2002) Performance of travelling gun irrigation machines. In: *Proceedings of the Australian Society of Sugar Cane Technologists*, pp. 235–340. PK Editorial Services; 1999.

Soil Association (2016) *Soil Association Organic Standards: Farming and Growing.* Soil Association, Bristol, UK.

Spiehs, M.J. (2005) *Nutritional and Feeding Strategies to Minimize Nutrient Losses in Livestock Manure.* University of Minnesota Extension Service.

Spliid, N.H., Helweg, A. and Heinrichson, K. (2006) Leaching and degradation of 21 pesticides in a full-scale model biobed. *Chemosphere*, 65(11), 2223–2232. DOI: 10.1016/j.chemosphere.2006.05.049

Strzepek, K., McCluskey, A., Boehlert, B., Jacobsen, M. and Fant, C. (2011) Climate variability and change: a basin scale indicator approach to understanding the risk to water resources development and management. *Water Papers 67338,* World Bank.

Suffolk Coast and Heaths AONB (2010) *Farm Reservoir Design Guide: A Guide to Good Planning and Design of Farm Reservoirs in the Suffolk Coast and Heaths Area of Outstanding Natural Beauty.* Suffolk Coast and Heaths AONB Unit, Melton, Woodbridge, Suffolk, UK.

Sutton, A. and Beede, D. (2003) *Feeding Strategies to Lower Nitrogen and Phosphorus in Manure.* Best Environmental Management Practices, Michigan State University Extension Service.

Tharme, R.E. (2003) A global perspective on environmental flow assessment: emerging trends in the development and application of environmental flow methodologies for rivers. *River Research and Applications*, 19(5–6), 397–441. DOI: 10.1002/rra.736

Tourism Emilia Romagna (2017) *Taro River Regional Park.* Official Tourist Information Site. Available at: http://www.emiliaromagnaturismo.com/en/apennines-and-nature/parks/taro.html (Last accessed: 16/10/17).

Tsakiris, G., Nalbantis, I. and Pistrika, A. (2009) Critical technical issues on the EU flood directive. *European Water*, 25(26), 39–51.

UN – United Nations (2014) *Concise report on the World Population Situation in 2014.* United Nations, Department of Economic and Social Affairs Population Division, Report number: ST/ESA/SER.A/354. United Nations, New York, ISBN 978-92-1-151518-3.

University of Warwick (2006) Sector: Livestock. In: *Technology to Improve Water Use Efficiency*

– *Report on Defra Project WU0123.* Warwick University, Coventry, UK.

Van Beilen, N. (2016) *Effects of Conventional and Organic Agricultural Techniques on Soil Ecolog.* Center for Development and Strategy, Getzville, New York.

Van der Molen, W.H., Beltrán, J.M., and Ochs, W.J. (2007) *Guidelines and Computer Programs for the Planning and Design of Land Drainage Systems,* Vol. 62. Food and Agriculture Organisation.

Wang, L., Butcher, A.S., Stuart, M.E., Gooddy, D.C. and Bloomfield, J.P. (2013) The nitrate time bomb – a numerical way to investigate nitrate storage and lag time in the unsaturated zone. *Environmental Geochemistry and Health,* 35(5), 667–681. DOI: 10.1007/ s10653-013-9550-y

Waterwise (2012) *Water – The Facts: Why do we need to think about water?.* Available at: http:// www.waterwise.org.uk/data/resources/25/ Water_factsheet_2012.pdf.#

Weatherhead, E.K., Knox, J.W., Daccache, A., Morris, J., Kay, M. and Groves, S. (2014) *Water for Agriculture: Collaborative Approaches and On-farm Storage – FFG1112 Final Report, March 2014.* Cranfield University, Cranfield, UK.

Weldon, J. and McBride, V. (2009) *Production*

Standard.: Organic Food Federation, Swaffham, Norfolk, UK.

Wheater, H. and Evans, E. (2009) Land use, water management and future flood risk. *Land Use Policy,* 26(S1), S251–S264. DOI: 10.1016/j. landusepol.2009.08.019

World Bank (2013). *Agricultural land (% of land area).* Available at: http://data.worldbank.org/ indicator/AG.LND.AGRI.ZS

World Bank (2015) *Rural population (% of total population).* Available at: http://data. worldbank.org/indicator/SP.RUR.TOTL.ZS

World Bank Group (2016) *Global monitoring report 2015/2016: development goals in an era of demographic change.* World Bank, Washington, D.C.

Wriedt, G., Van der Velde, M., Aloe, A. and Bouraoui, F. (2008) *Water Requirements for Irrigation in the European Union: A Model Based Assessment of Irrigation Water Requirements and Regional Water Demands in Europe.* Joint Research Centre, Ispra, Italy.

WVZ Maifeld-Eifel (2016) *Maifelder Landwirte kooperieren mit Wasserversorger.* Available at: http://www.wvz-maifeld-eifel.de/ info/346/Ueber-uns/News-_-Presse/News- -_-Pressearchiv/Maifelder-Landwirte- kooperieren-mit-Wasserversorger.html

CHAPTER 5

Resource management and productivity

5.1. Setting the scene

5.1.1. Introduction

Seeking improvement in agricultural productivity in a sustainable manner is an important aspect of agri-environmental policy across Europe. The aim is to secure farm incomes, improve rural development and economic growth and to help ensure Europe is food self-sufficient in the future without the need to rely heavily on imports. As discussed in Chapter 4, a fundamental driver of this is the huge growth in population anticipated over the next 30 or so years. Reports suggest that the current world population of 7.2 billion may increase by more than a third (UN, 2014). There are many reasons for this, including increasing numbers of people living to reproduction age and so greater fertility within the population, and longer life expectancies due to improved health care, hygiene, nutrition and medical advances. Accelerating migration is also adding to global population growth as movement away from areas prone to natural disasters, from war zones and poverty-stricken areas, increases survival rates and lifespan and the improved quality of life increases fertility. Similarly, increasing urbanisation, particularly in poorer countries, tends to

increase populations as birth rates rise and death rates fall due to access to better health care and nutrition. Education and government policy also play a vital role in determining birth and death rates in a country. Simple programmes in hygiene can impact upon death rates, while family planning and education can reduce birth rates. Along with a significant population expansion comes an equally significant increase in the demand for food. Projections indicate that to feed 9.6 billion people overall, food production needs to increase by around 70 per cent (FAO, 2009). According to the FAO (2010) there are close to a billion undernourished people in the world and so food security is a growing global issue. Therefore, it is easy to see that food security will be high on political agendas in the coming decades. This issue is also discussed in Chapter 4 in the context of demand for water.

The concept of 'food security' has been developing both scientifically and politically since the 1970s. The initial focus was concerned with food supply problems and the need to ensure the availability of basic foodstuffs at the international and national level. The World Food Summit, held in Rome by the UN in 1974, established a suite of institutional agreements and processes

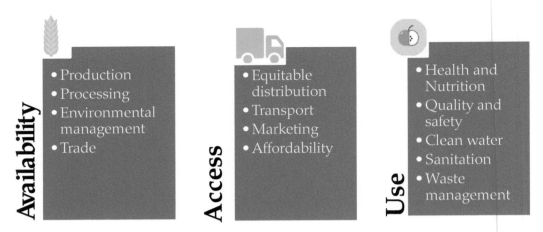

Figure 5.1: Three pillars of food security

covering information, resources for promoting food security and forums for dialogue on policy issues (ODI, 1997). Food security was defined during the Summit as 'availability at all times of adequate world food supplies of basic foodstuffs to sustain a steady expansion of food consumption and to offset fluctuations in production and prices'.

Thus, food security was described as having three pillars (Figure 5.1) as outlined below (FAO, 2003; Quisumbling *et al.*, 1995).

1. **Food availability**. All humans should have an adequate supply of food on a consistent basis.
2. **Food access.** This is related to economics and is the ability of nations, communities and individuals having sufficient financial or trade resources to obtain appropriate food to ensure a healthy, nutritious diet for all.
3. **Food use**. This is an acknowledgement that education plays a part in food security as there is a need for individuals to have a basic understanding of nutrition and sanitation, and to appreciate the consequences of waste.

In 1983, the FAO (2005) expanded this food security concept, adding a fourth pillar: 'stability', which incorporated into the original definition issues relating to price stability and securing adequate incomes for vulnerable people such that they can afford to purchase the food they need. Food supply can be affected by many different things such as weather variability, crop failures, price fluctuations and other economic factors, as well as political issues and by including stability in the food security definition, the FAO acknowledged the complexity of the problem. By the middle of the 1990s food security was a growing concern and was increasingly linked with food safety and nutritious diets such that during the 1996 World Food Summit, 'food security' was re-defined as 'a status where all people on planet Earth have, at all times, access to sufficient, safe, nutritious food to ensure a healthy and active life'.

Food security is now considered by most to be a complex sustainable development issue as it is linked to environment (production and consumption impacts), economics (commodity prices, trade, economic development) and societal well-being (health, malnutrition, food-borne diseases).

There is intense scientific and political debate, as well as some controversy, around food security. Many argue that there is enough food in the world to sustain current population levels as well as satisfy nutritional needs and that the problem is actually a distribution failure due to both political and logistical issues. Simply put, food does not always reach those that need it whilst in other areas food is so plentiful that it is casually wasted. The global rise in some food commodity prices coupled with natural disasters (such as floods and famine) and man-made disasters (such as war and mass migration) have resulted in some populations struggling to gain access to the food they need. In times of shortages, prices increase and it is the richer nations that have the greatest ability to secure food requirements from world markets – a luxury poorer nations do not have. It is obvious that the need to address food governance is just as pressing as the need to increase food production. Thus, action is needed to address the affordability and access to food, food distribution and to minimise food loss and wastage throughout the entire food supply and consumption chain (Lean *et al.*, 1990). Nevertheless, the predicted population growth does pose significant questions regarding the world's capacity to satisfy future food requirements. Therefore, taking steps to increase agricultural productivity and so the amount of food available is a sensible, forward-looking policy.

5.1.2. *Growing demand*

Increasing productivity is not just about producing more but also about producing more of the right foods. Over the next 30 years or so, the projected population increase is expected to drive up demand for cereals for food and animal feed by more than 40 per cent and potentially higher if global demand for biofuels also continues to grow. Cereals are still the mainstay of human diets but there is evidence to suggest that demand for protein-rich foods will increase in coming years as consumers in developing countries gradually adopt a more western diet, rich in protein, fats and sugars as opposed to starches. The demand for these types of commodities, especially meat, dairy and vegetable oil, tends to be linked to the growth in incomes of developing countries. Consequently, demand may increase much quicker for these commodities than that for cereals (FAO, 2009). Evidence continues to mount to support these conclusions and the OECD–FAO (2014) predicted an annual rise in global meat consumption of 1.6 per cent, with the demand for meat from China and India, for example, potentially rising by 80 per cent by 2022 due to a new (and growing) middle class. Africans are also starting to eat more meat, although their supply and demand is not yet comparable with other parts of the world. According to the UN, production of all the world's top ten major commodities increased significantly during the period 1969 to 2009 with the demand for soybeans increasing from 42 million tonnes to 223 million tonnes, an increase of 431 per cent; wheat and rice more than doubled in production during the same period (OECD–FAO, 2014).

Europeans are significant consumers of meat and are second after the United States in beef consumption and fourth behind the United States, Japan and Russia for beef imports. As demand for meat increases, there will be a concurrent increase in demand for livestock feed, with the OECD–FAO (2014) predicting that we will need to produce an extra 160 million tonnes by 2023. Figure 5.2 shows how the consumption of various agricultural commodities has increased within the EU-27 over the last two decades or so. However, as can be seen in the graph, consumption for some products appears to have somewhat levelled off. Indeed, in the UK,

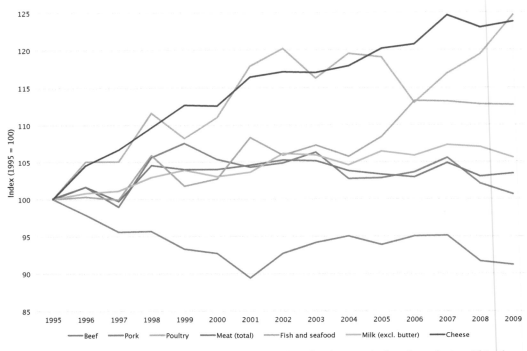

Figure 5.2: Changes over time in consumption per capita of various agricultural products within the EU-27

(Source: European Environment Agency)

meat consumption has fallen by around 13 per cent since 2007, although trends vary significantly depending on the type of meat. For example, consumption of chicken has increased whilst that of beef products has declined.

Although the increased demand for protein-rich foods is linked with increasing prosperity which can only be seen as a positive, it is expected to drive prices for these commodities higher (e.g. meat and dairy) compared to crop prices although the price of coarse grains and oilseeds used for feed will probably also see sharp increases.

5.1.3. Agricultural productivity

The main goal of every farmer, whether he farms a few olive trees, runs a small family farm or big industrial facility, is to have a productive business. From a farmer's perspective, productivity is crucial if the business is to stay viable and provide a reasonable income level. At a national and European level, productivity provides vital information on the competitiveness of the sector and on food security. Agricultural productivity is a measure of how efficient the sector, or indeed a farm, is at producing desirable products. However, measuring it in a meaningful way is not so straightforward. Using an economic definition, farm or agricultural productivity is the ratio of agricultural outputs (e.g. crops, milk, eggs, beef, etc.) to the agricultural inputs (e.g. fertilisers, pesticides, energy, water, land, labour, etc.) required to produce them. Based on this approach, the total factor productivity (TFP) index is often used at national and international level and, in this context, 'total production inputs' is

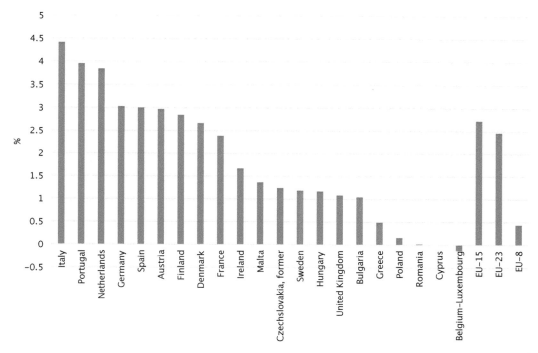

Figure 5.3: Productivity in the EU as measured using TFP
(Created from: Matthews, 2014)

usually taken to refer to land area, man-hours of labour, livestock numbers and the quantity of chemical fertilisers applied, these being estimated by one means or another. According to data produced by the United States Department of Agriculture for the EU, productivity growth during the period 2001–2010 in the new Member States (MSs) has been consistently lower than in the old MSs (Matthews, 2014).

At the farm level, productivity can be expressed more accurately as it is easier to quantify the inputs and outputs with a degree of certainty. However, productivity can also be expressed using other measures. For example, land is the most significant resource a farmer has and delimits the maximum productivity for a crop and variety being grown. Therefore, one of the key productivity indicators used at farm level is crop output (e.g. yield) per area of land, however,

this does not relate to farm economics which is a fundamental parameter in productivity and so the yield per hectare indicator is usually viewed side by side with a cost–efficiency indicator which compares the income from the crop with the cost of the inputs used to produce it.

Productivity can be taken further by investigating returns on investments and medium-term income forecasting, but whichever technique is used or whether it is at national or field level it is clear that the key to increasing agricultural productivity is to either increase production without increasing, or at least not to the same degree, the use of inputs, or to decrease the use of inputs but still achieve the same amount of production. Both are extremely difficult, particularly as productivity should not be increased at the expense of the environment. Consequently, considering these constraints, the aspiration of policymakers is to find ways to increase yields without adversely impacting on the environment and without the cultivation of more land, that is, we need to produce more food using less resources (e.g. land, energy, water, chemical inputs etc.).

Studies have shown that farming systems based on high levels of inputs tend to be considerably more productive (CWFT, 1999) than low-input systems. Farmers maximise their yields by protecting crops from pests and diseases, and by ensuring they have enough nutrients and water. With respect to livestock, yields are maximised by paying attention to their health and welfare requirements, and ensuring sufficient quality feed, nutrients, water and medicines are provided. If done effectively, the inputs, that is the fuel, time and capital invested, should be rewarded with greater yields whilst optimising the use of non-renewable resources and without damaging the environment, causing loss of biodiversity or increasing emissions of greenhouse gases.

A more productive agricultural sector delivers important benefits such as more efficient land use, climate change mitigation, biodiversity protection and social stability. According to a recent report (Noleppa *et al.*, 2013) each percentage point increase in agricultural productivity within the EU:

- Feeds more than 10 million humans per year.
- Increases the annual social welfare generated in European agriculture by approximately €500 million.
- Contributes €500 to the annual income of an average EU farmer.
- Reduces our net virtual land imports by about 1.2 million hectares.
- Acts to save 220 million tons of carbon dioxide emissions thus mitigating climate change.
- Preserves global biodiversity equivalent

to fauna and flora of up to 600,000 hectares of rainforest.

In Europe, agricultural productivity is amongst the highest in the world and so the probability of long-term food shortages is considered to be minimal, albeit there may be short-term pressures on a particular commodity due to natural disasters such as fire, flood or drought. Even in these instances, Europe is sufficiently prosperous to import all it needs. Nonetheless, there is the desire to further safeguard supplies as there are many uncertainties surrounding issues affecting European food security longer term.

5.1.4. *The pressures*

Much has already been done to increase agricultural productivity across Europe and indeed globally. Steady increases in European productivity per unit area were seen throughout the latter part of the 20th century due to:

- Optimal use of fertilisers through a better understanding of crop requirements, timely application and greater application precision.
- Optimal use of pesticides and increased availability and awareness of biopesticides coupled with the use of integrated pest management and pest forecasting techniques.
- Increased irrigation.
- Development of disease-resistant crops and high-yielding crop varieties and livestock breeds.
- Greater use of natural resources such as seeds, encouraging and protecting pollinators and beneficial fauna.
- Improved animal welfare and higher-quality animal feeds.
- Greater use of new technologies and mechanisation together with improved

infrastructure in poorer European nations.

Some of the conventional approaches for increasing productivity given above, such as increased fertiliser use and irrigation, are not without adverse environmental implications and may not be sustainable long term (see Chapter 4). Currently, productivity growth appears to have levelled off and, at least in the wealthier European countries such as Germany, France and the UK, there now seems to be limited opportunity for further increases in efficiency through conventional methods plus there are also a considerable list of pressures that could, if left unresolved, make maintaining current levels longer term difficult.

Planet Earth does not have the capacity to indefinitely support an expanding human population. Food cannot be produced without water, land and natural resources and so the current pressure levels on these will increase further as we strive to increase food production, as will the generation of waste. European productivity is amongst the highest in the world but its dependency on non-renewable resources (e.g. fossil fuels, phosphate) makes it unsustainable in the longer term. Many of the resources needed by agriculture are finite and so there is a pressing need for us to produce more food using less resources and ensuring those resources that are consumed are used effectively.

5.1.4.1. *Water security*

As can be reasoned from the information given in Chapters 3 and 4, there are equally strong demands for ample supplies of unpolluted water by both humans (for drinking, food production, industry, etc.) and biodiversity (for drinking, habitat and ecosystem services). However, ensuring both demands are met has in recent decades proved problematic and, as described in Chapter 4, water

security in the future is by no means certain. The UN defines water security as:

> the capacity of a population to safeguard sustainable access to adequate quantities of acceptable quality water for sustaining livelihoods, human well-being, and socio-economic development, for ensuring protection against water-borne pollution and water-related disasters, and for preserving ecosystems in a climate of peace and political stability.
>
> (UN, 2013).

Water security is a complex issue and inextricably linked to many other problems facing global society (see Figure 5.4 and Chapter 4). The UN explains this complexity succinctly:

> Many factors contribute to water security, ranging from biophysical to infrastructural, institutional, political, social and financial – many of which lie outside the water realm. In this respect, water security lies at the centre of many security areas, each of which is intricately linked to water.
>
> (UN, 2013)

Water shortages are already a fact of life in many parts of the world, including Europe, and the situation will undoubtedly be exacerbated by climate change which is expected to impact on both water availability and on water infrastructure. Populations, especially those in poorer European countries and those in southern Europe where the climate is hotter and dryer, may become increasingly more vulnerable to water insecurity in the future.

It has been estimated that global water consumption doubles every 20 years and this trend may accelerate in the coming decades due to greater direct demand for water from the domestic, industrial and municipal sectors (Population Institute, 2010). Future

Figure 5.4: Complexity
of water security

water demands from increasing population and agricultural consumption are expected to increase to around 5,500 km^3 per year, an increase of approximately 57% which will significantly enhance the impacts of climate change, especially in arid regions (OECD, 2012). At this rate of consumption and population growth, demand for water will outstrip availability by 56 per cent, with 1.8 billion people living in regions of water scarcity by 2025. In addition, human activities that depend on high water abstraction and use, such as irrigated agriculture, hydropower generation and use of cooling water, will be affected by changed flow regimes and reduced annual water availability. Currently, and considering Europe as a whole, there are adequate water resources but water scarcity and drought are becoming more frequent and further deterioration in the situation is

expected if temperatures rise and less rainfall is seen in future years due to climate change. Food cannot be produced without water and increasing productivity will depend heavily on water security. This is discussed in more detail in Chapter 4.

5.1.4.2.   Energy security

The International Energy Agency (IEA) is an intergovernmental organisation that works within the Organisation for Economic Co-operation and Development (OECD) to help deliver energy security for its 29 member nations. It defines energy security as 'the uninterrupted supply of energy sources at affordable prices' (IEA, 2014).

Energy is a fundamental part of all societal needs and is required for transport and for the distribution of food. It is also needed by the water industry to extract, pump, collect, clean and distribute water to agriculture, households and industry. Energy is needed to heat and cool our homes and to deliver services including hospitals and schools. It is also essential for the production of more energy. Across the world, 1.2 billion people live without access to electricity with 87 per cent of these living in rural areas (Băhnăreanu, 2015). Whilst in most parts of Europe access to energy is not considered a problem, this situation is not guaranteed and the ongoing threats can be divided into two main areas:

1. Insufficient capacity of electricity infrastructure to meet growing demands

As populations grow, energy demands will also grow and as economies correspondingly expand, so does the demand for energy-intensive goods and services. Within the EU-28, energy consumption between 1990 and 2013 increased by 2.2 per cent but actually showed a decrease of 7.0 per cent during the period 2005 to 2013 due to poor economic performance, greater energy efficiency by consumers and industry, and lower demands due to more favourable weather conditions (EEA, 2015). Nevertheless, demands are still relatively high on a per capita basis compared with, for example, China that only uses about 35 per cent of the OECD member nations' average albeit this is expected to increase significantly in the coming years due to the country's rapid economic growth (EEA, 2012). Concurrently, with large increases in demand, energy (oil, coal and gas) prices are also soaring and there is, in many European countries, a significant lack of investment in energy infrastructure. Consequently, there are fears that power supplies could be interrupted or rationed in the future. For example, in the UK, considerable investment is needed to maintain and update aging power stations and the national energy supply grid.

However, there is some evidence that where economies can afford them, technological improvements in devices and appliances, smart homes and white goods that optimise energy use, improved building materials and greater use of more effective insulation will significantly reduce energy demands in the future.

2. Supply interruptions due to external forces

There are many potential global situations that could slow or completely stop oil and gas supplies reaching the global markets, not least conflicts in oil-rich countries such as Iraq, political instabilities and societal unrest such as regime changes, labour strikes and protests such as those seen in the UK during 2000 to 2012 when lorry drivers blockaded refineries protesting against rising fuel prices. There are also examples where some nations have used oil and gas supplies as leverage to further their own aims and geo-political rivalries. For example, Russia has a history of using its oil and gas resources to reward its allies and punish countries that act against its wishes. Georgia and the Ukraine have, in recent years, experienced interrupted supplies and high prices when they appeared to favour the West, whilst the loyalty of Belorussia and Armenia were rewarded with ample supply and subsidised prices (Newnham, 2011). Currently, Europe relies heavily on Russian oil and gas. Major accidents and terrorist activities on oil fields, gas pipelines or distribution grids also have potential to disrupt supplies. For example, the explosion that occurred at an oil storage facility at Buncefield, Hertfordshire, UK in 2005 caused major concerns regarding fuel distribution in the south east of England. Another example is the explosion in 2011 at the Vasiliko power station in Cyprus that lead to daily power blackouts across the southern part of the island for many weeks.

Critical energy infrastructure has frequently been threatened by terrorist groups and such an attack would have major consequences for energy and economic security in most countries. To date, most of these types of attacks have been in the Middle East and North Africa and not in Europe but it does not follow that Europe is not vulnerable. Natural disasters and extreme weather conditions such as fires, floods and earthquakes can also interrupt gas and oil supplies, sometimes for long periods of time. For example, Hurricane Katrina that hit some American States in 2005 interrupted oil supplies and deliveries of some commodities for several weeks. In addition, extreme volatility in oil and gas markets are affected by both demand and concerns over supply interruptions.

From the above, it is clear that as the world's population grows, so will demand for energy along with food and water, and that the relationship between these three resources is highly complex and often referred to as the 'water–energy–food nexus' (Figure 5.5). As described above, water is a significant input for food production and is needed by the entire farm to fork food supply and consumption chains. Indeed, it is the limiting factor in increasing food productivity. Energy is a fundamental input for clean water being used to extract and pump water for ground and surface supplies including irrigation, for water treatment (raw and waste), for drainage, desalination and distribution. Countries rich in oil and gas need water as part of the energy production process but these activities can often cause water pollution. Growing demand for fresh water increases demand for energy and vice versa. Increases in food productivity will increase demand for both water and energy.

5.1.4.3.   Land use

One obvious way of improving agricultural production is to turn more land over to growing food crops and raising livestock.

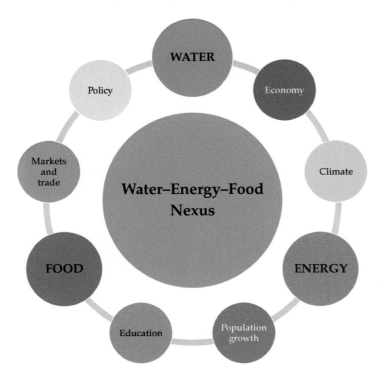

Figure 5.5: Water–energy–food nexus

However, land is in significant demand for many other purposes. As populations and economies grow, so does the need for housing and infrastructure, and so towns and cities expand, new roads, offices, shops, schools and hospitals, and so on, are built. Land pressures, such as urbanisation, production of non-food crops, soil degradation and conservation initiatives could mean that Europe's food-producing land decreases by 20 million ha by 2030 (Rienks, 2008). It has also been estimated that by 2020, approximately 80 per cent of Europeans will be living in urban areas which will have expanded into the countryside to accommodate them. This type of expansion, known as 'urban sprawl', refers to the expansion of poorly planned, low-density developments that spread out over large amounts of land. This puts long distances between residences, services and work places and creates a high level of segregation between residential and commercial uses with harmful impacts on the people living in these areas and the ecosystems and wildlife that have been displaced. Whilst urban sprawl offers society benefits and improves the quality of life of inhabitants, urbanisation comes at a cost as it does mean less land is available for food production and also for wildlife and their habitats. In addition, because of the scattered nature of developments, habitats become fragmented and when this occurs considerable impacts on biodiversity can occur (see Chapter 3). There are also many other consequences, such as increased car dependency, water pollution caused by hard cover run-off into surface waterbodies and a decrease in societal integration (EEA, 2011).

According to the EEA, across Europe during the period 1990 to 2000, large areas of land were lost to urbanisation. Germany lost around 200,000 ha, Spain around 180,000 ha and Italy around 150,000 ha (EEA, 2010). Satellite data used for land cover mapping by the UK's University of Leicester revealed that within the UK, during the period 2006 to 2012, 22,000 ha of green space was lost to housing, 7,000 ha of forestry was removed, 14,000 ha of farmland has been covered with concrete and 1,000 ha of wetlands drained for urban sprawl. As a consequence, Europe has seen significant impacts on biodiversity (Chapter 3) and increased flooding (Chapter 4) as well as the loss of land with food production potential.

In addition to demand for human food, crops are also grown to feed livestock and other animals. As demand for meat increases so will the need to grow crops for feed. In addition, there is a relatively small but growing industry in plant-based feed additives which are promoted for their benefits for livestock health and welfare and so farm productivity. These substances include essential oils from plants such as garlic, rosemary and peppermint. Some of these substances also offer environmental benefits such as reduced enteric methane from ruminants or reduced waste production due to improved nutrient absorption (Lewis *et al.*, 2015). Whilst there may be some productivity benefits it is unclear if these benefits outweigh the loss of productive land.

Some land that was once being used to grow food is now given over to non-food crops such as those that can be used to generate bioenergy, fibre and other valuable products such as pharmaceuticals, cosmetics, adhesives, plastics and industrial soaps. Table 5.1 shows the large range of non-food crops grown in Europe and their potential end use. A good example is lavender (*Lavandula* spp.), which is a small, aromatic shrub which has a wide range of applications in the cosmetic, fragrance, specialty food and alternative medicine industries (See Figure 5.6). It is grown commercially in various places across Europe, particularly in the UK, Ukraine and France. In many cases,

Table 5.1: Example non-food crops grown in Europe

End use	Example crops
Agrochemicals	Annual wormwood; caraway; pyrethrum; quinoa; spurge
Cellulose production	Wood; hemp; cotton
Construction materials	Cotton; common reed; flax; hemp; kenaf; miscanthus; sunflower; cork
Cords and sacking materials	Hemp; kenaf; nettle
Cosmetics and hygiene products	Almonds; amaranth; avocado; bugloss; borage; caraway; chamomile; coriander; evening primrose; jojoba, lavender; linseed; marigold; OSR/high erucic acid rape (HEAR)
Dyes, inks and stains	HEAR; madder; safflower; woad
Energy and fuels	Barley; cordgrass; elephant grass; maize; OSR; poplar; reed canary grass; sorghum; spurge; sugarbeet; sunflower; miscanthus; wheat; willow
Industrial raw materials	Castor; chicory; crambe; kenafe; OSR; sunflower
Lubricants, oils and waxes	Honesty; linseed; meadowfoam; OSR; rain-daisy; spurge
Paints, coatings and varnishes	Hemp; linseed; pot marigold; rain-daisy; Stokes' aster
Paper and pulp (incl. starch)	Flax; hemp; kenaf; miscanthus; potatoes; wheat
Pharmaceuticals and nutritional supplements	Amaranth; borage; caraway; chamomile; evening primrose; hemp; honesty; linseed; mallow; scabious
Plastics and polymers	Castor; honesty; meadowfoam; crops such as potatoes, wheat, maize for starch-based bioplastics
Resins and adhesives	Rain-daisy; Stokes' aster; Canada balsam; potato (to produce Coccoina); acacia tree
Soaps, detergents, solvents, surfactants and emulsifiers	Castor; coriander; gold of cuphea; hemp; OSR; poppy; quinoa;
Textiles	Hemp; flax; nettle

there may also be sustainability benefits from plant-based products, for example by replacing non-biodegradable or unsustainable materials, by helping meet renewable energy targets through reducing greenhouse gas emissions and from potentially less toxic substances. They can also offer biodiversity benefits by land use heterogeneity in intensive arable areas, providing nesting sites and plant architectures very different from typical arable crops. However, it should not be forgotten that land used for these crops is not available for food cropping although, with respect to bioenergy crops, there is the option to use less productive farmland such as that abandoned or fallow, and that unsuitable for food due to contamination, so as to avoid competing with food production.

5.1.4.4. Soils

Urbanisation and soil sealing are not the only concerns regarding land resources. Of equal concern is the degradation of arable soils and grazing land and this degradation can take several forms including erosion, loss of fertility, structural damage, salinisation and contamination. Estimates suggest that globally around a third of arable land has been lost due to these types of issues (Carey and Oetti, 2006). The FAO (2002) estimated that, when expressed as calorific value, 99.7 per cent of all human food is produced by the land, with only 0.3 per cent coming from aquatic bodies including the oceans. Therefore, it is obvious that we need to protect, conserve and, wherever possible, enhance Europe's soil resources.

The most serious form of soil degradation

Figure 5.6: English lavender
(Photo courtesy of: Dr John Tzilivakis)

is from erosion. There are two forms of soil erosion: wind and water. Both will reduce the quantity and the quality of soil and, consequently, will reduce its potential to undertake its ecosystem functions, which includes its ability to support food production. Wind erosion occurs when the action of wind on exposed soils causes soil particles to be dislodged and transported away from the site. Over time, soil losses can be quite significant. Water erosion, which happens more frequently than wind erosion, occurs when the energy of raindrops falls on exposed soils, dislodging them and thus providing an opportunity for them to wash away from the site.

There are a number of factors that affect erosion. First, soil structure is important. Soils with medium to fine textures and low

in organic matter are particularly vulnerable to both wind and water erosion. The finer and dryer the soil, the more vulnerable it is when exposed to wind. As exposed soils (see Figure 5.7) are more prone to erosion, those with vegetative cover, being crops, grass, shrubs, mulch or trees, are often protected because plant foliage absorbs much of the energy of raindrops and provides shelter from wind. In addition, plant roots help to bind soil together. Topography also plays a role. Sloping fields add to the risk of water erosion as the gradient encourages water to flow, washing the soil away. Flat, open fields are more prone to wind erosion as the wind speed and energy is unchecked by land undulations or trees which can act as barriers. Soil disturbance whereby vegetation is removed for whatever purpose or cause (e.g.

Figure 5.7: Recently ploughed and exposed soil
(Photo courtesy of: Dr John Tzilivakis)

excessive cultivation, building of roads and other infrastructure, earthquakes and land-slides) will increase the risk of erosion.

According to Pimentel (2006), around 90 per cent of the world's arable soils suffer erosion to some degree with 80 per cent experiencing moderate to severe levels. Over the last 40 years, erosion has caused around 30 per cent of the world's arable land to become unproductive with much of this being abandoned as it is no longer profitable to farm. Within Europe, erosion is a wide-spread problem with the Mediterranean area particularly affected (Van der Knijff *et al.*, 2000) due to drought conditions often being followed by heavy rainstorms on steeply sloping landscapes. It is also a severe problem in Iceland due to harsh weather conditions and landslides (Arnalds, 1987).

Erosion is obviously a serious concern for food security but the impacts are broader. For example, where water erosion occurs, run-off tends to be high and so less water is absorbed into the soil and less is available for crops and other plants. Eroded soil takes important plant nutrients such as nitrogen, phosphorus and potassium away from the site, depleting reserves available for vegeta-tion with this loss potentially ending up in polluting water bodies. Erosion reduces soil quantity and over time soil may become so depleted that there is no longer enough depth to sustain plant roots. Both wind and water erosion remove fine organic matter from the soil surface leaving behind the larger particles, stones and silt, degrading the soil structure and its fertility. Consequently, soil erosion will undoubtedly affect productivity.

Soil fertility is an important aspect of increasing productivity but across many parts of the world including Europe, it is in decline. Whilst erosion is one cause, high soil nitrogen is another. Usually nitrogen, in the context of crop productivity, is seen as ben-eficial as it promotes crop growth. However, high nitrogen inputs, especially over an extended period, cause excessive nutrient saturation which can adversely affect soil quality. This is a growing problem in many parts of Europe, caused by large volumes of manure from overgrazing livestock, excessive applications of chemical nitrogen fertilisers and, in some regions, high atmospheric nitrogen deposition. High soil nitrogen can result in soil acidification which in turn impacts on crop yields, forest growth and on soil organism populations. Changes in soil organism populations can change nitrogen mineralisation and denitrification, which can affect crop growth and so productivity. Excess nitrogen is eventually lost to the environment (water and air); although processes are complex this can, potentially, affect surface and groundwater quality, and contribute to climate change (Velthof *et al.*, 2011). There is some good news in this respect in that over the last decade, within Europe, nitrogen inputs into the soil have decreased due to strong agri-environmental policies and promotion of good fertiliser practices.

Soil Organic Matter (SOM) is often used as an indicator of soil fertility. Although in most soils the SOM content tends to be less than 10 per cent, typically between 1 and 6 per cent, its value for crop productivity is immense. SOM, however, is broken up into three fractions which can be referred to as 'the living, the dead, and the very dead'.

The 'living' fraction refers to the quantity of microbiological organisms, plant matter, manures and composts, and so on, the 'dead' fraction is the carbon reserves and the 'very dead' fraction is soil humus which is an amorphous material that has decayed so that it does not have the structure or characteristics of plants, microorganisms or animals (see Figure 5.8). The ratio of these three factions is used as an indicator of soil health (Snapp and Grandy, 2012). Soils rich in SOM tend to be more fertile, have good water-holding capacity and better structure, as aggregates are closer bound and so, more stable. Therefore, they are less likely to suffer from compaction and erosion. Poor SOM levels can impact on biological, chemical and physical soil processes such as nitrogen transformation, all of which help ensure high crop yields. Across many parts of Europe, low SOM is a problem; around 45 per cent of European soils have 2 per cent or less SOM and a further 45 per cent have 6 per cent or less (Velthof *et al.*, 2011). There are several causes, including the fact that SOM levels can decrease where low-residue crops, such as potatoes and soybeans, are

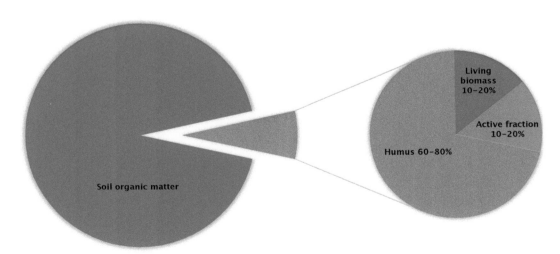

Figure 5.8: Composition of SOM
(Adapted from: Snapp and Grandy, 2012)

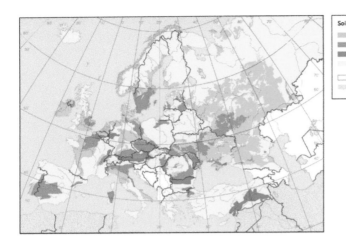

Figure 5.9: The extent of soil compaction in Europe (Source: EEA, 2010)

grown or if removal of corn for silage and of straw from small grains is a regular process, especially if organic matter is not regularly replaced with composts, manures or other soil conditioners.

Soil structure is a key factor in the functioning of soil, its ability to support plant and animal life, and perform ecosystem services including soil carbon sequestration and water quality (Bronick and Lal, 2005). Soil structure refers to the arrangement of the individual soil components, such as clay, silt and sand, organic matter, minerals, and how they are bound together. The arrangement of soil particles forming the soil can occur in different patterns resulting in a variable amount of pore space, which in turn, results in different soil structures. The specific soil structure will determine the behaviour including its water-holding capacity, water permeability, erosion potential, crusting, nutrient recycling, root penetration and crop yield. Soil degradation in the form of a decline in soil structure, known as compaction, occurs when the soil is compressed such that individual soil particles become packed closer together with less pore space, causing a reduction in the volume of air (Chan *et al.*, 2003). It has been defined as 'the densification and distortion of soil by which total

and air-filled porosity are reduced, causing deterioration or loss of one or more soil functions' (van den Akker and Hoogland, 2011). It is caused by poor land use and soil/crop management practices including using tillage equipment during soil cultivation in unsuitable (i.e. wet) conditions or from the frequent use of heavy field equipment. Most effects of compaction are detrimental and may include less plant root proliferation in the soil, lower rates of water permeability and less available to the crop, and restricted air movement. Slower internal soil drainage can result in poorer subsurface drain performance, longer periods of time when the soil is too wet for tillage following rainfall or water application, increased denitrification and decreased crop yields. Increased compaction also adds to the energy consumption by tractors for subsequent tillage. From Figure 5.9, it can be seen that soil compaction is a serious issue across Europe.

Salinisation, a form of soil contamination, refers to a situation where there is a build-up of soluble mineral salts in the soil to such an extent that soil fertility is reduced (Eckelmann *et al.*, 2006). A wide range of mineral salts can be responsible, including potassium (K^+), magnesium (Mg^{2+}), calcium (Ca^{2+}), sulphate (SO_4^{2-}), chloride (Cl^-) and sodium (Na^+).

Where mainly sodium salts are accumulated, the process is usually referred to as sodification. Salinisation can be a natural process caused by a high mineral salt concentration in the soil but more usually it is the result of inappropriate irrigation practices such as the use of salt-rich water, waste waters or where drainage is poor. Where it occurs, it can have significant impact on crop productivity and the situation may well be exacerbated by climate change. Mineral salt accumulation may disrupt plant development by limiting its nutrient uptake and reducing the quality of the water available to the plant. It may also affect the metabolism of soil organisms, leading to severely reduced soil fertility. High levels of salinity in soils can also cause plants to wilt and die off due to an increase in osmotic pressure and the toxic effects of salts. In extreme cases, it can lead to desertification. Approximately 4 million hectares of European farmland suffer from salinisation (Thiel-Bruhn, 2009) and the problem is most acute in very dry areas which are highly irrigated.

According to the EC's Joint Research Centre (JRC), there are potentially 2.4 million potentially contaminated soil sites across Europe (EC, 2014). Municipal and industrial waste disposal and treatment causes around a third of Europe's soil contamination problem with metal industries, mining and petrol stations also responsible for many contaminated sites. The most frequent contaminants are mineral oils, heavy metals and chemicals of various types. Often this means these areas are not suitable for food production, public amenities or house building until they are cleaned up. Soil contamination is not restricted to municipal and industrial sites: agriculture can also contribute to loss of productive land due to contamination caused by overapplication of pesticides and fertilisers, oil spills and on-farm waste disposal.

5.1.4.5. Waste

Society today creates huge volumes of waste. Waste management efficiency across Europe varies enormously. According to recent Eurostat data, Estonia, Slovenia and Belgium are amongst the best waste managers combining low levels of production with reasonable levels of recycling and composting such that they send, on a per capita basis, less to landfill and incinerators than other EU MSs (Eurostat, 2014). Nevertheless, most European countries realise that more needs to be done to reduce waste generation and often this is a socio-economic issue as well as an environmental one that affects the entire farm to fork system.

Farm waste is very diverse and includes natural wastes such as manures, slurries, crop residues, hedge trimmings and other green wastes, dead stock, silage effluent and waste from dairies, as well as more familiar wastes such as plastics, paper and card. In 2012, the combined agricultural sectors of the EU-27, Switzerland and Norway collectively produced 1.3 million tonnes of plastic waste (PlasticsEurope, 2012). Many of these wastes, especially the natural materials, can be economically valuable resources to the farm if managed correctly, that is, by actions to minimise and recycle using the waste management hierarchy (see Chapter 5.2). As well as representing a significant loss of resources (land, fertilisers and nutrients, pesticides, water, energy, labour and investment), improperly managed waste has the potential to cause land and water pollution.

Where wastes are sent to landfill or an incinerator there are potentially serious associated environmental and socio-economic impacts. Landfill is the least environmentally sound process, yet, whilst in some European countries significant amounts of recycling and incineration are achieved, as shown in Figure 5.10, some MSs such as Romania

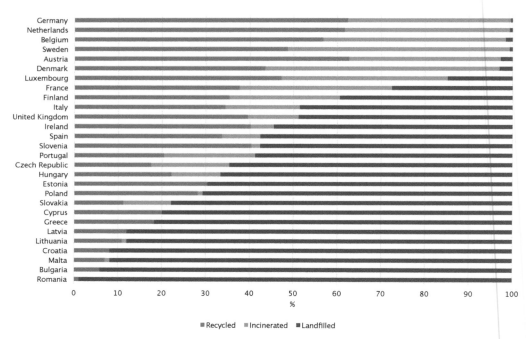

Figure 5.10: Waste disposal in European MS countries in 2011
(Source: Eurostat, 2014)

and Bulgaria, rely almost entirely on landfill for waste disposal. Landfill sites are essentially large holes in the ground. Some are excavated for the specific purpose of waste disposal whilst in other instances, old quarries or gravel pits may be used. Some waste will eventually rot, but not all, and during the degradation process it will generate methane, a greenhouse gas. In modern landfill sites, this gas may be collected and used as a fuel but this is not always the case. Landfill processes also produce a leachate which has potential to cause pollution and there have been many incidents of groundwaters being contaminated by landfill-generated pollutants across the world (e.g. Afzal *et al.*, 2000; Ahel *et al.*, 1998; Christensen *et al.*, 1998; Sangodoyin, 1993). Landfill sites can, and often do, have a significant adverse impact on the individuals who live and work close to them. Historically, they were located close to urban areas where the bulk of wastes

are produced and to reduce transport costs. However, unpleasant odours, increased volumes of large heavy vehicles, the associated noise and visual impacts can result in reductions in the quality of life for residents and the lowering of house prices. In many European countries, the process of commissioning a landfill is extremely complex as site selection (particularly considering the needs for hydrogeological assessments), environmental impact assessments (EIAs), licensing applications, public hearings, appeals and sometimes court cases, can often extend the process over several years. There is not enough land available to meet the demands of agriculture and the infrastructure needs of a growing population so continuously using land for landfill is unsustainable. Incineration is also not free from impacts. Hazardous wastes can generate toxic substances when burnt, such as dioxins, and ash residues can be high in heavy metals. This ash then needs to be disposed of and this is usually to land-fill. Incinerators are also rarely welcomed by residents due to the potential impact on the surrounding area, for example, aesthetic issues, noise and dust plus problems relating to traffic congestion, noise and air pollution. Consequently, these developments can also take a long time to progress through planning processes.

Much discussion and debate is ongoing regarding the amount of food waste generated by the food supply chain, particularly that from households, restaurants and food retailers. In fact, according to the World Bank (World Bank Group, 2014) almost a third of all food produced is never eaten. A significant proportion of waste food occurs at farm level. Crops and fresh produce are left in fields or on the ground in orchards and never harvested. This is sometimes due to adverse weather, pests and disease-causing crop damage such that it is below quality thresholds and so unsaleable. In other cases, this waste is due to market issues, greater supply than demand, cosmetic imperfections and a lack of sufficient cost-effective labour. It costs a farm money to harvest a crop and if the market value of that crop is less than the cost of harvesting it, the crop will not be harvested. This is also a waste of the valuable resources used to produce it, such as energy, water, fertilisers, pesticides and agricultural land. It also represents a real cost to the farmer and the consumer.

5.1.4.6. Depletion of natural resources
In the context of increasing agricultural productivity, the major concern with respect to natural resources is the use of fossil fuels and fertilisers, in particular phosphate. Fossil fuel use has been discussed extensively in Chapter 2 and under energy security above (Chapter 5.1.4). Whilst there is a need from a climate change perspective to reduce the use of fossil fuels, the planet cannot, as yet,

replace them entirely with other energy sources and so ensuring fossil fuels are used wisely is essential.

Phosphorous (P) is critical to the survival of all living things. In the human body, phosphorus helps to shape DNA, and strengthens bones and teeth, among many other functions. In agriculture, phosphorus acts as fertiliser to improve crop yields and is a major component of NPK (nitrogen-phosphorous-potassium) fertilisers. With respect to animals, phosphorus is an essential mineral and P-deficiency is the most commonly occurring mineral deficiency seen in livestock. Phosphorus, in the form of phosphate minerals, is very abundant in the natural environment, estimated to be 0.12 per cent of the Earth's crust, making it the eleventh most common element. However, currently, the only financially viable source of phosphate is mineral rock extraction and this is a non-renewable resource. The most common form is apatite and the related minerals, chlorapatite and fluorapatite, whilst the less abundant phosphate minerals include monazite and varasite. Calcium phosphate is a constituent of all fertile soil, having been supplied to the soil by the disintegration of rocks containing it.

Due to demand, principally for the production of mineral-based fertilisers, detergents, animal feed and other chemicals, the depletion of phosphorus reserves is receiving increasing interest and is perceived as a long-term threat to global food security (e.g. van Vuuren *et al.*, 2010; Cordell *et al.*, 2009; Gilbert, 2009; Koning *et al.*, 2008; Vaccari, 2008; Bartels and Gurr, 1994). Phosphate rock was added to the EU List of Critical Raw Materials in 2014. According to Cordell *et al.* (2009), at current consumption rates, phosphorus reserves will become exhausted within 200 years. This is of particular concern as demand is expected to increase significantly in the near future and current reserves are controlled by just a very few countries (Cordell *et al.*, 2009; Koning *et al.*, 2008) such as the USA, China and Morocco, controlling approximately 70 per cent of all global reserves (Cooper *et al.*, 2011). The large markets of the USA and China are largely satisfied by their own supplies meaning that much of the rest of the world relies on imports from the Moroccan, Sahara and other smaller suppliers. The Moroccan Bou Craa phosphate mine, situated in a remote region of the Western Sahara, is one of the world's largest reserves producing several million tonnes of phosphate-rich rock every year and currently supplies around 15 per cent of global demand. However, this area of the Western Sahara is occupied territory and has been under the control of Morocco since 1976. Ownership of the region is disputed by a rebel faction known as the Polisario Front; a claim largely supported by the UN. Should the region see political unrest in the future and the people of Western Sahara start a conflict to regain control of the region, significant uncertainty in global phosphate supplies would result. Longer term, the reserves in the USA and potentially China, will run out before those of Morocco and the Sahara meaning that global crop productivity may well be highly dependent on these potentially unstable regions. This would undoubtedly push prices very high, possibly to the point where some countries may not be able to afford them and result in a rapid decrease in the country's food productivity.

5.1.4.7. Impacts on ecosystem services

Ecosystem services are defined as the benefits humans receive from ecosystems and include, for example, the provision of food and water, provision of substances with pharmaceutical and manufacturing applications, the control of climate and disease, soil formation, oxygen production, wildlife and the countryside, and the provision of recreational and spiritual benefits. Some of the benefits seen are vital to crop productivity, for example nutrient cycling, pollination, genetic resources and pest regulation.

Nutrient cycling, to put it simply, is the movement of nutrients (e.g. nitrogen, phosphate, potassium, calcium) between the physical environment and living organisms, and back again to the physical environment (see Chapter 1). This continuous recycling of nutrients helps to sustain plant growth and so agriculture. For nutrient cycling to work effectively, a balance is required in the ecosystem but many anthropogenic activities have disrupted this balance. Historically, man extracted nutrients, as harvested crops, from the soil at roughly the rate nutrients, as human waste, animal manures and other organic wastes, were returned. However, nowadays human waste is managed via water sewage processes and large-scale, intensive farming has caused land degradation, erosion and loss of nutrients to the environment via, for example, run-off and leaching, which has resulted in more nutrients being removed from the soil than are being replaced. Flood management can also cause significant changes in nutrient cycling, as during flood events nutrient-rich river water and sediments are redistributed to land, whereas dams and flood water controls confine these to the river until the waters wash out to sea or sediments accumulate on the river bed. The consequences of these anthropogenic activities is an imbalance in the availability of nutrients: surface, ground- and marine waters are too rich in nutrients whereas soils are depleted and so less able to support agricultural processes.

In agricultural systems, pollination is an important part of ensuring a good crop yield, particularly for orchard, horticultural and forage crops as well as seed production for root and fibre crops. Without adequate pollinators, significant loss in productivity subsequently affecting food security would be expected. Within the UK, for example, 20 per cent of the UK cropped area is comprised of pollinator-dependent crops and a high proportion of wild flowering plants require insect pollination for reproduction. The UK National Ecosystem Assessment report (Watson *et al.*, 2011; NEA, 2011) concluded that pollination services are worth around £430 million per year to the UK. However, pollinators, such as bees and syrphids, are suffering from the same threats as the rest of biodiversity such as habitat loss, degradation and fragmentation (see Chapter 3). There is also evidence that various parasites (e.g. the Varroa mite), disease and agricultural pesticides (particularly the neonicotoids) are also having significant negative impacts. Whilst there are many initiatives ongoing (including research on bee health, surveillance measures and the establishment of partner initiatives such as the Pollinators Network) to address these issues, climate change is also a very significant threat and is likely to

cause a wide range of additional pressures. First, there is a risk of desynchronisation (spatial and temporal) between the life cycle of the pollinator and the life cycle of the flora it depends on. Earlier springs may cause a de-synchronisation between emergence of pollinators and crop/plant flowering times. This will vary with species and location. Pollinators may emerge before flowers are open or flowers may open before pollinators arrive resulting in a temporal mismatch in plant–pollinator interactions. Second, there may be direct climate effects. Pollinating insects (see Figure 5.11) are sensitive to temperature – the degree of their thermal tolerance and plasticity to temperature change is important for their survival. Temperature may also affect their foraging patterns. Higher temperatures and heavy rainfall may reduce the longevity of flowers reducing the availability of pollen and nectar production. Severe weather also tends to limit pollinators' foraging activity and this may also lead to reduced plant vigour, delayed plant maturation and a decline in nectar production,

Figure 5.11: Pollination (Photo courtesy of: Dr Doug Warner)

leading to changes in the natural synchronisation between pollinators and plant life cycles. There is evidence that many pollinators are likely to adapt but the rate of this genetic adaptation is unlikely to match the speed of climate change. There is a strong expectation that pollinators will migrate and that as some species are lost to more suitable climates, they will be replaced by others, but there is no guarantee that their pollination effectiveness will be equivalent to that of the species they replace. There are also concerns that some pollinator species will become extinct (see Chapter 3).

The regulation of pests and disease by the provision of natural biological control (predators and parasites), inherent defence mechanisms and genetic diversity (providing natural pest resistance) is yet another important ecosystem service. However, nature alone is not always able to solve all the pest and disease problems that food production faces. It has been estimated, for example, that black-grass (*Alopecurus myosuroides*) infestations in the UK can result in up to a 21 per cent decrease in cereal production, equivalent in financial terms to £532 million (Clarke *et al.* 2011). In addition, crop protection products offer other benefits such as reducing labour costs that would otherwise be required for pest management, and decreasing soil erosion by reducing the need for some cultivations. Nevertheless, they do pose risks to both human health and the environment. Pesticides have been linked to a wide range of human health problems, ranging from short-term impacts such as headache, nausea and skin irritation to chronic issues like cancer, reproduction problems, sensitisation and endocrine disruption. As they are applied in the open environment there is opportunity for them to reach non-target areas including residential and recreational areas and pollute water bodies, air and terrestrial areas. In turn, this

can result in poisoned wildlife and reduced populations of beneficial insects and earthworms which will impact on food productivity and soil fertility (see also Chapters 2 and 3). A study published in 2005 suggested that the total global loss of crops due to pests and diseases varies from around 50 per cent for wheat to more than 80 per cent for cotton, despite modern crop protection methods (Oerke, 2005). Therefore, if we are to increase food productivity we need to continue to use pesticides but, as discussed above, these substances can, in some instances, adversely affect ecosystem services such as pollination.

The challenge is to use pesticides in a way that balances the need to increase productivity whilst minimising the effects on non-target organisms that provide important ecosystem services (Forbes and Calow, 2013). However, it is not simply a case of using less due to the risk of pest resistance developing. Pesticide resistance develops due to a process of natural selection: the most resistant specimens survive a pesticide application and pass on their genetic traits to their offspring such that eventually the general population has that trait. It is associated with the repeated use of the same class of pesticide. There is no doubt that the problem is growing and has the potential to seriously damage agricultural production worldwide. One report suggests that resistance has been recorded in at least 546 species of arthropod pests, 218 species of weeds and 190 species of plant pathogens (Heap, 2013; Whalon *et al.*, 2008). In recent years, many pesticide-active substances have been removed from the market (under EC Regulation 1107/2009) due to a review of their risks to human health and the environment including those found to present an unacceptable risk to ecosystem services such as the neonicotinoid class of pesticides and the risk to pollinators (Cressey, 2013). As a result, the choice of pesticide products is often quite limited and this itself increases the risk of resistance developing as farmers may not have scope to use a range of different chemical classes. Should this happen, the risk of food production losses is likely to increase.

The number of species of plants, animals, aquatic life, invertebrates and microorganisms, the enormous diversity of genes in these species, the different ecosystems on the planet, such as deserts, rainforests and coral reefs, are all a part of a biologically diverse Earth. In terms of ecosystem services, the planet's genetic resources play a vital role in satisfying basic human food and nutritional requirements. In addition, these resources play a major role in the world's economy, its economic development, education, tourism and leisure, medical discoveries, cultural values and adaptive responses to new challenges such as climate change and drug resistance. However, due to excessive harvesting and predation, loss of habitats, inadequate nutrition, disease epidemics, extreme weather events, natural hazards and climate change, the diversity of the planet's genetic resources is under threat. Human activities also threaten this diversity. Over the last century or so, many livestock breeds and crop varieties have disappeared and been replaced with more productive varieties, many of which have common gene heritages, thus reducing gene stock. According to the FAO (1999), since the 1900s some 75 per cent of plant genetic diversity has been lost as farmers worldwide have abandoned local varieties and landraces for genetically uniform, high-yielding varieties. The FAO report also suggests that currently 75 per cent of the world's food is generated from only 12 plants and five animal species. Just nine crops (wheat, rice, maize, barley, sorghum/millet, potato, sweet potato/yam, sugar cane and soybean) account for over 75 per cent of the plant kingdom's contribution to human dietary energy.

The consequences of this are serious, not least for the agricultural sector and future food security. Overbreeding and dwindling genetic diversity could limit the ability of livestock populations to adapt to environmental changes, such as global warming and new diseases. The Irish Potato Famine of the 1840s is a dramatic example of the dangers of genetic uniformity. Potatoes were introduced into Ireland from South America in the 1500s and huge amounts were subsequently grown predominately from the original imported stock. Over time, potatoes became a significant part of the Irish rural population's diet. However, in the 1840s, a significant outbreak of potato blight (*Solanum tuberosum*) completely decimated the national crop. As all the plants were derived from the same genetic stock, there was little ability to respond and fight the pathogen and the consequences of this were mass starvation and a loss of 25 per cent of the Irish population. Another more recent example of the problems of genetic uniformity was seen in 1970 in the United States when the maize crop was found to be highly vulnerable to southern corn leaf blight, a fungal disease of maize caused by the plant pathogen *Bipolaris maydis*. An outbreak of the disease destroyed almost $1,000 million worth of maize and reduced yields by up to 50 per cent. Over 80 per cent of the commercial maize varieties grown in the United States at that time were susceptible to the virulent disease. Resistance to the blight was eventually found in an African maize variety called Mayorbella. A major catastrophe was averted by incorporating this resistance into commercial varieties. However, there is growing concern that not enough is being done to expand gene stocks to safeguard against similar catastrophes in the future.

5.1.4.8. Air pollution and crop production

There are numerous causes of air pollution including the burning of fossil fuels, power generation, industrial emissions, mining, vehicle exhausts and agriculture (see Chapter 2). The latter includes emissions of ammonia from fertiliser use, livestock buildings and animal wastes. With respect to agriculture and food production, air pollution can cause direct damage to crops including injury to foliage. This can range from visible markings such as scorching on crop leaves, to reduced growth and yield potentially leading to premature death. Impacts can be pollutant- and concentration-specific. For example, chronic effects of sulphur dioxide pollution include plant chlorosis and the development of brown spots and patches on foliage. Sulphur dioxide penerates the plant through the stomata and, to a lesser extent, via the cuticle with the amount penetrating dependent on prevailing environmental conditions, such as humidity and temperature. Most damage is done during daylight hours as stomata tend to close at night and so the amount of pollutant entering the plant is restricted. Recovery may be possible if the pollution concentration level decreases but it is probable that physiological processes are already damaged and crop yields will suffer as a consequence. It has also been reported that a significant proportion of sulphur dioxide in air is deposited to the soil, where it is converted to various acids (sulphuric and sulphurous) which can lead to soil acidification and loss of soil fertility (Shaibu-Imodagbe, 1991). As well as impacts on crops, sulphur dioxide can cause livestock health and welfare issues including respiratory impairment. Generally, the N in ammonia and nitrogen oxides compounds deposited from air can be used by plants, however, these pollutants can also induce soil acidification and be detrimental to animal health.

Air pollution can also induce growth disturbance due to changes in plant metabolism, reduced photosynthesis and increased susceptibility to weather, pests and diseases. Indirect effects of air pollution also have the potential to damage agricultural productivity. For example, changes in climatic conditions such as temperature and humidity, changes in the hydrological cycle affecting water availability and extreme events can all affect yields and crop quality. Impacts of air quality are discussed in detail in Chapter 2.

5.2. Policy and interventions

5.2.1. Introduction

Many of the issues discussed above have been on the world's political agenda for a long time, some for more than 50 years. However, it is only when they are considered collectively and in the context of food security that their true significance and severity can be properly comprehended. Increasing agricultural productivity was the main objective of agricultural policy during the Second World War and the various policy frameworks introduced subsequently (such as the 1947 Agriculture Act in the UK, the Treaty of Rome in 1957 and the introduction of the EU's Common Agricultural Policy in 1962; see Chapter 1) sought to maintain the momentum achieved. These policies focused predominately on increasing productivity through farm subsidies, intensification and modernisation to provide greater food security for Europe. They also sought to increase rural employment and introduce a level of countryside management, both of which were desperately needed following the destruction and neglect caused by the war. However, by the late 1960s and early 1970s, it was evident that whilst agricultural intensification had been hugely successful (actually leading to a massive overproduction of some products) it was having a destructive effect on the countryside. Many farms had lost their natural features, such as hedgerows, field boundaries and semi-natural habitats, to provide more productive land and, in turn, this led to large increases in the use of fertilisers and pesticides with increased incidences of pollution and impacts on biodiversity (Latacz-Lohmann and Hodge, 2003).

Around this time, agricultural and environmental policies, although intrinsically linked, were developing independently. Agricultural policies, dominated by the CAP, continued to evolve with environmental issues gradually gaining more and more importance. Eventually, under Agenda 2000 and the associated reform of the CAP, the link between subsidies and production was severed so that the latter was much more market-led and farm payments were linked directly to compliance with a set of standards on food safety, animal rights (cross compliance) and environmental concerns rather than productivity levels (Chapter 1).

Society now faces the challenge of increasing productivity to address future food security issues whilst still achieving the same high standards of environmental protection and animal welfare. As described above (Chapter 5.1), the concept of delivering a sustainable improvement in productivity at the farm level relies on ensuring sustainable access to resources, from land, water and energy to fertilisers and pesticides, as well as being able to sustainably handle wastes. As the resources required are very broad, there is no one single policy that addresses resource management but rather a whole spectrum of policies across all areas of the agri-environment and, more generally, the management of natural resources. A summary of the most important policies and legal instruments is given in Table 5.2. This table is not all-inclusive and excludes those related to other aspects such as climate change, biodiversity and water, discussed in separate chapters.

5.2.2. *Energy policies and agricultural productivity*

The importance of a secure energy industry in Europe has grown considerably in the last couple of decades such that it is now a core policy of the European Union. This growth in significance has had two main drivers. First, it has been a reaction to the rising concerns regarding climate change and the need to develop a low-carbon economy. Second, there has been a liberalisation of national energy markets and several actions and initiatives to create a barrier-free EU-wide energy market to help combat, through competition, high prices and limited choice for consumers as well as seeking to ensure secure supplies for all EU MSs.

Currently, the EU imports more than half of the energy it consumes. Many European countries are also heavily reliant on a single supplier; some rely entirely on Russia for their natural gas. The problem is also compounded by many countries having aged energy networks and infrastructures desperately in need of investment and modernisation. These issues equate to a situation that, in this uncertain world, may not be sustainable. In response to these concerns, the European Commission released its Energy Security Strategy in May 2014. The strategy aims to ensure a stable and abundant supply of energy for European citizens and the economy. Supporting the Energy Security Strategy is the EU's Energy Roadmap to 2050 which describes the path to not only assuring security of supplies but also to investing in infrastructure and developing a low-carbon energy industry. The roadmap will allow MSs to make the required energy choices and create a stable business climate for private investment.

The breadth and significance of all the risks facing the European energy sector means that realistically, eliminating them all is not possible as many are completely out

Table 5.2: The main policies and legal instruments relating to different aspects of agricultural productivity[1]

Policies and instruments	Description	Relevant areas						
		W	En	L	S	NR	Wa	Ec
EU Energy strategies 2050	EU strategy for the transition to a competitive, secure and sustainable energy system by 2050.	✓	✓			✓	✓	✓
Energy Roadmap 2050	Describes the pathway towards securing safe, secure, sustainable and affordable energy for Europe and for developing a low-carbon economy.	✓	✓			✓	✓	
Renewable Energy Directive (2009/28/EC)	Establishes an overall policy for the production and promotion of energy from renewable sources in the EU. It requires the EU to fulfil at least 20% of its total energy needs with renewables by 2020 – to be achieved through the attainment of individual national targets.	✓	✓			✓	✓	
Fuel Quality Directive (2009/30/EC)	Provides quality specifications for petrol and diesel and established criteria that must be met by biofuels if they are to count towards the greenhouse gas intensity reduction obligation.	✓	✓			✓	✓	
Directive to Reduce Indirect Land Use Change for Biofuels and Bioliquids (EU2015/1513) (ILUC Directive)	Growing biofuels on existing agricultural land can displace food and feed production. This Directive seeks to reduce the environmental impacts associated with this and prepares the transition towards advanced biofuels.	✓	✓	✓		✓	✓	
Biofuels Directive (2003/30/EC)	Promotes the use of biofuels and other renewable fuels, requiring MSs to take national measures to replace a proportion of fossil fuels used in transport with renewables.	✓	✓	✓		✓	✓	
Roadmap to a Resource-Efficient Europe	Describes the pathway to transforming Europe's economy into a sustainable one by 2050. It seeks to decouple economic growth from resource use.			✓	✓	✓	✓	
Thematic Strategy on the Sustainable Use of Natural Resources, 2005	Linked to the Thematic Strategy on Waste, this highlights the challenges for sustainable resource use and describes a framework for action.			✓	✓	✓		

Policies and instruments	Description	Relevant areas						
		W	En	L	S	NR	Wa	Ec
EU Action Plan for Sustainable Consumption and Production	A series of initiatives related to encouraging more sustainable production of goods and actions to raise consumer awareness of environmental performance and the green credentials of manufacturers.	✓	✓	✓	✓	✓	✓	
BioEconomy Action Plan	A plan to reconcile demands for sustainable foods and the sustainable use of natural resources with environmental and biodiversity protection.	✓	✓	✓	✓	✓	✓	
Thematic Strategy on the protection and Conservation of the Marine Environment, 2005	Aims to achieve good environmental status of the EU's marine waters by 2020 and to protect the resource base upon which marine-related economic and social activities depend.		✓					✓
Waste Framework Directive (2008/98/EC)	Describes the basic concepts and definitions related to waste management and provides the legislative framework for implementing waste management according to the waste hierarchy.		✓		✓	✓	✓	
Hazardous Waste Directive (1013/2006)	Introduces a legal definition for hazardous waste and how it is differentiated from standard wastes. Also provides a stricter control regime for hazardous wastes including measures for labelling, record keeping and monitoring.			✓		✓	✓	
Circular Economy Strategy/ Package	Policy and various measures to drive Europe towards an economy that is restorative and regenerative by design. It aims to keep products, components and materials in use for a maximum period of time to minimise waste and optimise the use of raw materials.	✓	✓	✓	✓	✓	✓	✓
Thematic Strategy on the Prevention and Recycling of Waste, 2005	Introduces an integrated approach to waste management based on the waste hierarchy and establishes the legislative framework. Closely linked to the Thematic Strategy on Sustainable Use of Resources and the Waste Framework Directive. Aims to move Europe towards a circular economy.			✓	✓		✓	

Table 5.2: (*continued*)

Policies and instruments	Description	Relevant areas						
		W	En	L	S	NR	Wa	Ec
Soil Thematic Strategy, 2006	Aims to protect soils against all risks including contamination, erosion, loss of fertility, salinisation, etc. Seeks to ensure all European soils are managed sustainably and provides a legislative framework for their protection and remediation.						✓	✓
Sewage Sludge Directive 86/278/EEC	Main objective is to protect soil from contamination when sewage sludge is used as a soil amendment and spread to agricultural land.			✓	✓	✓	✓	✓
Thematic strategy on air pollution, 2002	This policy sets objectives and targets for reducing certain pollutants. It reinforces the legislative framework for combating air pollution via two main routes: improving Community environmental legislation and integrating air quality concerns into related policies.		✓	✓	✓	✓	✓	✓
Industrial Emissions Directive (IED)	Seeks to ensure a sustainable approach to the prevention, control and mitigation of polluting emissions from industrial installations.	✓		✓	✓	✓	✓	
Thematic strategy on the sustainable use of pesticides	A policy to reduce the environmental and public health risks related to pesticide use such that their use becomes more sustainable whilst protecting end users from financial losses due to crop damage. Measures concern principally surveillance and research into pesticides, training and information for users, as well as on specific measures regarding their use.	✓		✓	✓			✓
Environmental Labelling (EU Eco-Label Regulation No 66/2010)	A voluntary approach to communicate environmental performance to consumers and to encourage improved sustainable production by competition. Main EU scheme is the Ecolabel but there are a number of other private schemes.	✓	✓	✓	✓	✓	✓	
Council Regulation (EC) 834/2007 on Organic Food	An internal market and consumer protection regulation, broadly describes the organic production standards and the control and labelling requirements.	✓	✓	✓	✓	✓	✓	✓

Policies and instruments	Description	Relevant areas						
		W	En	L	S	NR	Wa	Ec
Commission Regulation (EC) 354/2014	Lays down detailed rules for the implementation of Council Regulation (EC) No 834/2007 with regard to organic production, labelling and control.	✓	✓	✓	✓	✓		✓
International Treaty on the Sustainable Development of the Alps, 1995	Signed by the Alpine MS, this treaty seeks to ensure the sustainable development of the Alps including the protection of their biodiversity, water, natural capital and economic environments.	✓		✓		✓		✓
Environmental Liability Directive (2004/35/EC)	This Directive aims to establish a framework of environmental liability based on the 'polluter pays' principle, in order to prevent and remedy environmental damage.					✓	✓	✓
Environmental Impact Assessment – EIA (2011/92/EU) and Strategic Environmental Assessment Directive – SEA (2001/42/EC)	Environmental assessment is a procedure that ensures that the environmental implications of decisions are taken into account before the decisions are made. EIA tends to refer to individual projects whereas SEA is used for public plans and projects at national level. The common principle of both Directives is to ensure that plans, programmes and projects likely to have significant effects on the environment are subject to an EIA/SEA, prior to their approval or authorisation.	✓		✓	✓	✓		✓
Convention on Biological Diversity, 1992	Multilateral treaty for the conservation of biological diversity, sustainable use of biological materials and to ensure fair and equitable benefits from genetic resources.							✓

1. The CAP is described extensively in Chapter 1. Water policies and related legislative instruments are described in Chapter 4 with some additional coverage of the Water Framework Directive in Chapter 3.

2. Key: W – water, En – energy, L – land, S – soils, NR – natural resources, Wa – waste, Ec – ecosystem services

of the control of individual governments. Therefore, risk management and European energy polices generally concentrate on:

- Taking actions to reduce dependency on imports.
- Diversifying energy supply types to reduce dependency on any one single source (e.g. oil, gas). A fundamental part of most European energy security policies is to increase reliance on more localised, sustainable, renewable energy sources such as hydroelectricity, solar energy, wind energy, wave power, geothermal energy, bioenergy and tidal power.
- Increasing infrastructure and collaborative networks across Europe to rely more on internal markets and less on external ones.
- Implementing actions to increase each European country's ability to produce more of the fuel and electricity it needs by greater investment and sound energy security policies. To this end, investment in more sustainable and locally produced energy technologies are often prominent in many European countries' energy security policies.
- Becoming more energy-efficient with particular emphasis on buildings and industry.

At first glance, it may seem that these types of issues have little to do with the European agri-environment, however this is far from the truth. Agriculture is a significant consumer of energy, directly through transport, machinery use, heating, lighting, cooling and irrigation, and indirectly through the manufacture of fertilisers, pesticides and farm machinery. However, it is also a producer of energy via biofuels, on-farm wind and solar installations and energy from waste systems. Therefore, agriculture has a vital role to play in European energy security.

Energy crops used to create bio-alcohol fuels, such as wheat, corn, soybeans and sugar cane, can help decrease the amount of fossil fuels consumed (see Chapter 2). It is generally considered that these fuels are cleaner and less polluting to burn than fossil fuels, emitting fewer greenhouse gases and air pollutants. Biodiesel from vegetable oils offers similar environmental benefits. These types of fuels can be produced locally reducing the dependency of individual countries and, indeed, the EU, on expensive foreign fuel. Producing biofuels locally can also generate a significant economic stimulus as production centres need good, stable supplies of a suitable feed stock which increases the demand for suitable biofuel crops to be grown locally. This creates jobs at production sites, increases agricultural job security and helps boost rural economies. However, biofuels are not without their disadvantages. They have significant demands for land and water and so compete with food production. As biofuels have a low energy output compared with fossil fuels greater quantities are required to satisfy existing demands. Indeed, some analysts believe that when a more holistic, whole-system view is taken that considers the farm production and cultivation processes of the crops needed for biofuel production (which all need energy), the benefits are not so clear-cut. Some reports even suggest that ethanol produced from corn in the United States and in parts of Europe actually consumes more energy than it produces (Pimentel *et al.*, 2007; Ulgiati, 2001), although not all studies agree (e.g. Farrell *et al.*, 2006; Zah *et al.*, 2007). Considerable amounts of bio-alcohol are produced from sugar cane and there is some evidence that carbon-rich tropical forests are being destroyed to make way for sugar cane fields, thereby causing vast greenhouse gas emission increases (Righelato and Spracklen, 2007). The picture becomes even more of a concern if the full benefits to society of tropical forests are taken into consideration, such as the wildlife they support, hydrological functioning, soil protection and genetic resources that can benefit the pharmaceutical industries and so human health (e.g. Bala *et al.*, 2007; Albuquerque *et al.*, 2012).

As a consequence, the EU has, over the last 15 years or so, developed a comprehensive policy on biofuels as well as a number of legal instruments including the Renewable Energy Directive, the Fuel Quality Directive and the Biofuels Directive, 2003. With respect to biofuels, the Renewable Energy Directive ensures that the feedstock for biofuel production does not come from carbon-rich land or land that is biodiverse. This directive also gives MSs an incentive to create biofuels from waste, residues, non-food cellulosic material and lignocellulosic material as these will count twice towards the targets for using renewable energy in transport. Similarly, in some respects, the Directive to Reduce Indirect Land Use Change for Biofuels and Bioliquids (ILUC) limits the way MSs can meet the target of 10 per cent for renewables in transport fuels by placing a cap of 7 per cent on the contribution of biofuels produced from 'food' crops, and a greater emphasis on the production of advanced biofuels from waste feedstocks. Thus, this goes some way to protecting land suitable for food production from being lost to energy crops.

Policies that drive the production of biofuels from wastes are not limited to EU MSs. Indeed, most of the non-EU nations have similar policies. In Switzerland, for example, there is evidence of a growing trade in biofuels produced from used cooking oil (RSB, 2015). One particular company, Nidera, claims to be able to convert a wide variety of vegetable oils to a biodiesel that has the highest level of greenhouse gas savings. In Iceland, because of the harsh landscapes and climate, it is simply not feasible to produce a significant amount of biodiesel from crops, however, there are several policy-driven initiatives to produce biofuels from waste. A demonstration project is underway in Akureyri in north Iceland, utilising waste vegetable oil and animal fat as feedstock for biodiesel production. The biodiesel is blended with diesel and used to power Akureyri's public transport system (Orkustofnun, 2017). Not only does this help address the issue of energy security but it also reduces food waste.

Most of the world's oil and gas reserves are under oceans and so oceans are important for ensuring energy security. However, oceans are also essential for the proper functioning of the Earth's water cycle (see Chapter 1) and it has been estimated that over 48,000 fauna and flora species rely on the marine environment (Costello and Wilson, 2011). In addition, 41 per cent of the EU population, equating to around 206 million people live in coastal regions (EEA, 2014) relying significantly on the oceans for their food and often livelihoods. Oceans deliver many other ecosystem services. For example, they play an important role in removing carbon from the atmosphere and producing oxygen. The oceans are also a rich resource of valuable minerals and metals such as salt, sodium, potassium, magnesium, manganese nodules, phosphorites and deposits of gold, tin and titanium. Therefore, it is vital that oceans are protected from damage when oil and gas are extracted from the ocean floor. The Thematic Strategy on the Protection and Conservation of the Marine Environment, adopted in October 2005, is aimed at providing an integrated approach to the protection of the marine environment and setting clear, sustainable objectives to reduce pollution, including that from oil and gas facilities, and ensuring that all European coastal waters and marine environments achieve 'good environmental status' by 2021. Part of the policy process includes a socio-economic analysis of human activities that depend on these waters. In addition, the 1992 OSPAR Convention is a legal instrument that seeks to protect the marine environment of the North-East Atlantic from pollution.

5.2.3. Land and soils policies across Europe

Due to the broad range of soil functions and the wide range of pressures exerted upon them, and land in general, soil protective measures are spread over many different policies, which have various degrees of effectiveness. Few MSs have policies in place which protect soils from all risks. Whilst many European countries have policies and regulations to protect soil from contamination such as oil, chemical spills and waste as well as those relating to noxious emissions from industry (e.g. Industrial Emissions Directive [IED]), these are not sufficient to protect soil against all risks, in particular erosion, compaction and salinisation. Recognising this, in 2006 the EU adopted the Soil Thematic Strategy (COM(2006) 231/ Com(2012) 46 final) that seeks a broader sustainable approach to soil protection with emphasis on avoiding further degradation, preservation of soil functions and the restoration of degraded European soils. The strategy recognises that whilst agriculture can impact negatively on soil quality and function it can also be beneficial via sympathetic land management practices that enhance soil organic matter and reduce erosion and soil structure degradation. Therefore, this framework policy links directly to various other policies to further enhance soil protection. Not least of these is the CAP (Chapter 1) and, in particular, elements of cross compliance (Chapter 1) whereby good agricultural and environment conditions (GAEC) describe a minimum level of soil protection expected. For example, GAEC 4 is concerned

with providing minimum soil cover, GAEC 5 protecting against soil erosion and the aim of GAEC 6 is maintaining soil organic matter. The rural development programme (Pillar 2 of the CAP) also supports soil protection objectives via various agri-environment measures, in particular preventing soil erosion, degradation and boosting fertility.

The issue of contaminated agricultural land is also addressed by the Soil Thematic Strategy which includes the objective of restoring all EU contaminated sites. In terms of protecting soil from potentially polluting activities, there are specific legislative measures such as the Sewage Sludge Directive (86/278/EEC) which aims to encourage the use of sewage sludge in agriculture so as to benefit from its nutrient content, but recognises that its use needs to be regulated in order to prevent negative impacts on soil, vegetation, animals and humans. The use of untreated sludge on agricultural land is prohibited unless it is injected or incorporated into the soil, and sludge of any type must not be applied to soil in which fruit and vegetable crops are growing or grown, or less than ten months before fruit and vegetable crops are to be harvested. Grazing animals must not be allowed access to grassland or forage land less than three weeks after the application of sludge.

Soil protection is also offered through a range of broader laws and regulations that apply to any land and not just agricultural, such as the aforementioned IED Directive and those that seek to ensure the polluter pays principal for the clean-up of contaminated sites, such as the Environmental Liability Directive (2004/35/EC). Most EU MSs have enshrined the polluter pays principal into their national laws relating to contaminated land. Spain, for example, addresses contaminated land via Law 22/2011 on Waste and Contaminated Soils and Spanish Law 26/2007 on Environmental Liability requires operators to prevent and remedy, at their cost, any contaminated sites.

Besides directives and other legal instruments addressing soil protection, there are several other policies that include soil and land protection in their remit. For example, the Resource Efficiency Roadmap includes initiatives introducing sustainability in contaminated land management. Soil protection also features in various European planning laws covered by the Environmental Impact Assessment and Strategic Environmental Assessment Directive (2001/42/EEC). This directive seeks to ensure that the environmental consequences, including those for soil biota and fertility, are properly evaluated and considered before new developments proceed.

In addition to this strategy, there are a number of international agreements that also strive to protect soils, for example the Convention on Biological Diversity (CBD) was introduced at the Earth Summit in Rio de Janeiro which entered into force in 1993 (See Chapter 3). This convention identified soil biodiversity as an area requiring particular attention, following which an International Initiative for the Conservation and Sustainable Use of Soil Biodiversity was established. In July 2016 under CBD 101 countries worldwide including most of Europe submitted new national biodiversity strategies and action plans that set out the contributions of each nation to the achievement of the Aichi Biodiversity Targets, a set of 20 global targets to be achieved by 2020. Several of these targets are highly relevant to soil protection. For example, Target #7 seeks to ensure agriculture, aquaculture and forestry are managed sustainably. Target #15 seeks to ensure that ecosystem resilience and the contribution of biodiversity to carbon stocks are enhanced, through conservation and restoration. Other targets seek to address issues such as raising awareness and

knowledge sharing, implementing plans for sustainable production and consumption, and addressing the loss of habitats as well as including actions to reduce pressure on vulnerable ecosystem services and the protection of genetic diversity. Another relevant international treaty relates to the sustainable development of the Alps. This was signed in 1995 by eight European MSs with a vested interest in the region. This treaty seeks to protect the natural alpine environment whilst promoting its economic development and one of the specific measures adopted by the signatories is the Soil Conservation Protocol which seeks to preserve the ecological functions of soil, prevent soil degradation and ensure rational use of soil in that region.

Whilst many European countries do not have comprehensive national policies in place for protecting against the loss of fertile agricultural land and fragmentation, there are some that do. For example, in Germany soil protection is enshrined in many different laws and regulations including the Federal Soil Protection Act and the Federal Soil Protection and Contaminated Sites Ordinance. Both building and regional planning regulations also contain provisions relevant to soil. Germany also has a national target to reduce soil sealing, the covering of land for housing, roads or other construction work, from 120 ha day^{-1} to 30 ha day^{-1} by 2030. The Netherlands has comprehensive legislation in place that underpins their national soil policy. These include the Soil Protection Act (Wet bodembescherming – Wbb) and the Environmental Protection Act (Wet milieubeheer – Wm). Similarly, in Poland issues relating to soil protection and its quality are regulated by the Environmental Protection Act, 2001 (Ustawa o ochronie środowiska, 2001) and the Prevention and Remedying of Environmental Damage Act, 2007 (Zapobiegania i zaradzania szkodom wyrządzonym środowisku naturalnemu ustawy, 2007).

The contribution of soils towards climate change cannot be ignored as soils are a major carbon store that must be protected and enhanced. Carbon sequestration in agricultural soils by some land management practices can contribute to mitigating climate change. The Kyoto Protocol highlights this and under this protocol the European Climate Change Programme (ECCP) Working Group on Sinks Related to Agricultural Soils estimated this potential at equivalent to 1.5 to 1.7 per cent of the EU's anthropogenic carbon dioxide emissions during the first commitment period. This is discussed extensively in Chapter 2.

5.2.4. Policies to protect natural resources

Without exception, everything produced (such as food, fibre, fuels, constructions and consumables) uses natural resources. Everything is either animal, plant, mineral or a composite of these and consequently, natural resources are fundamentally important to the survival of all species, our quality of life and the development of economies the world over. In addition, ensuring ample affordable access to resources is critical to increasing agricultural productivity. The EU's ten-year economic growth strategy 'Europe 2020', clearly recognises that agricultural production faces a multitude of pressures but believes that by making good use of new technologies and management tools as well as optimising our use of natural resources, agriculture has the capacity to meet societal demands for food. The Roadmap to a Resource Efficient Europe (COM(2011) 571) takes the 'Europe 2020' strategy further and outlines how Europe's economy can be transformed into a sustainable one by 2050. It illustrates how policies interrelate and build on each other. The Roadmap provides

a framework and timetable in which future actions can be designed and implemented coherently.

A significant part of this policy is the Thematic Strategy on the Sustainable use of Natural Resources which was introduced in late 2005. It seeks to reduce the environmental impact associated with natural resource use particularly within the context of a rapidly growing population that will see an associated increase in demand for food, fibre and fuel. The driver for the introduction of this strategy was to see if the rate of resource use could be decoupled from economic growth such that society starts to see greater value from natural resources per unit of quantity used. Originally, there were plans to include quantitative targets for resource efficiency but this was eventually dropped due to the lack of data and appropriate scientific knowledge. Instead, the strategy concentrates on (i) raising awareness of the issues associated with depleting natural resources, (ii) gathering, evaluating and providing greater access to knowledge such as that on resource reserves and their depletion rates which included promotion of life cycle thinking and assessments, (iii) developing indicators to allow progress towards more efficient use to be monitored, and (iv) developing sustainable alternatives to natural resources such as biomass energy and safe sewage sludge products. In addition, an international panel and working groups were established to focus on the progress towards decoupling natural resource usage rates and economic growth.

Encompassed within this strategy are a number of key policies and initiatives which collectively seek to encourage sustainable use of resources. The principal approach is laid out in the EU Action Plan on Sustainable Consumption, Production and Industry. This action plan, adopted in late 2008, seeks to improve the environmental performance of products and increase the demand for more sustainable goods and production technologies. It also seeks to encourage EU industry to take advantage of opportunities to innovate. In respect to innovation, the EU's Bioeconomy Action Plan aims to achieve economic growth by promoting technological innovations that have low environmental impact. The bioeconomy in this context is defined as the production of renewable biological resources and their conversion into food, feed, bio-based products and bioenergy. It includes agriculture, forestry, fisheries, food, and pulp and paper production, as well as parts of the chemical, biotechnological and energy industries. There are also initiatives to encourage environmental labelling, particularly via the EU Ecolabel scheme, established through Regulation (EC) 66/2010, which seeks to identify products and services that have a reduced environmental impact throughout their life cycle compared with the norm. The objective of labelling schemes generally is to encourage consumers to make more informed purchasing choices and to encourage, via market competition, manufacturers to improve their production processes.

The EU Raw Materials Initiative was launched in 2008. Its objectives are to recognise and attempt to address issues relating to access, supply and costs of raw materials to the European Union. Sustainable access to raw materials is essential to economic growth. Considering the importance of phosphorus to agriculture and so food security (Chapter 5.1), it is not surprising that phosphorus is amongst the 20 critical raw materials identified by this initiative. However, as yet, there are not strong policy measures in place across Europe to ensure the resource is used sustainably. This mineral is widely used in agricultural fertilisers and within livestock feeds and so is vitally important for increasing productivity.

However, as described in Chapter 5.1, supplies are limited and access is vulnerable. A report commissioned by the European Commission concluded that policies and regulations throughout the EU and EEA did not take into account the threat to supplies of phosphorus and that there was a severe lack of global governance of the mineral. The report's authors also stated that the issue of phosphorus scarcity was not a global priority despite its importance for food security and it being a finite resource for which there is no substitute. In July 2013, the EC launched a consultation to identify the opportunities available for securing phosphorus supplies and for ensuring it is used more sustainably (Schröder *et al.*, 2010). Consequently, it is probable that policies for phosphorus will develop in the coming years.

5.2.5. *Waste management*
The EU's policy for waste management is based on the 'waste hierarchy', also known as Lansink's ladder after the politician who first proposed the approach, which ranks disposal options according to which offers the least environmental impact: prevention and reduction, reuse, recycling, recovery (including the generation of energy from waste) and, as the least preferred option, disposal (which includes landfilling and incineration without energy recovery; see Figure 5.12).

The development and implementation of EU waste policy is addressed by a number of wider EU policies and programmes and is closely linked to policies on the sustainable use of natural resources. A key part of the EU waste management policy is the Thematic Strategy on the Prevention of and Recycling of Waste (COM(2005)666) which was adopted in December 2005 and aims to avoid waste production, encourage recycling and to seek ways of making waste disposal more sustainable, by, for example,

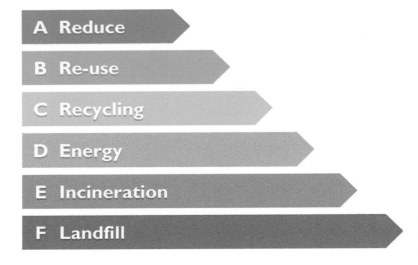

A Reduce

B Re-use

C Recycling

D Energy

E Incineration

F Landfill

Figure 5.12: Waste hierarchy or Lansink's ladder

encouraging waste to energy schemes and introducing life-cycle thinking into waste policy. The overall objective is to reduce the environmental impacts associated with resource depletion and waste management. There are a number of European regulations associated with this thematic strategy that relate directly or indirectly to agricultural resource use and productivity. The key legal instrument is the Waste Framework Directive (2008/98/EC) that provides the legislative framework for the collection, transport, recovery and disposal of waste, and includes a common definition of waste. It has a number of broad aims and objectives including the introduction of the 'polluter pays principle' and the 'waste hierarchy'.

The two main statutes relating to waste disposal are (i) the Landfill Directive (99/31/EC) that aims to prevent or minimise the environmental impacts of the disposal of waste to landfill and to harmonise controls and standards for the design, operation and management of such sites across the EU, and (ii) the Waste Incineration Directive (2000/76/EC) that seeks to prevent or minimise the environmental impacts of waste incineration, in particular, emissions to air, soil and water.

There are several policies and legal instruments concerned with waste from industrial installations. The majority of these are not greatly relevant to the agri-environment. The main exception to this is the Industrial Emissions Directive (IED Directive (2010/75/EU) which introduces an integrated environmental approach to the regulation, control and management of certain industrial activities including intensive pig and poultry units. Air, water (including discharges to sewer) and land emissions (including ammonia, noxious odours) and many other environmental effects must be considered together and an operating permit system is used to ensure a high level of environmental protection. These conditions are based on the use of the Best Available Techniques (BAT), which balances the costs to the operator against the benefits to the environment.

There are also various legal instruments for handling specific waste streams particularly those that are highly damaging to the environment or particularly harmful to human health, such as the Waste Electrical and Electronic Equipment Directive (2012/19/EU). Few of these will affect farm businesses to any great extent other than, perhaps, dairy farmers who rely on cooling equipment for milk storage and may need to consider the Ozone Depleting Regulations (1005/2009) that prohibit and control the production, use and waste management of ozone-depleting substances in order to reduce emissions in line with those agreed by the Montreal Protocol (an international agreement to combat the threat of damage posed to the ozone layer by ozone-depleting substances).

5.2.6. *The circular economy*

The circular economy is closely related to the conservation of resources and to sustainable waste management. Currently, the majority of the foods and products that are produced and consumed are part of what is often referred to as a 'linear economy', meaning that raw materials are used to make a product which is then eaten or used. By-products and waste from the production process are disposed of. Any food or other product no longer considered fit for purpose is then also disposed of. In contrast, a 'Circular Economy (CE)' aims to ensure that products are designed such that all the resources used in its production are kept in use for the maximum amount of time and the maximum amount of value is obtained from them. A perfect CE (shown in Figure 5.13) would mean that a product was designed, produced and used in a manner that produced no waste. The waste from one product would become a secondary raw material for the production of another product. Industry and society are a long way from achieving this utopia but nevertheless there is growing interest in this approach as it offers a wide range of benefits beyond waste minimisation, including the potential for greater resource productivity, a more competitive industry and reduced environmental impact.

Recently, the CE concept has become part of the 'Green Economy (GE)' concept which

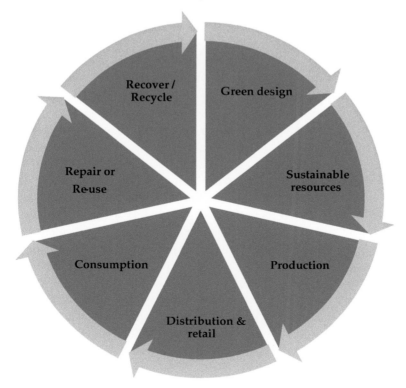

Figure 5.13: Circular economy

combines CE with various social issues to provide a means to protect future supplies of natural resources. GE was the central theme of the United Nations Conference on Sustainable Development (Rio+20) held in Rio in June 2012. The concept proposes a sustainable approach to growth and development promoting a triple bottom line of sustaining and growing economic, environmental and social well-being. UNEP describe the GE as one that is low-carbon, resource-efficient and socially inclusive (UNEP, 2011).

The CE concept is now embedded in many policies and strategies across Europe. Within the EU it is seen as conducive to many of the EU's policy objectives and is obviously closely linked to the thematic strategies on waste and natural resources. In 2015, the EC published its Circular Economy Strategy 'Closing the Loop' (COM(2015)614 final) which aimed to stimulate all industries, including the food sector, to adopt the concept and so help the European economy become more sustainable and competitive. The strategy includes a 'circular economy package' which incorporates a comprehensive action plan with measures aimed at all stages of a product's life cycle from the design phase, production, consumption and waste with significant emphasis on reducing landfill, increasing recycling and reuse. The strategy also includes plans to revisit existing related legislation to identify any opportunities for developing a CE including, for example, sustainable sourcing of raw materials, extending producer responsibilities, energy recovery and addressing food waste.

Adoption and implementation of the CE concept across Europe is variable but attention to sustainable waste management is a high priority on most political agendas. In Germany, the focus is on ensuring that waste is sustainably managed and waste production is decoupled from economic growth. Revised legislation (BMU, 2013) ensures that

the five-step waste hierarchy (see Figure 5.12) is applied. In Switzerland, the country has introduced various incentive schemes associated with clean production and the GE including, for example, a national waste management action plan which has three steps: avoidance, reduction (reuse and recycling) and schemes for environmentally responsible waste disposal. Developing a CE is a priority in the Netherlands with many initiatives seeking to make the country a leader in its implementation. Many of the projects that are supported by the 'Green Deal' initiative and the RACE (Realisation of Acceleration of a Circular Economy) programme go well beyond waste management and include those on circular product design, high-quality products based on secondary raw materials and analysing and removing the barriers to a CE. In Sweden, the country's 2017 budget proposals include reducing the rate of VAT payable on repairs to various goods in order to encourage people to be less wasteful and to increase jobs in the service industries.

Within the agricultural and forestry sectors, the CE is particularly pertinent due to the high dependency of these industries on natural materials and natural cycles. Failure by these sectors to address unsustainable practices threatens productivity, food security and the quality of life of many European communities. Therefore, policy tends to focus on issues such as making more of crop waste to avoid this going to landfill or being incinerated. Crop residues can be used as green manures or composted, thus returning nutrients back into the soil, or used for cellulose production for biofuel feed stocks and other products such as bioplastics. Anaerobic digestion can be used to convert biowaste into biogas (see Chapter 5.3). Food waste has potential for use as livestock feed, although currently the use of many food waste streams for this purpose is prohibited by many countries due to health issues associated with their mixed composition. However, potentially single-product food waste streams could be used for this purpose.

Another key focus is on making better use of grey water. In Cyprus, for example, treated waste water is already used for irrigating agricultural crops, landscapes, golf courses and green areas around hotels and is considered to be an indispensable resource. Italy uses treated waste water to irrigate over 4,000 ha of agricultural land. Portugal uses treated wastewater for irrigation and road construction and in Spain, several regions already rely on treated waste water to irrigate agricultural crops, recharge groundwater and for river flow augmentation (EU, 2013).

The CE offers other opportunities for waste water reuse. Globally, the key nutrients phosphorus (P) and nitrogen (N) are lost in vast quantities due to unsustainable industrial and farming practices. Loss of these nutrients via run-off, leaching and direct release to surface waters is the main reason why many EU MSs fail to achieve European water quality objectives. However, nutrients are often recoverable from many waste water streams using new nutrient recovery technologies that can yield value-added products such as environmentally-friendly fertilisers. This can be particularly beneficial in areas where excess manure applications have caused a build-up of soil P that can then be lost to surface waters. It will also reduce pressure on scarce natural resources. There are a number of different nutrient recovery technologies suitable for both waste water and liquid manure/slurries and the two attracting the most interest are:

- Struvite extraction, whereby P is precipitated as struvite (also known as magnesium ammonium phosphate hexahydrate – $MgNH_4PO_4.6H_2O$); the

resulting salt can be used as a slow-release fertiliser.
- Quick Wash combines chemical treatment and membrane separation to produce a granular form of calcium phosphate and low-phosphorus biosolids. The calcium phosphate can be sold as a secondary raw material and the biosolids can be safely spread on the land or sent to landfill without fear of causing pollution.

In support of the CE and due to concern over both water pollution and the increasing scarcity of P, the EU published a draft revision of the Fertiliser Regulations 2003/2003 in 2016. Seen as the first deliverable in the CE package, the aim of the revision was to ease the access of organic and waste-derived fertilisers to the EU single market, putting them on a level playing field with traditional, non-organic fertilisers. Regulations in this area are highly likely to become more and more stringent such that in the not too distant future, the sustainability of many livestock farms on P-rich land will depend upon their ability to meet new regulations requiring phosphorus management as well as nitrogen management.

5.2.7. Direct policies to enhance or protect productivity

As can be appreciated from the discussions in Chapter 5.1, the policies to address the need to increase agricultural productivity are, out of necessity, multiple and diverse. Those discussed above in Chapter 5.2.2 to 5.2.6 are concerned with management and regulation of issues relating to resources including soils, whilst Chapter 4.2 discusses water policy and Chapter 3.2 is concerned with polices relating to ecosystem services. In addition, there are a number of other policies that are more directly concerned with increasing productivity. These fall into a number of key areas:

FINANCIAL INCENTIVES, MARKETS, TRADE AND INNOVATION

The EU's Common Agricultural Policy has several major objectives, namely encouraging agricultural productivity, providing a fair standard of living for farmers, boosting rural economies and promoting stability in markets. With respect to this latter objective, the aim is to help stabilise imports and exports, secure food supplies and ensure quality produce at reasonable prices for EU consumers. Global agricultural markets are often turbulent and uncertain. The balance between production and prices is often finely tuned such that a shortfall in demand due to, for example, a failed crop caused by adverse weather conditions or a natural disaster can quickly increase costs. Price volatility makes planning for farmers and growers very difficult. In consequence, the 2013 reform of the CAP (see Chapter 1) focuses more than, perhaps, previous reforms, on improving food security, reducing price volatility, promoting sustainability, improving long-term competitiveness and achieving greater economic efficiency by better targeting of support. In addition to the fundamental CAP process of farm subsidies through cross compliance and grant aid, there are a number of other mechanisms and interventions including those to address price volatility and farm incomes, such as crisis management to deal with exceptional market situations and affordable farm income insurance. Competitiveness is currently driven through agricultural research and innovation as well as fostering partnerships between researchers and farmers. The EU's Horizon 2020 research and innovation programme is the main funding mechanism for these types of initiatives. By coupling research and innovation, Horizon 2020 aims to places an emphasis on science excellence, industrial leadership and tackling societal challenges such as food security and land use change. To support the delivery of new and innovative technologies, the CAP (under the Rural Development Programme) also offers a range of grants and initiatives for training and the upgrading of machinery that allows, for example, for improved mechanisation such as that which is appropriate for precision farming (e.g. satellite images and global positioning systems). The availability of these types of grants is usually subject to MS management and rules, and varies considerably from one year to the next.

Competitiveness is also increased by ensuring that consumer demands are met, not only in terms of access to a variety of quality produce at affordable prices but also by providing evidence that sustainable production methods have been used. In many European countries there are various schemes for assuring that quality food standards have been met. These can be operated by independent organisations such as the UK's Assured Food Standards (Red Tractor Schemes), LEAF Marque, the Dutch Milieukeur Agro and the RSPCA Welfare Standards, whilst others are operated by food retailers such as Carrefour's Filière Qualité, Tesco's Nurture, Sainsbury's 'Integrated Crop Management System' and GlobalGAP. Each of these schemes has their own objectives and focus. For example, Assured Food Standards and GlobalGAP are primarily concerned with food safety and quality; LEAF Marque is an environmental assurance system, whilst the RSPCA's standard is concerned with farm livestock health and welfare. However, it is often the case that delivering a scheme's specific objectives provides secondary productivity and sustainability benefits. For example, if protecting food quality requires sound pesticide use then this will deliver environmental benefits whilst high animal welfare standards will often offer productivity gains.

Most of these schemes communicate with consumers using distinctive labels that farmers, enrolled in the scheme, can display on their produce to show it has been grown to the scheme's standards. The schemes discussed above are all industry-driven and the standards embedded within each, in terms of their scope and how stringent they are, is entirely at the discretion of the scheme management. However, this is not the case with certified organic produce. All organic food sold and marketed in Europe must be grown according to legally enforceable standards and compliance is assured by inspections and certification by approved independent bodies. The organic farming philosophy and the standards are currently enshrined into European law by Council Regulation (EC) 834/2007 on Organic Food and Commission Regulation (EC) 354/2014 (which revises and corrects EC regulation 889/2008) which lays down the detailed operational rules and standards. The European regulations define the objectives and principles of organic production, as well as practices and inputs that may be used in farming, livestock production and processing. Farmers complying with the Organic Regulations will benefit from the CAP cross compliance farm payments normally without fulfilling any further obligations because of their overall significant contribution to environmental objectives. Support to organic farming can help to improve the competitiveness of agriculture and as a farm management system it contributes to improving the environment and the countryside. Currently, the European approach to organic farming is under review and so changes can be anticipated to these rules and standards in the coming years.

The EU also provides a wide range of different mechanisms for providing grants for farmers to improve productivity. These are usually subject to MS management and rules, and vary considerably from one year to the next. Typically, these grants are available for upgrading facilities or equipment to improve energy or water management, investment in equipment, technologies or processes to reduce waste or investment to modernise or mechanise production and increase productivity.

CROP PROTECTION

There is little doubt that the use of pesticides (see Figure 5.14) delivers a wealth of benefits to society. They significantly improve agricultural yields and help to ensure quality produce and so protect farm incomes. Without pesticides it has been estimated that global food production could fall by as much as 35 to 40 per cent (Oerke, 2005), increasing the cost of food and threatening food security. However, they do have the potential to cause significant harm to the environment and human health, and so a balance is needed between protecting crop productivity and protecting environmental quality and non-target species such as pollinators. Pesticide safety has been a primary objective of EU policy since the late 1970s and policies have continued to evolve and develop to address emerging concerns. However, it has proved difficult to control some issues such as levels in the aquatic environment and residue levels in fresh produce. Consequently, the Thematic Strategy on the Sustainable Use of Pesticides was introduced in 2006 to try and address these issues by providing a coherent and integrated approach. The strategy provides a set of controls on pesticide use and distribution, a stringent regulatory process which aims to evaluate risks and reduce the use of the more damaging substances, encouraging the use of safer alternatives, and establishes a comprehensive system for monitoring and reporting progress made towards the strategy's objectives.

Like other Thematic Strategies, there are a number of related policies, initiatives and regulations. The main one is the CAP which includes many measures to promote

Figure 5.14: Crop spraying
(Photo courtesy of: Pixabay.com)

the sustainable use of pesticides particularly within cross compliance and the Rural Development Programme. There are also a number of regulations relating to the protection of biodiversity (Chapter 3.2) and water pollution (Chapter 4.2) including the Water Framework Directive (2000/60/EC). In addition, the Drinking Water Directive (98/83/EC) establishes minimum standards for various substances in water. As discussed in Chapter 4.2.3, the maximum allowable concentration for individual pesticides in drinking water is 0.1 µg L^{-1} and the accumulated total of all pesticides must not exceed 0.5 µg L^{-1}. These limits have been set based on a precautionary approach and are not related to the actual toxicity or potential for harm that any one pesticide may have.

The Directive on the Sustainable Use of Pesticides (2009/128/EC) is the key policy that influences which pesticide active substances are available to farmers and how they may be used. It establishes a framework of EC-wide action to ensure pesticides are used sustainably. The main feature of the directive is that each MS should develop and adopt a national action plan and set up quantitative objectives, targets, measures and timetables to reduce risks and impacts of pesticide use on human health and the environment and to encourage the development and introduction of integrated pest management and of alternative approaches or techniques in order to reduce dependency on the use of pesticides.

Within the pesticide thematic strategy, the Plant Protection Products Regulation (1107/2009) provides the controls on which crop protection substances can be placed on the European market. Before any pesticide can be legally sold or used, it must first be approved at EU level which requires the risks to human and animal health, and the environment as well as the substances efficacy, to be assessed. Once approved at EU level, it can then be used in plant protection product formulations but these formulated products must then be authorised in each MS where they are to be sold and used.

Residues in food stuffs are regulated via the Maximum Residue Levels (MRL) Regulations (EC 396/2005 and amendments), which control the maximum amount of pesticide residue that is legally permitted in food or animal feed and is based on what would be expected if good agricultural practice has been undertaken. These regulations seek to provide a harmonised assessment scheme across the EU and set a general default MRL for all foodstuffs and pesticides at 0.01 mg kg^{-1}. However, this default is overridden by a more stringent value where scientific evidence suggests this is necessary.

LIVESTOCK HEALTH AND WELFARE

Much of the focus on improving agricultural productivity is on crop production and its use of natural resources. However, the livestock sector should not be ignored. Demand for animal products (e.g. meat, milk and eggs) is increasing globally and this trend is expected to continue as populations and economies grow. Therefore, improving the productivity of this sector is as important as that of the crop sector. This can be achieved by either increasing livestock numbers and/or increasing the health and welfare of the stock we keep to improve productivity. Livestock production is linked with significant environmental impacts contributing towards climate change, water and air pollution and impacts on biodiversity, for example. Livestock production is also the world's largest user of agricultural land, through grazing and the use of feed crops (Steinfeld *et al.*, 2006). Whilst, due to demand and market forces, this is probably inevitable, if we significantly increase the number of livestock we should expect similar increases in the environmental impact. Therefore, it is important that we seek to increase livestock productivity through health, welfare and sustainability measures as much as possible. Good nutrition is a fundamental requirement of all livestock and it is considered one of the biggest contributors to animal welfare. Obviously, poor nutrition will strongly affect productivity but it can also impact on the safety and quality of the food chain as it can increase susceptibility to diseases among animals, thus increasing the need for veterinary treatment. In turn, this can pose risks to consumers and can mean in some instances that the animal cannot be sold into the food chain, therefore decreasing profitability and endangering environmental sustainability of the livestock production systems and of the associated animal food chains. The EU Strategy on Animal Welfare focuses on enhancing knowledge among the many different stakeholders involved in the process. It also works to improve the competitiveness of European agricultural products by ensuring that markets and consumers recognise animal welfare as an added value.

There are also a number of EU regulations that seek to control feed quality such as Regulation (EC) No 882/2004 that describes how MS Competent Authorities should enforce compliance with feed and food laws including guidelines relating to animal health and welfare. Commission Regulation (EC) No 152/2009 addresses sampling and analysis of feed for control purposes, and the Marketing, Labelling and Composition of Feeds Regulation (767/2009) requires scientific evidence to support claims relating to the composition and function, establishes the minimum requirements for feed labels and introduces a procedure for applications for the authorisation of new nutritional purposes.

5.3. Farm-level management

5.3.1. Introduction

Much of what can be done at farm level to address the issues outlined in this chapter has been discussed in detail elsewhere, for example climate change (Chapter 2) and water (Chapter 4) and so the text below addresses those elements not previously discussed.

5.3.2. Energy

Energy efficiency is covered in Chapter 2.3 from the perspective of reducing emissions. However, at farm level there are also opportunities for reducing the consumption of non-renewable energy resources by taking action to produce energy/fuel on the farm.

For the last two decades or so, Europe has invested significantly in biofuels and energy crops and this has included using initiatives to encourage farmers to produce crops for energy rather than for food. Biofuels include solid biomass crops, liquid fuels such as biodiesel, bio-alcohol and various biogases. Data for 2008 shows that 66 per cent of European-produced biodiesel came from rapeseed (see Figure 5.15), 13 per cent from soybean and 12 per cent from palm oil (Junginger *et al.*, 2013). Bioethanol is an alcohol made by fermenting the sugar components of plant materials and it is made mostly from sugar and starch crops such as wheat, sugarbeet and corn. Using new technology, cellulosic biomass, such as trees and grasses, can now also be used as bioethanol feedstocks. Bioethanol can be used as vehicle fuel in its pure form but it

Figure 5.15: Rape for biodiesel production
(Photo courtesy of: Dr Doug Warner)

is usually used as a fuel additive to increase octane rating and improve vehicle performance (Shah and Sen, 2011). EU bioethanol production in 2005 exceeded 910 million litres, an increase of 73 per cent on the previous year. The majority of production is undertaken in Spain, Germany, Sweden and France. Across Europe, production of crops for solid biomass is undertaken at a much smaller scale, that is, in 2007 between 50,000 and 60,000 ha of land was given over to solid biomass energy crops including miscanthus, willow, poplar and reed canary grass. Much of this is grown in the UK, Sweden, Finland, Germany and Spain (Jyväskylä Innovation Oy and MTT Agrifood Research Finland, 2009).

In the EU, the ILUC Directive (see Table 5.2) and the food versus fuel debate (see Chapter 5.1) originally discouraged farmers from growing crops for biofuels. However, new drought-resistant varieties of oil crops that can be cultivated on marginal lands such that they do not compete with food crops went some way towards boosting production. Despite this, many EU MSs have not seen the widespread production of crops for energy that was anticipated by governments. There are several reasons why farmers have been discouraged from these crops, including the lack of secure and viable markets, the modest returns from power station contracts and the poor cash flow in early years. However, there are opportunities for farmers that can be win–win, providing financial benefits and helping issues relating to the use of non-renewable resources and the global energy security issue. In many cases, the best economical option is solid biomass crops that are grown for the purpose of generating energy for on-farm use such as heating farm buildings and the farmhouse. The savings made by replacing expensive fossil fuels, such as oil and LPG, especially where grant aid or incentives for using renewable heat are available, can turn willow, poplar and miscanthus into lucrative cash crops.

More and more farmers and landowners are using their own land to generate energy. For example, the energy output generated by a wind turbine installed on a farm (see Figure 5.16) can be fed directly into the farm's incoming supplies, reducing its need to draw from the distribution grid, and so reduce electricity costs. Similarly, solar farms are becoming increasingly popular, particularly in the UK and Germany. It is not unusual in some areas to see banks of ground-mounted photovoltaic panels on mixed and livestock farms. These provide farmers and landowners with a stable, long-term income whilst helping to reduce reliance on fossil fuels and diversifying energy sources. A downside of wind power and solar facilities is that, in many cases, the upfront costs can be quite considerable. Total project cost can vary significantly from one site to the next and from one country to the next. They depend on a range of factors, including the cost of the turbine or solar panels (and how many), the extent and scope of supporting environmental work for planning applications, the cost of any electrical distribution network connections and the cost of site works including access roads, foundations and cabling costs. In the UK, costs of installing these types of systems typically range from a few million pounds to several hundred million, depending on power output.

Another rapidly growing on-farm technique for sustainable energy generation is Anaerobic Digestion (AD). This is the controlled breakdown of organic matter, such as livestock slurry, without oxygen to produce a combustible biogas and nutrient-rich organic by-product (Gray *et al.*, 2008). As well as generating a gas that can be used for heating, thus reducing reliance on fossil fuels, helping mitigate climate change and

Figure 5.16: Wind turbine on farmland in Bedford, UK
(Photo courtesy of: Dr John Tzilivakis)

reducing fuel costs, it also offers farmers numerous other benefits including improved and more efficient handling and use of farm wastes and significant reductions in odour emissions. AD systems can be owned and located on a single farm, can be shared by farm groups and located conveniently, or farms can use privately-owned larger Centralised Anaerobic Digestion management facilities (CAD plants). Like wind and solar power, there can be significant costs associated with installing these facilities. The cost varies with the required capacity of the plant, the amount of environmental work required for planning applications and the cost of site works such as access roads and foundations. However, costs are significantly less than that for a wind turbine or solar fields and tend to be in the thousands rather than the millions of pounds. Whilst the capital cost for an AD plant is still relatively high, payback times are generally short because of the costs avoided in waste disposal. One disadvantage is that the management and

maintenance of an AD plant is often the responsibility of farm workers, although in a properly run facility this should generally be minimal. This is in contrast to wind and solar power installation, which require little maintenance from the farmer or landowner. If work is required, this usually needs to be done by a specialist contractor. In many EU MSs, financial support from the CAP's Rural Development Programme, via national governments, is available for establishing sustainable energy facilities on-farm.

5.3.3. Land and soils

In the case of land and soil as a resource, the issue is not consumption but maintaining the existing resource, that is, avoidance of erosion and various forms of degradation that can lead to a loss of productivity. At the farm level, soils can of course be directly inspected and monitored in the field for signs of erosion, damage or degradation.

The definition of 'quality' in terms of soil can be dependent on the functions the soil performs and judging quality is partially subjective. Soils are multifunctional, being used for growing plants (food, fuel, fibre, landscaping and decorative), providing habitat for fauna and flora, interacting with various natural cycles including water and nutrients (see Chapter 1) and for supporting our buildings, roads, and so on.

Due to this multifunctional nature of soils and their varied properties, there is no single tool for 'measuring' quality. There are, however, a wide range of indicators available covering physical, chemical and biological properties of the medium. The most appropriate indicator (or more likely, suite of indicators) is dependent on the purpose for which 'quality' is being assessed, although within the agricultural context, where the growth of crops (including grass) is the dominant soil function, it is likely that physical, chemical and biological factors will all have a role to play. In addition, the soil is a heterogeneous resource, such that the 'target' values of the relevant properties being used to define what is meant by 'good quality' will tend to vary (sometimes considerably), depending on the type of soil being evaluated. There are a significant number of soil properties which could be assessed to determine soil quality status at farm level (including those in Table 5.3), and a number of authors have endeavoured to identify those most suitable (Loveland *et al.*, 2002; Merrington *et al.*, 2006). Those likely to be of most importance (Merrington *et al.*, 2006) are likely to include pH (a measure of soil reaction, which affects nutrient availability for example), aggregate stability (soil structure), bulk density (compaction/loosening), total organic carbon (affects soil structure, fertility and carbon sequestration, for example), total N and extractable P (nutrient status) and heavy metal contamination (e.g. Cu, Ni, Zn, etc. – toxicity). This is however, by no means an exhaustive list (as evidenced by Table 5.3), but illustrates that the number of potential measures is significant, and assessing even just those deemed to be of particular importance would present a significant challenge.

In most instances on-farm activities to assess soil quality and preserve it as a resource will be limited to taking soil samples for analysis and relying on obervations and

Table 5.3: Potential indicators of soil quality

Indicators	Indicators
Above ground biomass	Extractable K
Soil water content at 1 m	Extractable Mg
Soil wetness characterisation	Extractable S
Soil storage capacity	Extractable Ca
Topsoil aggregate stability	pH
Bulk density	Total Zn, Cu, Ni, Cd
Aeration	Total Pb
Organic carbon	Extractable B, Cu, Mn, Se
Macroporosity	Biomass indicator
Soil (horizon) depth	Earthworms (total number)
Root penetration	Soil-borne diseases
Total N	Persistent Organic Pollutants
Wind throw	Agricultural Land Classification/land capability
Depth to waterlogged layer	Salinity (Electrical conductivity)/sodicity
Potentially Mineralised Nitrogen	Seed bank
Extractable P	Plastic glass/extraneous material

field walking. The aim will be to evaluate the current soil condition and then take suitable steps to mitigate any arising issues. This might mean replenishing nutrients and/ or organic matter or applying lime or some other soil conditioner to adjust pH. Rarely will anything more complex be undertaken. Soil erosion, for example, is rarely quantified at farm level with respect to the amounts of soil lost, other than under experimental conditions. However, there are several tools that apply at farm level and that may be used by consultants, advisors and for research impact assessment purposes.

QUALITY INDICES

As intimated above, to obtain a holistic impression of soil quality, multiple indicators are required, which has led a number of authors to propose the use of compound indices (e.g. Parr *et al.*, 1992; Doran and Parkin, 1994) of quality based on combinations of indicators and/or function characteristics. One of the first was that by Parr *et al.* (1992), which can be stated as:

Soil quality $= f$ (SP, P, E, H, ER, BD, FQ, MI)

Where:
SP = soil properties
H = human/animal health
FQ = food quality/safety
P = potential productivity
ER = erodability
MI = management inputs
E = environmental factors
BD = biological diversity

However, the authors made no attempt to define either how the factors could be calculated or how they should be integrated (Warwick, 2007), a problem which to a lesser or greater extent hinders many such systems, particularly when applied to a wide range of soil types and/or production

systems. Although Karlen *et al.* (2001) suggested that this may be less of a problem if more narrowly applied. Faced with this complexity, impact assessments have typically therefore tended to use the simplest and/or single measures of impact, for example soil loss/depletion, soil organic matter and soil compaction, each of which is discussed in more detail below.

SOIL LOSS/DEPLETION

There are a number of methods for estimating losses, although each has its limitations. These include the universal soil loss equation (USLE, Wischmeier and Smith, 1978) and the Morgan, Morgan and Finney model (Morgan *et al.*, 1984) amongst others.

The USLE, its derivatives, the Revised Universal Soil Loss Equation (RUSLE; Renard *et al.*, 1998) and Modified Universal Soil Loss Equation (MUSLE; Williams, 1975), are amongst the most established techniques for assessing soil loss, and have been used extensively around the world. Five component factors (R, K, LS, C and P) are multiplied together to compute the average annual sheet and rill erosion per unit area (the main differences being in the determination of the factors).

$$A = R \times K \times LS \times C \times P$$

Where:
R = annual erosivity
K = base soil erodibility
P = practices factor
C = cover management
LS = slope length, steepness, and shape

The USLE/RUSLE/MUSLE equations are relatively simple and so lend themselves well to being used in environmental assessment. However, they do have a number of limitations and shortcomings. First, although it has been used throughout the world, the model was initially designed for use in the USA east of the Rockies, and as a result it may not be truly applicable in some European situations. In addition, it can only give a very crude estimate of long-term expected soil loss; gully erosion (where this occurs, for example in southern Europe) is not taken into account, although the technique has been applied in Italy (Grimm *et al.*, 2003); and likewise, the effect of stones and rock fragments in the soil are not included. There are also uncertainties in the rainfall erosivity and soil erodibility factors. The Morgan, Morgan and Finney model (Morgan *et al.*, 1984) in contrast, was designed for use in UK situations (and therefore may again not be wholly suitable for some areas of Europe), and unlike the USLE is a process-based model of erosion. It calculates the ability of rainfall to detach sediments (based on the kinetic energy available in rainfall, and the detachability of the soils), and then the ability of overland flow to transport those sediments (based on run-off volumes, crop cover, management and slope). This too is generally applied at the field scale, although it can be used in small catchments.

More recently, there have been a number of models specifically designed for European situations (often funded by various arms of the EU), and although they are mainly designed to work at the catchment scale or larger, include concepts which may be applied at the farm scale, there being a continuum in scale between the farm and catchment scales. PESERA (pan-European Soil Erosion Risk Assessment) for example, is a model built around the partition of precipitation into a number of components relating to overland flow, evapo-transpiration, and soil moisture storage (factors which can be varied according to vegetation cover and soil properties) and soil erodibility (Kirkby *et al.*, 2008; Cerdan *et al.*, 2010). MESALES

(modèle d'evaluation spatiale de l'aléa erosion des sols – regional modelling of soil erosion risk; Cerdan *et al.*, 2010) similarly estimates erosion risk from a soil's sensitivity to erosion (soil, relief and land use-dependent) and climate, whilst G2 allows estimates of soil loss to be mapped from sheet and interril erosion caused by raindrop splash and surface run-off, on a month time step on a local to regional scale (Panagos *et al.*, 2014).

In addition, a number of the models, particularly in relation to water quality issues, include the ability to model soil losses from agricultural land, albeit from the standpoint of the impact on receiving waters, and their role in transporting attached pollutants (e.g. Alvarez *et al.*, 1997; Klein *et al.*, 2000).

SOIL ORGANIC MATTER (SOM) OR SOIL ORGANIC CARBON (SOC)

SOM/SOC is widely recognised as the best stand-alone indicator for soil quality even though it does not fully consider all aspects of soil functioning (Milà i Canals *et al.*, 2007) and has been described as the most promising indicator for use in situations in which the information will feed into management decisions (Wander and Drinkwater, 2000). Changes in SOM/SOC as a result of land use practices can impact upon soil functions including agricultural fertility, buffering and filtering capacity and (GHG emissions). In the context of impact assessment, SOM/SOC can be examined in a number of different ways. For example, Cowell and Clift (2000) propose simply that additions of organic matter to the soil be used as the indicator of impact, whereas Brandão *et al.* (2011) and Milà i Canals *et al.* (2007) use the change in SOM/SOC between the start and end of a land use activity, and also take account of the time it takes for the soil to return to its original SOM/SOC value following that land use (relaxation time). Sparling *et al.* (2003)

went on to use modelling to determine ideal ranges of SOM, which is attractive in that it allows for the setting of targets; however, it requires a detailed knowledge of the soils in question, which means it is only really applicable to soils that have already been modelled.

SOIL COMPACTION

Compaction of soil is an issue of particular relevance for the sustainability of agricultural production because it both affects yields of future crops and is a reflection of the production system (amongst other factors). In addition, subsoil compaction in particular is extremely difficult to treat once it has taken place, making its avoidance of even greater importance, and a number of authors have proposed methodologies for evaluating the risk of compaction under given conditions. Cowell and Clift (2000) for example, proposed a Soil Compaction Indicator (SCI) based on Field Load Index (FLI; Kuipers and van de Zande, 1994), as a means of assessing soil compaction. This involves taking the weight of vehicles and implements for each operation and multiplying by the time (hours/ha) taken to undertake the operation and the area (ha) on which the operation(s) is/are carried out. In the on-farm context this has the advantage of being directly linked to the process of crop production, and to data that is recordable on-farm. Although, clearly, actual soil compaction will be dependent on site-specific soil characteristics, relatively simple data could be used to provide an indicator of the potential for compaction. Similarly, precompression stress vulnerability (Horn and Fleige, 2003) is an indicator of the maximum stress from an external force, such as a vehicle, that a soil horizon is able to tolerate, and varies in response to soil classification and soil moisture, offering the potential to predict spatially, subject to the availability of spatially-referenced soil classification data, the susceptibility of arable sub-soils to compaction. Soils are classified for a given tolerance that can then be used to recommend exclusion of vehicles likely to exert a force beyond this threshold at certain times of year when soil moisture may exceed certain levels.

QUALITATIVE TECHNIQUES

In addition to a host of quantitative systems for assessing soil quality, a number of qualitative systems have been developed for use at the farm/field scale, including the visual soil assessment developed by the Soil Management Initiative and Väderstad (2005) for use in the UK, from that produced by Landcare Research for use in New Zealand (Shepherd, 2000; Shepherd *et al.*, 2000a, b, c). These techniques are based on the sort of assessments it is possible for farmers to do on their own farms (i.e. they require no laboratory analysis), and are therefore predominantly related to soil physical properties (e.g. structure, porosity, colour, the presence of pans, etc.), each of which is evaluated against a set of predefined photoraphs, although there is the capability to include some biodiversity indicators (e.g. earthworm counts), and plant indicators are also used (crop height, weed infestation, yields).

5.3.4. *Waste*

As outlined in Chapter 5.2.6, the world is evolving from a linear to a circular economy and what is one person's waste can often be another's resource. In many respects, at the farm level the circular economy has always been in existence. If a material has any value then it is usually reused or recycled, such as the use of animal wastes (manure and slurry) as a fertiliser. Waste management on farms should follow the waste hierarchy (see Figure 5.12), where, first, waste production should be minimised, then reused or recycled, used for energy, then either disposed of via incineration or landfill as a last resort. This hierarchy is usually reflected in the policy and legislative landscape (see Chapter 5.2) in which farms have to operate. The key aspect is to ensure any waste is minimised and, when it is generated, it should be managed using the best practical environmental and economically viable option that is permitted within the regulatory framework. For example, reuse is the second option within the waste hierarchy (see Figure 5.12), but there are instances where some wastes should not be reused due to potential hazards to people and/or the environment (e.g. empty pesticide containers.)

The types and amounts of waste generated on farms can be diverse and depend on the enterprises on the farm. Table 5.4 lists a number of different types of waste that can arise on farms along with their European Waste Catalogue (EWC) codes. The EWC is a hierarchical list of waste descriptions established by Commission Decision 2000/532/EC, the aim of which is to classify each waste stream according to its origin and its hazardous nature, thus helping to ensure that it can be appropriately managed.

In recent years, there has been a number of schemes and initiatives launched across Europe to encourage improved waste management on the farm. In the UK, for example, the National Farmers' Recycling Service, collects and recycles farm-generated plastic wastes such as that used to cover crops, from polytunnels and the stretch film used for wrapping straw bales. Similarly, in Germany there is a nationwide recycling scheme for crop plastics called Erntekunststoffe Recycling Deutschland (ERDE) which collects and recycles silage stretch films, net replacement films, underlay films and silo hoses. There are also similar collection and recycling schemes for other streams including, for example, used, dirty oils,

end-of-life tyres from agricultural machinery and wooden pallets. These schemes are not usually free of charge but the aim is to ensure that their charges are similar to or less than those for general waste disposal routes. In addition, several agricultural chemical manufacturers offer their products in returnable or refillable containers that reduces the volume of contaminated packaging that must be stored and managed.

5.3.5. *Natural resources*

As mentioned in Chapter 5.1.5.6, practices that affect the consumption of natural resources (such as fossil fuels, water and fertilisers) are largely covered elsewhere in the book (especially in Chapters 2 and 4). This section focuses on phosphorus, due to its importance to agriculture and food production and due to its non-renewable status.

Similar to energy, phosphate consumption can be monitored at the farm level. However, consumption does not reflect the impact in terms of resource depletion. Most efforts tend be focused on resource use efficiency to ensure phosphate is used to maximum economic benefit and to reduce the effects and impacts associated with losses of phosphate (such as losses to the aquatic environment and as a consequence eutrophication). Most EU MS provide their farmers with best practice guidelines and/or tools for advising on quite specific crop requirements ensuring that phosphate use is optimised (cost of input versus crop needs and yield improvements). Similarly, there are models and best practice advice for mitigating soil erosion as this will help maintain soil phosphate levels reducing the need for supplementation.

Assessing the consumption of phosphate in the context of resource depletion compared to other inputs is not typically done on farms. However, it may be undertaken in the context of an LCA for a product.

Table 5.4: Farm wastes

Type	Waste	EWC code
Vehicle and machinery	Brake fluids	16 01 13*
	Brake pads (no asbestos)	16 01 11
	Fuel oil and diesel	13 07 01*
	Lead/acid batteries	16 06 01
	Mineral-based, non-chlorinated hydraulic oils	13 01 10*
	Mineral-based, non-chlorinated waste engine, gear and lubricating oils	13 02 05*
	Most antifreeze	16 01 14*
	Oil filters	16 01 07*
	Petrol	13 07 02*
	Redundant vehicles and machinery (depolluted – liquids and hazardous components removed)	16 01 06
	Redundant vehicles and machinery (undepolluted)	16 01 04*
	Tyres	16 01 03
Packaging materials	Contaminated agrochemical containers and packaging, oil containers	15 01 10*
	Empty gas cylinders	15 01 11*
	Metal agrochemical containers, empty oil drums, empty paint tins, empty aerosol cans	15 01 04
	Paper and cardboard packaging, agrochemical bags, animal feed bags, animal health packaging, cores for silage sheet, silage wrap boxes, seed bags	15 01 01
	Plastic agrochemical packaging, animal feed bags, animal health packaging, fertiliser bags, miscellaneous packaging, seed bags, bale twine and net wrap, cores for silage wrap	15 01 02
	Wood – wooden pallets, crates	15 01 03
Building (construction or demolition)	Asbestos (including cement sheet asbestos)	17 06 05*
	Glass, e.g. window glass	17 02 02
	Mixed concrete, bricks and tiles and ceramics	17 01 07
	Plasterboard/gypsum, arising from construction and demolition	17 08 02
	Waste soil and stones arising from construction and demolition	17 05 04
	Wood, also scrap wood (e.g. fencing posts, telegraph poles)	17 02 01
Animal health products	Hazardous medicines	18 02 07*
	Infectious swabs and dressings, infected animal health care wastes	18 02 02*
	Non-hazardous waste medicines	18 02 08
	Other non-infectious dressings, etc.	18 02 03
	Swabs and dressings	18 02 03
	Unused syringes, used syringes	18 02 01
Other farm-specific wastes	Agrochemical wastes containing dangerous substances, e.g. sheep dip, pesticides	02 01 08*
	Agrochemical wastes not containing dangerous substances, e.g. some fertilisers	02 01 09
	Plastics (non-packaging), mulch film and crop cover, silage plastic, other horticultural plastics, tree guards, greenhouse and tunnel film	02 01 04
	Waste metal (non-packaging), e.g. fencing wires	02 01 10
Other general wastes	Discarded electrical equipment containing CFCs, e.g fridges	16 02 11*
	Fluorescent light tubes	20 01 21*
	Halogenated organic solvents	14 06 02*
	Miscellaneous cardboard and paper separated, e.g. office paper, fax, printer paper	20 01 01
	Paint (non-hazardous)	20 01 28
	Waste from a portable toilet	20 03 06
	Wiping cloths, protective clothing	15 02 03

* These wastes are classified as hazardous.

One approach, the method proposed by the Institute of Environmental Sciences (CML), Leiden University (Guinée, 2001), aggregates different resources using their Abiotic resource Depletion Potential (ADP), where antimony is used as a reference substance (ADP is expressed in kg antimony-equivalents), based on the scarcity of reserves. However, Brentrup *et al.* (2002) highlight that this neglects to consider that many resources are used for different purposes and are not equivalent to each other. Therefore, the depletion of reserves of functionally non-equivalent resources should be treated as separate environmental problems. Brentrup *et al.* (2002) develop the concept of grouping resources based on their function, for example, the use of oil, natural gas and coal as energy sources, and then expressing use of those resources in MJ, as a means of aggregating the impacts. The methodological uncertainties associated with these approaches tend to be related to the different definitions of 'resource depletion' and the assumptions made. For example, in some approaches, the cost of mining the raw material is considered to be a limiting factor whilst in others this aspect is ignored (Steen, 2006). Another assumption made in some models but not in others, is that for low-grade mineral reserves the cost of energy is the main limiting factor. These economic factors may well be dependent on the mining location and its economic status and energy markets. Brentrup *et al.* (2002) proposed a distance to target approach that determines a target rate of consumption based on an estimate of total resource reserves and the need to make the reserve last a certain period of time, for example, 100 or 1000 years. There are also approaches based on the future consequences of extraction that assume that unsustainable consumption now will create greater problems for the environment and world economics in future generations.

Various methods of quantifying this have been proposed, including assessing the impacts of over-extraction in other resource categories (Weidema, 2000) and considering impacts based on the increased energy requirements of future generations (Müller-Wenk, 1998). It is clear that these types of approaches are subject to huge uncertainties and inaccuracies, and lack transparency and scientific credibility. The importance of a specific resource is something that is not resolved fully in the scientific debate. This is significant, as the quantities of resources used and even improvements to resource use efficiency do not necessarily reflect the impacts of resource use, especially with respect to the use of significantly depleted reserves or where there are local issues, such as scarce water resources. In these instances, production can still have high efficiency, but still be drawing upon resources in an unsustainable fashion.

5.3.6. Sustainable agriculture, farming systems and productivity

Due to the economic and market pressures on farmers as well as the need to comply with CAP cross compliance rules, European farmers have, to a large extent, been 'encouraged' to improve the sustainability of their practices to help address the threats discussed in Chapter 5.1. There is no specific definition as to what constitutes 'sustainable farming', it will vary from farm to farm and it is certainly not a new idea having been around since the second half of the 20th century.

Sustainable agriculture has three main pillars: (1) environmental protection, (2) economic profitability and (3) social and economic equity. It is a process for addressing the nexus of complexity and often conflicting objectives facing the farming industry today, that is, producing more food from less land and resources, whilst still protecting the environment and remaining profitable. Academically, it is straightforward to describe what needs to be done but translating this into farm practice is complex. A whole-farm perspective is essential simply because few processes on the farm operate in isolation. For example, poor crop nutrition or drought conditions may result in a crop more susceptible to disease. This may mean more pesticides are needed which has an associated cost and potential environmental consequences. In addition, yields might also be affected which will impact on farm incomes. Taking a systems perspective will enable all the consequences of farming practices on other farm activities, farm incomes, local communities and the environment to be considered.

Implementing sustainable agriculture is not straightforward at farm level and there is not a single prescriptive plan to follow. Each and every farm is different from others in terms of its geographical location, soils, weather, crops, animals raised and farming practices and so the way forward for any specific farm will be unique to that farm. Usually the transition to a more sustainable form of farming is a process introduced over time. There is also no endpoint as the farm activities and the local environment will be constantly in flux, changing as markets and stakeholders present different demands, as crops and rotations are modified, and as the environmental status adjusts to these variable conditions. Therefore, the sustainable farming plan may also need to be adjusted in view of these changes and pressures. In addition, the farm economics and farmers or landowners' personal interests will also influence how fast progress is made and what, if any, new approaches and technologies are implemented.

Economics inevitably plays a huge role. Agricultural investment consistently delivers social and economic benefits and can be correlated with increased productivity and low food prices. In the wealthier EU MSs, agricultural production is already high due to the significant investments in selecting productive breeds, the use of high energy- and protein-rich animal feeds, the selection of high-yielding crop varieties, more efficient machinery and modern technologies. Whereas in many poorer MSs, where farm incomes and agricultural investments are low, such luxuries are often beyond the reach of many farmers.

Sustainable intensification (SI) is the general term used to describe the objective of obtaining more productivity from less resources. SI seeks to simultaneously increase farm output and competitiveness, whilst protecting the countryside and enhancing the environment. 'Integrated Farm Management' (IFM) is a recognised farming approach to help deliver SI. IFM (see Chapter 7) is a whole-farm business approach for improving the economic, environmental and social performance of a farm. IFM considers all the resources on the farm (including soils and waterbodies, labour, machinery, capital, biodiversity and wildlife habitats, natural heritage and archaeological features) and seeks to optimise their use to maximise the overall benefits. Its successful implementation requires the understanding of beneficial husbandry principles and traditional methods of farming, being aware of scientific advances and adopting appropriate innovation whilst complying with regulation and the demands of other stakeholders including, for example, retailers and quality assurance schemes. IFM will help deliver greater productivity by, for example, the use of greater-yielding crop varieties, more productive livestock breeds, the use of new technologies and innovation and greater capital/labour investments. IFM also includes a risk management approach that involves the development of a general farm management plan that describes the actions required to achieve the farm's objectives while protecting water and other natural resources. The planning process takes into consideration the size of the farm, soil types and their condition, topography of the land, proximity to waterbodies, pest and disease pressure, type of livestock or crops, manure and waste management, resources such as machinery or buildings and available finances. The management plan should also identify how emergencies and natural events which might disrupt farm activities (e.g. failed harvest; flood; drought; extreme weather events; biosecurity issues) will be handled.

Traditionally, IFM is associated with conventional farming rather than organic farming. Organic production is an approach to farming that seeks to work with natural cycles and ecosystems, adopting legally-defined practices that help to ensure the minimum environmental impact and the least risk to human health. Organic farming practices include the effective use of crop rotations, strict limits and restrictions on chemical inputs such as pesticides, synthetic nutrients and livestock feeds, and measures to maintain healthy soils and the responsible use of natural resources. The extent to which organic farming is sustainable is the focus of much debate. Organic production undoubtedly seeks to protect the natural environment and produce quality food for the benefit of both producers and consumers (Sundrum, 2001; Hole *et al.*, 2005). Studies have shown that the nutritional value of organic foods compared with non-organic is marginal at best (Dangour, *et al.*, 2010). Organic products can often fetch higher prices in the marketplace which means they could contribute to economic sustainability but only if the higher price more than compensates for the lower yields usually achieved by organic production compared with conventional farming (Leake, 1999, 2000; Seufert *et al.*, 2012). Studies have also

shown that demand for organic produce is highest when economic growth is strong. In times of economic stress, demand can rapidly decrease as consumers switch to cheaper alternatives and organic farmers can suffer economic hardship as a result. Indeed, studies have also shown that when it comes to choosing organic produce over traditionally-grown foods, the decision-making process is very complex and price is often a barrier (Padel and Foster, 2005; Krystallis and Chryssohoidis, 2005; Leake, 2000). Organic farming is more labour-intensive than traditional farming, which accounts for some of the higher production costs and will, inevitably, mean higher GHG emissions. However, it can also have socio-economic benefits, for example, by having greater employment opportunities.

Generally, the scientific literature suggests that organic farming offers a range of biodiversity benefits (e.g. Fuller *et al.*, 2005; Gabriel and Tscharntke, 2007; Hole *et al.*, 2005; Higginbotham *et al.*, 2000) as well as benefits for soil fertility (Mäder, *et al.*, 2002; Stockdale *et al.*, 2002). There are three broad management practices that are particularly beneficial for farmland wildlife. These are (1) the prohibition or reduced use of chemical inputs such as pesticides and inorganic fertilisers, (2) the sympathetic management of non-cropped habitats and (3) preservation of mixed farming. However, whilst these are part of the embedded philosophy of organic farming they are not exclusive to it and a review by Hole *et al.* (2005) highlights the fact there is insufficient evidence to suggest that organic farming would be superior in terms of biodiversity benefits compared to a conventional farming system that carefully targeted inputs via, for example, the agri-environment schemes or precision farming.

Precision farming seeks to make the most of advances in science and technology to maximise productivity and farm incomes (see

Chapter 8.4.2) and often uses 'geomapping' to produce farm and field maps of soil type, nutrient levels, soil organic matter, and so on. Global positioning systems (GPS) can be used to record the position of the field using geographic coordinates (latitude and longitude) and locate and navigate agricultural vehicles within a field according to these geomaps, with a two-centimetre accuracy, to optimise the delivery of inputs such as fertiliser and soil amendments taking account of natural variabilities within the field. Sensors and remote sensing can be used to collect data from a distance, relating to soil and crop health. Variable Rate Technology (VRT), helps to reduce overseeding, spraying and spreading on farms by tracking the equipment with GPS location and preventing redundant use of product.

Adoption of precision farming has been quite slow to date, in part due to the cost of the equipment and also due to the lack of training opportunities in some areas across Europe. It is also, therefore, not surprising that it is the larger farms that have the greatest uptake (Takács-György *et al.*, 2013). The focus has been, predominately, on the site-specific application of fertilisers and so the resulting cost advantages have been quite small and the payback time for equipment lengthy, although there are undoubtedly environmental benefits (Chavas, 2008; Gutjahr *et al.*, 2008). The uptake of precision crop protection, which meets the requirement of environmental and economic sustainability, is still under debate but one study at the EU level suggests that the savings in pesticide use following the adoption of precision plant protection could be as much as 30,000 tonnes (calculated using the current dose levels) per annum, which would be equivalent to reducing the environmental burden by 10 to 30 per cent (Takács-György *et al.*, 2013).

Taking the targeted application further, nano-technology has the potential to enable pesticides and fertilisers to be delivered in tiny quantities but in formulations shown to be highly active. This offers significant opportunities for sustainable intensification of farming and improving competitiveness (Parisi *et al.*, 2015). This is discussed further in Chapter 8.

References

Afzal, S., Ahmad, I., Younas, M., Zahid, M.D., Khan, M.H.A., Ijaz, A. and Ali., K. (2000) Study of water quality of Hudiara drain, India-Pakistan. *Environment International*, 26(1–2), 87–96. DOI: 10.1016/S0160-4120(00)00086-6

Ahel, M., Mikac, N., Cosnovic, B., Prohic, E. and Soukup, V. (1998) The impact of contamination from a municipal solid waste landfill (Zagreb, Croatia) on underlying soil. *Water Science and Technology*, 37(8), 203–210. DOI: 10.1016/S0273-1223(98)00260-1

Albuquerque, U.P., Ramos, M.A. and Melo, J.G. (2012) New strategies for drug discovery in tropical forests based on ethnobotanical and chemical ecological studies. *Journal of Ethnopharmacology*, 140(1), 197–201. DOI: 10.1016/j.jep.2011.12.042

Alvarez, J., Guirao, J., Herguedas, A. and Atienza, J. (1997) Evaluation of the PRZM2 model for transport of metamitron in undisturbed soil monoliths. Water pollution IV. Modelling, measuring and prediction. *Computational Mechanics Publications 1997*, pp. 67–74.

Arnalds, A. (1987) Ecosystem disturbance in Iceland. *Arctic and Alpine Research, Restoration and Vegetation Succession in Circumpolar Lands*: Seventh Conference of the Comité Arctique International, 19(4), 508–513.

Băhnăreanu, C. (2015) *Risks and threats to strategic energy resources in the contemporary world. Discourse as a form of multiculturalism in literature and Communication*, Section: Sociology, Political Sciences and International Relations, ISBN: 978-606-8624-21-1.

Bala, G., Caldeira, K., Wickett, M., Phillips, T.J.,

Lobell, D.B., Delire, C. and Mirin, A. (2007) Combined climate and carbon-cycle effects of large-scale deforestation. *Proceedings of the National Academy of Sciences*, 104(16), 6550–6555. DOI: 10.1073/pnas.0608998104

Bartels, J.J. and Gurr, T.M. (1994) Phosphate rock. In: Carr, D.D. (ed.) *Industrial Minerals and Rocks*, 6th Edition. Society for Mining, Metallurgy, and Exploration, Inc., Littleton, Colorado.

Brandão, M., Milà i Canals, L. and Clift, R. (2011) Soil organic carbon changes in the cultivation of energy crops: implications for GHG balances and soil quality for use in LCA. *Biomass and Bioenergy*, 35(6), 2323–2336. DOI: 10.1016/j.biombioe.2009.10.019

Brentrup, F., Küsters, J., Lammel, J. and Kuhlmann, H. (2002) Impact assessment of abiotic resource consumption conceptual considerations. *The International Journal of Life Cycle Assessment*, 7(5), 301–307. DOI: 10.1007/BF02978892

Bronick, C. J. and Lal, R. (2005) Soil structure and management: a review. *Geoderma*, 124(1–2), 3–22. DOI: 10.1016/j.geoderma.2004.03.005

Bundesministerium für Umwelt (BMU) (2013) *Naturschutz und Reaktorsicherheit, Bericht 'Abfallwirtschaft in Deutschland 2013'* (Fakten, Daten, Grafiken), Available at: https://secure.bmu.de/fileadmin/Daten_BMU/Pools/Broschueren/abfallwirtschaft_2013_bf.pdf

Carey, C. and Oetti, D. (2006) *Determining links between agricultural crop expansion and deforestation*. Report to the WWF Forest Conversion Initiative.

Cerdan, O., Desprats, J-F, Fouché, J., le Bissonnais, Y., Cheviron, B., Simonneaux, V., Raclot, D. and Mouillot, F. (2010) Soil erosion modelling of the Mediterranean basin in the context of land use and climate changes. *Geophysical Research Abstracts*, 12, EGU2010-8679-1.

Chan, K.Y., Heenan, D.P. and So, H.B. (2003) Sequestration of carbon and changes in soil quality under conservation tillage on light-textured soils in Australia: a review. *Animal Production Science*, 43(4), 325–334. DOI: 10.1071/EA02077

Chavas, J. P. (2008) On the economics of agricultural production. *Australian Journal of Agricultural and Resource Economics*, 52(4), 365–380. DOI: 10.1111/j.1467-8489.2008.00442.x

Christensen, J.B., Jensen, D.L., Grøn, C., Filip, Z. and Christensen, T.H. (1998) Characterisation of the dissolved organic carbon fraction in landfill leachate-polluted groundwater. *Water Research*, 32(1), 125–135. DOI: 10.1016/S0043-1354(97)00202-9

Clarke, S.A., Green, D.G., Bourn, N.A. & Hoare, D.J. (2011) *Woodland Management for Butterflies and Moths: A Best Practice Guide*. Butterfly Conservation, Wareham.

Compassion in World Farming Trust (CWFT) (1999) *Factory Farming and the Environment*. CWFT, Cork, Republic of Ireland, ISBN 1-900156-11-3.

Cooper, J., Lombardi, R., Boardman, D. and Carliell-Marquet, C. (2011) The future distribution and production of global phosphate rock reserves. *Resources, Conservation and Recycling*, 57, 78–86. DOI: 10.1016/j.resconrec.2011.09.009

Cordell, D., Drangert, J-O. and White, S. (2009) The story of phosphorus: global food security and food for thought. *Global Environmental Change*, 19(2), 292–305. DOI: 10.1016/j.gloenvcha.2008.10.009

Costello, M. J. and Wilson, S. P. (2011) Predicting the number of known and unknown species in European seas using rates of description. *Global Ecology and Biogeography*, 20(2), 319–330. DOI: 10.1111/j.1466-8238.2010.00603.x

Cowell, S. J. and Clift, R. (2000) A methodology for assessing soil quantity and quality in life cycle assessment. *Journal of Cleaner Production*, 8(4), 321–331.

Cressey, D. (2013) Europe debates risk to bees. *Nature*, 496(7446), 408. DOI: 10.1038/496408a

Dangour, A.D., Lock, K., Hayter, A., Aikenhead, A., Allen, E. and Uauy, R. (2010) Nutrition-related health effects of organic foods: a systematic review. *The American Journal of Clinical Nutrition*, 92(1), 203–210. DOI: 10.3945/ajcn.2010.29269

Doran, J.W. and Parkin, T.B. (1994) Defining and assessing soil quality. In: Doran, J.W.,

Coleman, D.C., Bezdicek, D.F. and Stewart, B.A. (eds.) *Proceedings of a Symposium on Defining Soil Quality for a Sustainable Environment* (Minneapolis, 1992). Wisconsin: Soil Science Society of America/American Society of Agronomy.

EC – European Commission (2014) *Progress in the management of contaminated sites in Europe*. Joint Research Centre Reference Reports Series, Institute for Environment and Sustainability, EUR 26376, Luxembourg: Publications Office of the European Union, ISBN 978-92-79-34846-4.

Eckelmann, W., Baritz, R., Bialousz, S., Bielek, P., Carré, F., Hrušková, B., Jones, R.J., Kibblewhite, M., Kozak, J., Le Bas, C. and Tóth, G. (2006) *Common Criteria for Risk Area Identification According to Soil Threats*. Office for Official Publications of the European Communities.

EEA – European Environment Agency (EEA) (2010) *The European environment — State and Outlook 2010: Land Use*. European Environment Agency, Copenhagen, Denmark.

EEA – European Environment Agency (EEA) (2011) *Analysing and Managing Urban Growth*. European Environment Agency, Copenhagen, Denmark.

EEA – European Environment Agency (EEA) (2012) *Today in energy: economic growth continues to drive China's growing need for energy*. September 21, 2012, Available at: http://www.eia.gov/todayinenergy/detail.cfm?id=8070

EEA – European Environment Agency (EEA) (2014) *Marine messages. Our seas, our future –moving towards a new understanding*. Available at: http://www.eea.europa.eu/publications/marine-messages

EEA – European Environment Agency (EEA) (2015) *Final energy consumption by sector and fuel*. Available at: http://www.eea.europa.eu/data-and-maps/indicators/final-energy-consumption-by-sector-9/assessment

EU – European Union (2013) *Service Contract for the support to the follow-up of the communication on water scarcity and doughts*:

Updated report on waste water reuse in the European Union. Report from TYPSA Consulting Engineers and Architects, 7452-IE-ST03_WReuse_Report-Ed1.

Eurostat (2014) Eurostat News Release: Environment in the EU28. *In 2013 42% of treated municipal waste was recycled or composted*. STAT/14/48 25 March 2014.

Farrell, A.E., Plevin, R.J., Turner, B.T., Jones, A.D., O'Hare, M. and Kammen, D.M. (2006) Ethanol can contribute to energy and environmental goals. *Science*, 311(5760), 506–508. DOI: 10.1126/science.1121416

FAO – Food and Agriculture Organisation (1999) *Women: Users, Preservers and Managers of Agrobiodiversity*. Women and Population Division, FAO, Rome.

FAO – Food and Agriculture Organisation (2002) *The Salt of the Earth: Hazardous for Food Production*. Food and Agricultural Organisation of the United Nations; 2002. World Food Summit.

FAO – Food and Agriculture Organisation (2003) *Trade Reforms and Food Security: Conceptualising the Linkages*. Food and Agriculture Organisation, Commodity Policy and Projections Service, Rome.

FAO – Food and Agriculture Organisation (2005) *Voluntary Guidelines to Support the Progressive Realization of the Right to Adequate Food in the Context of National Food Security*. Rome.

FAO – Food and Agriculture Organisation (2009) *Global Agriculture Towards 2050. High-level Expert Forum: How to Feed the World 2050*. Rome, 12–13 October 2009.

FAO – Food and Agriculture Organisation (2010) *Global Hunger Declining but Still Unacceptably High*. FAO Economic and Social Development Department, Rome, Italy. Available at http://www.fao.org/docrep/012/al390e/al390e00.pdf

Forbes, V. E. and Calow, P. (2013) Use of the ecosystem services concept in ecological risk assessment of chemicals. *Integrated Environmental Assessment and Management*, 9(2), 269–275. DOI: 10.1002/ieam.1368

Fuller, R.J., Norton, L.R., Feber, R.E., Johnson, P.J., Chamberlain, D.E., Joys, A.C., Mathews, F., Stuart, R.C., Townsend, M.C., Manley, W.J.

and Wolfe, M.S. (2005) Benefits of organic farming to biodiversity vary among taxa. *Biology Letters*, 1(4), 431–434. DOI: 10.1098/rsbl.2005.0357

Gabriel, D. and Tscharntke, T. (2007). Insect pollinated plants benefit from organic farming. *Agriculture, Ecosystems & Environment*, 118(1), 43–48. DOI: 10.1016/j.agee.2006.04.005

Gilbert, N. (2009) The disappearing nutrient. *Nature*, 461, 716–718. DOI: 10.1038/461716a

Gray, D.M.D., P. Suto, and C. Peck (2008) *Anaerobic Digestion of Food Waste*. East Bay Municipal Utility District, Oakland, California.

Grimm, M., Jones, R.J.A, Rusco, E and Montanarellaeu, L. (2003) *Soil Erosion Risk in Italy: A Revised USLE Approach*. European Soil Bureau Research Report No.11, EUR 20677 EN, (2002). Office for Official Publications of the European Communities, Luxembourg, 28 pp.

Guinée, J.B. (2001) *Life Cycle Assessment: An Operational Guide to the ISO Standards*. Centre of Environmental Science (CML), Leiden University, Netherlands.

Gutjahr, C., Weiss, M., Sökfeld, M., Ritter, C., Möhring, J., Büsche, A., Piepho, H.P. and Gerhards, R. (2008) Erarbeitung von Entscheidungsalgorithmen für die teilfl ächenspezifi sche Unkrautbekämpfung [Development of decision-making algorithms for site-specific weed control]. *Journal of Plant Diseases and Protection Special Issue XXI*, 143–148.

Heap, I. (2013) *International Survey of Herbicide Resistant Weeds*. Available at: www.weedscience.org

Higginbotham, S., Leake, A.R., Jordan, V.W.L. and Ogilvy, S.E. (2000) Environmental and ecological aspects of integrated, organic and conventional farming systems. *Aspects of Applied Biology*, 62, 15–20.

Hole, D.G., Perkins, A.J., Wilson, J.D., Alexander, I.H., Grice, P.V. and Evans, A.D. (2005) Does organic farming benefit biodiversity? *Biological Conservation*, 122(1), 113–130. DOI: 10.1016/j.biocon.2004.07.018

Horn, R. and Fleige, H. (2003) A method for assessing the impact of load on mechanical stability and on physical properties of soils. *Soil and Tillage Research*, 73(1–2), 89–99. DOI: 10.1016/S0167-1987(03)00102-8

International Energy Agency (IEA) (2014) *Energy Supply Security*. IEA, Paris, France.

Junginger, M., Goh, C. S. and Faaij, A. (eds.) (2013) *International Bioenergy Trade: History, Status & Outlook on Securing Sustainable Bioenergy Supply, Demand and Markets* (Vol. 52). Springer Science & Business Media, Dordrecht.

Jyväskylä Innovation Oy & MTT Agrifood Research Finland (2009) *Energy from Field Energy Crops – A Handbook for Energy Producers*. Jyväskylä Innovation Oy, Finland.

Karlen, D.L., Andrews, S.S. and Doran, J.W. (2001) Soil quality: current concepts and applications. *Advances in Agronomy*, 74, 1–40. DOI: 10.1016/S0065-2113(01)74029-1

Kirkby, M.J., Irvine, B.J., Jones, R.J.A., Govers, G. and the PESERA team (2008) The PESERA coarse scale erosion model for Europe. I. – Model rationale and implementation. *European Journal of Soil Science*, 59(6), 1293–1306. DOI: 10.1111/j.1365-2389.2008.01072.x

Klein, M., Hosang, J., Schäfer, H., Erzgräber, B. and Resseler, H. (2000) Comparing and evaluating pesticide leaching models: results of simulation with PELMO. *Agricultural Water Management*, 44(1–3), 263–281.

Koning, N.B.J., Van Ittersum, M.K., Becx, G.A., Van Boekel, M.A.J.S., Brandenburg, W.A., Van Den Broek, J.A., Goudriaan, J., Van Hofwegen, G., Jongeneel, R.A., Schiere, J.B. and Smies, M. (2008) Long-term global availability of food: continued abundance or new scarcity? NJAS – *Wageningen Journal of Life Sciences*, 55(3), 229–292. DOI: 10.1016/S1573-5214(08)80001-2

Krystallis, A. and Chryssohoidis, G. (2005) Consumers' willingness to pay for organic food: factors that affect it and variation per organic product type. *British Food Journal*, 107(5), 320–343. DOI: 10.1108/00070700510596901

Kuipers, H. and van de Zande, J.C. (1994) Quantification of traffic systems in crop production. In: Soane, B. D. and Ouwerkerk,

C. (eds.) *Soil Compaction in Crop Production.* Elsevier, Amsterdam, pp. 417–45.

Latacz-Lohmann, U. and Hodge, I. (2003) European agri-environmental policy for the 21st century. *Australian Journal of Agricultural and Resource Economics*, 47(1), 123–139. DOI: 10.1111/1467-8489.00206

Leake, A.R. (1999) A report of the results of CWS Agriculture's organic farming experiments 1989–1996. *Journal of the Royal Agricultural Society of England*, 160, 73–81.

Leake, A.R. (2000) Weed control in organic farming systems. *Farm Management*, 10(8), 499–508.

Lean, G., Hinrichsen, D. and Markham, A. (1990) *Atlas of the Environment.* Arrow Books Ltd, 192 pp, ISBN 0-09-984620-9.

Lewis, K.A., Tzilivakis, J., Green, A. and Warner, D.J. (2015) The potential of feed additives to improve the environmental impact of European livestock farming: a multi-issue analysis. *International Journal of Agricultural Sustainability*, 13(1), 55–68. DOI: 10.1080/14735903.2014.936189

Loveland, P. J., Thompson, T.R.E., Webb, J., Chambers, B., Jordan, C., Stevens, J., Kennedy, F., Moffat, A., Goulding, K.W.T., McGrath, S.P., Paterson, E., Black, H. and Hornung, M. (2002) *Identification and development of a set of national indicators for soil quality.* R and D Technical Report P5-053/2/TR, Environment Agency, Swindon.

Mäder, P., Fliessbach, A., Dubois, D., Gunst, L., Fried, P. and Niggli, U. (2002) Soil fertility and biodiversity in organic farming. *Science*, 296(5573), 1694–1697. DOI: 10.1126/science.1071148

Matthews, A. (2014) *What is happening to EU agricultural productivity growth.* CAP-Reform. eu, Available at: http://capreform.eu/what-is-happening-to-eu-agricultural-productivity-growth/

Merrington, G., Fishwick, S., Barraclough, D., Morris, J., Preedy, N., Boucard, T., Reeve, M., Smith, P. and Fang, C. (2006) *The development and use of soil quality indicators for assessing the role of soil in environmental interactions.* Environment Agency Science Report

SC030265. ISBN: 1844325466, Environment Agency March 2006.

Milà i Canals, L., Romanyà, J. and Cowell, S.J. (2007) Method for assessing impacts on life support functions (LSF) related to the use of 'fertile land' in Life Cycle Assessment (LCA). *Journal of Cleaner Production*, 15(15), 1426–1440. DOI: 10.1016/j.jclepro.2006.05.005

Morgan, R.P.C., Morgan, D.D.V. and Finney, H.J. (1984) A predictive model for the assessment of soil erosion risk. *Journal of Agricultural Engineering Research*, 30, 245–253. DOI: 10.1016/S0021-8634(84)80025-6

Müller-Wenk, R. (1998) *Depletion of abiotic resources weighted on the base of virtual impacts of lower grade deposits used in future.* IWOE discussion paper no 57, St Gallen, Switzerland.

National Ecosystem Assessment (NEA) (2011) *UK National Ecosystem Assessment: Technical Report.* United Nations Environment Programme World Conservation Monitoring Centre.

Newnham, R. (2011) Oil, carrots and sticks. Russia's energy resources as a foreign policy tool. *Journal of Euroasian Studies*, 2(2), 134–143. DOI: 10.1016/j.euras.2011.03.004

Noleppa, S., von Witzke, H., and Cartsburg, M. (2013) *The value of agricultural productivity in the European Union.* HFFA Working Paper 03/2013. Humboldt Forum for Food and Agriculture e. V.

ODI – Overseas Development Institute (1997) Global hunger and food security after the World Food Summit. ODI Briefing Paper 1997 (1) February. Overseas Development Institute, London.

OECD (2012) *OECD Environmental Outlook to 2050: The Consequences of Inaction.* OECD Publishing, Paris.

OECD–FAO (2014) Agricultural Outlook 2014, ISBN: 9789264211742, OECD Publishing, Paris. DOI: http://dx.doi.org/10.1787/agr_outlook-2014-en

Oerke, W.–C. (2005) Crop losses to pests, *Journal of Agricultural Science*, 144(1), 31–43. DOI: 10.1017/S0021859605005708

Orkustofnun (2017) Biodiesel. Orkustnun, Iceland's National Energy Authority website.

Available at: http://www.nea.is/fuel/ alternative-fuels/biodiesel/ (Last accessed: 16/10/17).

Padel, S. and Foster, C. (2005) Exploring the gap between attitudes and behaviour: understanding why consumers buy or do not buy organic food. *British Food Journal,* 107(8), 606–625. DOI: 10.1108/00070700510611002

Panagos, P., Karydas, C.G., Ballabio, C. and Gitas, I.Z. (2014) Seasonal monitoring of soil erosion at regional scale: an application of the G2 model in Crete focusing on agricultural land uses. *International Journal of Applied Earth Observations and Geoinformation,* 27(Part B), 147–155. DOI: 10.1016/j.jag.2013.09.012

Parisi, C., Vigani, M. and Rodriguez-Cerezo, E. (2015) Agricultural nanotechnologies: what are the current possibilities? *Nanotoday,* 10(2), 124–127. DOI: 10.1016/j.nantod.2014.09.009

Parr, J.F., Papendick, R.I., Hornick, S.B. and Meyer, R.E. (1992) Soil quality: attributes and relationship to alternative and sustainable agriculture. *American Journal of Alternative Agriculture,* 7(1–2), 5–11. DOI: 10.1017/S0889189300004367

Pimentel, D. (2006) Soil erosion: a food and environmental threat. *Environment, Development and Sustainability,* 8(1), 119–137. DOI: 10.1007/s10668-005-1262-8

Pimentel, D., Patzek, T. and Cecil, G. (2007) Ethanol production: energy, economic, and environmental losses. In: *Reviews of Environmental Contamination and Toxicology.* Springer, New York, pp. 25–41.

Plastics*Europe* (2012) *Plastics – the facts 2012. An analysis of European plastics production, demand and waste data for 2011.* Association of Plastic Manufacturers: Plastics*Europe*, Brussels, Belgium.

Population Institute (2010) *Population and Water.* Available at: https://www.populationinstitute. org/external/files/Fact_Sheets/Water_and_population.pdf

Quisumbling, A.R., Brown, L.R., Feldstein, H.S., Haddad, L. and Pena, C. (1995) *Women: the key to food security.* Food Policy Report. International Food Policy Research Institute, Washington, DC.

Renard, K.G., Foster, G.R., Weesies, G.A., McCool, D.K. and Yoder, D.C. (1998) Predicting soil erosion by water: a guide to conservation planning with Revised Universal Soil Loss Equation (RUSLE). In: *Agriculture Handbook,* no. 703, USDA–ARS, Washington DC.

Rienks, W.A. (2008) *The Future of Rural Europe: An Anthology Based on the Results of Euralis 2.0 Scenario Study.* Wageningham University and the Netherlands Environmental Assessment Agency.

Righelato, R. and Spracklen, D.V. (2007) Carbon mitigation by biofuels or by saving and restoring forests? *Science,* 317(5840), 902. DOI: 10.1126/science.1141361

RSB – Roundtable on Sustainable Biofuels (2015) *RSB Standard for the certification of biofuels and biomaterials based on end-of-life-products, by-products and residues.* Reference code: RSB-STD-01-010 (version 1.7), www.rsb.org

Sangodoyin, A.Y. (1993) Considerations on contamination of groundwater by waste disposal systems in Nigeria. *Environmental Technology,* 14(10), 957–964. DOI: 10.1080/09593339309385370

Schröder, J.J., Cordell, D., Smit, A.L. and Rosemarin, A. (2010) *Sustainable Use of Phosphorus.* Report for EU Tender ENV.B.1/ETU/2009/002. Plant Research International, Wageningen UR. Available at: http://ec.europa.eu/environment/natres/pdf/phosphorus/sustainable_use_phosphorus.pdf.

Seufert, V., Ramankutty, N. and Foley, J.A. (2012) Comparing the yields of organic and conventional agriculture. *Nature,* 485(7397), 229–232. DOI: 10.1038/nature11069

Shah, Y.R. and Sen, D.J. (2011) Bioalcohol as green energy – a review. *International Journal of Current Scientific Research,* 1(2), 57–62.

Shaibu-Imodagbe, E.M. (1991) The impact of some specific air pollutants on agricultural productivity. *Environmentalist,* 11(1), 33–38. DOI: 10.1007/BF01263196

Shepherd, T.G. (2000). *Visual Soil Assessment. Volume 1. Field Guide for Cropping and Pastoral Grazing on Flat to Rolling Country.* Landcare Research, Palmerston North, New Zealand.

Shepherd, T.G., Ross, C.W., Basher, L.R. and Saggar, S. (2000a) *Visual Soil Assessment, Volume 2. Soil Management Guidelines for Cropping and Pastoral Grazing on Flat to Rolling Country*. Landcare Research, Palmerston North, New Zealand.

Shepherd, T.G., Ross, C.W., Basher, L.R. and Saggar, S. (2000b) *Visual Soil Assessment, Volume 3. Field Guide for Hill Country Land Uses*. Landcare Research, Palmerston North, New Zealand.

Shepherd, T.G., Janssen, H.J. and Bird, L.J. (2000c) *Visual Soil Assessment, Volume 4. Soil Management Guidelines for Hill Country Land Uses*. Landcare Research, Palmerston North, New Zealand.

Snapp, S. and Grandy, S. (2012) *Advanced soil organic matter management*. Michigan State University Extension Bulletin E-3137.

Sparling, G., Parfitt, R.L., Hewitt, A.E. and Schipper, L.A. (2003) Three approaches to define desired soil organic matter contents. *Journal of Environmental Quality*, 32(3), 760–766. DOI: 10.2134/jeq2003.7600

Steen, B. (2006) Abiotic resource depletion – different perceptions of the problem with mineral deposits. *International Journal of Life Cycle Assessment*, 11(Supplement 1), 49–54. DOI: 10.1065/lca2006.04.011

Steinfeld, H., Gerber, P., Wassenaar, T.D., Castel, V. and de Haan, C. (2006) *Livestock's Long Shadow: Environmental Issues and Options*. Food & Agriculture Organisation, Rome, Italy.

Stockdale, E.A., Shepherd, M.A., Fortune, S. and Cuttle, S.P. (2002) Soil fertility in organic farming systems–fundamentally different? *Soil Use and Management*, 18(s1), 301–308. DOI: 10.1111/j.1475-2743.2002.tb00272.x

Sundrum, A. (2001) Organic livestock farming: a critical review. *Livestock Production Science*, 67(3), 207–215. DOI: 10.1016/S0301-6226(00)00188-3

Takács-György, K., Lencses, E. and Takács, I. (2013) Economic benefits of precision weed control and why its uptake is so slow. *Studies in Agricultural Economics*, 115(1), 40–46. DOI: 10.7896/j.1222

Thiele-Bruhn, S. (2009) Agricultural soils in Europe – special demands related to intensive agriculture in an industrialised environment. *Agricultural Sciences*, Volume II, 28.

Ulgiati, S. (2001) A comprehensive energy and economic assessment of biofuels: when 'green' is not enough. *Critical Reviews in Plant Sciences*, 20(1), 71–106. DOI: 10.1080/20013591099191

United Nations (UN) (2013) *Analytical Brief on Water Security and the Global Water Agenda*, United Nations University and Institute for Water, Environment and Health, ISBN 978-92-808-6038-2.

United Nations (UN) (2014) *Concise report on the World Population Situation in 2014*. United Nations, Department of Economic and Social Affairs Population Division, Report number: ST/ESA/SER.A/354, United Nations, New York, ISBN 978-92-1-151518-3.

United Nations Environment Programme (UNEP) (2011) *Towards a Green Economy: Pathways to Sustainable Development and Poverty Eradication*. Version 02.11.2011, ISBN: 978-92-807-3143-9.

USGS, in *Nature*'s Special report, Courtland, R (2008) Enough water to go around? *Nature*. DOI:10.1038/news.2008.678. See http://www.nature.com/news/2008/080319/full/news.2008.678.html

Vaccari, D.A. (2008) Phosphorus – a looming crisis. *Scientific American*, 300, 54–59.

Väderstad (2005) *Visual Soil Assessment*. Soil Management Initiative, Chester, UK.

van den Akker, J.J.H. and Hoogland, T. (2011) Comparison of risk assessment methods to determine the subsoil compaction risk of agricultural soils in the Netherlands. *Soil and Tillage Research*, 114(2), 146–154. DOI: 10.1016/j.still.2011.04.002

Van der Knijff, J.M., Jones, R.J.A. and Montanarella, L. (2000) *Soil Erosion Risk Assessment in Europe*. European Soil Bureau, Joint Research Centre and Space Applications Institute, EUR 19044 EN.

van Vuuren, D.P., Bouwman, A.F. and Beusen, A.H.W. (2010) Phosphorus demand for the 1970–2100 period: a scenario analysis of resource depletion. *Global Environmental*

Change, 20(3), 428–439. DOI: 10.1016/j.gloenvcha.2010.04.004

Velthof, G., Barot S., Bloem, J., Butterbach-Bahl, K., de Vries, W., Kros, J., Lavelle, P., Olesen, J.E. and Oenema, O. (2011) Nitrogen as a threat to European soil quality. In: M.A. Sutton, C. M. Howard, J.W. Erisman, G. Billen and A. Bleeker (eds.) *The European Nitrogen Assessment.* Cambridge University Press, New York.

Wander, M.M. and Drinkwater, L.E. (2000) Fostering soil stewardship through soil quality assessment. *Applied Soil Ecology,* 15(1), 61–73. DOI: 10.1016/S0929-1393(00)00072-X

Warwick H.R.I. (2007) *Farm Practice and Soil Health.* Defra Project Report OF0370, Defra, London.

Watson, R., Albon, S., Aspinall, R., Austen, M., Bardgett, B., Bateman, I., Berry, P., Bird, W., Bradbury, R., Brown, C. and Bullock, J. (2011) *UK National Ecosystem Assessment: Understanding Nature's Value to Society. Synthesis of Key Findings.* United Nations Environment Programme World Conservation Monitoring Centre.

Weidema, B. (2000) *Can resource depletion be omitted from environmental impact assessment?*

Poster presented at SETAC World Congress, May 21–25, Brighton, UK.

Whalon M., Mota-Sanchez, D., Hollingworth, R. and Duynslager, L. (2008) *Global Pesticide Resistance in Arthropods.* Cabi, Oxfordshire, UK.

Williams, J.R. (1975) Sediment-yield prediction with universal equation using runoff energy factor. In: *Present and Prospective Technology for Predicting Sediment Yields and Sources,* ARS-S-40, USDA-ARS.

Wischmeier, W.H. and Smith, D.D. (1978) Predicting rainfall erosion losses. In: *Agricultural Handbook* 537. USDA, Washington DC.

World Bank Group (2014) Food loss and waste a barrier to poverty reduction. Press Release. Available at: http://www.worldbank.org/en/news/press-release/2014/02/27/food-loss-waste-barrier-poverty-reduction (Last accessed: 16/10/17).

Zah, R., Böni H.G.M., Hischier, R. and Lehmann M.W.P. (2007) Ökobilanz von energieprodukten: ökologische bewertung von biotreibstoffen. EMPA. [Електронний ресурс]. Available at: http://www.news.admin.ch/NSBSubscriber/message/attachments/8514.pdf

CHAPTER 6

Cultural heritage within the agri-environment context

6.1. Setting the scene

6.1.1. Background

Cultural heritage is a description of the way a community lives, passed on from one generation to another. It includes both intangible and tangible assets. The intangible assets are the community's customs, traditions, spiritual and religious beliefs, practices, places and values, whilst tangible assets are the built and natural environment, and the community's artefacts such as literary, musical, audiovisual and other artistic works. Tangible assets can also be described as immovable and movable heritage. Immovable heritage includes places, buildings, monuments, biodiversity, landscapes and archaeological sites, whereas moveable heritage refers to assets such as artefacts, paintings and furniture. Whilst the majority of movable assets are not likely to be under the management of farmers or land managers, the agri-environment contributes significantly to Europe's non-tangible and immovable assets.

The precise definition of what constitutes cultural heritage can differ from one country to another and tends to be described by the laws of that country which have, themselves, developed based on local culture and values.

History, having left widespread material traces of ancient buildings and characteristic structures, such as monuments, fortress ruins, burial sites, standing stones, wells, mills, fountains, dry stone walls and terraces, clearly demonstrates that Europe's cultural heritage is both extensive and impressive which is best described using the broadest definition.

Europe's cultural heritage contributes towards its economic, social and environmental well-being. Closely linked to the tourism industry, the economic value of Europe's cultural heritage is exceptional, generating billions of euros annually from the millions of visitors attracted to its monuments, historical city centres, archaeological sites, museums and sites of natural beauty. According to the European Commission, cultural heritage generates revenue of approximately €335 billion annually for European economies (JPI, 2010). The same report estimated that the heritage conservation market alone was €5 billion annually. Tourism revenues contribute to more than 15 per cent of gross domestic product (GDP) in most countries and a buoyant tourism industry creates jobs, increases local spending and income, encourages social cohesion and generally enhances a region's attractiveness and competitiveness. Indeed, during the economic crisis that began in 2008, a growing number of EU MSs (e.g. Greece, Cyprus and Spain) were, and still are, heavily dependent on tourism for economic recovery and growth. Over 300,000 jobs in the European tourism sector are linked directly to cultural heritage and it has been estimated that for each direct job a further 26.7 indirect jobs are created in sectors such as transport, restaurants, other food and drink outlets and retail (CHCfE, 2015; KEA, 2006). As well as boosting the local labour market and economy, cultural heritage helps to create a local identity that relates to the historical and heritage background, local traditions and unique characteristics of an area. It also provides a focus and stimulant for educational activities and learning. This encourages community ownership, participation, social cohesion and a 'sense of belonging'.

The importance of cultural heritage for human well-being is illustrated by its inclusion, in terms of values and identity, amongst the cultural ecosystem services defined by the millennium ecosystem assessment (MEA, 2015). This initiative aimed to assess the consequences of ecosystem change for human well-being and the scientific basis for action needed to enhance the conservation and sustainable use of those systems. The other related categories being: spiritual services (sacred, religious, or other forms of spiritual inspiration); inspiration (use of natural motives or artefacts in art, folklore, etc.); aesthetic appreciation of natural and cultivated landscapes; and recreation and tourism (Hassan *et al.*, 2005).

6.1.2. *Cultural heritage and rural areas*

As around 50 per cent of the total EU-28 land area belongs to agricultural holdings, it is not surprising that a significant proportion of Europe's cultural heritage is found on farmland or within a rural setting.

Farmed landscapes provide a vital repository of the European cultural inheritance not only from its landscapes and biodiversity but also in the form of traditional buildings, mills, wells, historic features, terraces, stone walls, distinctive settlements, monuments, earthworks, burial mounds and other buried archaeological remains as well as local customs, traditions and produce. Tracing the history of our farming ancestors through these assets aids the understanding of what life was once like and how humans interacted with and harnessed natural resources in their environment over time, documenting how they adapted to ongoing societal, economic, industrial and technological change.

The landscape features, structures and buildings of archaeological or historical importance help create an area's 'sense of place' and its local distinctiveness. The history of agriculture within an area is often encapsulated in traditional buildings and structures, as these help to demonstrate how agriculture has developed over time. For example, the German tithe barn shown in Figure 6.1 was typical of those used across northern Europe in the Middle Ages and these were used to store the 'tithes' – one tenth of the farm's produce that was due to the local rector. There are still a few surviving medieval tithe barns across Europe and many of these have been restored and now have alternative tourism, community and commercial uses.

Wind-powered pumps were once used to help drain land of excess water to enable crop production or to move water for irrigation and to satisfy livestock needs. For example, they were used to drain polders, low-lying areas of land in the Netherlands and to convert the Fens, a naturally marshy and very wet area in eastern England, to fertile, productive land. Although quite common in parts of Europe, there are now very few

Figure 6.1: An ancient tithe barn in Jesberg, Germany, typical of those used across northern Europe in the Middle Ages
(Photo courtesy of: Axel Hindemith)

remaining wind pumps. One that has survived (Figure 6.2), and which is now a busy tourist attraction, can be found on Wicken Fen, England. This is the only working wooden wind pump remaining on the Fens, although there were once many thousands of them. Today the area is drained by electric pumps and pipes under the fields that carry the water to a network of drainage ditches.

Agricultural land may also be home to historical monuments, ancient earthworks and standing stones. These are usually nationally important heritage providing links to the historic past and helping to understand the lives of our ancestors and how they have influenced today's society. Well-known examples include Stonehenge and Avebury, both in Wiltshire, southwest England; the Jelling stones in Denmark; Ale's Stones in Sweden; the Carnac stones in Brittany, France; and the Odry megalithic stone circle in northern Poland. Many of these sites were once privately owned and were on farmland but are now often owned and managed by the state. Nevertheless, there are still many lesser known stones and monuments still on farmland, meaning that farmers and

landowners are often caretakers of this part of our cultural heritage.

Many farming practices have long historical connections and traditions. For example, 'biodynamic farming' is a spiritual and ethical form of farming that is similar to organic farming in many ways and rich in folklore and old traditions, that dates back to 1924 in Germany and so has a place in Germany's agricultural cultural heritage. Some of the practices adopted by biodynamic farmers may seem out of place in the modern agricultural setting, such as burying ground quartz stuffed into the horn of a cow in order to harvest 'cosmic forces in the soil'; timing many farming activities, such as planting and harvesting, according to the astronomical calendar; and using a wide range of herbal and mineral soil amendments (Turinek *et al.*, 2009). The Biodynamic Association claims this approach 'awakens and enlivens co-creative relationships between humans and the earth, transforming the practice and culture of agriculture to renew the vitality of the earth, the integrity of our food, and the health and wholeness of our communities'. Nevertheless, there is a niche market

Figure 6.2: The last functioning wind pump on Wicken Fen, England
(Photo courtesy of: Benjamin Hall)

for such foods often being available in 'whole food' outlets. Food produced by the Biodynamic Association protocols are often certified against organic standards.

Looking back through time at how farm machinery and technology has changed is also indicative of how society, particularly rural ones, have changed. In the 18th century, oxen and horses were used for power; cultivations were undertaken by hand with hoes and sickles; and threshing with flail. Tractors were not used until the early 1900s; the combine harvester was first introduced around 1938; and diesel engine machinery was not commonplace until the 1950s. Across Europe there are many museums that specialise in rural life and agricultural history which display old farming equipment and use it to illustrate agricultural heritage relevant to the particular area.

Over time, selective breeding has developed crops and livestock with enhanced and specific traits that are more productive than the traditional types or more suited to current farming methods and these have, consequently, displaced the old, traditional ones. However, traditional breeds and crop varieties are indicative of how our forefathers farmed and their continued existence is not only important as cultural heritage but also to ensure genetic diversity, as this will provide the genetic resources necessary to respond to disease threats that might emerge in the future in modern breeds and crop varieties. To conserve the world's food seeds for the future, the Global Crop Diversity Trust has built the 'doomsday vault', the first global seed bank, housed in a frozen bunker buried under the Norwegian island of Spitzbergen, near Longyearbyen, 1,300 km from the North Pole. In addition, there are many NGOs and national initiatives aimed at preserving old varieties, such as the UK's Heritage Seed Library. Traditional livestock breeds, usually referred to as heritage breeds, are as equally important as traditional crop varieties as they are also needed to preserve genetic diversity. The list of traditional European livestock breeds is extensive and includes cattle, sheep, poultry and many other species. Many of these breeds are not considered economically viable within modern livestock systems but, nevertheless, often have important attributes such as meat quality, resistance to particular diseases and the ability to adapt to specific environments. For example, the Varesina, a rare breed of sheep from the province of Varese in Lombardy, northern Italy, is a breed adapted for the alpine environment and which is prized for its meat. The Mangalica pig (see Figure 6.3), a traditional Hungarian hairy breed, is equally happy in hot, dry summers and snowy, cold winters. Mangalica pork is valued for its high-quality fat obtained through continuous exercise even in bitter Hungarian winters where temperatures routinely fall to -30°C. Another example is the Crèvecœur, a rare breed of black chicken that originates from Normandy, which is thought to be one of the oldest poultry breeds in France. The breed was developed principally for the quality of its meat. Breast

Figure 6.4: Cypriot 50 cent coin depicting the Mouflon
(Photo courtesy of: Prof. Kathy Lewis)

meat is exceptionally white, whilst leg meat is dark, almost the colour of duck flesh.

Not all traditional breeds are farmed animals, yet these are still important to our cultural heritage. The Cyprus Mouflon, known locally as the Agrino (*Ovis gmelini ophion*, previously *Ovis orientalis* ophion), is a sub-species of wild sheep endemic to the forests of Paphos, Cyprus. Although now quite rare, it is considered feral derived from ancient domestic stock by the International Union for Conservation of Nature (IUCN). Nevertheless, it is protected by Cypriot law and its primary habitat on the island, Paphos Forest, is a designated nature reserve and a Natura 2000 site (see Chapter 3) for the protection of the Mouflon. It is considered of vital importance to the Cypriot cultural heritage and is one of the national emblems of the island. Its place in history is demonstrated by its imagery appearing on excavated mosaics of the Hellenistic–Roman period. It also appears on the back of several Cypriot coins (see Figure 6.4) and is the symbol of the Cypriot national airline (Cyprus Airways) and the national rugby team.

Over time, differences in local farming practices, the large variety of different crops grown and different livestock reared have helped create a rich variety of landscapes that has shaped European rural character and these are often highly distinctive of a

Figure 6.3: The traditional Hungarian Mangalica hairy pig
(Photo courtesy of: Simon Meyer, Nienetwiler)

region. This distinctiveness will also be due to regional climates, the highly variable geology and topography as well as diverse biodiversity and habitats. UNESCO (2009) defined 'cultural landscapes' as those where human interaction with natural systems has, over a long period of time, formed a distinctive landscape. Similarly, the term 'landscape' has been defined within Article 1 of the 2004 European Landscape Convention as a zone or area as perceived by local people or visitors, whose visual features and character are the result of the action of natural and/or cultural factors. In this context, 'human interactions' might have resulted in ancient reservoirs and waterways, agricultural terracing, networks of dry stone walls, plantings of woodlands and forestry and the installation of drainage. For example, the extensive networks of dry stone walls in the UK's Lake District National Park are highly characteristic of the region as are the terraced agricultural fields of many Mediterranean landscapes.

The uniqueness of fauna, flora and their habitats in an area can also add to its distinctiveness and so attract considerable levels of tourism. For example, rare plant species such as the pink and white Calypso Orchid (*Calypso bulbosa*), the alpine species Pohjanailakki (*Silene furcata*) and the Marsh Saxifrage (*Saxifraga hirculus*), are amongst the main tourist attractions of the Oulanka National Park, Finland. As well as being valuable to our cultural/natural heritage, plants and animals provide key ecosystem services (e.g. food, fibre, medicines, carbon sequestration, oxygen production) and genetic resources, and each species provides value to others as habitat or food. Consequently, it is vital that they are protected. Often public attention to species loss is focused on animals and birds, particularly those that look appealing such as the Iberian lynx (*Lynx pardinus*), the Bavarian pine vole (*Microtus bavarius*),

the dormouse (*Muscardinus avellanarius*), the black-tailed godwit (*Limosa limosa*) and the slender-billed curlew (*Numenius tenuirostris*). However, of the 83,000 species on the IUCN Red List, there are 8000 plant species and over 1700 insect species that are considered threatened worldwide (IUCN, 2015).

Ancient dry stone walls (see also Chapter 3.3.3.8) are a characteristic part of several European landscapes, particularly in northern England and Scotland, as well as their modern-day equivalent. These types of walls are a common feature in the UK upland areas such as the Pennines, the Lake District, North Yorkshire Moors, Dartmoor and Exmoor. In these areas, the network of walls clearly defines the landscape providing perspective and distinction. They are also found to a lesser extent in the English lowlands in, for example, the Cotswolds, West Midlands and Cornwall. Their history in the UK can be traced back to the Iron Age brochs of northern and western Scotland, however they are not unique to the UK and are also found in other parts of Europe. For example, in Greece, Cyprus (see Figure 6.5), Italy and Spain they are a common landscape feature in rural areas where they form boundaries and traditional terraces. Acting as field boundaries, they enclose livestock and provide shelter. They are also unique and valuable habitat supporting many different species of birds, insects, small mammals, ferns, mosses and lichens (DSWA, 2011). However, they are often neglected and if not properly maintained they can soon deteriorate and crumble.

Artificial terracing, described in Chapter 3.3.3.8, has long been used to ameliorate a steep slope in a terrain in order to bring such areas into cultivation, taking advantage of or minimising the impact of key local environmental factors such as the topography, geology, climate and soil type. Such structures are very distinct features of the

Figure 6.5: Old dry stone wall, Cyprus
(Photo courtesy of: Prof. Kathy Lewis)

agricultural landscape across Europe, especially alpine, Mediterranean and sub-Mediterranean regions offering both historical and aesthetic value; this has been recognised by UNESCO which has many World Heritage Sites containing large areas of terraces, including several in Europe such as the Upper Middle Rhine valley in Germany, Costiera Amalfitana in Italy and the Tokaj wine region in Hungary. They are used for a variety of crops such as vegetables, pulses and citrus trees with low-input traditional plantations of olives, almonds or vines commonly seen. Some of these terraces have a very long history, for example in Spain, the

first terraced fields dated from the second millennium B.C. (Asins, 2006). Globally, there are also examples of terraces with prehistoric origin and others dating as far back as the Iron Age (Christopherson *et al.*, 1996).

Veteran trees are an often forgotten aspect of our cultural heritage yet they are indicative of past land management and use (Garner, 2004). In addition, many species of tree, characteristic of a specific landscape, are also in decline. Over 8000 tree species, representing 10 per cent of the planet's stock, are threatened with extinction due to the degradation or destruction of woodland and forest habitat or due to unsustainable timber production (IUCN, 2001). As well as being vital to the character of rural landscapes, trees provide a large number of ecosystem services, helping to regulate water, stabilise soils and sequester carbon as well as providing food and habitat for many other plants, insects, birds and animals. Within the UK, there are 15 endangered species. There are less than 20 surviving examples of the Ley's Whitebeam (*Sorbus leyana*) in the wild and all are found on the steep limestone cliffs in the Brecon Beacons, Wales, in the UK (Forestry Commission, 2010). Another example is the cork oak tree (*Quercus suber L.*) which is undoubtedly part of the Mediterranean cultural heritage. One of the first references to the cork oak was by the Greek philosopher Theophrastus in the 4th century BC, who marvelled at the tree's ability to renew itself after the bark had been removed. A cork oak tree is not harvested until it is at least 25 years old and then harvesting is typically undertaken every nine or ten years such that the tree produces a constant supply of raw, natural material. It also offers climate change mitigation potential as cork oaks store carbon in order to regenerate their bark. A harvested cork oak tree absorbs up to five times more carbon dioxide than an unharvested tree. Cork forests have been exploited in a sustainable way for over 3000 years providing jobs, income and valuable products. However, the current move away from cork bottle stoppers towards plastic alternatives and screw tops means that the distinctive cork forests of many Mediterranean areas, including the Iberian Peninsula and Catalonia, Spain are under threat. The key to their survival lies with there being a market incentive to manage them sustainably (Bugalho *et al.*, 2011). In recent years, the response to this has been a huge diversification into other products including cork bags, hats, belts and kitchen products such as heat-resistant mats.

Farm incomes can significantly benefit from local cultural heritage. In many areas across Europe, agriculture is disadvantaged due to natural handicaps such as difficult topographies, remoteness, extreme climates or low soil fertility. These areas often have great natural beauty supporting high levels of biodiversity and such culturally rich landscapes can be economically exploited by landowners by encouraging visitors. Despite the potential of receiving additional CAP grant aid, this extra income may be vital to farmers in these areas as they often have relatively low agricultural productivity and high production costs. Traditional barns and farm buildings also represent a financial asset as they can be put to new uses such as visitor centres, farm shops, offering holiday retreats or simply as the farm office. Local identity can be exploited in effective marketing strategies to promote locally produced goods and services which in turn boost local economies.

6.1.3. The risks to cultural heritage

Cultural heritage is considered by many as one of Europe's most important assets and an integral part of what it means to be 'European'. Considering that much of our cultural heritage is valuable and irreplaceable as well as being both fragile and

vulnerable, it is vitally important that we take the necessary steps to protect and preserve it. The threats to our natural heritage and landscapes are described in detail in Chapter 3.1, so the focus here is largely on the built environment and moveable assets.

Culturally-valuable built structures on agricultural land are exposed to many threats including those from biodeterioration, various chemical and physical processes, natural aging, weather, environmental pollutants, human negligence, vandalism and general wear and tear. Biodeterioration was defined by Hueck (1968) as 'any undesirable change in the properties of material of economic importance brought about by the activities of living organisms'. The living organisms include those from the plant world, mammals, insects and microorganisms. A summary of the causes and threats is provided in Table 6.1. Many of these threats can also damage moveable assets such as art works, books and documents.

6.1.3.1. Biological and chemical threats

Threats from the plant world are wide-ranging and can be serious. Microbial colonisation by a range of phototrophic organisms, typically algae and cyanobacteria, is usually the first stage of biodeterioration. Algal growth, lichen and mosses tend to be superficial and do not generally cause significant levels of structural damage and can add to the natural-aged appearance and aesthetic appeal of stone. However, in some instances they have been known to cause pitting of surfaces and they are sometimes unsightly. In addition, dense growth can mean that inscriptions and engravings cannot be seen clearly, thus reducing their aesthetic quality.

Fungal attack on timber-based structures is common, especially in areas with high atmospheric moisture content (usually above 20–25%). A number of different fungi can cause wet rot, in particular the

Table 6.1: Cultural heritage: built structure damage

Decay type	Examples of causal agent	Target	Environmental factors
Plant-based biodeterioration	Fungi, e.g. wet and dry rots Lichens and mosses Algae Invasive vegetation	Timber Brick Masonry Plaster	Moisture and humidity Lack of air circulation Temperature Sunlight
Insect-based biodeterioration	Wood-boring beetles Termites Ants Masonry bees	Timber Brick Mortar	Moisture and humidity Temperature Food source
Microorganism-based biodeterioration	Cyanobacteria Chemolithoautotrophic bacteria Actinomycetes	Timber Concrete Mortar Brick	Moisture and humidity Temperature
Chemical	Acids Alkalis Salting	Timber Metal Stone Plaster	Pollution Moisture levels Rainfall
Physical	Mechanical abrasion Cultivations Stock and wildlife damage General handling, wear and tear Vandalism, metal detecting Erosion (wind, rain, animal)	Timber Brick Mortar Stone Plaster Soil	Wind, rain and flooding Tourism Accidental and deliberate damage Grazing livestock Wildlife burrowing
Climatic	Heat/temperature Humidity/moisture Frost Rainfall Excessive irrigation UV radiation	Timber Brick Mortar Metal	Location Topography

cellar fungus (*Coniophora puteana*) and the pore fungus (*Fibroporia vaillartii*), both found widely across Europe. Dry rots such as that caused by *Serpula lacrymans* (previously known as *Merulius lacrymans*) are also common causes of damage to building structures. The development of fungi on stone and masonry requires the presence of an organic substrate, such as wind-deposited soil or decaying plant matter, for it to grow. In addition, fungi, and indeed some lichens, also produce organic and inorganic acids and various pigments which have potential to cause structural damage. Many crustose

lichens generate acids such as oxalic and gluconic acids in abundance and these can etch and erode stone and other surfaces over time, whilst the pigments can cause discolouration and staining (Wilson and Jones, 1983). Several studies have shown that UV radiation is a predominate factor in the development of lichens and so they can be prolific in the warm climates of Southern Europe (Edwards *et al.*, 1998).

Moulds, which are sometimes called mildews, are a type of multicellular filament fungus that can develop under humid conditions especially when a suitable nutrient source is available. On exterior surfaces, this could be any organic substrate such as wind-blown soil and dusts, or decaying plant material. Inside buildings, materials rich in cellulose, such as textiles, paper and wood, provide the required nutrients. Moulds produce powdery spores that are released into the atmosphere and when settled can form new colonies exacerbating the problem. As well as eventually damaging the building structure, they may constitute a health hazard aggravating allergies, asthma and other respiratory conditions. Certain moulds are toxigenic producing mycotoxins. The exact nature of the mycotoxin and its toxic effect varies significantly depending on the specific fungus and the sensitivity of the organism exposed. For example, *Stachybotrys chartarum* is a green-black mould that can be found in soils and in stored grain but is also found in damp buildings especially where cellulose-rich materials are present. Several of the mycotoxins produced by *S. chatarum* belong to the trichothecenes class and these are well known for their inhibition of DNA, RNA and protein synthesis (Ueno, 1980). *S. chartarum* has been associated with irritation of the nose and airways, coughs, wheezing and idiopathic haemorrhage syndrome. The toxic nature of mycotoxins was first reported in the 1930s in Russia, when researchers reported livestock mortalities following the ingestion of mould-contaminated hay (Hussein and Brasel, 2001).

Cultural heritage structures can become discoloured and may gradually deteriorate through the activity of microorganisms due to the formation of surface-coating colonies known as biofilms. These biofilms are extensive clusters of interconnected cells that are often embedded in an extracellular polymeric slime-like material, composed primarily of polysaccharides and proteins, which is secreted by the microorganism itself. Biofilms can be produced by a wide range

of microorganisms such as bacteria (including *actinomycetes* and cyanobacteria) as well as other organisms such as fungi, algae and lichens. They can lead to greater water permeability and can accelerate the accumulation of atmospheric pollutants. *Actinomycetes* are a specific type of filamentous bacteria and are akin to fungi with respect to their structure. As well as causing structural damage similar to that of moulds, *Actinomycetes* can be the cause of respiratory problems in humans and diseases such as 'lumpy jaw' in livestock. Cyanobacteria can colonise a wide range of different building materials. They damage structures by exerting pressure from within the material by the uptake of water, expansion of cell mass and the precipitation of various salts (e.g. carbonates and oxalates). As cracks and fissures expand, soil and dust particles, insects, small creatures and even opportunistic plants can enter increasing the pressure further. Chemolithoautotrophic bacteria can release acids such as nitrous, nitric and sulphuric acid (e.g. *Nitrosomonas* spp., *Nitrobacter* spp., *Acidothiobacillus* spp.), lowering the surface pH and subsequently causing biocorrosion. The released acids can react with constituents of the building material to form unslightly sulphate crusts and chelate with metal ions; both processes can eventually undermine the strength of the structure. The intensity of the bacterial attack on stone surfaces has been associated with lower pH and metabolic acid formation (Griffin *et al.*, 1991).

Higher plants can also cause significant damage to structures. Invasive plants, tree stems and roots can penetrate open joints, grow through mortar and damage foundations, causing displacement of brick and weakening structures. Trees and large shrubs growing close to masonry monuments can also cause structural damage and disturbance by root movement and abrasion. Plant material and debris can block building drains leading to water leaks and a build-up of moisture causing additional problems. Creeping ivy is frequently a problem and can result in significant damage to historic buildings and monuments although it is not without its benefits. Mats of dense ivy growth form dense barriers protecting the structure from the weather and air-borne pollutants as well as providing excellent habitat for wildlife. However, on older, weaker structures, ivy aerial roots can invade cracks and crevices, displacing masonry and eventually leading to structural failures.

Insect pests can cause continuous and often undetected damage to our cultural heritage. Woodworm is a generic description for the larvae of many wood-eating beetles, particularly those commonly known as the furniture beetle (*Anobium punctatum*), the house longhorn beetle (*Hylotrupes bajulus*) and the deathwatch beetle (*Xestobium rufovillosum*). In nature, these insects play a vital role in the decomposition of forest timber as well as providing a vital food source for other species such as birds and small mammals, however, when they infest building timber they can inflict significant damage. Although termites are known to cause very severe economic and structural damage to buildings in places such as Australia, Malaysia, India, Vietnam, China and the USA, across most of Europe they are not responsible for significant levels of damage. However, established subterranean populations of *Reticulitermes* spp. have damaged historic sites in Portugal, Italy and France, and reports are increasing at quite an alarming rate in Cyprus and Greece. As an example, in the northern Italian town of Bagnacavall, a large infestation of *Reticulitermes urbis* caused extensive damage to the old brickwork structure as well as the furniture and artefacts of several churches and monasteries of historical importance (Ferrari *et al.*, 2011). According to Kambhampati and Eggleton (2000), around

150 different termite species have been recorded as being responsible for building damage and subterranean termites account for 80 per cent of the economically important species. The annual economic cost of termite damage and termite prevention worldwide is estimated in the billions of US dollars (Ahmed and French, 2005). Once termites have invaded a building, they will damage, to the point of complete destruction, timber and other building materials rich in cellulose.

Termites are not the only destructive insect. Masonry bees are solitary bees, belonging to the genus *Osmia*, that nest in holes in mortar, brick and stones as well as in the ground. Whilst they often utilise existing orifices, for example, nail holes, gaps in wood joints and those in air bricks and vents, they are able to burrow into soft materials and may also make existing holes larger. Most buildings will easily tolerate a few holes, however, they have been known to create an extensive network of small tunnels which can undermine structural strength, especially if they fill with water which can freeze and expand causing further deterioration.

Salt crystallisation ('salting') can be devastating to old stone, masonry and brickwork. It is the production of efflorescence which involves the formation of soluble secondary mineral salts via the reaction of anions from acids (potentially excreted from bacteria and lichens and from the atmospheric deposition of acidic pollution) with cations in the building material. It can be a significant problem where there is high humidity and when the building material is porous, as the pores provide a channel along which water can flow and the formed salts thus move to the surface. Mineral salt solubility is related to temperature and changes in temperature can cause salts to dissolve as well as form. Fluctuations in relative humidity and temperature can, therefore, mean cycles of efflorescence and deliquescence occurring and so salt damage is usually the result of multiple (rather than single) events. As well as being unsightly, moisture beneath the crust cannot evaporate and more salts will then crystallise beneath it which eventually weakens the structure by blistering, flaking and a gradual disintegration.

Structural weakening problems can also be caused by the direct deposition of acidic and alkaline pollutants – wet or dry. Sulphur dioxide, carbon dioxide and nitrogen oxides can all attack stonework. When dissolved in rainwater these pollutants form highly damaging acidic rain (i.e. sulphurous, carbonic and nitric acids). Dry deposition of these chemicals is also highly damaging. The amount of direct deposition of 'dry' pollutants onto a building surface varies with factors such as wind speed and direction, building/surface shape and orientation and the amount of humidity. For example, the reaction of sulphur dioxide on limestone is significantly enhanced if humidity is greater than 80 per cent. Some calciferous building materials widely used in monuments and ancient structures, such as limestone and marble, are highly susceptible to damage from acid rain and the dry deposition of acidic pollutants. Porous and cracked surfaces are also vulnerable to the action of chemical salts as these can penetrate deep into the material, expand and so exert stress on the material from the inside.

Ancient buildings and structures are often attacked by many of these biological and chemical threats simultaneously and severe damage can occur if such structures are neglected. Those that are in the hands of the state or NGOs are, perhaps, somewhat safer than those on private agricultural land where they can be out of sight, neglected due to ignorance and lack of funds. An example of this is Harmondsworth Great Barn in the village of Harmondsworth, London Borough of Hillingdon (previously Middlesex). This barn is the largest timber-framed building in England and a prime example of medieval carpentry. It was originally used to thresh and store grain from the manor farm. It was still used for agricultural purposes up to the late 1970s and kept in a reasonable condition but when it fell into new ownership it became severely neglected. By 2009, the roof was in a deleterious condition, there had been significant ingress of rainwater, plants had invaded the floor and walls, and there was woodworm damage. English Heritage began legal action to enable the compulsory purchase of the building on the basis of extensive cultural heritage value. A private settlement meant that English Heritage was able to restore the barn and it is now managed by a local conservation group on behalf of English Heritage.

6.1.3.2. Physical disturbance

One of the major threats to historically important monuments, standing stones and circles is physical disturbance and abrasion. Loss of soil around these types of structures can weaken foundations and increase water penetration which itself can lead to further damage from biodeterioration. Whilst wind, rain and trampling by tourists and visitors can cause this type of damage, agricultural production and forestry is also a serious threat. Ineffectively planned forest developments including new tracks, digging drains and sediment traps can also cause irreversible damage.

Darvill *et al.* (1998) highlighted that agriculture had been the cause of the outright destruction of 10 per cent of the recorded archaeological resource base, and a further 30 per cent has been affected by piecemeal loss mainly due to ignorance, neglect and unsympathetic management. Arable cultivation is considered to be a major cause as it can result in damage to sensitive sites, particularly where the site slopes or where the soil conditions are challenging. Tillage and other cultivations will encourage erosion weakening supporting materials and foundations which may lead to the structure toppling over (Figure 6.6). Cultivations may also lead to the gradual deposition of soil around the base of ancient remains and standing stones, burying them deeper or covering them totally. When out of sight, these structures can be completely destroyed during future cultivations passing over them. For this reason, many sites of historical importance are under permanent grass. Currently, the equivalent of 9 per cent of the national total of scheduled monuments is still being actively ploughed in the UK.

Livestock can cause erosion around the base of monuments and standing stones from the continuous and extensive trampling of adjacent grass and soil. This can cause a number of problems. Hooves cause compaction of the soil surface, leaving depressions which can be 10 to 12 cm deep. If this happens close to a monument or standing stone, hollows can form at the base. The action of hooves can also displace supporting materials and encourage erosion which may also lead to hollows forming, which ultimately means that the structure is less stable and liable to tumble. It is then more vulnerable to damage by farm vehicles, livestock and biodeterioration.

As well as livestock, wildlife such as badgers, foxes and rabbits can also cause problems by burrowing beneath structures causing the surrounding support materials to loosen, making them unstable. The deposition of bird faeces can cause surface damage due to its highly acidic nature. Whilst in an open field this tends not to be a serious problem as it rapidly biodeteriorates or is washed away by rain, but should birds get inside a building and are left unchecked, very serious damage can occur. For example, in 2015, 25 tonnes of pigeon faeces, 3 feet deep, were removed from inside an ancient monument in the coastal town of Rye in England. The monument, the 14th-century Landgate Arch, was not open to the public and the faeces had accumulated unnoticed over a number of years. The cost to the taxpayer was significant, however, no permanent damage to the structure was incurred.

6.1.3.3. Natural hazards

Constant exposure to harsh weather such as wind, rain and extreme heat will also damage structures over time. Even ultraviolet (UV) radiation contained in the sun's rays will gradually affect the appearance of many building materials by bleaching and fading

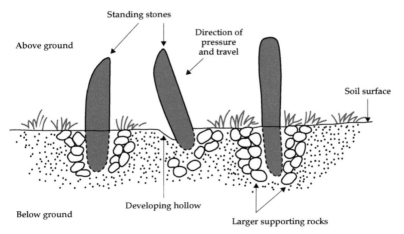

Figure 6.6: Stability factors for standing stones

the original colour. Timber can especially be affected by UV attacks and will turn into a grey, dull and unattractive state if not protected. In a hot Mediterranean climate, the process is accelerated and often accompanied by the timber drying out and cracking. It can be seen from Table 6.1 that, with the exception of damage caused by physical means, all other cultural heritage building damage mechanisms are significantly affected by the local moisture levels and this is particularly the case with old buildings. Biodeterioration of many building materials such as brick, timber, stone and plaster is more likely to occur under damp and humid conditions, which in turn are affected by temperature and the amount of ventilation and air circulation. These factors regulate the presence and activity of building-attacking organisms. If moisture and humidity levels exceed that which can be tolerated by the specific building material, mould growth and decay is also likely to occur. Many of the wood-boring insects generally require more water than is usually found in a dry and well-maintained building. Similarly, Cyanobacteria occur in abundance where there is water but are much less prolific in dry places.

It is inevitable that the effects of climate change (see Chapter 2) will increase the threat to Europe's cultural heritage, particularly longer term. Increased frequency and intensity of rain events and storms, stronger winds and increased temperatures will accelerate weathering and are likely to increase relative humidity, soil moisture content and the performance of existing rainfall disposal and collection systems. There will also be enhanced abrasive effects of more intense rainfall events, more powerful winds and storms, and increased levels of salt- and rainwater-carried pollutants. More intense UV radiation will also exacerbate the fading of timber whilst increased temperature will encourage greater expansion and shrinkage

of building materials leading to material fatigue. Climate change is also likely to affect plant growth patterns and increase the abundance of damaging insects such as termites. Consequently, the main biodeterioration control mechanism is to mitigate the amount of moisture in the building. Water leakage from internal water tanks and pipes, rain infiltration, convection of damp air and moisture condensation, rising damp from the ground and moisture accumulation in the structure can all cause significant damage if left unchecked (Figure 6.7). Timber-based agricultural buildings, such as those traditionally used in many European countries including Norway, Finland, Italy and the UK, are probably the most vulnerable to this type of decay.

Fire is a significant threat to the forestry and wooded cultural heritage areas of Europe and, in particular, southern areas due to their hot, dry climate and strong summer winds. According to a European Parliament report (2007), more than 450,000 ha of forestry in France, Greece, Italy, Portugal and Spain suffered serious damage due to fire between 2000 and 2006, and data for 2007 showed a worsening of the situation. The abandonment of agricultural land has been reported to enhance fire risk due to increased scrub and flammable vegetation increasing vegetation continuity (Benayas *et al.*, 2007; Moravec and Zemeckis, 2007) although some studies dispute this (Ricotta *et al.*, 2012). In many reports, humans are cited as the main protagonist in fire events, some being accidental and others criminal damage. Tourists generate a great deal of flammable litter and often leave it behind. Campers may cook with campfires which, if not properly extinguished, can easily lead to a fire, as can a carelessly discarded cigarette butt. Forest fires have potentially severe implications. These include (i) landscape damage, (ii) impact on communities

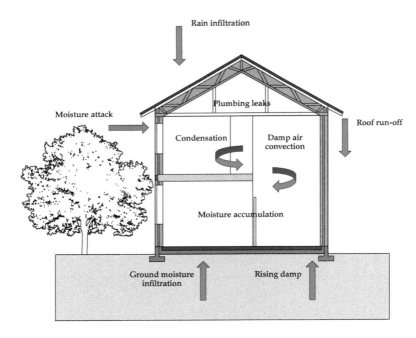

Figure 6.7: Moisture pathways in built structures

(mortalities and injuries, associated health impacts, loss of infrastructure), (iii) financial implications (firefighting, clean-up, restoration, loss of tourist income) and (iv) significant environmental impact which includes emitted carbon dioxide during the fire, increased atmospheric pollution, deterioration in soil quality and increased erosion, displacement of and damage to wildlife populations and habitat loss. Research indicates that the risk of fires will increase with climate change. In 2009, a fire seriously damaged the Dunnet Head SSSI which is part of the North Caithness Cliffs Special Protection Area in Scotland, one of the largest seabird colonies in the area with nesting populations of kittiwakes, guillemots, puffins and cormorants amongst others and a regionally valued landscape. A total of 750 ha of coastal land was damaged which included vegetation of national importance, cliffs were eroded and nesting habitat destroyed. The fire was caused when moor burning by local landowners to manage overgrown shrubs got out of control.

6.1.3.4. Anthropogenic threats

Characteristic landscapes are a highly valued part of a nation's cultural heritage but in many regions, the threat from tourists and land use change are very significant. Tourism is the largest industry in the world and with this comes a wealth of both positive and negative consequences. As discussed previously, it does have a positive impact on society, creating jobs and boosting local economies. However, nature can only tolerate the presence and activities of humans up to a point before negative impacts occur and these impacts happen when the level of visitors is greater than the local environment's capacity to accommodate them.

Tourists also contribute to other risks to the natural heritage, often out of ignorance of the damage they are causing. There have been incidences where the public have illegally removed rare plants damaging local populations and where stones from dry stone walls have been displaced or removed to the point that the wall has become unstable. Tourism is both dependent on fresh water resources and an important factor in fresh water use. Tourist demand for fresh water, exacerbated by swimming pools, open green spaces and golf courses, is in many places outstripping supply, placing extreme pressure on natural habitats (Gössling *et al.*, 2012). The longer and warmer summers anticipated due to climate change will undoubtedly significantly exacerbate the problem (Moriondo *et al.*, 2006).

Large numbers of tourists can also disturb wildlife and livestock which can deter nesting and lead to population decreases. For example, in both Greece and Cyprus, marine turtles have had coastal nesting sites disturbed by the intensive use of beaches by tourists and by sand extraction for beach refurbishment and other purposes (Warner, 1999; Davenport and Davenport, 2006). However, the EU and many Mediterranean countries now have legislation in place to protect sensitive species such as turtles, dolphins and seals, and their habitats (see Chapter 6.2) such that with sensitive and proactive management, tourists can enjoy natural areas without causing damage. For example, Lara in Cyprus is a nesting site of green turtles (*Chelonia mydas*) and loggerhead turtles (*Caretta caretta*). In 1976, with government and WWF funding, a conservation station and turtle 'hatchery' were established which seeks to manage the use of beaches by tourists at sensitive times but also to provide an information and education service.

Land use change associated with developments, construction and deforestation can also damage landscapes not only by consuming agricultural land and open spaces, thus destroying wildlife habitat, but also by increasing noise and pollution and ultimately encouraging more human activity within an area. Developments related to the tourism industry (hotels, roads and related amenities) are often focused along the coast, destroying natural areas and severely

impacting on fragile, valuable coastlines and marine ecosystems. For example, three quarters of the sand dunes on the Mediterranean coastline from Spain to Sicily have disappeared mainly as a result of urbanisation linked to tourism development (WWF, 2004). Of course, development is not just tourism-related and urban sprawl is also a growing concern to our natural heritage and, indeed, to agriculture. Land is a limited resource and as towns and cities expand into the countryside across Europe to accommodate a rapidly expanding population, landscapes and habitats are lost forever. As discussed in Chapter 3.3, developments, roads and railways cause fragmentation of habitat into smaller parcels which damage wildlife populations and species richness by reducing their ability to migrate and blocking their access to new areas and food sources (EEA, 2011).

One of the causes of neglect to stone walls and terracing is land abandonment and there are many areas across Europe where this is a severe problem. For example, of the terraced farm areas in the Iberian Range, Spain, 95 per cent have been abandoned since the middle of the 20th century (Lasanta *et al.*, 2013). The reasons for this are complex and include natural constraints and natural hazards, land degradation, socio-economic factors and demographic structure including rural depopulation and, in Eastern Europe, the transition process of integrating into the global economy (FAO and European Commission, 2006; Keenleyside *et al.* 2005). Land abandonment can often cause considerable negative social, economic and environmental impacts (Moravec and Zemeckis, 2007) including an enhanced risk of fire, increased erosion, desertification, detrimental effects on landscapes and declining biodiversity due to the loss of semi-natural habitats (FAO and European Commission, 2006; Keenleyside *et al.*, 2005; Lesschen *et al.*, 2008).

Europe's cultural heritage within the agri-environment (as well as more generally) is being lost at a disquieting pace, not only due to biodeterioration and human impacts (including tourism and conflict) but also as a result of climate change and natural hazards. Climate change is probably the most significant challenge currently facing our society and, as discussed previously, weather is a major factor influencing the rate of damage. The predicted increased frequency and intensity of extreme weather events together with risks associated with natural hazards, such as fire and floods, present an added challenge for the sustainable management and conservation of cultural heritage in Europe, calling for improved adaptation and mitigation strategies in this vulnerable sector. Undoubtedly, preservation and conservation of our cultural heritage must be a high priority.

6.2. Policy and interventions

6.2.1. The general policy arena

Awareness of the importance of protecting and preserving cultural heritage is not new and can be long traced back in most national histories. For example, the Greek historian Polybius, the earliest extant author, criticised the wartime plundering of art in the 2nd century BC and in 1891 Edward Dodwell, an Irish painter and author of archaeological works, stated that historic monuments such as the Parthenon should be the 'property of all nations'. Between 1943 and 1944 General Eisenhower, the 34th President of the United States, issued two directives relating to cultural heritage; one for the protection of historical monuments and the second concerned with cultural property. In recent times, the development and implementation of European cultural heritage policy has been influenced by several major global events that have, consequently, led to

cultural heritage policies and interventions being driven by a number of different stakeholders and not governments alone. These events include the following.

- The world's financial crisis that began in 2008 has meant that national budgets across Europe for cultural heritage have been slashed and cultural tourism has seen a drop in revenue. This is yet to be reversed and, consequently, there is pressure to use available resources more wisely and to the maximum benefit.

- Despite budget cuts, tourism has experienced significant growth and diversification in recent decades. Modern tourism is linked to development including, for example, new destinations; large-scale infrastructure; hotels; and shopping facilities. In addition, there has been strong growth in eco-tourism with visitors to archeological sites, caves, areas rich in biodiversity, rural communities and areas with strong cultural backgrounds. Whilst this growth has delivered economic benefits, these have been coupled with adverse environmental issues. Even though sustainable tourism is a growing approach, tourists still need resources such as water and accommodation.

- The acceptance that climate change is fact and not theory has emphasised the urgency to protect cultural heritage from a more damaging climate. It has highlighted the need to ensure that protection from climate change effects is an intrinsic part of management policies and planning.

- Globalisation, due to massive increases in world trade and cultural exchange, has meant that there is increasing awareness of cultural heritage uniqueness and its value in defining identity. Simultaneously, however, globalisation has also contributed to the dissemination of culture and social habits. Therefore, the need to protect what we have has been given greater priority.

- A rapidly expanding world population places greater pressure on cultural and natural heritage through increasing demands for land, food, fuel, fibre, water and natural resources, greater urbanism and higher levels of tourism.

Much of the world's cultural heritage policy is driven by multilateral agreements between nations and international conventions; the most significant of these are summarised in Table 6.2. Across Europe, policy development and implementation is largely driven by two major actors – The United Nations Education, Scientific and Cultural Organisation (UNESCO) and the Council of Europe (CoE). It is the EU's relationship with these two bodies that shapes, to a large extent, cultural heritage policy in Europe. It is interesting to note that whilst natural heritage is an important part of cultural heritage, it tends to have separate policies (see Chapter 3). However, built structures, moveable assets and intangible heritage that are part of the agri-environment all fall under the policies and international agreements of UNESCO and the CoE as implemented in national laws and regulations.

6.2.2. Cultural heritage policy at the global level

UNESCO, based in Paris, France, is probably the most important global organisation with respect to the preservation of cultural heritage. It is involved with all aspects of cultural heritage both tangible and intangible, moveable and immovable. It was founded in 1945 as an agency of the United Nations and seeks to deliver its objectives through education, the sciences, culture and communication. It has a number of tools and services available to aid the preservation of cultural heritage.

These include, for example, the provision of training and technical assistance, and a comprehensive database of national laws relating to the protection of cultural heritage both directly and indirectly by addressing issues such as theft, illicit trade and the degradation of cultural property.

However, with respect to cultural heritage, the most important aspect of its work is the establishment of standards, largely aimed at management and preservation, via a process of multilateral international conventions in which documents that detail potential measures and recommendations are drafted by experts. These drafts are formally proposed to the UNESCO membership for discussion and amendment. Once agreed, they are then signed and ratified by national governments on a voluntary basis. In order to be legally binding, the details of the agreement must be transposed into national law. The most important UNESCO conventions regarding cultural heritage are summarised in Table 6.2.

In addition to the delivery of multilateral treaties, UNESCO helps fund and promote worldwide initiatives related to cultural heritage. Its World Heritage Sites are landmarks officially recognised as being of cultural, historical or scientific importance or as having other significance for cultural heritage. These sites are legally protected by international treaties such as those given in Table 6.2. Globally, there are over 1000 World Heritage Sites, of which around 80 per cent are cultural, the remainder being designated due to their natural heritage. These sites include many well-known tourist attractions such as the Acropolis in Athens, Blenheim Palace in the UK and Centennial Hall in Wrocław, Poland. However, many are also part of Europe's agri-environment such as the Alto Douro Wine Region, in Portugal recognised for the remaining archaeological evidence of

Table 6.2: A summary of the most important cultural heritage conventions[1]

Convention	Description
1954 Convention for the Protection of Cultural Property in the Event of Armed Conflict	This UNESCO convention covers both immovable and moveable cultural heritage. It seeks to establish peacetime measures and emergency planning to protect cultural heritage during armed conflict.
1969 European Convention on the Protection of Archaeological Heritage and revisions	Commonly known as the Malta Convention and together with the Granada Convention (1885) and Valletta Convention (1992) revisions, this CoE multilateral treaty aims to protect European archaeological heritage with an emphasis on its use for historic and scientific study. In addition, it aims to preserve archaeological heritage 'as a source of European collective memory'.
1970 Convention on the Means of Prohibiting and Preventing the Illicit Import, Export and Transfer of Ownership of Cultural Property	The aim of this UNESCO Convention is to set up an international cooperation framework to help prevent theft and fraudulent imports, and to establish a range of preventative measures.
1972 Convention concerning the Protection of the World Cultural and Natural Heritage	Known more generally as the 'The Wold Heritage Convention', it is probably the most important and undoubtedly the best known of the UNESCO conventions. It evolved when two separate movements – one on the protection of cultural sites and another on the conservation of nature - were merged. It provides for the designation of World Heritage Sites and as of July 2015, there were 1031 listed sites globally of which 802 were classified as 'cultural', 197 as 'natural' and the remaining 32 as 'mixed sites'. The Geoparks initiative offers another dimension focusing on earth science.
2001 Convention on the Protection of Underwater Cultural Heritage	This UNESCO convention seeks to improve the effectiveness of measures at international, regional and national levels for the preservation of underwater cultural heritage including sites, structures, buildings, artefacts and human remains, together with their archaeological and natural context.
2003 International Treaty on Plant Genetic Resources for Food and Agriculture (the Seed Treaty)	The objectives of this FAO multilateral agreement is the conservation and sustainable use of plant genetic resources for food and agriculture and the fair and equitable sharing of the benefits arising from them.
2003 Convention on the Safeguarding of Intangible Cultural Heritage	The aim of this UNESCO Convention is to safeguard the uses, representations, expressions, knowledge and techniques that a community, group and, in some cases, an individual, recognise as an integral part of their cultural heritage.
2004 European Landscape Convention	Commonly known as the Florence Convention, this CoE treaty seeks to promote the protection, management and planning of European landscapes. Signed in 2000 and coming into force in 2004, it also serves as a vehicle for European cooperation on landscape issues. It was the first international treaty that was exclusively concerned with all aspects of European landscapes.
2005 Framework Convention on the Value of Cultural Heritage for Society.	Usually referred to as the Faro Convention, this CoE intiative is diverse in its application to cultural heritage and seeks to strengthen the links between cultural heritage, the quality of life, identity and sustainable development. It has a strong ethos that individuals and communities have the right to benefit from cultural heritage.

[1]. Excludes conventions relating to natural heritage as these are covered in Chapter 3.

wine making dating back to the 3rd century AD. Another example, is the agricultural landscape of Southern Öland in Sweden, designated a World Heritage Site because of its biodiversity, the Gettlinge Gravefield standing stones and limestone plains. Also in Sweden are the decorated farmhouses of Hälsingland which are extraordinary examples of a traditional Swedish timber-based construction technique unique to the region.

UNESCO also recognises sites with an earth science interest by the establishment of geoparks. The geopark initiative was first established by UNESCO in 2015 to satisfy demand for an international initiative to promote sites of geological significance. A geopark is an area identified for a specific geological heritage which has particular importance in terms of its scientific quality, rarity, aesthetic appeal or educational value. As such, geoparks often also have archaeological, ecological, historical or cultural value. The idea, due to its potential to increase tourism, has been adopted in many different countries and many geoparks now exist that are independent from UNESCO. Many of these sites occur within World Heritage Sites or Natura 2000 sites and there are many scattered across Europe including one in the Troodos Mountains of Cyprus, internationally recognised for its stratigraphic completeness and well-preserved, well-exposed plutonic, intrusive, volcanic rocks and chemical sediments. The area has been exploited since antiquity for its rich sulphide mineral reserves. Another example is the Reykanes geopark in Iceland located on a peninsula with a wide variety of different landscapes, including fissures, lava fields and geothermal activity. Many of these geoparks are in regions that do not readily support product agriculture due to harsh climates, rough terrains and infertile soils, however, this is not always the case and the

Shetland geopark, in the UK, which is comprised of over 100 small islands, is a prime example. The Shetland Isles, recognised as a geopark for its geological diversity within a small area, also has exceptional biodiversity and a thriving agricultural community, particularly for sheep famous for their fine wool.

The FAO also has a role to play in cultural heritage, although this has a much narrower focus than that of UNESCO and is only concerned with food and agriculture aspects. Globally Important Agricultural Heritage Systems (GIAHS) are similar in concept to UNESCO World Heritage Sites, however the FAO GIAHS are food production systems that have evolved over millennia in harsh and remote landscapes and often in extreme climates largely due to indigenous people. Currently, there are not any recognised GIAHS sites in Europe although the FAO do provide additional support to European government bodies by promoting policies and incentives that support conservation. Perhaps the most significant FAO initiative that affects European agriculture is the 2003 International Treaty on Plant Genetic Resources for Food and Agriculture. The objectives of this are to assure the conservation and sustainable use of plant genetic resources for food and agriculture and the fair and equitable sharing of the benefits arising from them. It requires all signatory parties to create an inventory of plant genetic resources, consider their status and degree of variation in existing populations, including those that are of potential use and, wherever feasible, assess any threats to them. It also requires signatories to put in place mechanisms for promotion and support to manage and conserve genetic resources. The FAO also organises various conferences, seminars and workshops related to traditional agriculture and its products.

6.2.3.   Cultural heritage policy in Europe

6.2.3.1.   The Council of Europe (CoE)

The Council of Europe (CoE) is an intergovernmental organisation that was founded in 1949 by the Treaty of London. It is a separate body from the European Union although there are undoubtedly strategic and collaborative links between the two. The aims of the CoE are to seek a greater unity between its members in order to protect common heritage and facilitate socio-economic development. The CoE has no legislative powers, instead it seeks to promote its ideals via voluntary agreements between its membership nations via multilateral conventions and it has facilitated a number related to the preservation and protection of cultural heritage. The first was the European Culture Convention signed in Paris in 1954. One of the principal objectives of this was to facilitate cultural cooperation across Europe by promoting the mobility and exchange of people as well as cultural goods. There are now 50 national signatories to this Convention although they did not all commit at the same time. The United Kingdom, Belgium, Denmark, France and Germany were among the first countries to sign in 1954. Serbia became a signatory in 2001 and Montenegro in 2006. Whilst most of the signatories are members of the CoE, the Convention does have some non-European members, that is, Belarus, and Kazakhstan. Since the signing of the 1954 Convention there have been several others including the European Landscape Convention, 2000 and the Framework Convention on the Value of Cultural heritage for Society, 2005. These are summarised in Table 6.2.

The CoE also facilitates a number of different cultural heritage activities. These include the HEREIN programme that seeks to stimulate cooperation between membership nations responsible for heritage management and the CoE, and the organisation of conferences related to various aspects of European cultural heritage. These included the 2009 conference that addressed the need for a sustainable model for cultural tourism and the 2015 Namur conference that considered European heritage strategy. The CoE, often in collaboration with the European Commission, organises and financially supports various regional and local initiatives to promote and manage local cultural, natural and human heritages as sustainable resources and a common good, particularly in European countries that have socio-economic difficulties or which have been damaged by conflict.

6.2.3.2.   The European Commission

The European Union is a signatory to all the conventions summarised in Table 6.2 but the current cultural heritage policy approach really began with the 1993 EC Treaty (Article 167, formerly known as the Treaty of Rome) that specified that safeguarding moveable and immovable cultural heritage of European significance must be treated as a priority for the EU. This treaty is the legal basis for the policy framework, preservation programmes and research initiatives on cultural heritage. Currently, the EU bases its cultural heritage policies and programmes on the Lisbon Treaty (Article 3.3; originally known as the Reform Treaty). This treaty was signed by EU MSs in 2007 and entered into force on 1 December 2009. It states that the EU 'shall respect its rich cultural and linguistic diversity, and shall ensure that Europe's cultural heritage is safeguarded and enhanced'. As such, the Lisbon Treaty imposes on the European Commission the task of encouraging and contributing to the protection and development of culture within individual MSs whilst respecting their diversity, bringing 'the common cultural heritage to the fore' (Article 167 of the Treaty on the Functioning

Figure 6.8:
Maintenance of dry stone walls can be supported by the Countryside Stewardship Scheme (EU RDP) in the UK
(Photo courtesy of: Dr John Tzilivakis)

of the European Union). As a result of this requirement and because of the diversity of Europe's cultural heritage, the European Commission does not work alone but seeks to work with, complement and assist the actions of individual MSs as well as the other policy-driving organisations such as UNESCO and the CoE. As a consequence, the principal responsibility for the sustainable preservation of cultural heritage is with individual European nations, and their regional and local authorities. However, the EU has a number of relevant policies and programmes, and the Commission also supports and promotes policy collaboration between MSs and heritage stakeholders. In addition to the highly focused policies, there are also those concerned specifically with natural heritage (discussed more fully in Chapter 3) and EU policies in other areas also take increasing account of heritage.

With respect to the agri-environment, one of the most important European policies relevant to cultural heritage is the Common Agricultural Policy (CAP; see Chapter 1). It is the second pillar of the CAP via the Rural

Development Programme (RDP) that is the most relevant. The focus is undoubtedly natural heritage, that is, protection of biodiversity, rather than the broader concept, but it does cover cultural assets. Biodiversity is a vital part of Europe's cultural heritage and it is the EU Biodiversity Strategy to 2020 that currently focuses on its preservation and protection (see Chapter 3). For example, environmental stewardship is a scheme that offers funding to farmers in return for environmental land management. Its focus is the protection of natural heritage, however, there are various land management options within the initiative that aid the protection of the rural historic environment including hedgerow and stone wall maintenance (see Figure 6.8), maintenance of traditional farm buildings, scrub management on archaeological features, managing earthworks and the removal of archaeological features from cultivation.

There are also various European Policy interventions designed specifically for solving a recognised problem, for example land abandonment. In this area, European policy has developed to encourage farmers and landowners not to abandon their land due to the significant risks to natural and cultural heritage (Benayas *et al.*, 2007), and under the current rural development programme compensation payments are given by many MSs to farmers and landowners in areas designated 'less favoured' (LFA) to improve farmers' incomes and discourage land abandonment. The scale and location across Europe of LFAs is shown in Figure 6.9. In some instances, however, land abandonment is not necessarily detrimental to biodiversity or natural heritage (see Chapter 3.3), for example it can help reconnect highly fragmented landscapes and increase woodland area which will be of benefit to some species (Sirami *et al.*, 2007; Gehrig-Fasel *et al.*, 2007). However, it can be highly detrimental to built cultural heritage such as traditional farm buildings as these quickly deteriorate without sympathetic and ongoing management.

There are also policies at EU level that seek to protect rare and traditional livestock breeds although the objective is predominately to preserve genetic diversity rather than cultural heritage per se, albeit these two aspects are closely related. Via the CAP Pillar 2 RDP, most EU MSs offer financial support to breeders of endangered native breeds. This takes the form of a subsidy per animal which is often co-funded by national schemes managed by local breeding associations or by national genetic resources monitoring programmes. There are also opportunities for support for the preservation of traditional cropping varieties. Traditional varieties are referred to by a host of other names including landraces, conservation varieties, peasant varieties and ancestral varieties but essentially refer to locally adapted traditional varieties of crops at risk of genetic erosion. There is as yet little direct legislation on this at EU level, the exception to date being Directive 2008/62/EC, but this only applies to certain agricultural species and is largely concerned with seed trade. Directive 2008/62/EC provides certain derogations from the normal listing and marketing requirements for varieties

Less favoured areas

Mountain/hill areas

Less favoured areas in danger of depopulation

Areas with specific handicaps

Non less favoured areas of EU-15

Lakes

Non EU-15

Figure 6.9: Less Favoured Areas as of 2012 (Source: European Environment Agency [EEA])

of seed, including seed potatoes, which have some conservation value. The directive provides for less prescriptive listing and certification regimes to encourage the preservation and use of older varieties which may not match contemporary varieties in terms of yield and disease resistance but do have value in sustaining cultural and traditional practices. The provisions are optional in as much as no seed producers are compelled to make applications for listing conservation varieties or to market them. The directive simply facilitates the legal marketing of such seed. There are also several options within the CAP agri-environment schemes that are relevant to the protection and maintenance of traditional orchards expecially those that are managed to offer wildlife and historic landscape benefits and/or where traditional varieties of fruit trees are grown.

Many areas across the EU seek to economically exploit the uniqueness of their areas, be that specific breeds, crops or characteristic landscapes, by promoting locally produced agricultural products and foodstuffs to enhance rural incomes. Regional branding of goods and services is increasingly recognised as a valuable tool for boosting sales in an increasingly globalised market. Indeed, the reputation and quality of regionally unique produce can be protected under European law (EU Regulation No. 1151/2012) using one of three EU schemes concerning geographical indications and traditional specialities (see Chapter 6.2). These schemes, which prescribe stringent requirements guaranteeing the quality and origin of certain foods and drinks, aim to prevent inferior products being marketed under the same name and provide consumers with information about the nature and origin of the product. The three related schemes are:

- Protected Designation of Origin (PDO). This scheme applies predominately to agricultural products which are produced, processed and prepared within a defined geographical area and use unique, recognised, particular knowledge, skills and expertise.
- Protected Geographical Indication (PGI). This scheme links agricultural products with a distinct area. The scheme requires at least one part of the production, processing and preparation stages to be uniquely carried out in that region.
- Traditional Speciality Guarantee (TSG). The aim of the TSG scheme is to highlight products of a traditional nature. This can refer to either its composition or its production processes.

Perhaps the most famous protected product is champagne, a sparkling wine produced from grapes grown in the Champagne region of northeast France. There are also many other well-known, legally protected products that reflect the cultural heritage of a region, for example, Parma in the Italian region of Emilia-Romagna is famous for its prosciutto (ham) and its Parmesan cheese. Within the UK, many unique cheeses are protected including Buxton blue, White stilton and 'West country farmhouse cheddar'. The Black Forest area of southwestern Germany also benefits from having a number of protected products. The two most famous are Black Forest ham (*Schwarzwälder Schinken*), a dry-cured smoked ham produced from the thigh and rump of the haunch of a pig or boar, and Black Forest gateau which includes kirsch alcohol and sour cherries sourced from the region.

One of the key objectives of the European Commission is to raise public awareness and, subsequently, the value the general public places on cultural heritage; there are a number of EU initiatives specifically concerned with this issue. These include: European Heritage Days, EU Prize for

Cultural Heritage, the European Heritage Label, European Capital of Culture and the Europe for Citizens Programme. With respect to the agri-environment, the European Commission launched the Natura 2000 Award in 2013 and awarded the first prize on Natura 2000 Day in 2014. The annual award aims to raise awareness about the Natura 2000 network, showcase excellence and encourage networking between people working on Natura 2000 sites. In addition to the Natura 2000 award, European Heritage Days, the EU Prize for Cultural Heritage and the European Heritage label are also applicable to the agri-environment.

Launched in 1985, the European Heritage Days have been organised since 1999 as a joint initiative of the EU and the CoE, and involve all signatories of the 1954 European Culture Convention. Annually, the entire EU takes part in an event which seeks to increase public awareness of Europe's cultural heritage by encouraging visitors to monuments and important cultural sites including the opening of some sites that are usually closed to the public. The monuments and sites open include a diverse mix of castles, historic buildings, industrial heritage sites, agricultural areas, ancient farm buildings, windmills, natural landscapes, places of religious importance, public spaces and architectural highlights. For example, in 2013, a Heritage Day held in Hedmark County, Norway coordinated events and demonstrations at the regional museum showing life as a local mountain farmer. The Netherlands began to recognise the importance of cultural heritage and public appreciation as early as 1986. Their first event 'Open Monumentendag' coincided with major changes in how monuments and other cultural heritage structures were managed in the Netherlands. A new Monuments Act was proposed which would decentralise their management and protection with duties of the Netherlands central

government being transferred to municipalities. This Act came into force in 1988 and as part of their duties each municipality was required to establish a body responsible for promoting the region's own cultural heritage. European Heritage Days are now one of the main responsibilities of these bodies. In the Netherlands, Heritage Days are one of the biggest cultural events with between 3000 and 4000 monuments and sites opening. The European Heritage Day initiative has been a huge success and now several non-European countries are involved, including Taiwan and Kazakhstan. European Heritage Days receive €200,000 in support from Creative Europe, the EU's programme for the cultural and creative sectors, and a further €200,000 from the CoE. Many events are also funded with national or regional backing.

The EU Prize for Cultural Heritage was launched in 2002 by the European Commission with support from Creative Europe. It is organised and managed by Europa Nostra which is a pan-European Federation for cultural heritage comprised of 250 member heritage associations and 150 associated bodies which includes governments, local authorities and various other stakeholders. The Prize is awarded to up to 30 individual projects and initiatives which demonstrate and promote best practices related to heritage conservation, management, research, education and communication. The aim of the award is to boost public awareness and create a stronger public recognition of cultural heritage as a strategic resource for Europe's society and economy. Awards are divided into four categories: (i) conservation, (ii) research, (iii) education, training and awareness and (iv) the recognition of the service to cultural heritage by individuals and organisations. Since its inception, awards have been given to a diversity of different projects and initiatives, many of which relate to the agri-environment. For example, one of the 2015 awards was given to a conservation project located in Salt Valley of Añana, Basque country, Spain. The large-scale initiative sought to recover the characteristic landscape, traditional architecture and the salt industry, having seen a serious deterioration of the area and the loss of ancient 'know-how' and traditions. The project had been highly successful, rejuvenating the landscape of the entire valley. Also in 2015, an award was given to an Estonian museum for an education and training programme directed towards the owners of old rural buildings. In Estonia, ancient farm structures are not generally treated as listed monuments. Their conservation and preservation is the responsibility of the owner and there was a serious need for the provision of advice and training to prevent their degeneration. In an example from 2012, an education award was given to a Danish project that sought to revive and update traditional skills of thatching farmhouses on Laesoe Island with seaweed.

The European Heritage label is administered by the European Commission and the CoE. The current scheme, which built on a previous one by the same name, was formally established in 2011 with the first sites being awarded the label in 2013. Sites are carefully selected such that they have played a role in European history and clearly symbolise European heritage ideals and values. It is their European symbolic value rather than their beauty or architecture that is of interest to the scheme and it is this that separates the European Heritage Label from the UNESCO World Heritage Sites, although some sites will have both labels. Heritage sites participating in the scheme are offered a number of promotional benefits such as being able to display a plaque bearing the EU logo, gain marketing and exposure via the EU's Heritage label communication and promotion strategy, and benefit from networking opportunities with other heritage sites. However, they are not offered direct financial support. Whilst open to agricultural sites of cultural significance, few have so far joined the scheme and, to date, most labels have been awarded to structures and places such as the Acropolis and surrounding archaeological sites in Greece, the historic Gdańsk Shipyard in Poland and the Archaeological Park Carnuntum in the east of Austria.

The European Commission has a history of underpinning cultural heritage conservation and protection with grant funding and research opportunities. During the period 2007 to 2013 the EU invested around €4.5 billion in heritage largely from its European Regional Development Fund and European Agricultural Fund for Rural Development. In the current economic funding period 2014–2020, cultural heritage is again set to receive significant investment including research and innovation funding from its Horizon 2020 programme. Support for cultural heritage research is not new as the first research projects in the field were funded in 1986; since then cultural heritage research has been an integral part of the EU's environmental research programmes and investment is undoubtedly increasing even under difficult economic conditions. Funding and grant opportunities are summarised in Table 6.3.

It is probably LEADER (Links between Actions for Rural Development) that provides the most benefits for cultural heritage. LEADER is an initiative aimed at supporting local projects that seek to revitalise rural areas and provide other socio-economic benefits such as jobs and improved infrastructure by mobilising and delivering rural development in local rural communities. It can provide grants for increasing farm or forest productivity, for establishing small business or farm diversification activities and provision of rural services. Cultural heritage

Table 6.3: EU grant funding and research mechanisms for cultural heritage

Title	Support type	Programme description	Types of agri-environment projects funded
European Structural and Investment Funds (Growth programme)	Investment Research	Five EU funds work together to support economic development across the EU: European Regional Development Fund (ERDF), European Social Fund (ESF), Cohesion Fund (CF), European Agricultural Fund for Rural Development (EAFRD) and the European Maritime and Fisheries Fund (EMFF).	• Tourism • Promoting employment and social inclusion • Education and skill development • Agricultural competitiveness • Environment and countryside
Horizon 2020	Research	A large research and innovation programme with a budget of almost €80 billion up to 2020 with an emphasis on science, industrial leadership and tackling social challenges.	Calls for specific projects and challenges are announced periodically relating to the development of multinational, transdisciplinary and demonstration projects that explore and showcase the potential of cultural heritage for urban and rural regeneration in Europe.
Creative Europe	Individual project support	A framework programme to support the culture and audiovisual sectors. In 2015, it had a budget of €1.46 billion aimed at encouraging digitalisation, growth in employment and social cohesion and the development of new markets and audiences.	Not greatly relevant to the agri-environment as funding is mainly directed towards artists, cultural professionals, cinemas, films and book translations.
Erasmus+	Capacity-building projects	Scheme that seeks to improve the skill base and employability of Europeans. The main opportunities of relevance to the cultural heritage sector are under Key Action 2 – Cooperation for innovation and the exchange of good practices.	• Strategic transnational partnerships to pilot, implement and promote innovative practices • Capacity building in higher education to improve competencies and skills • Knowledge alliances between higher education and business partners to foster and strengthen innovation
Europe for Citizens	Individual project support	Programme aims to contribute to citizens' understanding of EU history and diversity, to raise awareness of EU history and values and encourage societal and intercultural engagement.	Priorities for funding change annually. In 2015, projects funded included those relating to the 70th anniversary of WWII and those associated with democratic engagement and civic participation.

Title	Support type	Programme description	Types of agri-environment projects funded
LIFE programme	Research Innovation Education	This is the EU's funding instrument for the environment and climate action. The general objective of LIFE is to contribute to the implementation, updating and development of EU environmental and climate policy and legislation by co-financing projects with European added value. The total budget for 2014-2017 is €1.1 billion under the sub-programme for Environment and €0.36 billion under the sub-programme for Climate Action.	Predominately projects relating to natural heritage conservation projects including halting biodiversity loss. • Environment and resource efficiency • Nature and biodiversity • Environmental governance and information • Climate change mitigation and adaptation • Climate change governance and information
Pillar 2 of the CAP, e.g. LEADER	Grants and project support	LEADER supports multi-sectoral community-based development. It helps individuals, communities and businesses to come together to design and implement Local Development Strategies.	Projects include those to: • Support micro and small businesses and farm diversification • Boost rural tourism • Increase farm or forestry productivity • Provide rural services • Provide cultural and heritage activities
Pillar 2 of the CAP, e.g. Agri-Environment Schemes	Farm-level support scheme – farm payments and grants	National schemes co-funded by the EU Rural Development Programme and Member States (MSs) offering farm level financial support in return for the adoption of various land management options including those relating to habitat management, woodland, forestry, dry stone walls, preservation of genetic diversity, ancient monuments and archaeological sites.	Farmers and landowners enter into medium- to long-term agreements with government in return for land management that is sensitive to the environment, natural, cultural heritage and other environmental services.

projects have featured significantly in past LEADER programmes. These have established small rural heritage trails and heritage centres with such projects contributing to increased tourism, local jobs and economic development. For example, a LEADER-supported project sought to establish a mini mill (Griffiths Mill, Derbyshire, UK) to process locally produced fibre and fleeces. Mill products including carded fleece and fibres, spun yarns and knitted articles are sold via a website, at farmers' markets and in a small mill shop. The project created a small business and provided a new outlet for a local product (fleece) whilst also adding value to it as a local rural product and will, overtime, create jobs.

6.2.4. Cultural heritage policy at the national and regional level

Smaller sites of cultural heritage value often fall outside the protection of the major initiatives of Europe and bodies such as UNESCO and rely on the protection of national regulation. Indeed, action plans, policies and strategies are often best developed at regional level as they are better placed to address local needs and priorities. Often the establishment of legal instruments is driven by a country ratifying a particular convention, as commitment effectively means that that country will abide by the principles of the convention's charter. For example, many EU MSs have specific laws relating to historic buildings and monuments embedded in their planning and development procedures which, in part, satisfy their obligations under the Valletta Convention, albeit many European countries had the legal foundations in place before the convention was ratified. France has more than 43,000 protected structures and under the French Heritage Code VI these protected buildings are divided into two categories: 'classé' (classified) and 'inscrits au titre des monuments historiques' (registered/

included on the list), these classifications being originally enacted in the French Law of 2 May 1930. The type of legal protection conferred on the structure depends on its category and all listed structures are subject to special provisions relating to maintenance, restoration or modification which seeks to ensure their cultural interest is protected. A similar approach is in place in England whereby nationally important structures and sites are protected within the panning system by 'scheduling' and 'listing'. Monuments are scheduled under the provisions of the Ancient Monuments and Archaeological Areas Act 1979 (as amended). This Act makes it an offence to disturb a scheduled monument, either above or below ground, without having obtained prior permission from the Secretary of State. If a scheduled monument occurs on agricultural land, the landowner may be subjected to restrictions regarding the agricultural activities that are permitted. Buildings of special architectural or historic importance are placed on a statutory list of buildings and said to be 'listed' under the Planning (Listed buildings and conservation area) Act 1990. Listed buildings (see Figure 6.10) cannot be demolished, extended or altered without special permission from the local planning authority. In Belgium, a regional approach is taken with slight variations in legislation. In the Walloon region, as in France and England, a listing approach embedded in the planning system was first established in the early 1990s and then embedded in the Walloon Spatial and Town Planning and Heritage Code, 1999. However, in the capital (Brussels) and the Flemish region, the laws concerning architectural heritage are not directly linked to the planning system and discrete regulations must be considered concurrently with other related legislation when cultural heritage is affected by a planning process (Pickard, 2001; Guštin and Nypan, 2010).

Figure 6.10: Bodiam Castle, East Sussex, England (scheduled monument and Grade I listed building) (Photo courtesy of: Dr John Tzilivakis)

Regional approaches are common regarding the protection of cultural heritage relating to landscapes, habitats and biodiversity. The development of regional strategies is vitally important to ensure the uniqueness of an area has adequate protection – something that may not be easy within generic national approaches. For example, within Europe there are over 470 national parks, areas protected for their natural and cultural heritage, and the UK alone has 15. The Peak District National Park in the UK is rich in cultural heritage, however, only 5 per cent of its cultural heritage has any form of designated legal protection and this is mainly related to listed buildings, scheduled monuments and specific conservation areas. The remaining 95 per cent relies heavily on farmers and their participation in agri-environment schemes as well as formal legal processes, consequently the Peak District Management Plan is vital in ensuring that all stakeholders are aware of the issues and how they are being managed (PDNPA, 2012). The UK's Lake District is the largest national park in England with 14,650 archaeological sites and monuments (see Figure 6.11), 1760 listed buildings and 23 conservation areas. However, only 1.3 per cent have any legal designated protection (i.e. conservation areas, listed buildings, scheduled monuments, registered parks and gardens, and World Heritage Sites) and its management plan, which sets out to define how its cultural heritage will be protected, is vitally important for the park (LDNPP, 2015). The Lake District National Park is a prospective UNESCO World Heritage Site.

Figure 6.11: Castlerigg stone circle, Lake District, England
(Photo courtesy of: Dr Douglas Warner)

In addition to the EU-level funding for cultural heritage, there are usually funding opportunities at the national level. There are many examples of national charities and heritage societies offering grants and other forms of support such as the Society for the Preservation of the Greek Heritage, the Scandinavian Heritage Foundation and Hispania Nostra, Spain. Some national governments are also quite innovative in encouraging the public to invest. For example, in the UK, the National Heritage Lottery Fund offers funding to a wide range of heritage-related projects that aim to sustain and transform the UK's heritage including parks, historic places, natural environment and cultural traditions. The fund was established in 1994 under the provisions of the National Lottery etc. Act, 1993. In Italy, tax benefits are offered to private investors of Italy's extensive cultural heritage and this is seen as invaluable whilst the country recovers from the current economic crisis.

Specific, dedicated legislation at EU level to protect and enhance genetic resources and the cultural heritage they represent is largely absent. At national level an obligation does exist to conserve and protect genetic resources if that country is a signatory to the Convention on Biological Diversity and the International Treaty on Plant Genetic Resources for Food and Agriculture (see Chapter 3.2, Table 3.1). Germany, for example, is a signatory to both the convention and the treaty and in order to meet its obligations, places the conservation of genetic resources into the context of a broader overarching concept for conserving agricultural biodiversity, rather than addressing it separately. This approach is typical of many EU MSs, however there is, in a few European countries, specific legislation. Italian regional legislation is one of the few operational examples where there is a legal driver for protecting and enhancing the genetic resources for food and agriculture. The Tuscan Regional Law 50/97 on the protection of autochthonous genetic resources was underpinned by an awareness that Italy was rapidly losing its local and old livestock breeds and crop varieties. This regional law was followed soon after by similar laws in the regions of Lazio, Umbria and Emilia Romagna amongst several others (Bertacchini, 2009).

Public support for cultural heritage is invaluable and therefore if an obvious link can be made between goods that are purchased and the countryside, agriculture and traditions that have produced it, this can have a very positive effect on consumer habits and thus the economic vitality of a region. In turn, this can have benefits for how the land is managed within the region, help preserve and maintain traditions and will often lead to further opportunities for tourism and leisure. Across Europe, many nations have developed initiatives to promote specific regions that have a food or agricultural heritage. For example, in England a very successful initiative called 'Eat the view' was launched in 2000 whereby the government works in partnership with a wide range of organisations to improve the market for products which come from sustainable forms of farming that enhance or protect natural heritage and/or which strengthen the sense of place of an area and so 'add value' to the product. For example, 'Tastes of Anglia' is a regional food group from the East Anglian region of England that successfully markets speciality food and drink from the region, such as distinctive fruit juice and locally produced meat, poultry and game. The group also promotes 'food tourism' and work to help develop consumer awareness not only of speciality food and drink, but of a much broader availability of local products from the East of England. Sud de France is a brand name used to promote regional products and tourist destinations from the Languedoc-Roussillon region in the South of France. The brand objective is to bring the wine, food and wellness produce of the region together. Its produce range is very broad including, for example, anchovy puree, sweet Cévennes onion confit, cured meats and artisan sausages as well as a variety of fruits and vegetables. The brand is also well known for its wines, olives and regional delicacies such as rousquille doughnuts, marshmallows, liquorice, gingerbread and various honeys.

The food and culinary tourism industry is a rapidly growing concept in Europe. Most food tourists are interested in local food culture, rather than gourmet foods in particular as local cuisine gives travellers a direct and authentic connection with their destination. They experience local heritage, culture and people through food and drink. Activities can range from tasting local food and drink to more adventurous and active experiences such as participating in local harvests, visiting farms, farmers' markets and taking part in food and drink tours. These types of initiatives are often instigated through local partnerships and can benefit from enterprise funding from various sources including local and national governments.

6.2.5. The role of other stakeholders in cultural heritage policy

Whilst the EU, UNESCO and CoE are key players in developing cultural heritage policies, the role played by farmers should not be overlooked. Agriculture is seen as both a threat and a caretaker of cultural heritage and farmers and landowners are the ultimate deliverers of cultural heritage policy outcomes at farm level. This is recognised in some European countries but not all. Norway is an example whereby the responsibilities of farmers and landowers as guardians of cultural heritage is recognised in policy. The belief is that cultural heritage is best protected within an active and profitable farming system. Cultural heritage is promoted as having economic value with a market for goods and services which can benefit the farm business but this benefit can only be realised long term if the heritage is sustainably maintained. Across Europe, where agriculture is conducted on an 'industrialised' scale and highly mechanised, many farm businesses are only concerned with their own production enterprises and are largely divorced from local cultures. However, this is not the case in Norway where most farms are small family enterprises, handed down from one generation to the next.

Whilst not necessarily enshrined in law or policy, the concept of recognising farmers as the bearers of knowledge and traditions is true in many rural regions of Europe, especially where agriculture can no longer be relied on for jobs or as the main source of regional or even farm income. Instead, farm incomes are supplemented by offering recreation activities, access to the countryside and its heritage such as landscapes, monuments and standing stones, tourism and farm shops. Hence, farmers gain many benefits from a multifunctional business in which culture plays a strong part. If exploited it can be both profitable and beneficial for sustainable development of the sector. For example, the survival and protection of rare livestock breeds is only guaranteed if they are continuously maintained by farmers within their local biological, cultural and socio-economic

context (Veteläinen *et al.*, 2011). This, in turn, offers the opportunity for state subsidies to help support the animals; these animals can be a lure for tourists and food produced from them can often attract high prices. In addition, keeping rare breeds helps maintain and protect genetic diversity.

The role of the private sector in the conservation of cultural heritage should also not be ignored. Whilst many businesses depend upon cultural heritage for their income and survival, for example tourism and related services such as food and drink, transport and accommodation, they, along with many small businesses that create and/ or sell unique and characteristic local goods and produce, aid the economic stability of rural areas, job creation and social cohesion which helps provide the impetus to continue protecting local cultural heritage.

In addition to the action of the EU and individual countries, there are also the activities of a range of NGOs, charities, trusts, conservation bodies and the business sector. Amongst these is the 'European Heritage Alliance 3.3' which is an informal collaboration and cooperation platform launched in 2011 and comprised of various cultural heritage networks and organisations including European Landowner's Organisation, International National Trusts Organisation and Heritage Europe. The '3.3' is a link to Article 3.3 of the consolidated Lisbon Treaty that refers specifically to safeguarding and enhancing cultural heritage. Europa Nostra coordinate and manage the alliance. The European Cultural Foundation was established in 1854 with a mission to initiate and support cultural expression and cultural interactions. The foundation is a registered charity and operates various funding programmes for cultural heritage projects. There are also many at MS national and regional level. For example, in England, the Campaign to Protect Rural England

(CPRE) is a lobbying organisation that seeks to influence policies to protect landscape character and diversity. In Germany, the Deutsche Stiftung Denkmalschutz (German Foundation for Monument Protection) is a private initiative that works for the preservation of cultural heritage and to promote the idea of cultural heritage management in Germany.

6.2.6. *Monitoring the effectiveness of cultural heritage policy*

Monitoring the effectiveness of any policy is an important part of its continual development. It helps identify which approaches work best, where they can be improved, identifies best practice and ensures that money is well spent. Monitoring and evaluation activities are widely recognised as important elements of demonstrating performance, accountability and identifying future policy direction. This process is increasingly seen as a component of cultural heritage policies. However, it can be one of the most difficult aspects of a policy mainly because it is not always possible to identify a sound cause and effect relationship between a policy measure and an outcome, especially where 'intangible' aspects are concerned which can be highly subjective and where there are many different stakeholders influencing those outcomes. For example, increased public awareness of a heritage site might be due to a particular policy but could also be influenced by numerous other factors, even adverse media attention. Another problem is that a particular policy objective may not be measureable, for example the value of a landscape or a monument as this can be highly subjective.

Different organisations tend to develop their own monitoring and evaluation programmes. These may be simple reporting processes, measurement towards certain targets and goals using milestones, or benchmarking the use of best practice or the use of indicators. UNESCO, for example, require World Heritage Site managers to undertake reactive monitoring of the site's conservation status and report back describing any ongoing or new issues and what mitigation measures are being implemented. In addition, a periodic reporting process seeks to monitor conservation progress and changes in World Heritage Sites over time to assess the effectiveness of the management programme. This is achieved, in part, using key quantitative and qualitative monitoring and evaluation indicators.

Indicators are a common approach used in policy monitoring and evaluation to provide valuable information on the success and functionality of a policy. However, they are not without their limitations as they are often a simplification of reality, based on limited or poor data and are potentially open to differing interpretations. They are also usually used to measure progress in policy delivery and not an outcome. UNESCO is also currently developing an indicator approach to measure and monitor sustainable management of cultural heritage. The research initiative aims to establish a set of indicators highlighting how culture contributes to development at the national level, fostering economic growth and helping individuals and communities to expand their life choices and adapt to change. UNESCO's Institute for Statistics published in 1968 and revised in 2009 a methodology for measuring cultural participation and the economic contribution of cultural industries. Indicators are also the tool of choice for the European Commission and the EC have developed indicator sets for a variety of policies related to cultural heritage. For example, a European tourism indicator system for sustainable management at destination level, and a suite of Agri-Environmental Indicators (AEIs) to monitor the integration of environmental concerns into the CAP at European, national and regional level. Eurostat provides statistical data to the European Union from all EU MSs including statistics on the cultural sector and these are used in its monitoring and evaluation activities.

6.3. Farm-level management and protection

6.3.1. Background

As such a large proportion of European cultural heritage lies on agricultural or rural land, and as a relatively small proportion of this area has any designated legal protection, it is farmers and landowners that act as guardians. Over time, farming practices have led to major changes in the farmed landscape and in the design of many farmsteads. Large machinery and the need for greater levels of productivity has, in the past, meant that hedges were removed and tracks made wider, and bigger buildings were required for housing the larger equipment. Improved standards in quality, environment, energy efficiency, hygiene and animal welfare required higher standards in animal housing and farm utility buildings, and it was often more economical to build new than convert and upgrade the existing buildings. Consequently, over time many old farm buildings have become redundant and in danger of deterioration. However, these buildings illustrate how farms once operated and demonstrate regional differences as well as the building techniques of the period. Similarly, agricultural terraces and dry stone walls are indicative of farming in the past and it is important, therefore, that the value of these assets in terms of both their historical importance and their economic value is recognised at the farm level and wherever possible that they are kept in active use. Other cultural heritage and archaeological sites and structures such as monuments,

standing stones and circles also need careful management when on agricultural land and any work undertaken on a legally protected structure usually requires prior permission from the appropriate authority.

As custodians of our cultural and natural heritage, the European agricultural sector relies heavily on financial support. Without this, especially in economically stressed regions, many farmsteads would be abandoned and in these situations it would be difficult to maintain many traditional values, upkeep ancient and historically valuable farm buildings or maintain non-productive areas such as woodlands and other habitats. As discussed in Chapter 6.2, grants and support are available under Pillar 2 of the CAP (See Chapter 1) and at farm level and with respect to cultural heritage, two of the most important support mechanisms for farmers and landowners are LEADER and the agri-environment schemes. Within the rural development programme, many MSs, for example Austria, Germany and France, include measures that are targeted towards enhancing the quality of life in rural areas (e.g. village renewal, protection and enhancement of rural heritage) by improving and promoting cultural heritage.

6.3.2. On-farm practice

The on-farm conservation of natural heritage, landscapes and features such as dry stone walls and ancient terraces, is covered in detail in Chapter 3.3.

ANCIENT FARM BUILDINGS

It is important that farm buildings, especially those that are legally protected, are cared for appropriately and in compliance with national and local regulations. One of the best ways of ensuring that these buildings are protected is to keep them in active use as there is then good incentive to keep them maintained and in reasonable condition.

Left unused, problems may go unnoticed and rapidly get to a stage where repair is expensive. If the building has no active use, its value may be perceived to be low and so maintenance may not be a high priority.

Ideally, old farm buildings should be retained for agricultural use as this is less likely to fundamentally alter their character. Regular inspection and maintenance is essential. Ongoing general repairs are often relatively simple and cost-effective to undertake, considering the value of the building to the farm, if noticed early. Repairs should be done using appropriate materials and methods in keeping with the building's character. The most common problems which may be encountered include dry or wet rot, insect attack, damage by vermin, lost tiles from roofs and weather damage. If the building or the area in which it is located is legally protected then prior permission from the appropriate authority may be required before such repairs are undertaken. In some cases, especially where damage is extensive, effective maintenance and/or restoration may require specialist advice from architects, surveyors or structural engineers as well as the local planning authority.

Before repairs to traditional farm buildings are undertaken it is important to also consider the potential impact on wildlife. Many old barns, for example, provide habitat to birds and animals and in EU MSs, it is illegal to disturb protected species under the Habitat and Bird Directives which are normally enforced under national law (see Chapter 3). For example, in England the Conservation (Natural Habitats) Regulations, 1994 provides extensive protection to bats and their roosts and so any roofing work undertaken on old barns needs the consent of the responsible authority, which in the UK is Natural England. Similarly, in Germany, bats and their roosts are protected under the Federal Nature Conservation Act, 2009 and in Italy,

bats have had legal protection since 1939 by Royal Decree. Nowadays, they are protected under the 'Legge quadro in materia di fauna e attività venatoria' (L. 11 February 1992, n. 157). Bats and their habitats are also protected under the 1994 Agreement on the Conservation of Populations of European Bats (EUROBATS) which was set up under the Convention for the Conservation of Migratory Wild Species (Bonn Convention).

In some instances, there may be a desire for ancient farm buildings to be converted for alternative uses such as offices, farm shops or even residential use. It is important, when considering alternative uses that the building's character is maintained as much as possible and the landscape context is considered. In these instances, planning and building regulations will almost certainly apply.

ARCHAEOLOGICAL SITES AND MONUMENTS ON AGRICULTURAL AND FOREST LAND

The phrase 'archaeological site' is quite broad in its definition and can include anything from an internationally important site to a stone circle or single standing stone. In forests and woodland it is not uncommon to find earthworks and burial mounds. Most of the large sites that are nationally or internationally important will probably have legal protection and will belong to the state or a trust and be under the auspices of a government agency or a professional body of some sort which will manage its care, maintenance and public access. However, whilst some small sites may also have protected status, many will not and a significant proportion of these will occur on agricultural or forestry land. Indeed, many more are yet to be discovered.

As these sites can be fragile and especially vulnerable to any local ground disturbance, sensitive land management around such sites is essential. Many such agricultural sites are still actively cultivated and suffer damage from ploughing and other operations such as sub-soiling, and even growing root crops can be problematic. These activities can damage, disturb and displace buried remains, weaken structures and their foundations and erode protective soil around them. Forestry tends to be much less intensive in terms of cultivations but heavy machinery used in forestry operations can be very damaging. When new forestry areas are being established cultivation, road building and installing drainage can pose a risk to hidden archaeological remains. In mature forestry areas, tree thinning and clear felling can also cause damage and the trees themselves can cause problems due to deep roots.

Known sites that do have legal protection will normally need approval from the appropriate designated authority before any maintenance or renovation work can be undertaken. However, many sites may also get protection under an agreement within a national agri-environment scheme which would prescribe explicit management practices to prevent damage. For example, in 2009 over 6000 archaeological features on farmland were protected under the English agri-environment scheme (Countryside Stewardship) and this included 59 per cent of scheduled monuments, 62 per cent of undesignated monuments and 54 per cent of registered battlefields (Natural England, 2009). The scheme includes the option for farmers to take historic and archaeological features out of cultivation if these features are on arable land or temporary grassland and in return farm payments were given on a per hectare basis.

Specific land management practices for archaeological and historic sites and features on agricultural land will often depend on the characteristics of the land and the local topography. Archaeological areas prone to soil erosion require preventative actions as erosion can have a highly damaging impact on the survival of historical remains as, over time, it removes the protective layer of soil around or over the site. Light soils can be more easily eroded especially when they are deeply cultivated or on a slope. Peat soils can shrink when they dry out exposing remains and are also subject to wind erosion. Consequently, the aim is to avoid any activity that causes soil shifting. Ideally, keeping the area under permanent grass or reverting arable land to grass will be the best management option as it will mitigate erosion and such areas are not actively cultivated. Where this is not possible, no-till, mulching and direct sowing will also all offer a good level of protection. The current plough depth should not be increased in order to avoid any new damage. Sub-soiling, de-stoning and drainage operations should be avoided in such areas. There are also certain crops that should be avoided. For example, root crops such as sugar beet and potatoes can cause physical damage especially during harvesting and some energy crops such as short rotation coppice (SRC) or *Miscanthus* may also lead to damage through root action, water depletion and from grubbing out old SRC stools and the harvesting of *Miscanthus rhizomes*.

It is also important to ensure that the closely surrounding agricultural area is carefully managed. In particular, scrub growth and areas of vigorous vegetation should be controlled and actions should be taken to avoid animal burrowing close to any historic remains. With respect to sites in forestry areas, these can often be sub-surface and preparing land areas for planting new forests can cause significant damage where historical records show that a site is within an area of potential archaeological interest; the services of a professional archaeologist may be required. Often, national forestry authorities will have established procedures to protect historic sites and remains including those that must be followed if an unrecorded site or remains are found. In order to protect a site it may be necessary to establish an exclusion zone with defined boundaries for forestry operations around the site and to build purpose-built access routes from the nearest road or track to prevent random access across potentially important areas.

References

Ahmed, B.M. and French, J.R.J. (2005) Report and recommendations of the national termite workshop, Melbourne. International Biodeterioration. *Biodegradation Journal*, 56, 69–74. DOI: 10.1016/j.ibiod.2005.05.001

Asins, V.S. (2006) Linking historical Mediterranean terraces with water catchment, harvesting and distribution structures. *The Archaeology of Crop Fields and Gardens*. Edipuglia. Bari, Italia, pp. 21–40.

Benayas, J.R., Martins, A., Nicolau, J.M. and Schulz, J.J. (2007) Abandonment of agricultural land: an overview of drivers and consequences. *CAB Reviews: Perspectives in Agriculture, Veterinary Science, Nutrition and Natural Resources*, 2, 1–14. DOI: 10.1079/PAVSNNR20072057

Bertacchini, E. (2009). Regional legislation in Italy for the protection of local varieties. *Journal of Agriculture and Environment for International Development*, 103(1/2), 51–63. DOI: 10.12895/jaeid.20091/2.24

Bugalho, M.N., Caldeira, M.C., Pereira, J.S., Aronson, J. and Pausas, J.G. (2011) Mediterranean cork oak savannas require human use to sustain biodiversity and ecosystem services. *Frontiers in Ecology and the Environment*, 9(5), 278–286. DOI: 10.1890/100084

CHCfE (2015) *Cultural Heritage Counts for Europe – the Project*. International Cultural Centre, Krakow, Full Report ISBN 978-83-63463-27-4, Executive summary available at: http://www.encatc.org/culturalheritagecountsforeurope/wp-content/uploads/2015/06/CHCfE_REPORT_ExecutiveSummary.pdf

Christopherson, G.L., Guertin, D.P. and Borstad, K.A. (1996) GIS and archaeology: using ArcInfo to increase our understanding of ancient Jordan. In: *Proceedings of 1996 ESRI European User Conference*, London.

Darvill, T., Fulton, A.K., Bell, M. and Russell, B. (1998) *MARS: The monuments at risk survey of England*, 1995: main report. School of Conservation Sciences, Bournemouth University, Poole.

Davenport, J. and Davenport, J.L. (2006) The impact of tourism and personal leisure transport on coastal environments: a review. *Estuarine, Coastal and Shelf Science*, 67(1), 280–292. DOI: 10.1016/j.ecss.2005.11.026

DSWA – Dry Stone Walling Association of Great Britain (2011) *Dry stone walls and wildlife*. Dry Stone Walling Association, Cumbria.

Edwards, H.G.M., Holder, J M. and Wynn-Williams, D.D. (1998) Comparative FT-Raman spectroscopy of Xanthoria lichen-substratum systems from temperate and Antarctic habitats. *Soil Biology and Biochemistry*, 30(14), 1947–1953. DOI: 10.1016/S0038-0717(98)00065-0

EEA – European Environment Agency (2011) *Landscape fragmentation in Europe*. A joint EEA-FOEN report. No 2/2011. Publications Office of the European Union, Luxembourg, ISBN 978-92-9213-215-6.

European Parliament (2007) *Forest fires: causes and contributing factors in Europe*. Report by the DG Internal Policies, Policy Department A: Economic and Scientific Policy, Report number; IP/A/ENVI/ST/2007-15.

FAO and European Commission on Agriculture (2006) *The role of agriculture and rural development in revitalizing abandoned/depopulated areas*. 34th Session, Riga, Latvia, 7 June 2006. 15 pp.

Ferrari, R., Ghesini, S. and Marini, M. (2011) Reticulitermes urbis in Bagnacavallo (Ravenna, Northern Italy): a 15-year experience in termite control. *Journal of Entomological and Acarological Research*, 43(2), 287–290. DOI: 10.4081/jear.2011.287

Forestry Commission (2010) *A survey of*

Community Woodland Groups. A Combined Case Study Report by Wavehill Consulting for the Forestry Commission Wales. Available at: http://llaisygoedwig.org.uk/wp-content/uploads/2013/12/Wavehill-Report-%E2%80%93-Case-studies-of-Welsh-Woodlands-Sept-2010.pdf

Garner, A. (2004) Living history trees and metaphors of identity in an English forest. *Journal of Material Culture*, 9(1), 87–100. DOI: 10.1177/1359183504041091

Gehrig-Fasel, J., Guisan, A. and Zimmermann, N.E. (2007) Tree line shifts in the Swiss Alps: climate change or land abandonment? *Journal of Vegetation Science*, 18(4), 571–582. DOI: 10.1111/j.1654-1103.2007.tb02571.x

Gössling, S., Peeters, P., Hall, C. M., Ceron, J. P., Dubois, G. and Scott, D. (2012) Tourism and water use: supply, demand, and security. An international review. *Tourism Management*, 33(1), 1–15. DOI: 10.1016/j. tourman.2011.03.015

Griffin, P.S., Indictor, N. and Koestler, R.J. (1991) The biodeterioration of stone: a review of deterioration mechanisms, conservation case histories and treatment. *International Biodeteroration*, 28, 87–207.

Guštin, M. and Nypan, T. (eds.) (2010) *Cultural Heritage and Legal Aspects in Europe*. Institute for Mediterranean Heritage and Institute for Corporation and Public Law Science and Research Centre Koper, University of Primorska.

Hassan, R., Scholes, R. and Ash, N. (2005) *Ecosystems and Human Well-Being: Current State and Trends: Findings of the Condition and Trends Working Group*. Millennium Ecosystem Assessment Series. Island Press, Washington DC.

Hueck, H.J. (1968) The biodeterioration of materials-an appraisal. Biodeterioration of materials. Microbiological and allied aspects. *Proceedings of the 1st International Symposium*. Southampton, 9th–14th September, 1968. Elsevier Publishing Co.

Hussein, H.S. and Brasel, J M. (2001) Toxicity, metabolism, and impact of mycotoxins on humans and animals.

Toxicology, 167(2), 101–134. DOI: 10.1016/S0300-483X(01)00471-1

International Union for Conservation of Nature (IUCN) (2001) *IUCN Red List Categories and Criteria*. Species Survival Commission, IUCN.

International Union for Conservation of Nature (IUCN) (2015) *2015 IUCN Red List of Threatened Species. A Global Species Assessment*. IUCN Publication Services Unit, Cambridge, UK.

Joint Programming Initiative (JPI) (2010) *The joint programing initiative on cultural heritage and global change: a new challenge for Europe*. Vision document. Version 17 June 2010.

Kambhampati, S. and Eggleton, P. (2000) Taxonomy and phylogeny of termites. In: T. Abe, D.E. Bignell and M. Higashi (eds.) *Termites: Evolution, Sociality, Symbiosis, Ecology*. Kluwer Academic Publishers, Dordrecht, Netherlands, pp. 1–23.

KEA European Affairs (2006) *The Economy of Culture in Europe*. Study conducted on behalf of the European Commission, pp. 147–155 and pp. 303–306.

Keenleyside, C., Veen, P., Baldock, D. and Zdanowicz, A. (2005) Background document from the Seminar on Land Abandonment and Biodiversity in the New Member States and Candidate Countries in Relation to the EU Common Agricultural Policy held in Sigulda, Latvia, on October 7–8 2004. 64 pp.

Lasanta, T., Arnaez, J, Flano, P.R. and Monreal, N.L.R. (2013) Agricultural terraces in the Spanish mountains: an abandoned landscape and a potential resource. *Boletin de la Asociacion de Geografos Espanoles*, 63, 487–491, ISSN 0212-9426.

LDNPP – Lake District National Park Partnership (2015) *The Partnership's Plan – The management plan for the English Lake District 2015–2020*. Available at: https://app.box.com/s/uudp141pgugi1fi73pld0uj0r6tk0qz1

Lesschen, J.P., Cammeraat, L.H. and Nieman, T. (2008) Erosion and terrace failure due to agricultural land abandonment in a semi-arid environment. *Earth Surface Processes and Landforms*, 33(10),1574–1584. DOI: 10.1002/esp.1676

MEA (2015) *Millennium Ecosystem Assessment (MEA)*. Available at: http://www.millenniumassessment.org

Moravec, J. and Zemeckis, R., (2007) *Cross compliance and land abandonment*. Deliverable D17 of the CC Network Project, SSPE-CT-2005-022727, pp. 6–16.

Moriondo, M., Good, P., Durao, M., Giannakopoulous, G. and Corte-Real, J. (2006) Potential impact of climate change on fire risk in the Mediterranean area. *Climate Research*, 31(1), 85–95.

Natural England (2009) *Farming with Nature: Agri-Environment Schemes in Action*, Natural England, ISBN 978-1-84754-183-3.

PDNPA – Peak District National Park Association (2012) *A summary of the Peak District National Park Management Plan 2012–2017. A partnership for progress*. Available at: www.peakdistrict.gov.uk/npmp

Pickard, R. (ed.) (2001) *Policy and Law in Heritage Conservation*. Spon Press, London. ISBN 0-419-23280-X.

Ricotta, C., Guglietta, D. and Migliozzi, A. (2012) No evidence of increased fire risk due to agricultural land abandonment in Sardinia (Italy). *Natural Hazards and Earth System Science*, 12(5), 1333–1336. DOI: 10.5194/nhess-12-1333-2012

Sirami, C., Brotons, L. and Martin, J.L. (2007) Vegetation and songbird response to land abandonment: from landscape to census plot. *Diversity and Distributions*, 13(1), 42–52. DOI: 10.1111/j.1472-4642.2006.00297.x

Turinek, M., Grobelnik-Mlakar, S., Bavec, M. and Bavec, F. (2009) Biodynamic agriculture research progress and priorities. *Renewable Agriculture and Food Systems*, 24(02), 146–154. DOI: 10.1017/S174217050900252X

Ueno Y. (1980) Trichothecene mycotoxins mycology, chemistry and toxicology. *Advances in Food and Nutrition Research*, 3, 301–353. DOI: 10.1007/978-1-4757-4448-4_10

UNESCO – World Heritage Centre (2009) *World Heritage Cultural Landscapes: A Handbook for Conservation and Management*. World Heritage Papers 26. UNESCO, Paris, France. ISBN: 978-92-3-104146-4.

Veteläinen, M., Negri, V. and Maxted, N. (2011) A second look at the European strategic approach to conserving crop landraces. In: N. Maxted, M.E. Dulloo, B. Ford-Lloyd, L, Frese, J, Iriondo, M. Pinheiro de Carvalho (eds.) *Agrobiodiversity Conservation: Securing the Diversity of Crop Wild Relatives and Landraces*, Chapter 24, pp. 181–186, CAB International.

Warner, J. (1999) North Cyprus: tourism and the challenge of non-recognition. *Journal of Sustainable Tourism*, 7(2), 128–145. DOI: 10.1080/09669589908667331

Wilson, M.J. and Jones, D. (1983) *Lichen Weathering of Minerals: Implications for Pedogenesis*. Geological Society, London, Special Publications, 11(1), 5–12.

WWF – Worldwide Fund (2004) *Freshwater and Tourism in the Mediterranean*. WWF Mediterrean Programme, Rome, Italy.

CHAPTER 7

Integrated perspectives

7.1. *Introduction*

One of the key challenges faced in environmental management in any sector is to find solutions to problems in one area that does not cause new issues or exacerbate existing ones in other areas. This involves taking a whole-system perspective and considering all activities, effects and impacts. This is not always easy to do, and indeed this book illustrates that we often have to break down whole systems into discrete parts in order to examine them in detail and explore management solutions and policy approaches.

In reality, agriculture and the farming landscape is an open continuum with geological, biological, chemical and physical processes at work, with flows of materials and energy within a global system. Figure 7.1 illustrates the flows of materials and energy within an agricultural system, whilst also showing how the system can be broken down into a set of farm system components through which the materials and energy flow. It is a process of abstraction in order to 'paint' the best scientific picture to make sound decisions, be that at European, national or farm level.

Identifying different farm system components also helps to identify key intervention points, for example where in the system are the greatest GHG emissions, and these intervention points can then become the focus for developing mitigation options. Farms are also part of a broader system and are consumers of inputs, the manufacture of which also has environmental impacts, for example the production of inorganic nitrogen fertiliser. This is a Life Cycle Thinking (LCT; Brentrup *et al.*, 2004) approach to understanding the system, and as such encompasses both direct and indirect activities and associated effects and impacts.

There can be an inherent conflict between managing the individual parts and optimising the whole-system performance. However, this issue is not impossible to overcome, and taking an integrated perspective is an explicit approach to managing the system as parts and as a whole.

This chapter presents two elements for developing a more integrated/holistic perspective: at the policy (strategic) level; and at the farm level. There is perhaps also a third element, one of vertical integration with respect to the delivery of policy and transfer of knowledge to farms, but this will be dealt with under the policy/strategic level.

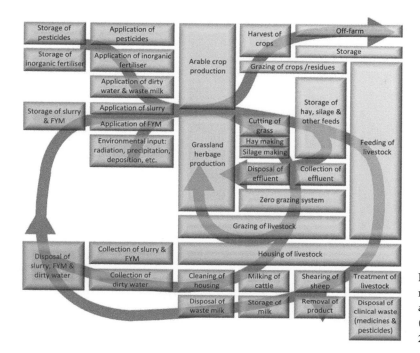

Figure 7.1: Flows of materials and energy in a farm system (Source: Tzilivakis *et al.*, 2010)

7.2. Integrated policy formulation and delivery

7.2.1. Introduction

There is little doubt that the generic process of policy formulation is complex. It has often been described as a process of 'muddling through' (Lindblom, 1959; 1979) and is far from perfect. The process is inherently tied up with the democratic and governmental systems that have evolved in society, which express the values and interests that set the aims, objectives and priorities for public policy. Consequently, as values and interests evolve over time, governments change and the policies and instruments for delivery also change to accommodate different perspectives and priorities. It is a dynamic landscape that generally tends to evolve slowly, but can sometimes occur very rapidly (as is currently being observed with the decision of the United Kingdom to leave the EU). As has been explored in the respective

chapters in this book, many policies focus on single issues (e.g. the Nitrates Directive focusing on a specific substance and the Soil Thematic Strategy focusing on a specific media). However, as scientific knowledge and understanding has grown, the need to take a more holistic and integrated approach has become more evident. In environmental management it is important to ensure that we do not solve one problem at the expense of another and that applies at both the farm and policy levels.

In addition to the need to take a whole-systems perspective, which could be regarded as a 'horizontal', policy formulation and development also needs to take a 'vertical' perspective to integration with respect to delivery and knowledge transfer. Policies and knowledge need to be successfully delivered and transferred to the farm level to be effective and achieve the desired outcomes at national or European level. This involves the development of effective delivery instruments, education and knowledge transfer initiatives, which integrate with other related activities and initiatives at the farm level (as opposed to conflicting with such activities and initiatives).

This sub-chapter explores some of the strategic challenges that arise when formulating agri-enviromental policy that highlight the need to take a holistic and integrated perspective. It explores some examples of how policies evolve and can be made more holistic/integrated and how policies can be delivered on farms within the context of knowledge transfer.

7.2.2. *Strategic challenges*

There are two key strategic challenges that need to be tackled to develop more integrated policy formulation. First, policies that aim to address specific issues need to do so in a way that does not negatively impact upon other objectives (that are tackled by other policies). Ideally, policy solutions need to tackle all issues positively (win–wins), but this may not always be the case and there may be instances where trade-offs between objectives need to be considered. Second, partly related to the first challenge, is the need to account for displacement effects, that is, where impacts are lowered in one region, but at the expense of increasing them in another region, effectively exporting effects and impacts to other parts of the country, Europe or elsewhere around the globe. Regional variability can be a key factor for some impacts (e.g. GHG emissions), thus it is important to account for this and not assume that a blanket policy, for example, has the same benefits and burdens in all locations. Such strategic aspects cannot be tackled by farm-level decision-making, thus policymakers need to manage such issues to ensure that there is an overall reduction in negative impacts and food production evolves in a more sustainable direction.

Strategic decision-making inherently requires a holistic perspective and environmental management demands that this perspective is broadened. As illustrated in Figure 7.1, the farm system that is affected by policy is interconnected, and a change in one part of the system can affect the other parts. Policies essentially provide a regulatory framework for the management of this system. It may restrict or prohibit some activities and practices and encourage others, and in so doing influence the effects and impacts that arise. Unintended consequences of policies are undesirable, but it is always a risk when policies are developed to tackle single issues. Within the EU, there are now legal requirements to undertake a Strategic Environmental Assessment (SEA) for new policies before they are implemented (under the SEA Directive 2001/42/EC). This is a systematic process to identify and mitigate negative effects and impacts to a level where they are acceptable for the stakeholders involved. This is not necessarily equivalent to taking a more integrated approach, as each policy is dealt with individually and SEA as a technique has a number of limitations and weaknesses (Sadler and Dusík, 2016; Thérivel and Paridario, 2013). However, it has the potential to be adapted so that a suite of more integrated policies can emerge (Brown and Thérivel, 2000; Lenschow, 2002; White and Noble, 2013).

Policies by their nature tend to be 'blanket' in approach in order to provide a common set of regulations and a 'level playing field' for all concerned. This may be deemed to be a fair approach but the benefits and burdens of a policy can vary from one location to another, and the burdens and benefits may also vary over time. For example, Tzilivakis *et al.* (2017) analysed the climate change benefits and burdens of 130 measures and operations that are supported by the EU Rural Development Programme (RDP) over different time horizons (1 to 250 years) for

1,281 NUTS3 regions in the EU. The 'NUTS' classification (nomenclature of territorial units for statistics) is a hierarchical system for dividing up the economic territory of the EU for statistical and analytical purposes. NUTS3 is a relative fine scale referring to an area with a population generally between 150,000 and 800,000. The analysis, undertaken by Tzilivakis *et al.* (2017), revealed the RDP operations tended to fall into the following categories:

- Long-term benefit: There is a net decrease in emissions across all time horizons.
- Short-term burden, long-term benefit: There is a net increase in emissions in the first year, followed by a net decrease in subsequent years.
- Medium-term burden, long-term benefit: There is a net increase in emissions over 50–100 years, followed by a net decrease in subsequent years.
- Variable benefit/burden: Emissions depend on location and time horizon.
- Long-term burden: There is only a net increase in emissions.

This highlights both the spatial and temporal variability in the benefits and burdens that can arise as a consequence of the implementation of policy.

Such regional variation presents an additional strategic challenge for agri-environmental policy with respect to ensuring that environmental impacts are reduced overall (at regional, national, European and global levels). The displacement of impacts is an issue that can be easily overlooked. This can happen at the farm level by shifting the burden from one field to another, and it can also happen at regional, national, European and global levels. For example, a policy may be implemented within a region to reduce GHG emissions from farms by developing more extensive (lower-yielding) systems of production. This may result in an increase in imports (to replace production lost due to lower yields) and those imports may have equivalent or higher GHG emissions – thus the GHG emissions have effectively been exported and have not been reduced overall. Efforts at the farm level to reduce GHG emissions or other impacts cannot account for this sort of displacement as it is a strategic issue that needs to be tackled by policy.

7.2.3. *Policy evolution and examples of integration*

If society were to start with a blank sheet of paper there is little doubt that the policies we have today would be very different to those that are currently in place. Policies could be drafted that are specifically tailored to tackle the issues society faces, including addressing multiple issues in an integrated fashion. In reality though, the policy landscape is an evolutionary one (Medina and Potter, 2017) and generally new policies are an incremental adaptation of existing policies – keeping the parts that do work and discarding those which have failed or are no longer fit for purpose. Ultimately, policy is the public administration of a complex, dynamic and adaptive system, in which the fitness landscape is constantly changing and consequently policies need to adapt accordingly (Perrot *et al.*, 2016; Teisman and Klijn, 2008). The evolutionary process is partly driven by growth in scientific knowledge and understanding, especially with respect to environmental systems, processes and cause–effect relationships. As scientific understanding improves, policies, instruments and practices are able to adapt and become more 'fit for purpose' and ultimately deliver the outcomes society desires.

The evolution of agri-environment policies, regulations and instruments is already apparent in existing legislation. Biological conservation, for example, has progressed from species-based to ecosystem-focused approaches (Franklin, 1993; Wallington *et al.*, 2005). The phrase 'ecosystem approach' was first coined in the early 1980s, but found formal acceptance at the Earth Summit in Rio in 1992. It underpins the Convention on Biological Diversity, where it is described as 'a strategy for the integrated management of land, water and living resources that promotes conservation and sustainable use in an equitable way' (CBD, 2017). The shift from species to ecosystem, and more integrated approaches is reflected in the legislation and regulations that have emerged in the past. For example, in 1979 the Birds Directive came into force (Council Directive 79/409/EC on the conservation of wild birds of 2 April 1979) which provides a framework for the conservation and management of, and human interactions with, wild birds in Europe. Then, in 1992 the Habitats Directive came into force (Council Directive 92/43/EEC of 21 May 1992 on the conservation of natural habitats and of wild fauna and flora), thus introducing a habitats and ecosystems approach to biological conservation (see Chapter 3.2 for further details).

In Europe, there is a clear recognition that biological conservation policies need to be integrated. The bulk of wildlife populations and biodiversity exist amongst landscapes managed for the production of goods and services (such as agriculture and forestry) rather than in dedicated reserves protected from any harmful activities. The latter have an important role, that is, to protect the most valuable and vulnerable areas, but this needs to be coupled with approaches to enhance and conserve biodiversity in productive landscapes. This demands an integrated approach to meet the needs of biological conservation, production and socio-economic viability. The ecosystem services framework aims to bring all needs

under one 'umbrella', recognising that they are all inherently linked. In many respects, the concept of ecosystem services is not new (Cairns and Pratt, 1995; Costanza and Daly, 1992; Ehrlich and Mooney, 1983; Gómez-Baggethun *et al.*, 2010), but its application in the context of agri-environmental management policy is relatively new (Hauck *et al.*, 2013; Landis, 2017; Potts *et al.*, 2016; Schulp *et al.*, 2016; van Zanten *et al.*, 2014) and it is still a developing topic. A key tool for many policies is the implementation of payment schemes to incentivise activities that protect and enhance ecosystem services. Consequently, in recent years the term 'Payments for Ecosystem Services' (PES) has emerged. The fundamental concept is not new, as for many years farmers have received payments for undertaking environmental activities as part of stewardship schemes under the EU rural development programme. However, there is now more emphasis on understanding and quantifying ecosystem services so that PES schemes can reflect this in the payments made. However, quantifying ecosystem services is not easy. There are many studies that attempt to do this (Farley and Costanza, 2010; Moore *et al.*, 2017; Scheufele and Bennett, 2017; QuESSA, 2017; Swinton *et al.*, 2007) but an established or standardised approach is yet to emerge.

Another approach to integration (in the context of evolving existing policies) is one of mainstreaming. An example of this can be observed with respect to climate change policy in Europe. In 2010, the Directorate-General for Climate Action (DG CLIMA) was created, taking over responsibility for climate change issues from the Directorate-General for Environment (DG ENV). In addition to the formulation and implementation of specific climate policies and strategies, a key element of the work of DG CLIMA is to facilitate the evolution and adaptation of other EU policies managed by other DGs. This involves the integration of climate change objectives, relating to both mitigation and adaptation, into other policy areas. This is the process of mainstreaming and has involved integrating mitigation and adaptation actions into all major EU spending programmes, in particular cohesion policy, regional development, energy, transport, research and innovation and the Common Agricultural Policy (CAP).

7.2.4. Policy delivery and knowledge transfer

As mentioned above, in addition to horizontal integration (tackling multiple environmental objectives), policy formulation and development also needs to take a 'vertical' perspective to integration with respect to delivery and knowledge transfer. For any policy to achieve its aims and objectives, there needs to be effective delivery mechanisms. This may include a suite of regulations and instruments with various incentives and penalties, but crucially these need to be coupled with education, awareness raising and knowledge transfer activities (Anderson and Feder, 2004; Fazey *et al.*, 2013; Juntti and Potter, 2002). This is apparent in the EU rural development programme where training and education measures are presented alongside measures incentivising agri-environmental management. Policies also need to integrate with initiatives and activities at the farm level (Buizer *et al.*, 2015; Raymond *et al.*, 2016), they need to be synergistic and it is essential they are not contradictory.

Farming is a business like any other and so needs strategic planning and careful management to continue to produce food and remain profitable. As such, farmers would not knowingly do anything that would reduce their crop yields, livestock productivity, damage their land or environment upon which they depend or waste resources for

which they have paid. The key here is the use of the term 'knowingly' and how successful a farmer is at achieving maximum yields from a minimum amount of resources whilst protecting the environment will depend upon their depth of farming knowledge and access to advice, information and training. Across Europe, the farming community is hugely diverse in terms of the nature of the farmed area and farming enterprises, their wealth and profitability, farming practices, attitudes and capabilities and so knowledge transfer needs and the types of training programmes required will be equally diverse. Consequently, over time a wide range of different approaches to knowledge transfer and training have developed across Europe.

Under the CAP, MSs must operate a system for advising farmers on land and farm management with an emphasis on sustainable production. Some financial aid is available under Pillar 2 (rural development programme) for both setting up the scheme and supporting farmers to access it. However, the CAP does not specify what form this advisory system should take and various approaches are evident. Whilst this flexibility means the service can be built to meet specific needs, it inevitably also means huge variability in the effectiveness of the service across different MSs. In some countries, a specific service has been established for the provision of information using either public or private finance. In other countries, a support service has been integrated within other existing services.

Knowledge is also transferred in a variety of different ways. As well as governmental bodies, private consultancies and NGOs offer a wealth of advice. For example, in the UK, LEAF (Linking Environment and Farming) is a UK registered charity that seeks to promote sustainable food and farming. The organisation achieves this via a range of interconnected initiatives that collectively act as a farm information and knowledge exchange service. LEAF has, over a number of years, developed its own sustainable farming review process that farmers can use to identify priorities and develop site-specific action plans. A set of environmental standards has also been developed and the LEAF marque is a food labelling scheme that allows consumers to identify produce grown to these standards in-store. The charity also organises farm open days, conferences and has a network of demonstration farms and innovation centres that promote sustainable farming and integrated farm management. In Denmark, the main agricultural advisory service (Danish Agricultural Advisory Service – DAAS) is owned and operated by farmers themselves. As part of DAAS, a knowledge centre has been established that acts as the national research and knowledge facilitator disseminating information from governmental authorities, food businesses and scientific institutions.

Industry also has a role to play and provides large amounts of advisory support, usually as part of a deal to sell a product or service. However, it must be recognised that there is always the risk that the farmer's best interests might not be perfectly aligned with the company's desire to sell their goods or services.

As well as advisory services offered by third parties, the role of the Internet and dedicated advisory and information websites should not be underestimated. Across much of Europe, these are easily accessed by farmers and can be done at a time convenient to the individual. In addition, 'peer to peer' knowledge exchange, demonstration farms and events are also efficient means of transferring information around the farming community.

7.3. Integrated farm management

7.3.1. Introduction

As shown in Figure 7.1, there are flows of materials and energy through agricultural systems, involving physical, chemical and biological processes, influenced by management decisions and actions. This inevitably means that changing one part of the system can affect other parts of the system. For example, insufficent nutrients can make plants less resilient to drought, pest and disease; or using less chemical pesticides may require the use of more mechanical approaches which will use more fuel and increase gaseous emissions. Consequently, a more holistic and integrated approach is needed, which is known as Integrated Farm Management (IFM).

Defining what constitutes an integrated approach, other than being holistic, is quite difficult. However, this chapter aims to do this by first exploring the history of the evolution of IFM, including some existing definitions. Second, it explores farm management processes and how these are adapted in order to develop a more integrated approach. Third, some examples are explored to illustrate integrated management.

7.3.2. History and definitions of integrated farm management

The basic concept of IFM can be traced back to Integrated Pest Management (IPM), which emerged in the 1950s. However, Leake (2000) explains that IPM can be traced back to the 18th Century when pests were controlled and managed using cultural, biological and mechanical techniques, and then chemical pesticides started to emerge, including heavy metals and plant extracts. In the 1940s, organochlorine, organophosphate and carbamate pesticides revolutionised crop protection, however, they were so good that their consequent widespread use led to several pest crises in the 1950s, such as with cotton crops in Peru, the spotted alfalfa aphid (*Therioaphis trifolii*) in California and the peach potato aphid (*Myzus persicae*) on glasshouse chrysanthemums in the UK (Van Emden and Peakall, 1996; Van Emden and Service, 2004), whereby resistance developed allowing the pest to attack the crop unchecked. In the 1950s it was realised that complete eradication of pests was not the answer and that careful management of the crop was a better approach, to keep pest levels below economic injury levels. In 1959, Stern *et al.* (1959) found, following this principle, that reduced rates of pesticide resulted in reduced populations of pests, but beneficial insects (biological control agents) were not killed off and would then reduce the pests further. This was a landmark study that clearly demonstrated the benefits of an integrated approach.

In the 60s and 70s, it became apparent that the intensification of agriculture was giving rise to a number of environmental problems, especially as a consequence of pesticide and nutrient use in arable agriculture, and the publication of *Silent Spring* (Carson, 1962) was a landmark book recognising these issues. Consequently, the concepts within IPM were expanded to encompass all aspects of arable cropping and the term integrated crop management (ICM) emerged. This concept is well described in the objectives of the LIFE project (Jordan *et al.*, 1997) that ran from 1989 to 1995, which were the manipulation and integration of farming practices; to maintain or enhance the chemical, physical and biological condition of the soil; increase biodiversity and reduce the impact of pest diseases and weeds in order to lower the requirement for agrochemicals.

In the 1990s, the concept of ICM was extended to the whole farm and the term

Integrated Farming Systems (IFS) emerged; for example, from 1992–1997 there were the LINK:IFS projects (research on production systems which would be profitable, meet environmental requirements, ensure greater conservation of resources and meet consumer requirements; Ogilvy *et al.* 1994). The concept and term, Integrated Farm Management (IFM), evolved alongside IFS but did not really enter the mainstream until the 2000s, being promoted by organisations such as LEAF and EISA (European Initiative for Sustainable Agriculture).

In the past, IFM has been viewed as a third system of farming (Morris and Winter, 1999; Parra-López *et al.*, 2006; Pacini *et al.*, 2003; Reganold *et al.*, 2001), with the others being conventional and organic, the idea being that IFM provides a third, middle way between conventional and organic. However, integrated farming is not half conventional and half organic. Intensive and extensive production systems can both be subject to integrated farm management, and in many respects the goal of IFM is one that aligns with the concept of sustainable intensification (i.e. maintaining or increasing production whilst reducing or minimising negative environmental and social impacts). This is apparent in the following three definitions of IFM:

> An holistic pattern of land use which integrates natural regulation processes into farming activities to achieve maximum replacement of off-farm farm inputs and to sustain farm income.
>
> (El Titi, 1992)

> Integrated Farming offers a whole farm policy and whole systems approach to farm management. The farmer seeks to provide efficient and profitable production, which is economically viable and environmentally responsible, and delivers safe, wholesome and high quality

food through the efficient management of livestock, forage, fresh produce and arable crops whilst conserving and enhancing the environment.

(EISA, 2010)

IFM is a whole farm approach that combines the best of traditional methods with beneficial modern technologies, to achieve high productivity with a low environmental impact.

(LEAF, 2016)

These definitions highlight how IFM attempts to balance the economic, environmental and social objectives. Efficient, highly productive, modern and profitable production are key elements – farms need to be economically viable in order to deliver environmental and social objectives as well.

A common element of all the definitions above is that IFM is holistic and so encompasses multiple issues. To visualise this holistic approach, there are some common 'wheel'-style diagrams in much of the IFM literature, for example from EISA (2012) and LEAF, the latter shown in Figure 7.2.

Figure 7.2 illustrates the numerous issues presented as circles around integrated farming including: organisation and planning, soils, crop protection, animal husbandry, energy, pollution, and landscape and wildlife. LEAF includes a node of PR and marketing as part of IFM (in addition to those of EISA), introducing social elements into the equation.

It is undoubtedly a bit of a juggling act to manage all these objectives in an integrated fashion. There is no prescribed approach and, although there are tools and approaches that can help, it fundamentally comes down to knowledge and understanding of:

- The objectives for the farm (and the functions it performs).
- How farm components achieve those objectives and functions.

Figure 7.2: LEAF: Integrated farm management

- Interactions between farm components.
- Synergies and trade-offs in outcomes.

A continuous learning process is inherently required to achieve multiple objectives in a dynamic and complex system. This process has been commonly termed 'continuous improvement'. It acknowledges that we live in a dynamic world where the environment changes around us, and objectives, priorities, markets and societal values evolve and change over time. Consequently, farming businesses, and their practices, need to adapt and evolve accordingly.

It is perhaps also important to acknowledge that although the term 'continuous improvement' is used, this does not necessarily mean infinitely improving. It is more about the process of evolution. Evolving and adapting, hopefully to improve, but at the very least to stay in the same place – recognising that if farming systems do not evolve, then they can get left behind as everything changes around them. Farming systems need to have the capacity to adapt and evolve (increasing resilience) to address changing demands, priorities and objectives, including a changing environment, and pressures such as climate change or emerging pollutants.

7.3.3. Integrated management processes

7.3.3.1. Introduction
As mentioned above, knowledge and expertise is fundamental to any management, and very much so for IFM. It is the decision-making and management processes that govern and steer activities on a farm and therefore the consequent impacts. Farm management decisions are taken over different time horizons, be it day to day (sometimes hour to hour), month to month, season to season, or year to year, or longer, for example five-yearly. For an integrated approach to be

adopted, the management and decision processes need to be adapted. Thus, the first step is to understand what those processes are.

Continuous improvement is at the heart of IFM (and any environmental management system). It is an iterative process of analysis and understanding; modelling and planning; action; and metrics and reporting, which in turn feed into further analysis and understanding, and so on – this basic concept is shown in Figure 7.3.

Building on the basic concept, some components with respect to farm management interactions with the environment can be added, including: 'actions'; which have 'effects and impacts'; 'information and data' about those effects and impacts, which can then be combined with 'knowledge and expertise', to take more integrated actions. This concept can be built upon further by adding four key management processes: 'options appraisal'; 'actions plans' (and their implementation); 'measures and indicators'; and 'performance assessment'. These management interactions and processes are overlaid on the cycle of continuous improvement concept, as shown in Figure 7.4.

The following sub-chapters explore the four management processes in order to provide an overview of how management and decision-making can evolve towards a more integrated approach.

7.3.3.2. Options appraisal
Options appraisal is about identifying suitable management options that meet the objectives and priorities of the farm in an integrated way. In many instances, management decisions involve choosing between different options. For example, it could be different pesticide products to treat a pest problem. Options appraisal is about selecting the option that best meets the objectives. Thus, for example, purely financial objectives

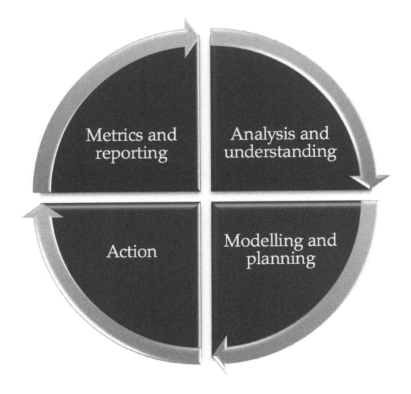

Figure 7.3: The cycle of continuous improvement concept

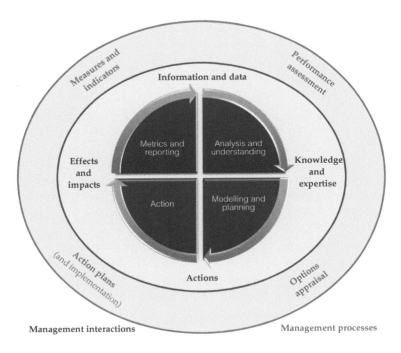

Figure 7.4: Farm management interactions, processes and continuous improvement

means selecting the economic optimum (e.g. cheapest pesticide with the greatest efficacy), which is usually a 'relatively' simple decision. However, the multiple objectives of an integrated approach make options appraisal more complex.

In some respects, options appraisal can be straightforward. In the example mentioned above, selecting an economically effective pesticide can be done using a spreadsheet that combines the costs and likely effectiveness to derive a cost–benefit ratio, thus flagging up those products that are the most cost-effective. However, when additional objectives and parameters are introduced, the complexity increases as a function of each objective added, for example the number of permutations/options can exponentially increase with each criterion. There is also a danger that to simplify the process, in an effort to take a more integrated approach, decisions are made to tackle each objective in isolation. Thus, issues and objectives, for example crop protection, crop nutrition, animal health and welfare, water, air and soil protection and biodiversity, are all managed on the farm, but not necessarily in an integrated way. The key is to understand the linkages between components and objectives, so that when decisions are made on crop nutrition, for example, the effects and impacts on crop protection, water, soil, air and biodiversity are understood and the decision-making process accounts for these accordingly.

Those undertaking the options appraisal process, such as farm managers, often know how to appraise options and make decisions. They are usually familiar with the farm, the conditions and circumstances, and have a range of skills and expertise. However, there are also instances when decision makers need to seek advice and guidance or draw upon tools to aid options appraisal, and when there are multiple objectives to tackle,

the need for such guidance and tools can increase. There are many tools that can aid options appraisal. They include: direct advice from experts (such as agronomists) either face-to-face or using other means of communication; books, guides, decision trees and codes of practice (hard copy or online); record keeping and/or calculation tools, such as spreadsheets, databases or bespoke software applications (often referred to as decision support tools); or a combination of two or more of the above. Currently, there are very few tools that offer support for truly integrated decision-making. The tools tend to be focused on single issues, for example tools for calculating nutrient requirements; tools for pesticide selection; carbon calculators; biodiversity and wildlife advice. These are still useful for providing/generating useful information to feed into an integrated decision-making process, but the integrated part of the process needs to be undertaken by the decision maker and/or consultants. This reflects the complexity of the issues involved and past approaches to tackling them. However, this is recognised and there have been recent efforts (e.g. Wynn, 2012) to develop tools and guidance that provide more integrated perspectives.

Finally, it is important to recognise that in the options appraisal and decision-making process, there comes a point where the decision maker has to take ownership of the process and make decisions based on options and information available. It is important that any guidance or tools available do not try to be overly prescriptive – they are tools for decision support, not tools that make the decision. The complexity of issues often dictates that there is no single answer – there should be a range of options with a number of advantages and disadvantages, and the decision maker can then take these into account, appraise the options and make a decision.

7.3.3.3. Management/action plans and implementation

A management/action plan is usually no more than a list of actions to be implemented on a farm. Actions arise from the options appraisal process and/or the results of performance assessment, and thus relate back to the objectives and priorities of the farm. Consequently, action plans sometimes can become a benchmark for judging performance in the next iteration of the cycle. Perhaps of more importance are the tools and instruments for implementing action plans. If these are not adequate, then actions may not be fully and/or correctly implemented. For example, sometimes additional actions may be required to ensure the actions in the plan are implemented correctly, such as the training and awareness raising of staff.

Management/action plans in the context of IFM can cover almost anything, including: crop protection; crop nutrition; biodiversity; water; energy; waste; soil; carbon (GHG emissions); and animal health and welfare. It is quite common to have issue-specific plans, for example waste management plan, carbon management plan, water management plan, and so on. As mentioned above, single-issue action plans can be contradictory to an integrated approach, as issues get tackled in isolation and not together. The following are some examples of typical management plans:

• Pests: Integrated Pest Management (IPM) is an ecosystem-based strategy that focuses on long-term prevention of pests or their damage through a combination of techniques such as biological and physical control, habitat and pest life-cycle manipulation, modification of cultural practices and use of resistant varieties. A crop protection management plan is developed that identifies the farm approach. It may include, for example, suitable rotations to limit disease pressures and/or cultural techniques to ensure that pests and weed populations are controlled without the need for chemical assistance. Ploughing and the maintenance of a fine seedbed, for example, have been shown to have a beneficial impact in reducing slug populations (e.g. Speiser *et al.*, 2001; Davey 2014; Dreves *et al.*, 2016), such that the need for molluscicides can be reduced. So, this might be a prority where slugs are a serious problem, however, such techniques can be at odds with the promotion of soil health and require considerable energy inputs (Howlett, 2012), meaning that they may be contrary to other on-farm environmental objectives, such as reducing erosion and minimising greenhouse gas emissions. Therefore, the crop protection management plan needs to work in harmony with other farm management plans. It should also recognise that the relationship between preventative techniques and the resulting pressures from pests and diseases may not be straightforward and may depend on species, soil type, climatic conditions and crop type (e.g. Davey, 2014), making the efficacy of cultural techniques both spatially and temporally variable, and difficult to predict. Indeed, what is beneficial in terms of one problem may make another worse. This will be highly site-specific and, for this reason, the observation of pests (e.g. by trapping), weeds, diseases and the damage they cause, together with the use of threshold values to determine whether the problem is sufficiently serious to warrant treatment, is essential within IPM for avoiding the prophylactic application of crop protection chemicals, something which has implications both for water quality and the build-up of pesticide resistance. If monitoring and cost–benefit interpretation indicates that there is a need for treatment, biological controls (e.g. parasitic nematodes, predatory insects, bacteria, viruses and biochemical pesticides) should be the first consideration and chemical methods the last option, at which point the chemical pesticide should be selected and applied in a manner that minimises risks to human health and the environment.

- Soil: Integrated Soil Management (ISM) refers to a set of agricultural practices adapted to local conditions to maximise the efficiency of nutrient and water use and improve agricultural productivity. Like other IFM-type approaches, ISM involves the development of a soil management plan which will describe the farm's soil protection strategy. It may focus on the combined use of mineral fertilisers, locally available soil amendments (such as lime and phosphate minerals) and organic matter (crop residues, compost and green manure) to replenish lost soil nutrients and improve soil organic matter status. This improves both soil quality and the efficiency of fertilisers and other farm inputs. The soil management plan may also include actions to reduce tillage and erosion. Reducing tillage has both advantages and disadvantages. The most important benefit of a conservation tillage system is that it can help prevent both wind and water soil erosion. It will also reduce fuel and labour requirements. However, tillage is often seen as an alternative to chemical weed control and can be very effective at reducing weed populations and so minimum tillage systems can lead to increased reliance on chemical herbicides.

  With respect to erosion, one of the most effective means of mitigation is to provide a green surface cover. Foliage will absorb much of the impact of raindrops and wind, thus protecting the soil. Vegetation will reduce the speed of water flowing over the land, providing time for the water to infiltrate and so also help to reduce soil loss. Erosion risk is significantly reduced when there is more than 30 per cent soil cover. Total cover is achievable for many grazing and cropping systems. Other options for reducing erosion include practices to reduce run-off such as avoiding row crops, farming across the contour on sloping sites and terracing, as well as providing windbreaks. Therefore, on farms where erosion is a problem, the soil management plan would be expected to fully describe the farm's erosion minimisation strategy.

- Water: Integrated Water Management (IWM) is a management approach to ensuring that on-farm water use is efficient and that agricultural activities such as irrigation, drainage and land reclamation do not disturb the natural water balance. Various practices can be implemented to ensure that agriculture uses water more efficiently. These include changing the timing of irrigation so that it closely follows crop water requirements, adopting more efficient techniques such as using sprinkler and drip irrigation systems, and implementing the practice of deficit irrigation. In addition, changing crop types can reduce water demand or shift peak demand away from the height of summer when water availability is at a minimum. As with other water-saving approaches in agriculture, providing advice (Viala, 2008), information and education to farmers will enhance their impact significantly. Suitable IWM practices are discussed in Chapter 4.3.

- Waste: Integrated Farm Waste Management (IFWM) seeks to ensure all waste streams on the farm are managed sustainably such that resources are not wasted and pollution is avoided. The approach includes the development of a farm waste management plan that will ensure the effective use of manures to realise their nutrient value and help the farm save on fertiliser costs. Plans to store and spread manures and slurry such that water pollution is avoided are also produced. The management planning process may include nutrient budgeting which is a procedure for quantifying the amount of nutrients imported to and exported from the entire farm system (Watson *et al.*, 2002). The budget is considered in balance if inputs and outputs are equal, however its main purpose is to identify if there is an imbalance such that actions to sustainably manage an excess or deficiency can be foreseen and addressed. Consequently, the farm waste management plan will be closely linked to the soil management plan. Ideally, the IFWM plan should also address non-natural waste streams such as pesticides, oils, plastics and packaging, animal health products, machinery and building waste. It should clearly distinguish between hazardous and non-hazardous wastes and adopt the waste hierarchy approach for managing these.

Actions within management plans can include almost anything, but may typically involve: investment in new equipment or infrastructure (e.g. new machinery); strategic adjustments (e.g. crop rotations); adjustments to activities and features on the ground (e.g. tillage techniques, creation of habitats, etc.); processes adjustments (e.g. emergency procedures); and training, awareness and education of staff.

For those actions involving spatial elements, the use of maps can play a crucial role in plans and implementation, for example when different activities are taking place on the same unit of land and need to be planned in an integrated way. Maps not only serve as useful planning tools but can also aid implementation. For example, cab cards in tractors can provide information to operators with respect to implementation of items in the action plan (especially useful if contractors are involved as they may not be aware of farm action plans). Another level of sophistication is the use of precision farming technology and GPS, for example automatically adjusting inputs based on crop requirement, but also based on action plans for specific fields, for example no spray zones.

7.3.3.4.   Metrics and performance

Following the cycle of continuous improvement and management (see Figure 7.3), options are appraised, decisions made, and actions planned and implemented. In order to continue the process, it is important to know if the decisions taken and actions implemented have achieved the objectives. Therefore, information and data are needed to judge this. Measures and indicators (often referred to as metrics) can be used to assess how well a farm has performed in relation to its objectives. This might be the performance of the farm against those objectives from year to year, or it might be performance of the farm in relation to other farms or standards (sometimes known as benchmarking). Metrics cover aspects of the farm and the environment that can be measured in some way. Some aspects are directly measurable, for example the amount of water or electricity used; while other aspects are not directly measureable or not easy to measure, for example habitat quality or GHG emissions.

For the majority of farm businesses, the

fundamental indicators are financial, for example farm profit or the status of the balance sheet – this is the 'bottom line'. However, for IFM there are wider objectives and the concept of the 'triple bottom line' (Elkington, 1998; 2008) applies, that is, the three pillars of sustainability: economic, environment and social. Measuring and judging environmental and social performance is typically more difficult than economic performance due to the complexity of the parameters involved.

Fundamentally, there are two types of metric: outcome-based and practice-based. Both have advantages and disadvantages, depending on their end use. Outcome-based metrics are tangible and measureable end points of desired objectives. They tend to be relatively straightforward for economic objectives and complicated for social and environmental objectives. For example, economic metrics might include enterprise/farm profit, net worth, gearing, and so on; social metrics might include rural employment, worker welfare, and so on; and environmental metrics might include increased bird populations, improvements in water quality, and so on. The social and environmental outcome-based metrics can be costly and/or impractical for a farm business to gather, and some may argue this should be the role of government, not the farm.

Practice-based metrics are tangible and measureable metrics relating to actions on the farm. They can be viewed as surrogate measures or indicators for outcomes. The concept is that changes in farm practices should result in progress towards desirable outcomes. For example, 'improved accuracy in calculation of crop fertiliser requirements' as an indicator of reduced nutrient pollution. This action should reduce excess nutrients and associated impacts when they are lost to the wider environment (see Chapters 2 and 4), but the farm does not measure the actual

nutrient losses, pollutant loads or water quality. These metrics tend to be cheaper and more practical for a farm business to gather.

With regard to the practicalities of data collection, a lot of useful data is already collected within most business. Good record-keeping is good practice on any farm and is often a legal requirement for some aspects (e.g. pesticide use). Additionally, monitoring use of resources and inputs, for example energy use (fuel and electricity bills); water use; fertiliser and pesticide use; and so on, should be part of any good business management and provide data for monitoring use over time. A step on from record-keeping and monitoring is the use of audits. Audits are an explicit data collection task that can encompass record-keeping and monitoring, but are usually more detailed. They can be quantitative and qualitative. For example, an energy audit is not just the total energy used by the farm, it involves establishing where the energy is being used (how much and where), with the aim of improving energy efficiency. Qualitative audits tend to relate to practices and/or actions being implemented on the farm, for example the practice could be to have a nutrient management plan in place. Audits can be a one-off exercise, but usually most value is obtained from them when they are undertaken on a regular basis, for example annually.

The ultimate point of judgement for performance is to determine whether the objectives/desired outcomes being achieved and/or progress is being made towards them. For example: is farm profit being maintained or increased; is water quality improving; are GHG emissions decreasing; is biodiversity being enhanced; are populations of wildlife species being maintained/increased; is soil erosion decreasing; and so on. Many of these are costly and impractical to measure and monitor on farms. Consequently, practice-based metrics tend to become the basis for performance assessment.

At a basic level, we can judge performance against action plans. This can be a list of actions, highlighting which ones have or have not been implemented/achieved. It is essentially a simple checklist approach, but can be useful. This approach can be made more sophisticated by adding scores and/or weightings to actions or practices in relation to desired outcomes and objectives.

Another option for utilising practice-based metrics for performance is to use models and calculators to assess potential impacts. Whilst some of these types of tools may well be used by the farming community, many may be more suited to consultants and advisors or for research purposes due to their complexity. One example is the use of carbon calculators in combination with practice-based metrics (activity data) to determine GHG emissions and/or a carbon balance for the farm (Bochu, *et al.*, 2013; Hillier *et al.*, 2011; Tzilivakis *et al.*, 2014). This can be useful for identifying options and issues on the farm, for example to highlight key emission sources and options for mitigation. However, their use is questionable for uses beyond the farm, for example for benchmarking or labelling, as they are modelled metrics and not actually measured.

Assessment of farm performance is useful for on-farm management as it can help identify areas where objectives have not been achieved and thus where resources can be targeted for improvement. It can also be beneficial to assess performance in relation to benchmarks or defined standards. This can be done in two ways: (i) against the performance of other farms; (ii) against a set of industry or best practice standards.

Comparing a farm to other farms can be a strong driver. If the farm is not performing as well as others, then the farmer may wish to take action to correct this. Comparing farms is fine for non-quantitative benchmarks, but when data are quantitative it is important to ensure that like is compared with like, or else the benchmark is meaningless. For example, comparing the GHG emissions of one farm to another can be meaningless as there are many variables that may give rise to different emissions, even for farms of the same type (e.g. different soil types could make a big difference). Therefore, it is important to ensure benchmarks and standards are meaningful.

There are many types of industry standards, some defined by standards organisations such as International Organisation for Standardisation (ISO) and British Standards Institution (BSI), others by industry organisations. In this instance, the most common standards are those which are referred to as 'assurance schemes' and/or 'labelling schemes'. Assurance schemes are largely practice-based. Farms can become certified by achieving all the required standards and this is independently verified by an auditor. Standards consist of critical failure points (CFP) and recommended lists of actions. Failure to meet CFPs usually results in suspension from the scheme until corrective action has been taken. The use of practice-based metrics in assurance standards is quite widespread. However, their use as a means of putting labels on food produce is increasingly being questioned (Lewis *et al.*, 2010; Tzilivakis *et al.*, 2012), principally because they are not measuring outcomes and this could potentially be misleading to consumers. One of the few exceptions is perhaps Conservation Grade (also known as 'Fair to Nature'; Conservation Grade, 2017) which requires farms to put aside 10 per cent of land for biodiversity conservation and, every five years, the farm is visited by independent assessors to assess habitat quality, to ensure high-quality habitats are being delivered, thus there is some measure of the outcomes.

7.3.4. Ilustrating/demonstrating integrated farm management

7.3.4.1. Introduction

IFM acknowledges and embraces the concept of farms being multifunctional. Every farm is different, and so their objectives and priorities will be different. The local circumstances will influence objectives and priorities (e.g. if there is no surface water, then environmental objectives relating to protecting surface water do not apply), and local features will affect environmental sensitivities and interactions.

Objectives and priorities drive thought processes in IFM. Clearly 'win–win' solutions are sought, but there is often a need to balance competing objectives. A 'sustainable balance' needs to be struck – albeit acknowledging that the term sustainable is used with some caution, as defining what is or is not sustainable is even more complex than defining IFM, and is often a contentious issue.

The following sections explore some specific examples where practices are implemented to tackle multiple objectives. Some are simple examples, some are more complex. The following objectives are used for these examples:

- Air quality: to reduce emissions of harmful substances to the air from the farm, such as ammonia, nitrous oxides and particulates, directly and indirectly.
- Biodiversity: to maintain or enhance biodiversity and wildlife populations on the farm.
- GHG emissions: to reduce emissions of greenhouse gases from the farm (directly and indirectly) and increase carbon sequestration on the farm.
- Livestock health and welfare: to ensure that livestock are in good health and have high welfare standards.

- Profitable production: to ensure that farm enterprises are economically profitable.
- Soil conservation: to ensure that soils on the farm are in good condition (physically, chemically and biologically) and that any erosion issues are minimised.
- Waste minimisation: to minimise the generation of wastes, and when they are generated, maximise reuse and recycling where appropriate.
- Water protection: to ensure that surface and groundwaters on or near the farm are protected from any potentially polluting/harmful substances and activities.
- Worker health and welfare: to ensure that farm workers are in good health, any health risks are minimised, working conditions are acceptable, and so on.

This is not an exhaustive list, but aims to illustrate the sort of objectives that might exist and thus how they can be addressed. Given these objectives, the aim for the farm is to satisfy as many as possible, with respect to achieving these objectives.

7.3.4.2. Soil management and crop nutrition

Soil is a fundamental component that interconnects with almost everything else. It can impact upon: crop nutrition (plant health and thus crop protection), yield and profitability; leaching of nutrients and pesticides; gaseous emissions (including GHGs); soil erosion (and sediment pollution); and habitats for biodiversity. Therefore, soil management is crucial to IFM.

OPTIONS

There are numerous options that can be considered during soil management and crop nutrition decisions. These might include, for example, aspects such as the choice of fertiliser (e.g. fertiliser with lower GHG emissions during manufacture; see Chapter 2.3.2.2); the amounts applied; the timing of application; and application techniques. These options need to be appraised to select those that optimally meet the objectives.

ACTIONS

There are many actions that could emerge from the options appraisal process. This first example is relatively simple and involves calculating soil nitrogen supply (SNS) in order to improve fertiliser use and minimise excesses of nitrogen in the soil (see Chapter 2.3.2.4). We can use tools (such as the fertiliser manual; Defra, 2010) which provide guidance on calculating SNS and adjustments to crop fertiliser application rates accordingly. Accurate calculation of SNS means that application rates of fertiliser can be adjusted to better meet crop requirements. This is not only good for profitability, but also reduces excess N which can be lost to water via leaching and run-off and gaseous emissions of ammonia and nitrous oxide. It could be argued that such an approach is not integrated, and indeed many conventional farmers use such tools to maximise production efficiency. Thus, it could be driven purely to achieve economic objectives. However, from an environmental and integrated perspective it is a win–win solution and it can result in positive effects for both economic and environmental objectives.

METRICS

Losses of nutrients to the wider environment are unlikely to be measured/monitored on most farms. Therefore, from an environmental perspective, outcome-based metrics are not viable and practice-based metrics will form the basis for performance assessment. These might include determining metrics such as nitrogen use efficiency and/ or productivity, to ensure that fertiliser is efficiently used, thus minimising the potential losses to the environment and maximising profitability. There may be more indirect

outcome-based metrics for environmental performance that could be considered, for example evidence of aquatic or terrestrial eutrophication, albeit these potentially overlap with performance metrics for biodiversity.

OBJECTIVES ACHIEVED

- Air quality improved through less emissions.
- GHG emissions reduced through less N-fertiliser use.
- Profitable production maintained or improved.
- Water protection improved through reduced risks and pollution.

7.3.4.3.　Pest management

As highlighted in the history of IFM (see Chapter 7.2), pest control was one of the first areas in which integrated approaches were taken, with the emergence of Integrated Pest Management (IPM). For enterprises to be profitable, damage caused by pests and diseases needs to be minimised. This can first be done via prevention and second, via control (i.e. mechanical, biological or chemical intervention) when prevention does not work.

OPTIONS

In this instance, the options appraisal process involves weighing up the cost–benefit of different options for prevention and control. For example, are prevention measures more costly (economically and environmentally) than control measures and vice versa. In essence, the options appraisal process should result in a crop protection strategy that aims to achieve the objectives. It is likely that the priority will be to ensure economic viability, and then amongst the economically viable options, those which best meet the other objectives can then be determined.

ACTIONS

There are many actions that could emerge when considering options for crop protection. The following examples provide a selection of integrated approaches to prevention:

- Crop rotations (avoiding monocultures) can help prevent the build-up of pest populations, especially soilborne pathogens. The use of rotations by farmers goes back many thousands of years, so its effectiveness as a technique is well understood. It remains an effective technique, despite the invention of synthetic fertilisers and pesticides which are unable to fully substitute the effect of a good crop rotation (Bullock, 1992).
- Varietal choice: The choice of crop variety is usually determined by yield, however, yield in itself can be dependent on a range of traits including resistance of the plant to pests, its drought tolerance and its nutrient use efficiency. These can all impact upon yield and profitability, but will also impact upon other objectives. Crop varieties that are more resistant to pests may reduce the need for chemical, biological or physical control; and drought-tolerant crops not only reduce water demand, but also ensure the crop remains strong during water deficit and thus less susceptible to pests and diseases.
- Management of biodiversity on the farm also includes natural predators, thus these can keep pest levels below threshold levels.

Reducing the need for control can reduce operations and inputs and associated GHG emissions. It improves profitability (as input costs can be reduced) and can be beneficial for biodiversity. Avoidance of handling of chemicals can be better for worker welfare and there can also be benefits for water protection in some instances.

In the event that pest levels go above economic thresholds, then intervention may be required to control those pests in order to ensure the crop remains economically viable. In many instances, there are choices, for example mechanical, biological or chemical control (and then choices within these). The choices will most likely depend on local and specific circumstances. For example, if there is water close to the crop or sensitive habitats, then avoidance of certain pesticides may be desirable. Consequently, mechanical control, such as flame weeding, could be considered but this has higher GHG emissions. In the absence of sensitive habitats, then the chemical control option may be the optimal decision in terms of meeting the objectives. There are trade-offs to consider across multiple objectives – there are not always win–win situations, but sometimes optimal solutions or best practicable environmental options (BPEO) based on local knowledge can be determined.

METRICS

Similar to soil management and crop nutrition, practice-based metrics are more pragmatic than outcome-based metrics. This might be minimising practices that have potential negative impacts (e.g. minimising the use of harmful substances) whilst maintaining yields and profitability. Observing any changes in negative impacts on biodiversity species and populations could be undertaken and regarded as an outcome-based metric, albeit directly attributing impacts on biodiversity to specific crop protection actions can be difficult, and again such metrics overlap with performance metrics for biodiversity.

OBJECTIVES ACHIEVED

- Biodiversity improvements through reduced pollution and lower risk.
- GHG emissions have both positive

and negative aspects due to better pest control but possibly more fuel usage.

- Profitable production maintained or improved through more efficient operations.
- Water protection improved through less pollution.
- Worker health and welfare improved due to less use of potentially toxic substances.

7.3.4.4. *Irrigation and efficient water use*

The consumption of fresh water for food production is a growing issue, more so with climate change increasing the scarcity of water in some locations (see Chapter 4.1). It is not just about efficient use of water resources, but also protection of waterbodies on the farm and downstream, and about managing water in terms of surplus and deficits to meet objectives. It integrates with many other aspects on the farm including wildlife and biodiversity, nutrients, pesticides, and so on.

OPTIONS

This can be split into options for managing situations where there is too much water (flooding) on the farm or downstream and those where there is deficit (drought) leading to drought stress in crops and livestock.

- Flooding: There may be instances where water needs to be retained on the land. For example, restoration of moorland peat bogs serves not only to enhance biodiversity, but also retains water and acts as a buffer, reducing flood peaks downstream. A balance needs to be struck between not retaining too much water on the farm (to the detriment of other objectives) and not releasing too much water too quickly, which can cause flooding downstream.
- Drought: When crops become stressed due to lack of water, they can be irrigated, but rather than using techniques

such as rain guns, more water-efficient techniques can be used, such as trickle irrigation, which supplies water more accurately and when the crop needs it. This not only saves water, but saves money, boosts yield and can reduce GHG emissions (as energy is required to pump water). Avoiding water stress can improve efficient crop nutrition and crop health/resistance to pests, thus reducing the need for pesticides – all of which can improve both yield and gross margins. There are also techniques such as partial root zone drying (see Chapter 8.4.3.2), which can trick the plant into using less water whilst maintaining or increasing yields and profitability. With regard to livestock, access to fresh water is important for their health and welfare, but this has to be managed in balance with managing the soil. For example, ensuring that river bank erosion does not occur when livestock have access to water courses (e.g. fencing off the most vulnerable parts of river banks).

ACTIONS

Those actions which help retain water might include blocking of drains on moorland or changing the land use (e.g. afforestation), both of which can reduce the flow of water into the catchment. With respect to drought, actions might include chosing drought tolerant plants and/or using more efficient irrigation equipment.

METRICS

With regard to flooding, clearly flooding incidents and the severity of flooding downstream of the farm are key performance metrics, albeit unless the farm covers the whole catchment such flooding may not be totally attributable to the farm. Related to this, water discharge rates and flood peaks could be measured, albeit these are not necessarily something the farmer could directly measure themselves (it is more the remit of the water management authority). With regard to drought, water use efficiency is a key metric – ensuring that production utilises water to the greatest effect. The concept of embedded water (see Chapter 1.5.3.4) could be applied to the products produced on the farm, albeit this embedded water needs to account for the levels of water stress within the region, which to date has been lacking from such metrics.

OBJECTIVES ACHIEVED

- Biodiversity improvement through better-quality habitats.
- GHG emissions through less energy use.
- Livestock health and welfare improvements.
- Profitable production maintained or improved.
- Soil conservation has both positive and negative aspects depending on how water flows are controlled.
- Waste minimisation achieved.
- Water protection achieved through greater efficiency.

7.3.4.5. Animal husbandry

This is a broad subject covering many issues including: health and welfare; feed and water; housing; and environmental impacts. There are significant impacts associated with livestock production (including emission of GHGs, NH_3, NO_3^-, pathogens [e.g. *Cryptosporidium parvum*], high BOD effluent, veterinary medicines, soil damage, etc.). It is important that these are not mitigated at the expense of livestock health and welfare, and consequently profitability – so an integrated approach is required. There are numerous examples detailed within this book on how animal husbandry can impact on other objectives (see Chapters 2.3, 3.3 and 4.3), however, the following provide

examples and how they interact with other objectives.

OPTIONS

Animal husbandry to meet multiple objectives is inherently a bit of a balancing act and there are inevitably trade-offs to consider, and thus an integrated approach is essential to meet multiple objectives. Options appraisal can take place at various stages including breeding; design of housing; health care; access to land; choice of diets; and so on. For example:

- Healthy animals with good welfare standards are essential for a profitable farm. If an animal is unwell or poorly looked after, its production will be lower. This results in a lower yield, often with no decrease in environmental impact, so the impact per tonne of output increases.
- Livestock can have direct impacts on the land, such as erosion of river banks or the compaction of soil, which in turn can impact upon soil erosion, biodiversity, water pollution and GHG emissions.
- The impact of what goes in (feed) and what comes out (excreta). There are significant impacts embedded within livestock feeds that are accrued during their production. Livestock waste can result in the pollution of water and air, from both housed and grazing livestock, thus it needs careful management.

ACTIONS

The options appraisal process should identify where there are trade-offs and circumstances where these occur. This should then aid decisions and the selection of optimal actions. For example, if the land is vulnerable to N leaching, then it is probably best to avoid build-up of excess N or high N in manure. This may lead to greater CH_4 emissions but this may be an acceptable trade-off for the given circumstance and vice versa (i.e. if N is not vulnerable to leaching, then increasing the N in manure to decrease CH_4 emissions could be an acceptable option).

METRICS

Livestock productivity and yield are key performance metrics, along with measures of livestock health and welfare (albeit the latter is partly embedded in productivity measures, as unhealthy livestock will not be very productive). The emissions of N or CH_4 are unlikely to be measured directly, therefore practice-based metrics (possibly in combination with emission factors) are likely to be the main metric to assess environmental performance.

OBJECTIVES ACHIEVED

- Air quality improvements due to less emissions.
- Biodiversity improvement due to reduced physical impacts from livestock.
- GHG emissions due to less emissions.
- Livestock health and welfare improvements through better management.
- Profitable production maintained or improved.
- Soil conservation due to reduced physical impacts from livestock.
- Waste minimisation due to efficient use of inputs.
- Water protection has positive and negative aspects depending on how livestock access to water is managed and any run-off from grazed fields.
- Worker health and welfare can be improved if livestock health and welfare is improved (e.g. zoonotic diseases).

7.3.4.6. *Farm habitat and biodiversity management*

This overlaps with many other areas (as already seen in Chapter 3), which is not surprising given that biodiversity is part of and/or interwoven into the farm landscape, thus an integrated approach is essential. There needs to be space for wildlife and biodiversity and the impact of farm operations needs to be minimised. This requires some understanding of the local flora and fauna, and relative benefits of different features and practices given the local circumstances. A few examples are explored below.

OPTIONS

Generally, the options appraisal process relates to the protection and management of existing habitats and biodiversity on the farm, to ensure that they are optimally managed to maintain or enhance biodiversity and/or minimise any negative impacts from agricultural operations. However, there may also be situations where there are opportunities for habitat creation, and thus options appraisal may involve determining how to create habitat that maximises the benefits for biodiversity whilst minimising any impacts on the productivity of the farm.

ACTIONS

There are many actions that can be taken to maintain and enhance biodiversity on the farm, whilst maintaining farm production. For example, the creation of beetle banks can provide a habitat for ground beetles across the middle of large fields with minimal disruption to farming. The tussocky grass and raised bank allow very large numbers of predatory ground beetles to become established. Farmers who maintain beetle banks rarely need to use a summer insecticide. Thus, although there can be an economic loss (from the reduction in cropped land), there can be economic benefits with reduced input costs. Beetle banks provide good nesting cover for gamebirds and, when planted with long stemmed grasses like cocksfoot (*Dactylis glomerata*), they have proved to be a perfect habitat for harvest mice (*Micromys minutus*). They can also impact on carbon sequestration (increased soil organic matter), soil erosion and water protection (by increasing infiltration and reducing run-off).

Some habitats can also be beneficial to livestock. For example, hedgerows and trees can provide shelter for livestock during extreme weather (shelter from wind, rain, sunshine and heat).

METRICS

In the past, metrics of performance were largely practice-based, minimising practices that have negative impacts on biodiversity, and to some extent this is still the case, especially with respect to indirect impacts of nutrient and pesticide use. However, there is an increasing need for more outcome-based metrics, especially with respect to biodiversity to ensure that practices that are designed to protect and enhance biodiversity do actually achieve this. Consequently, metrics that involve assessing the diversity, quality and quantity (populations) of habitats and species are becoming more important in order to demonstrate positive outcomes.

OBJECTIVES ACHIEVED

- Biodiversity improved through enhanced habitats.
- Net GHG emissions improved through greater carbon sequestration.
- Livestock health and welfare improved through a diversity of habitats.
- Profitability can be impacted upon positively and negatively, for example reduced need to use insecticide due to beetle banks can reduce input costs, but taking land out of production may reduce output.
- Soil conservation improved through enhanced soil OM.
- Water protection enhanced due to improved management.

7.3.4.7. Pollution management

Pollution management is integrated with many other aspects (see Chapters 2–5). Inputs wasted (lost as pollution) are lost profit, so their efficient use is essential for profitable production (and in cases of negligence, there can be financial penalties). In some instances, pollution management is about containment, and in other instances it is about efficient use of inputs and knowledge of the system into which they are input and from which they can be lost (be it from point or diffuse sources). A few examples are explored below.

OPTIONS

The appraisial process in this instance is related to pollution risk assessment and understanding the source–pathway–receptor (SPR) model (see Chapter 1.5.2.2). It is important to understand where the greatest risks of pollution can occur. There can be 'classic' point source pollution incidents where there is a loss of containment. For example, in the past, point sources have included effluent from silage or manure storage facilities. Such incidents may be due to failures in infrastructure, poor/old design, misuse, extreme conditions or a combination of all of these. It is not necessarily always old facilities that are the cause of such incidents, new infrastructure can fail as well; for example, there have been incidents where new anaerobic digestion plants (designed to reduce pollutants) have sprung a leak (e.g. see Tasker, 2014). There may also be losses of inputs from the field, for example nutrients and pesticides, which may be lost by drift, drainage and leaching or run-off. In some circumstances, a beetle bank or a buffer strip might help reduce run-off, erosion and losses and pollutants and sediments, and also provide benefits for biodiversity. There can also be losses of pollutants from biological processes, for example

CH_4 emissions from cattle. These need to be understood and managed in an integrated way, so that one pollutant is not decreased at the expense of increasing another, or where there are trade-offs, the local circumstances can be taken into account. For example, decreasing CH_4 emissions from cattle, but at the expense of increasing nitrogen in manures; if the local area and/or features are sensitive to nitrogen, then it may be desirable to avoid this, but if they are not sensitive then this could be an acceptable trade-off.

ACTIONS

These usually relate to the source and pathway elements of the SPR model, as these provide intervention points to reduce pollution risks (as usually there is not much that can be done with respect to the receptor element). Therefore, actions may try to remove the source of pollution, for example using substances that are less hazardous to the end receptor (such as a pesticide with different or lower toxicity). Actions that address the pathway may include different application techniques (which lower the risk of losses) or using mitigation options to block the pathway of pollutants, such as the use of buffer strips to filter pollutants and/or increase infiltration to prevent them reaching the receptor.

METRICS

Water quality is a key metric, but it is not something that is likely to be directly measured by the farmer. Pollution incidents may be a metric that can be used, but this may not account for the severity of the incident or the consequent effects and impacts. Therefore, performance metrics are likely to be largely practice-based metrics, where actions that are known to lower the risk of pollution are measured and monitored, and form the basis for performance assessment.

OBJECTIVES ACHIEVED

- Air quality generally improved by less emissions but these may be offset by greater fuel use.
- Biodiversity improvements due to less pollution.
- GHG emissions can be impacted upon positively and negatively, for example reduced methane emissions from livestock by changing their diet, but at the expense of increased nitrogen in manures and vice versa.
- Livestock health and welfare can be improved by reducing exposure to pollutants.
- Profitable production can increase and decrease depending on pollutants that need to be managed. Generally, efficient use of inputs should increase profit and reduce pollution, but some pollution control measures may be costly and thus impact on profit.
- Soil conservation is improved as many measures to mitigate diffuse pollution may also have benefits for soils.
- Waste minimisation is improved and pollution is essentially waste.
- Water protection should be generally improved, but in some instances there may be a trade-off between water and air pollution, for example.
- Worker health and welfare is improved by reducing exposure to pollutants.

7.4. Conclusions

A truly integrated approach to 'management' is perhaps still lacking. Management of multiple issues can often result in water management plans, energy management plans, nutrient management plans, and so on. This is probably unavoidable in tackling specific issues, but it is important to find ways to connect the management plans. Performance assessment is largely practice-based. However, increasingly an outcome-based approach is required, especially if produce is being sold with a label. A more outcome-focused approach may provide scope to make progress towards a more integrated approach. For example, it is conceivable that different management plans could share the same outcome-based performance metrics and consequently, via the cycle of continuous improvement, both action plans can become tuned to improving the same performance metrics. For example, biodiversity is impacted upon by multiple activities on the farm. Outcome metrics might include biological diversity and/or the population of some key indicator species; management plans for nutrients, water and pests all have the potential to impact on biodiversity, thus these plans could have common biodiversity performance metrics and so evolve in an integrated fashion, albeit in a more indirect way.

If integrated farming systems (or any farming system) provide benefits, economically, socially and environmentally, then this needs to be shown. The outcomes need to be evident (not necessarily measured by the farmer) and the cause–effect mechanisms need to be understood (not assumed based correlated data sets). If objectives and outcomes are not being achieved, then we need to understand why and adapt our systems accordingly. This does not apply just to farms, but society as a whole will need to evolve. For example, consumers may need to be prepared to pay more for food if multiple objectives are to be achieved, be it at the point of sale, or in taxes which are then used by the government to pay for ecosystem services ('public goods') provided by farms. As our scientific knowledge grows and technical capabilities increase, a more integrated approach to agri-environmental management will and is becoming more feasible from the policy level down to the farm and field level.

References

Anderson, J.R. and Feder, G. (2004) Agricultural extension: good intentions and hard realities. *World Bank Research Observer*, 19(1), 41–60. DOI: 10.1093/wbro/lkh013

Bochu, J.-L., Metayer, N., Bordet, C. and Gimaret, M. (2013) *Development of carbon calculator to promote low carbon farming practices: methodological guidelines (methods and formula)*. EC-JRC-IES, Ispra, Italy.

Brentrup, F., Küsters, J., Kuhlmann, H. and Lammel, J. (2004) Environmental impact assessment of agricultural production systems using the life cycle assessment methodology I. Theoretical concept of a LCA method tailored to crop production. *European Journal of Agronomy*, 20(3), 247–264. DOI: 10.1016/S1161-0301(03)00024-8

Brown, A.L. and Thérivel, R. (2000) Principles to guide the development of strategic environmental assessment methodology. *Impact Assessment and Project Appraisal*, 18(3), 183–189. DOI: 10.3152/147154600781767385

Buizer, M., Arts, B. and Westerink, J. (2015) Landscape governance as policy integration 'from below': a case of displaced and contained political conflict in the Netherlands. *Environment and Planning C: Politics and Space*, 34(3), 448–462. DOI: 10.1177/0263774X15614725

Bullock, D.G. (1992) Crop rotation. *Critical Reviews in Plant Sciences*, 11(4), 309–326.

Cairns, J. and Pratt, J.R. (1995) The relationship between ecosystem health and delivery of ecosystem services. In: Rapport, D.J., Gaudet, C.L. and Calow, P. (eds.) *Evaluating and Monitoring the Health of Large-Scale Ecosystems*. NATO ASI Series (Series I: Global Environmental Change), vol 28. Springer, Berlin, Heidelberg.

Carson, R. (1962) *Silent Spring*. Houghton Mifflin Harcourt, USA.

CBD (2017) *Ecosystem Approach*. Convention on Biological Diversity website: Available at: https://www.cbd.int/ecosystem/

Conservation Grade (2017) *Fair to Nature*

Farming. Conservation Grade. Available at: http://www.conservationgrade.org/

Costanza, R. and Daly, H.E. (1992) Natural capital and sustainable development. *Conservation Biology*, 6(1), 37–46. DOI: 10.1046/j.1523-1739.1992.610037.x

Davey, A. (2014) *Evidence review of catchment strategies for managing metaldehyde – Report Ref. No. 13/DW/14/7.* UKWIR, London.

Dreves, A.J., Andreson, N. and Sullivan, C. (2016) Slug control. In: C. S. Hollingsworth (ed.) *Pacific Northwest Insect Management Handbook*. Oregon State University, Corvallis, OR.

Ehrlich, P.R. and Mooney, H.A. (1983) Extinction, substitution, and ecosystem services. *BioScience*, 33(4), 248–254. DOI: 10.2307/1309037

EISA (2010) *European Integrated Farming Framework. A European Definition and Characterisation of Integrated Farming as Guideline for Sustainable Development of Agriculture*. European Initiative for Sustainable Development in Agriculture (EISA). Revised version August 2010.

EISA (2012) *European Integrated Farming Framework. A European Definition and Characterisation of Integrated Farming (IF) as Guideline for Sustainable Development of Agriculture*. European Initiative for Sustainable Development in Agriculture (EISA) February 2012.

Elkington, J. (1998) Accounting for the triple bottom line. *Measuring Business Excellence*, 2(3), 18–22. DOI: 10.1108/eb025539

Elkington, J. (2008) The triple bottom line. Sustainability's accountants. In: Russo, M.V. (ed.) *Environmental Management: Readings and Cases*, 2nd Edition. Sage Publications, Inc. ISBN: 1412958490, pp. 663.

El Titi, A. (1992) Integrated farming: an ecological farming approach in European agriculture. *Outlook on Agriculture*, 21(1), 33–39.

Farley, J. and Costanza, R. (2010) Payments for ecosystem services: from local to global. *Ecological Economics*, 69(11), 2060–2068. DOI: 10.1016/j.ecolecon.2010.06.010

Fazey, I., Evely, A., Reed, M., Stringer, L., Kruijsen, J., White, P., Newsham, A., Jin, L., Cortazzi, M., Phillipson, J., Blackstock, K., Entwistle, N., Sheate, W., Armstrong, F., Blackmore, C., Fazey, J., Ingram, J., Gregson, J., Lowe, P., Morton, S. and Trevitt, C. (2013) Knowledge exchange: a review and research agenda for environmental management. *Environmental Conservation*, 40(1), 19–36. DOI: 10.1017/S037689291200029X

Franklin, J.F. (1993) Preserving biodiversity: species, ecosystems, or landscapes? *Ecological Applications*, 3(2), 202–205. DOI: 10.2307/1941820

Gómez-Baggethun, E., de Groot, R., Lomas, P.L. and Montes, C. (2010) The history of ecosystem services in economic theory and practice: from early notions to markets and payment schemes. *Ecological Economics*, 69(6), 1209–1218. DOI: 10.1016/j.ecolecon.2009.11.007

Hauck, J., Görg, C., Varjopuro, R., Ratamäki, O. and Jax, K. (2013) Benefits and limitations of the ecosystem services concept in environmental policy and decision making: some stakeholder perspectives, *Environmental Science and Policy*, 25, 13–21. DOI: 10.1016/j.envsci.2012.08.001

Hillier, J., Walter, C., Malin, D., Garcia-Suarez, T., Mila-i-Canals, L. and Smith, P. (2011) A farm-focused calculator for emissions from crop and livestock production. *Environmental Modelling and Software*, 26(9), 1070–1078. DOI: 10.1016/j.envsoft.2011.03.014

Howlett, S.A. (2012) Terrestrial slug problems: classical biological control and beyond. *CAB Reviews*, 7(51), 1–10.

Jordan, V.W.L., Hutcheon, J.A., Donaldson, G.V. and Farmer, D.P. (1997) Research into and development of integrated farming systems for less-intensive arable crop production: experimental progress (1989–1994) and commercial implementation. *Agriculture, Ecosystems and Environment*, 64(2), 141–148. DOI: 10.1016/S0167-8809(97)00032-7

Juntti, M. and Potter, C. (2002) Interpreting and reinterpreting agri-environmental policy: communication, trust and knowledge in the implementation process. *Sociologia Ruralis*, 42(3), 215–232. DOI: 10.1111/1467-9523.00212

Landis, D.A. (2017) Designing agricultural landscapes for biodiversity-based ecosystem services. *Basic and Applied Ecology*, 18, 1–12. DOI: 10.1016/j.baae.2016.07.005

LEAF (2016) *LEAF's Integrated Farm Management (IFM)*. Available at: http://www.leafuk.org/leaf/farmers/LEAFs_IFM.eb

Leake, A. (2000) The development of integrated crop management in agricultural crops: comparisons with conventional methods. *Pest Management Science*, 56, 950–953. DOI: 10.1002/1526-4998(200011)56:11<950::AID-PS234>3.0.CO;2-5

Lenschow, A. (2002) *Environmental Policy Integration: Greening Sectoral Policies in Europe*. Routledge, Oxon., UK, pp. 241.

Lewis, K.A., Green, A., Tzilivakis, J. and Warner, D.J. (2010) The contribution of UK farm assurance schemes towards desirable environmental policy outcomes. *International Journal of Agricultural Sustainability*, 8(4), 237–249. DOI: 10.3763/ijas.2010.0495

Lindblom, C.E. (1959) The science of 'muddling through'. *Public Administration Review*, 19(2), 79–88. DOI: 10.2307/973677

Lindblom, C.E. (1979) Still muddling, not yet through. *Public Administration Review*, 39(6), 517–526. DOI: 10.2307/976178

Medina, G. and Potter, C. (2017) The nature and developments of the Common Agricultural Policy: lessons for European integration from the UK perspective. *Journal of European Integration*, 39(4), 373–388. DOI: 10.1080/07036337.2017.1281263

Moore, D.W., Booth, P., Alix, A., Apitz, S.E., Forrow, D., Huber-Sannwald, E. and Jayasundara, N. (2017) Application of ecosystem services in natural resource management decision making. *Integrated Environmental Assessment and Management*, 13(1), 74–84. DOI: 10.1002/ieam.1838

Morris, C. and Winter, M. (1999) Integrated farming systems: the third way for European agriculture? *Land Use Policy*, 16, 193–205. DOI: 10.1016/S0264-8377(99)00020-4

Ogilvy, S.E., Turley, D.B., Cook, S.K., Fisher,

N.M., Holland, J., Prew, R. and Spink, J. (1994) Integrated farming – putting together systems for farm use. *Aspects of Applied Biology*, 40(1), 53–60.

Pacini, C., Wossink, A., Giesen, G., Vazzana, C. and Huirne, R. (2003) Evaluation of sustainability of organic, integrated and conventional farming systems: a farm and field-scale analysis. *Agriculture, Ecosystems and Environment*, 95(1), 273–288. DOI: 10.1016/S0167-8809(02)00091-9

Parra-López, C., Calatrava-Requena, J. and de-Haro-Giménez, T. (2006) A multi-criteria evaluation of the environmental performances of conventional, organic and integrated olive-growing systems in the south of Spain based on experts' knowledge. *Renewable Agriculture and Food Systems*, 22(3), 189–203. DOI: 10.1017/S1742170507001731

Perrot, N., De Vries, H., Lutton, E., van Mil, H.G.J., Donner, M., Tonda, A., Martin, S., Alvarez, I., Bourgine, P., van der Linden, E. and Axelos, M.A.V. (2016) Some remarks on computational approaches towards sustainable complex agri-food systems. *Trends in Food Science and Technology*, 48, 88–101. DOI: 10.1016/j.tifs.2015.10.003.

Potts, S.G., Imperatriz-Fonseca, V.L., Ngo, H.T., Biesmeijer, J.C., Breeze, T.D., Dicks, L.V., Garibaldi, L.A., Hill, R., Settele, J. and Vanbergen, A.J. (2016) *Summary for policymakers of the assessment report of the Intergovernmental Science-Policy Platform on Biodiversity and Ecosystem Services on pollinators, pollination and food production*. Report. Intergovernmental Science-Policy Platform on Biodiversity and Ecosystem Services, Bonn, Germany. pp. 36. ISBN 9789280735680

QuESSA (2017) *Quantification of Ecosystem Services for Sustainable Agriculture (QUESSA) Project*. Available at: http://www.quessa.eu/

Raymond, C., Reed, M., Bieling, C., Robinson, G. and Plieninger, T. (2016) Integrating different understandings of landscape stewardship into the design of agri-environmental schemes. *Environmental Conservation*, 43(4), 350–358. DOI: 10.1017/S037689291600031X

Reganold, J.P., Glover, J.D., Andrews, P.K. and Hinman, H.R. (2001) Sustainability of three apple production systems. *Nature*, 410, 926–930. DOI: 10.1038/35073574

Sadler, B. and Dusík, J. (2016) *European and International Experiences of Strategic Environmental Assessment: Recent Progress and Future Prospects*. Routledge, Oxon., UK, pp. 380

Scheufele, G. and Bennett, J. (2017) Can payments for ecosystem services schemes mimic markets? *Ecosystem Services*, 23, 30–37. DOI: 10.1016/j.ecoser.2016.11.005

Schulp, C.J.E., Van Teeffelen, A.J.A., Tucker, G. and Verburg, P.H. (2016) A quantitative assessment of policy options for no net loss of biodiversity and ecosystem services in the European Union. *Land Use Policy*, 57, 151–163. DOI: 10.1016/j.landusepol.2016.05.018

Speiser, B., Glen, D., Piggott, S., Ester, A., Davies, K., Castillejo, J. and Coupland, J. (2001) *Slug Damage and Control of Slugs in Horticultural Crops*. IACR, Long Ashton, UK.

Stern, V.M., Smith, R.F., Van den Bosch, R. and Hagen, K.S. (1959) The Integrated Control concept. *Hilgardia*, 29, 81–101. DOI: 10.3733/hilg.v29n02p081

Swinton, S.M., Lupi, F., Robertson, G.P. and Hamilton, S.K. (2007) Ecosystem services and agriculture: cultivating agricultural ecosystems for diverse benefits. *Ecological Economics*, 64(2), 245–252. DOI: 10.1016/j.ecolecon.2007.09.020

Tasker, J. (2014) *AD plant collapses at Harper Adams*. Farmers Weekly, 30 May 2014. Available at: http://www.fwi.co.uk/news/ad-plant-collapses-at-harper-adams.htm

Teisman, G.R. and Klijn, E-H. (2008) Complexity theory and public management. *Public Management Review*, 10(3), 287–297. DOI: 10.1080/14719030802002451

Thérivel, R. and Paridario, M.R. (2013) *The Practice of Strategic Environmental Assessment*. Routledge, Oxon., UK, pp. 224.

Tzilivakis, J., Lewis, K. A., Green, A. and Warner, D. J. (2010) *The climate change mitigation potential of an EU farm: towards a farm-based integrated assessment*. Final Report for European Commission Project ENV.B.1/ETU/2009/0052. September 2010.

Tzilivakis, J., Green, A., Warner, D.J., McGeever, K. and Lewis, K.A. (2012) A framework for practical and effective eco-labelling of food products. *Sustainability Accounting, Management and Policy Journal*, 3(1), 50–73. DOI: 10.1108/20408021211223552

Tzilivakis, J., Green, A., Lewis, K.A. and Warner, D.J (2014) Identifying integrated options for agricultural climate change mitigation. *International Journal of Climate Change Strategies and Management*, 6(2), 192–211. DOI: 10.1108/IJCCSM-09-2012-0053

Tzilivakis, J., Warner, D.J., Green, A. and Lewis, K.A. (2017) Spatial and temporal variability of greenhouse gas emissions from rural development land use operations. *Mitigation and Adaptation Strategies for Global Change*, 22(3), 447–467. DOI: 10.1007/s11027-015-9680-x

Van Emden, H.F. and Peakall, D.B. (1996) *Beyond Silent Spring. Integrated Pest Management and Chemical Safety*. Chapman and Hall, London, UK, p. 38.

Van Emden, H.F. and Service, M.W. (2004) *Pest and Vector Control*. Cambridge University Press, UK, pp. 349.

van Zanten, B.T., Verburg, P.H., Espinosa, M., Gomez-y-Paloma, S., Galimberti, G., Kantelhardt, J., Kapfer, M., Lefebvre, M., Manrique, R., Piorr, A., Raggi, M., Schaller, L., Targetti, S., Zasada, I. and Viaggi, D. (2014) European agricultural landscapes, common agricultural policy and ecosystem services: a review. *Agronomy for Sustainable Development*, 34(2), 309–325. DOI: 10.1007/s13593-013-0183-4

Viala, E. (2008) Water for food, water for life. A comprehensive assessment of water management in agriculture. *Irrigation and Drainage Systems*, 22(1), 127–129.

Wallington, T.J., Hobbs, R.J. and Moore, S.A. (2005) Implications of current ecological thinking for biodiversity conservation: a review of the salient issues. *Ecology and Society*, 10(1), 1–16. DOI: 10.5751/ES-01256-100115

Watson, C.A., Bengtsson, H., Ebbesvik, M., Løes, A-K., Myrbeck, A., Salomon, E., Schroder, J. and Stockdale, E.A. (2002) A review of farm-scale nutrient budgets for organic farms as a tool for management of soil fertility. *Soil Use and Management*, 18(s1), 264–273. DOI: 10.1111/j.1475-2743.2002.tb00268.x

White, L. and Noble, B.F. (2013) Strategic environmental assessment for sustainability: a review of a decade of academic research. *Environmental Impact Assessment Review*, 42, 60–66. DOI: 10.1016/j.eiar.2012.10.003

Wynn, S. (2012) *Integrating advice on climate change mitigation and adaptation into existing advice packages to achieve multiple wins*. Final Report for Defra project FF0204, undertaken by ADAS, INNOGEN, RAND and AHDB.

CHAPTER 8

Future perspectives

8.1. Introduction

In the previous chapters, current knowledge and understanding of agri-environmental management has been explored, including the policies and practices that have been implemented to address the challenges society has faced to date. This chapter aims to consider the future of farming, however, there are many visions for what may lie ahead. Some scenarios can be quite wild, involving futuristic technologies, such as automated machinery and robotics whereby the human element is almost completely removed. Other scenarios might focus on addressing food security and include vertical farming in urban areas, as shown in Figure 8.1, rooftop production areas or even growing underground in controlled and automated environments.

Some of these visions are starting to be realised, for example Davies (2016) reported that there are plans in Japan to build a lettuce farm run by robots and Growing UNDERGROUND is a company in London that grows salad crops using hydroponics 33 metres below street level (Growing UNDERGROUND, 2017). There are also innovative companies like AeroFarms (AeroFarms, 2017), Sky Greens (Sky Greens,

2017) and VertiCrop (VertiCrop, 2017), all of which use forms of vertical farming. However, these concepts are fairly niche and tend to be aimed at either high-value crops and/or urban areas, rather than mainstream production in rural areas.

In the absence of a crystal ball or a time machine, it is difficult to gain a perspective or vision of what mainstream farming in rural areas will look like in the future. However, this chapter will attempt to provide some future perspectives by:

- Understanding the system, its components and stakeholders, how they interact and thus how they might evolve in the future.
- Identifying and exploring the key pressures and drivers that are likely to steer farming and food production systems.
- Examining potential responses (policies, actions and practices) that may emerge in response to the key pressures and drivers.

By understanding the system and the key pressures and drivers, it is possible to ascertain the likely direction of 'travel' for farming systems and consequently explore some of the responses that may emerge, and thus

Figure 8.1: Visions of the future of farming: vertical farming
(Photo courtesy of: Except Integrated Sustainability)

provide a perspective of future farming and agri-environmental management.

8.2. Understanding the system: the simple and the complex

Before exploring the pressures and drivers that will steer the future of farming, it is important to understand the system, starting with the OECD's Pressure–State–Response (PSR) framework from 1993 (Figure 8.2). It shows how human activities exert pressure on the environment, such as through pollution, and also draws upon the environment for resources; these activities give rise to changes in the state of the environment (and our ecosystem services), sometimes to an extent where they become an environmental problem. Information about the pressure and state, feeds our governance mechanisms and results in a response, sometimes economic, sometimes regulatory, sometimes individual or organisational, or all of these combined – these responses, in turn, have

an impact on the pressure, which impacts on the state, and so on.

The PSR framework is, of course, a simplification of the system. In reality, it is more complex. In 2010, the Government Office for Science (GOS) in the UK produced a landmark document exploring land use futures (GOS, 2010). This document, and its appendices, provide some excellent system diagrams of our land use and global food systems. They are too large and complex to present here, but they do start to neatly reveal and illustrate the complexity. There is agriculture and land use; ecosystem services; environmental issues; and economic, social and political elements, some of which govern the system. The perspectives may also differ at different levels, albeit they are connected, for example between the strategic (societal/government) level and the farm (local) level. At the strategic level, efforts can focus on managing multiple societal objectives and managing the bigger picture. This might, for example, include tackling issues

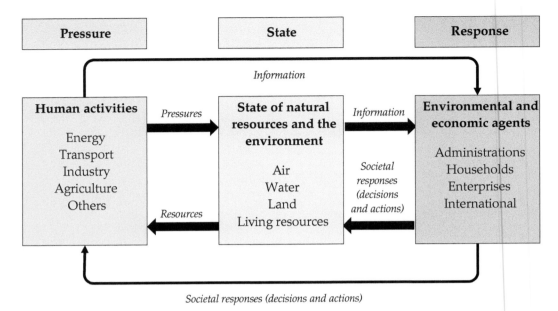

Figure 8.2: Pressure-State-Response (PSR) framework
(Adapted from: OECD, 1993)

that are not addressed by the market and/or become externalised (market failure), as has been the case with many environmental issues. Responses may then include creating interventions and infrastructure to bring about outcomes that society desires, for example regulations to reduce pollution (e.g. emissions of GHGs); tackling the impacts of climate change (e.g. water shortages); ensuring food and energy security; and strategic initiatives to conserve biodiversity. At the farm (local) level there are more immediate business concerns to address. These may include ensuring the farm is profitable; meeting the requirements that society demands both in terms of produce and regulations imposed by government; managing relationships with the local community, and, of course, personal objectives – many farmers take great pride in how they farm and see themselves as stewards of the environment.

When considering the bigger/strategic picture, there are, of course, some serious concerns and challenges that lie ahead in the future. Many of these were neatly summed up by Professor Sir John Beddington (the UK's chief scientific advisor in 2009) when he described the challenges as the 'perfect storm' of food, water and energy shortages in 2030 (Beddington, 2009). The key elements and factors of this perfect storm are outlined in Table 8.1 and discussed in Chapter 5.

In relation to food alone, Sheeran (2012) also neatly summed up the challenge with the statement that 'over the next 40 years, we need to produce more food than the last 8,000 years combined!'.

Clearly, these are major challenges that encompass a number of pressures and drivers that are likely to steer the evolutionary direction of our food and farming systems. These are explored in the next section within this chapter.

Table 8.1: Key elements and factors of the 2030 perfect storm

Elements	Description
Population	The world population will rise to 8 billion.
Food	Demand for food will increase by 50%.
Water	Demand for water will increase by 30%.
Energy	Demand for energy will increase by 50%.
Factors	
Climate change	There is a risk that climate change will have drastic effects on food production – for example, by killing off the coral reefs (which about 1 billion people depend on as a source of protein) or by either weakening or strengthening monsoon rains. Also, some scientists are predicting that the Arctic will be ice-free by 2030, which could accelerate global warming by reducing the amount of the sun's energy that is reflected back out of the atmosphere.
Urbanisation	Not only is the world's population predicted to grow (until the middle of the century, at least) but more people are moving to live in cities. The growth of cities will accelerate the depletion of water resources, which in turn may drive more country dwellers to leave the land.
Increasing prosperity	As people become wealthier in some parts of the world, such as China and India, their diets are changing. They are consuming more meat and dairy products, which take more energy to produce than traditional vegetable diets. Like city dwellers, prosperous people also use more energy to maintain their lifestyle.
Biofuels	The more land that is devoted to growing biofuels, in response to climate change mitigation, the less can be used for growing food.

8.3. Pressure and drivers

There are number of pressures and drivers on farming which have been discussed in detail in previous chapters, including:

- Resources: water, land, nutrients, energy.
- Environment: GHG emissions, biodiversity, and so on.
- Social: consumer demands/societal preferences.
- Economic: market prices and gross margins.
- Responses as drivers and pressures: technology and regulation.

These are all current pressures and drivers that are likely to increase in the future, as outlined in the 2030 perfect storm scenario (see Table 8.1).

8.3.1. Resources

WATER

There is no shortage of water in the world, but often we do not have it in the right quantity in the right place (floods and droughts) for food production – and this is likely to get worse with climate change and population growth (as outlined in the perfect storm scenario). Food production has a 'drink problem'. Consequently, there needs to be more efficient use of water in farming. The embedded water in a food product (see Chapter 1.5.3.4) needs to be minimised, especially from regions that are water-stressed.

LAND

Land is a limited resource, of variable capability, under pressure from multiple demands (food production, urbanisation, conservation, recreation, etc.). Any new agricultural

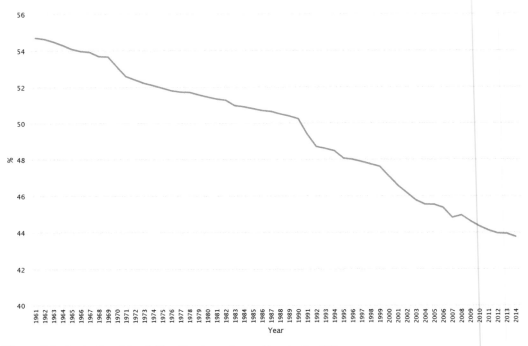

Figure 8.3: Agricultural land (% of land area) over time in the EU
(Source: World Bank, 2016)

land can only be gained at the expense of another ecosystem service – which is generally undesirable. The amount of land available to farm is relatively small, so it needs to be used efficiently. Figure 8.3 shows how, in the EU, agricultural land area has decreased over the past 50 years, from 54.7 per cent in 1961 to 43.7 per cent in 2014, while globally it has remained at around 36–37 per cent (World Bank, 2016).

NUTRIENTS

As discussed in previous chapters (1.4 and 2.3), agricultural production is dependent on some key nutrients: nitrogen (N), potassium (K) and phosphorus (P). The environmental impacts of nitrogen are well documented, both in terms of those arising from its application (pollution of water, GHG emissions), and energy used in its manufacture – so it needs to be used more efficiently. Phosphorus

also has impacts arising from use (e.g. loss to water, resulting in eutrophication). However, its consumption as a resource is less well documented, and there are increasing concerns as to the sustainability of its consumption in the longer term (see Chapter 5). For example, Vaccari (2009) highlights that, globally, about 90 years' worth of reserves remains.

ENERGY

As has been discussed in Chapters 1.4, 2.1 and 5.1, agricultural production is highly dependent on fossil fuels (and associated GHG emissions). Energy is used for the production of inputs (e.g. fertilisers, pesticides, feeds, etc.); farm operations (e.g. machinery, lighting, heating, drying, etc.); and transport (food miles).

To reduce the use of fossil fuels, efforts have been made in recent decades to develop and use biomass and biofuels, such as short-rotation coppice (willow, poplar), grasses (miscanthus), oilseed rape for bio-diesel, and sugar beet, wheat and sugarcane for bioethanol. However, these energy crops require land, which competes with land for food production.

8.3.2. *Environment*

The environmental impacts of farming are well documented, including within this book. They include:

- Climate change (GHG emissions and carbon sequestration): Farming needs to have a lower carbon footprint, and be able to adapt to the effects of climate change, for example through the development of drought-tolerant/water-efficient plants.
- Pesticides: Farming needs to effectively control pests and diseases without significant impacts on the environment.
- Nutrients: Farming needs to maintain or increase yields, but using nutrients more efficiently – reducing losses.
- Biodiversity: Farming needs to protect and enhance the biodiversity that is part of the agricultural landscape.
- Landscape and cultural heritage: Farming needs to conserve the landscape and heritage in line with what society desires.

8.3.3. *Social*

Farmers are suppliers to retailers and consumers, so they need to meet market demands (consumer purchasing behaviour, retailer marketing or both) with regard to price and quality, and also in relation to social, ethical and environmental issues, such as animal welfare, eco-friendly, and worker welfare requirements.

Although quality and price are still key decision points for purchasing, consumers are becoming more aware of wider issues

and governments are demanding that consumers should have the information made available to them to be able to make more informed purchasing decisions, such as through greater and more transparent information on food labels.

Where the market fails, government intervention in the form of regulation attempts to ensure that societal desires are met, such as regulations to reduce pollution, protect landscapes, cultural heritage, and so on.

8.3.4. *Economic*

This is very much related to social pressures; indeed, this can often be referred to as socio-economics, in the context of consumer demand and industry supply. At the farm level, the market price obtained for a product is a fundamental driver both in terms of the choice of farm enterprises and how they are produced (gross margin: input costs and output revenue). Market prices are very much driven by supply and demand, plus numerous social and economic factors. This is very apparent in the food price inflation and fluctuations that have been seen in recent decades.

Figure 8.4 shows the world food price index, with large spikes in 2008 and 2010. The 2008 spike was caused by the global financial crisis and increases in the price of oil (as described in Chapters 1.4, 2.1 and 5.1, agriculture is highly dependent on fossil fuels). The 2010 spike was caused by numerous factors including wildfires in Russia, commodity speculators and drought in the southern USA.

Clearly, the changes in prices have significant consequences for both farmers and consumers, affecting farm revenues and cost of living for consumers. Economic viability is one of the three core pillars of sustainability and thus agri-environmental management is not isolated from this – the economics need to work for agri-environmental management

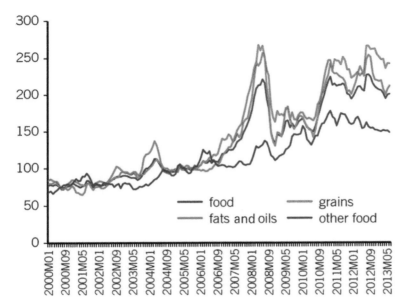

Figure 8.4: World food price index (Source: World Bank, 2013)

to be able to deliver the desired environmental objectives.

It also needs to be acknowledged that the consequences of such spikes can be very serious from a broader societal perspective. Figure 8.5 shows the same chart as Figure 8.4, except this time, the red lines point to the dates indicating the start of food riots and protests, and, sadly, the numbers in brackets next to each country is the number of people who died in those riots and protests, thus highlighting the very serious consequences of spikes in food prices.

8.3.5. Responses

Finally, societal responses can become drivers in themselves, especially at the farm level, where regulations drive the industry to operate in particular ways, or new technologies can make new enterprises economically viable. For example, banning some pesticides may make some crops economically unviable, as pests and diseases can no longer be economically controlled – so new cropping enterprises must be sought. Conversely, the development of new technology may make

some crops economically viable in areas or situations where previously production was not feasible, for example via improved efficiency and/or lower input costs.

This is a dynamic and evolutionary process and one which has been clearly observed over the past few decades, where there is a variety of responses to a range of drivers and pressures, including the need for reduced impacts on the environment or more extensive forms of production. In the 1970s and 80s, the CAP and advances in technology and biology drove the industry to overproduce – leading to 'food mountains' and 'wine lakes'. Intensive farming practices also resulted in many environmental impacts, for example water pollution and damage to wildlife habitats and biodiversity. Consequently, the response was to identify and encourage farming systems with lower environmental impact; land was taken out of production (set-aside); and more extensive production systems were considered. In relation to the latter, organic farming is probably the most well-known extensive system of production but it is

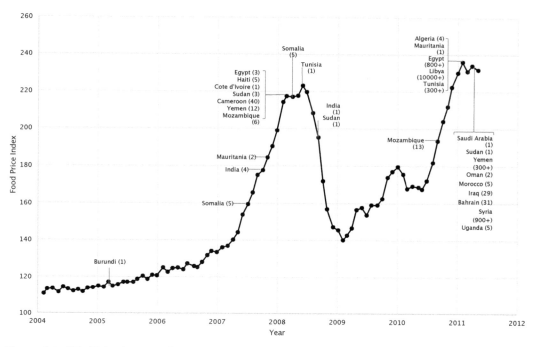

Figure 8.5: World food price index and food riots
(Adapted from: Lagi *et al.*, 2011)

not without its flaws. It is an inherently lower-yielding system, and consequently this conflicts with the demand for land (as more land is needed to produce the same amount of food) and in some instances the environmental impact per tonne of product may be higher. Consequently, its uptake has been limited, and, now that increasing production is higher up the agenda again, extensive solutions are looking less like a solution.

Sustainable intensification is the new 'buzz word' and concept associated with food security. The theory is to produce more food using the same amount of land (or less), whilst reducing (or avoiding) negative environmental and social impacts. Whilst this concept has been discussed in Chapter 5, it has yet to be fully defined or realised in practice, albeit some examples are starting to emerge and these will be explored in Chapter 8.4.

In summary, the sections above have outlined a number of pressures and drivers that present challenges for food security and sustainable production and consumption. In many instances, the direction of travel is also negative, that is, the pressures are increasing, for example diminishing resources, environmental pressures, socio-economic fluctuations, dynamic responses, and so on. The following sections explore some potential future responses that may help society address and overcome the challenges that lie ahead. Many of these responses could be regarded as 'near future' as they are based on evolutions of current/recent developments.

8.4. Future responses

8.4.1. Introduction

The societal responses that may emerge in the light of the drivers and pressures fall into four main areas: technological, biological (and biotechnology), management/economic and political/policy responses. In reality, these overlap to some extent as, for example, advances in technology need to be understood in terms of the impact they have on plants and the cost of the technology in relation to the bottom line of the farm business. The situation is further complicated by the fact that responses are also often interconnected with the environmental and socio-economic drivers and pressures. There are various synergies and conflicts between all of these. For example, small family farms in an aesthetic environment may be desired, but cheap/affordable food is also needed, and the latter tends to drive the industry towards being large-scale and highly efficient, rather than small and aesthetically pleasing.

In the following sections within this chapter, first some specific examples of current and near future responses are explored, with respect to technology, biology, management/economic and policy (Chapters 8.4.2–8.4.5), and then integrated responses (Chapter 8.4.6) are discussed, where multiple issues are tackled in an integrated manner through various combinations of technology, biological knowledge, management/economic and policy approaches.

Many of the responses featured below could fall under the banner of precision farming or smart farming, which combines technology with livestock and crop science to improve agricultural practice, improve efficiency, reduce inputs, reduce emissions and increase margins and profitability. Precision or smart farming is not covered in an explicit section below, as it effectively covers everything. However, Chapter 8.4.6 covers the combination of technologies, knowledge and approaches to address multiple objectives in an integrated fashion.

8.4.2. Technology

8.4.2.1. Introduction

In this instance, technology is used in the sense of conventional technology and not biotechnology (which is covered in Chapter 8.4.3.4) and includes mechanical innovations, computers and IT, and so on. Technological innovation is perhaps one of the oldest activities in farming. Over the centuries, farmers have been particularly innovative when seeking to tackle the challenges of farming and improve the efficiency of production, resulting in a huge diversity of machines, tools and devices.

Technological advances are often combined with better biological knowledge, and are commonly driven by the need to be more efficient, as well as addressing environmental concerns (which often go hand in hand). Technological innovations can cover a broad range of issues and activities but they can be broadly split into the following areas:

- Mechanisation and automation: replacing human labour with mechanical and automated devices.
- Sensing: remote and in-situ; detecting issues to enable them to be addressed efficiently.
- Information technology.
- Bioinformatics.
- Nanotechnology: techniques and technology at the molecular level.

8.4.2.2. Mechanisation and automation

Prior to the industrial revolution, farming was an exclusively hand-and-horse-powered process, coupled with a variety of tools and implements. The industrial revolution introduced mechanisation and enabled farming to become much less labour-intensive and more productive. This process continued and advanced rapidly during the 20th century, and the digital age advanced automation alongside mechanisation. This process will, undoubtedly, continue into the future as solutions are sought to improve efficiency and reduce costs, especially labour. Two key focus areas in the future will be to increase efficiency with respect to resource use and to ensure scarce resources are used more effectively whilst minimising emissions of pollutants. For example, there are numerous efforts (e.g. Flourish, 2017; SWEEPER, 2017) to develop robots for agricultural use. Their role will be to automatically roam fields and use sensors to detect pests and apply pesticides minimally where needed, using renewable (e.g. solar power) rather than non-renewable fossil fuels. This will reduce fuel use and optimise the use of pesticides. The use of robots will also reduce manpower requirements and enable the farm to be productive with a much smaller work force. However, this could result in socio-economic disadvantages as it will reduce employment opportunities.

In relation to resource use and efficient production, another key area (as it has been in recent years) is the development of machinery that can variably apply inputs on an 'as needed' basis. This can include seeds, nutrients, pesticides and water, which in turn can result in reduced costs, fuel/energy use and associated emissions. Variable application rate machinery is not new, but it is being further advanced, most notably in combination with other technologies such as improved sensors (see Chapter 8.4.2.3), mapping technologies (see Chapter 8.4.2.4) and precision guidance systems. The latter of these draws upon the network of satellites that form the Global Positioning System (GPS). Basic GPS equipment, such as that found in smartphones, is accurate to within 10 metres. This is not accurate enough for most field operations, so farm GPS equipment tends to use real time kinematic (RTK) systems which employ two or more GPS receivers, connected by radio, to refine positioning accuracy. With RTK, one GPS receiver operates as a base station while the other is on the moving vehicle. RTK can provide positioning accuracy down to a couple of centimetres. This spatial accuracy facilitates precise management and intervention. It also enables the development of automation and self-guiding machinery and robots.

Automation already exists on many farms, for example automated milking systems have been commercially available since the early 1990s. Cows can enter the milking parlour at any time and an ID sensor identifies the cow, then automatic feeding, teat cleaning, milking cup application, milking and teatspraying takes place, followed by milking. The process reduces labour, improves milking frequency and consistency, can lower stress and has advantages for herd management. There is little doubt the process of automation will continue in the future, especially in combination with other technologies, to develop a range of agricultural robots. In addition to milking, automation is being developed for a range of tasks including crop protection and nutrition; irrigation; harvesting; and herding of livestock. Some examples are provided in Table 8.2.

8.4.2.3. Sensing: remote and in-situ
The use of sensors (devices that measure a physical parameter and then convert the measurements into signals which can be read, interpreted and used by other devices), both remotely and in-situ (on or below the ground; in or around the crop), is another area of technological development which is likely to become more commonplace in the coming years. Observation is critical to modern farming, both in cropping and livestock enterprises. Like many aspects in human health, early detection of issues such as water stress or pest build-up, can be vital

Table 8.2: Examples of mechanisation and automation in the agri-environment

Area	Description	References
Crop protection	The Australian Centre for Field Robotics (University of Sydney) has developed RIPPA (Robot for Intelligent Perception and Precision Application). RIPPA is a solar-powered, four-wheeled device that identifies weeds in fields of vegetables and treats them individually. It does this with precise doses of herbicide, but there is also scope to use a beam of microwaves or a laser, which would allow the crops to be recognised as organic.	Bogue, 2016
Crop nutrition	Rowbot Systems (a company in Minneapolis) is developing a robot that can travel between rows of partly grown maize plants, allowing it to apply supplementary dressings of fertiliser without crushing the plants. In the future, it may also be possible to match the dose to the plant in farms where the needs of individual plants have been assessed by airborne multispectral cameras.	Cavender-Bares and Lofgren, 2016
Harvesting	AGROBOT have created the SW6010, which uses a camera to recognise strawberries and determine which are ready to be picked. Those that are have their stems severed by blades and are caught in baskets before being passed on by a conveyor belt for packing.	Hannemann, 2016
Harvesting	Wageningen University (Netherlands) are working on a robot harvester for larger produce such as peppers, known as SWEEPER.	SWEEPER, 2017
Multiple	The Flourish project aims to develop an adaptable robotic solution for precision farming by combining the aerial survey capabilities of a small autonomous multi-copter Unmanned Aerial Vehicle (UAV) with a multi-purpose agricultural Unmanned Ground Vehicle (UGV). The system will be able to survey a field from the air, perform targeted interventions on the ground, and provide detailed information for decision support, all with minimal user intervention. The system can be adapted to a wide range of farm management activities and different crops by choosing different sensors, status indicators and ground treatment packages.	Flourish, 2017

to ensure they are efficiently addressed with minimum economic cost. In many instances, sensors have the capability to do this cost-effectively, with minimum labour require-ments and often more efficiently than if the job was done manually.

The use of sensing equipment within agri-culture is not new, having been discussed in the literature since the early 1980s (e.g. Warrick *et al.*, 1983; Dragg *et al.*, 1984) but up until the last decade or so, few applications reached the market and even fewer were used on a regular basis to inform on-farm decision-making. Those that were used on a regular basis performed simple operations

such as monitoring greenhouse conditions and automatically providing ventilation or shading when needed. However, since 2005, this area of agricultural technology has rapidly gained popularity both as stand-alone applications, especially within green-houses and for speciality crops (Lee *et al.*, 2010), as well as within an entire precision farming enterprise and there is every reason to presume that this trend will continue such that in the short to medium term, sensor use on-farm will become an everyday practice.

Sensors use a wide range of different technologies and can undertake many differ-ent operations. A summary of these is given in Table 8.3. Some of the systems described are already available but costs can be prohibitive to many farmers. However, within other industrial sectors, costs have dropped rapidly and there are signs that this is now also happening within agriculture. Other systems described in Table 8.3 are currently being used for research or being piloted, and are yet to reach the commercial market but are likely to in the short to medium future.

Sensing techniques can be broadly applied in three main different ways:

1. **Ground-based sensing:** Sensors can be used in-situ, that is, on or below ground, in or around the crop and placed in a specific, strategic position. They can be hard-wired or use wireless technologies. These types of sensors can also be used in hand-held devices including smart phones.
2. **Airborne sensing:** This is a form of remote sensing that is done using either piloted aircrafts or on unmanned aerial vehicles (UAV, drones). UAVs are typically low cost, lightweight and low airspeed aircrafts that are well suited for remotely sensed data gathering. They can also be used in conjunction with a Global Positioning System (GPS).
3. **Satellite sensing:** This is another form of remote sensing that uses satellite imagery to monitor large land areas. Whilst these types of sensors have been used for many years, they have been limited in application as they tended to provide only low-resolution images and were often affected by weather conditions. However, this is an area that is rapidly improving and recent technological advances are now providing higher-resolution images (Wójtowicz *et al.*, 2016).

Sensor use offers a host of benefits. First, sensors can provide data with a high spatial resolution that is often more detailed and more accurate than could be provided by manual measurement unless significant manpower was utilised, thus they can save significantly on labour costs. They can be used in situations which can be hazardous, such as inside silos and so offer improved worker safety. The data they produce can be used to ensure that crop inputs, such as water, nutrients and pesticides are optimised such that they can increase crop productivity whilst simultaneously reducing costs and environmental impact. In-field wireless sensor technology has a range of additional advantages, particularly in terms of easy installation and mobility. Sensor technology within smart devices coupled with information technology will undoubtedly deliver multiple benefits to the farm business and to consumers through increased productivity and quality.

8.4.2.4. Information technology

Advances in technology are often coupled with an increase in the information that is available. We live in an age when there is a whole world of information available at our fingertips via the Internet. As much as this can be valuable, it can also be a burden with issues such as information overload or not being able to locate reliable information and knowledge – to use an appropriate analogy, it is about being able to 'sort the wheat from the chaff'. The correct information and data can aid decision-making and improve efficiency. Thus, advances in information technology may help with this in the future.

Information technology can serve in a variety of ways. First, it can provide an interface to other technologies, to manage and utilise them. Second, it can provide a 'brain' in the form of the calculations, knowledge and information that is embedded within a range of applications (apps) available on

Table 8.3: A summary of sensor technologies and potential applications

Sensor type	Technological description	Example applications	Example references
Electromagnetic sensors	A soil's electromagnetic properties are partly influenced by soil texture, salinity, organic matter and moisture content. When using these sensors, the soil becomes part of an electromagnetic circuit and changing local conditions immediately affect the signal recorded which can subsequently lead to an automated response. They are usually used in-situ or on land-based mobile machinery as in conjunction with GPS.	These types of sensors have a wide variety of different agricultural applications such as to generate field maps of particular soil properties (e.g. soil salinity, electrical conductance). They can also be used in a wide range of automated activities such as simply turning on irrigation equipment or the variable signal can be used to adjust variable rate applications of, for example, nitrogen fertiliser in real-time.	Doolittle and Brevik, 2014; Guo et al., 2015; Sun et al., 2014; Gutierrez et al., 2014.
Optical sensors	These sensors utilise camera technologies to analyse images to perform appearance, colour, character and positioning inspections. They can be used in-situ, on handheld devices or they can be used remotely on land-based vehicles or mounted on UAVs. This type of sensor includes satellite imagery which can detect visible and near-infrared wavelengths of light, reflected from land below, thus providing information on the spectral properties of the crop.	They can be used to identify when fruit is harvest-ready, evaluate harvest quality and identify the presence of weeds, pests and disease. For example, they can analyse foliage colour to provide an estimate of crop chlorophyll content which can subsequently be used to adjust nitrogen fertiliser applications. They can also be used to measure crop canopy reflectance which in turn can provide information on crop development variability across the field and to adjust fertiliser applications as needed. Satellite image data can be collated at different spatial, spectral and temporal resolutions for agriculture mapping and crop assessment, crop health analysis, landscape mapping, yield determination and soils analysis.	Sumriddetchkajorn, 2013; Rios do Amaral and Molin, 2014; Gottschalk et al., 2008; Peteinatos et al., 2016; Rembold et al., 2013; Yang et al., 2013; Esch et al., 2014; Quemada et al., 2014.
Biometric sensors	These sensors are closely related to optical sensors as they use imagery to analyse biological traits such as body shape and voice.	Few current agricultural applications but these types of sensors have been used in research studies to monitor animal health and for identifying species.	Park and Moon, 2016; Guarino et al., 2017.
Spectroradiometers	These are devices used to measure the spectral power of a light source. They can be used with or without a connected computer so acting as a spectrometer. They can be used in-situ or on handheld devices as well as remotely.	Spectroradiometers can be used for foliage (e.g. quality, ripeness, disease) or soil measurements (e.g. soil structure and composition) or to measure, for example, crop canopy reflectance.	Mat et al., 2014. Gautam et al., 2016. Brown and Pervez, 2014.
Electrochemical sensors	These sensors are based on ion-selective electrodes which detect the activity of specific ions (e.g. nitrate, potassium, fluoride, hydrogen) and so provide information on soil chemistry. They are usually used in-situ or on handheld devices.	They can be used to measure soil nutrient and macro/micronutrient levels, soil pH and concentrations of soil contaminants.	Kim et al., 2013. Moeinian et al., 2016. Shaw et al., 2013.

Sensor type	Technological description	Example applications	Example references
Pyrgeometer sensors	These sensors measure near-surface infra-red radiation spectrum at a wavelength 4.5 to 100 µm. These types of sensors are usually used remotely and can be mounted on UAVs.	Their most common application in agriculture is for measuring climatic and weather data such as air temperature, thermal radiation or humidity.	Borg et al., 2014; Chiaradia et al., 2015.
Airflow sensors	These sensors can be used to measure soil–air permeability but currently are largely limited to research.	Research has demonstrated potential for distinguishing between various soil types, moisture levels, and soil structure/compaction.	Guo et al., 2014. Chen et al., 2014.
Thermal imaging	This is a technique to convert the invisible radiation pattern of an object into visible images for feature extraction and analysis. This technology can be used on handheld devices or can be coupled with optical systems mounted on an UAV, airplane or, potentially, by satellite.	It has many applications in the agricultural sector including identifying water stress in crops, irrigation scheduling, disease and pathogen detection in plants, predicting fruit yield, identifying harvest-ready crops and crop quality analysis (e.g. detecting bruising in fruits).	Vadivambal and Jayas, 2011; Alchanatis et al., 2010; Baranowski et al., 2015.
Airborne multispectral and hyperspatial sensors	These sensors are used to measure reflected energy within several specific bands of the electromagnetic spectrum. Multispectral sensors usually have between 3 and 10 different band measurements in each pixel of the images they produce. Hyperspatial sensors record data at a high spatial resolution (<1 m). They can be used from piloted aircraft or mounted on UAVs.	These types of sensors have been used in recent research to estimate and characterise the spatial variability of microclimate conditions for risk assessment and site-specific management of vector-borne diseases and crop pests.	Yang et al., 2014. Contreras et al., 2014. Quemada et al., 2014.
Acoustic sensors	An acoustic wave sensor is an electronic device that can measure sound waves. Any changes to the characteristics of the surrounding area though which the sound wave moves will affect the velocity and/or amplitude of the wave. These changes can be correlated to the corresponding physical quantity being measured and so provide information for monitoring.	Within agriculture, acoustic sensors have currently been limited to mainly research studies where they have been used for determining soil texture, for monitoring crop height and so estimate yield and to capture livestock sound data for behaviour and health monitoring.	Sharma et al., 2016. Chuanzhong et al., 2017. Hakkim et al., 2016.
Motion sensors	These sensors are in common use in many industrial sectors but are relatively new to agriculture. They respond to any kind of motion including vibration. They can be based on a number of different technologies including optical (light), microwaves or passive infra-red (PIR). They can be mounted in collars or tags and are often used in conjunction with GPS.	These sensors can be used to track, manage and possibly control animal movements and behaviour.	Lubaba et al., 2015; Bishop-Hurley et al., 2014.

different devices. Often, it is the combination of these two aspects, and in combination with machinery (Chapter 8.4.2.2) and sensors (Chapter 8.4.2.3), that can lead to the realisation of valuable tools that make a significant difference on the farms of the future and their environmental impacts.

Information technology encompasses a range of devices including desktop and laptop computers, tablets, smartphones, other handheld devices and other devices that are running software and/or storing and transmitting data. The applications that are most commonly employed on farms include office applications; record keeping and accounting; mapping tools; decision support tools (e.g. databases and calculators); or a combination of any of these. It is beyond the scope of this book to describe every available application, but with respect to agri-environmental management, two key application areas are the use of geographical information systems and decision support tools.

Geographical Information Systems (GIS) are software applications that provide the infrastructure to capture, store, analyse, integrate, transform and visualise spatially referenced data (essentially integrates databases with mapping tools). Data can be points, lines or polygons that have geospatial referenced data so they can be mapped, with each item having associated environmental data as attributes (i.e. database fields and records). They are commonly stored as GIS layers (e.g. rasters and vectors). Having data structured this way can support analysis and communication. For example, it is possible to overlay and/or combine data in different layers as a means of analysis, for example combining spatial data emissions of pollutants with spatial data on vulnerable receptors. Farms inherently have a lot of spatial data and the amount of this data is growing with technological developments

such as GPS and sensors (see Chapters 8.4.2.2 and 8.4.2.3). To realise the full value of this spatial data, mapping solutions are required and GIS has the capability to meet this demand. The maps that emerge can be used for a variety of agri-environment applications, for example strategic planning (e.g. location of buffer strips or conservation areas); day to day management (e.g. no-spray zones marked on maps in tractor cabs); or precision application (e.g. maps of soil nutrient status or plant health used to target inputs accordingly). The advances in other technologies now and in the future are likely to broaden the role of the GIS.

Decision support tools can encompass a broad range of applications that can be used during decision-making and management processes (see Chapter 7.3). Many mainstream tools are usually simple calculators or databases, such as tools for calculating crop nutrient requirements or for selecting pesticides approved to treat specific pests on specific crops. In some instances, especially in the digitial age and with the plethora of information available on the Internet, decision support tools are essentially advanced search mechanisms that can help signpost, filter or tailor information. An important aspect for many tools is that they should be designed to 'support' the decision-making process and not be a tool that makes the decision. Tools that are overly prescriptive, for example the answer to issue Y is X, do not necessarily help decision-making, unless they are extremely robust and reliable. Tools that present a range of potential options to solutions (along with their various advantages and disadvantages) are likely to be more pragmatic within the context of decision-making. Such tools are more transparent and the decision maker can evaluate the options presented within their specific context (the details of which may be beyond the capability of the tool).

In the context of integrated farm management, there is an increasing need to find solutions and make decisions that will address multiple objectives on the farm (see Chapter 7), which inherently involves seeking optimal solutions. In the past, optimal solutions were about, for example, applying an economically optimal amount of fertiliser or pesticides that reduced crop damage at least cost. In the context of agri-enviromental management, optimal solutions are sought, but with additional objectives of minimising nutrient pollution, greenhouse gas emissions and impacts on biodiversity. This makes the decision-making process more complex and thus there is a demand for tools to support such complex decisions.

There have been many efforts over the last few decades to develop decision support tools to help with agri-environmental management decision-making. However, there are few, if any, that have made it to become mainstream (commercial) tools – they are all largely prototypes, academic or scientific applications. Many of these have also tended to focus on specific issues in isolation (e.g. nutrients, pesticides, greenhouse gas emissions, water use, etc.) and few have attempted to tackle all the issues in an integrated fashion (and those that have attempted this are often superficial and/or unreliable). There is still a demand for such tools, so there is still scope for their development in the future – some of the prototypes that exist today may evolve or be adapted to integrate with mainstream/commercial applications (such as record-keeping systems) so that eventually agri-environmental management is seamlessly integrated into the commercial management of the farm. The development of other technologies for gathering and managing agri-environment data may help this evolution, such as the development of sensors (see Chapter 8.4.2.3) and the Internet of Things (IoT).

IoT is a relatively new area that is emerging (facilitated by advances in IT) and is likely to grow in the future. There is an ever-increasing number of appliances that can be connected to the Internet, including mobile phones, cars, power sockets, lights, refrigerators, televisions, and so on and this will increase in the future as costs decrease. Accordingly, the use of IoT is likely to increase in the agricultural sector, especially in combination with developments in sensors (see Chapter 8.4.2.3). For example, IoT sensors in fields or buildings could feed data in real time to the farm office, which can then be used to support decision-making, or in a more advanced scenario it may trigger a fleet of robots (see Chapter 8.4.2.2) to travel out to a field to automatically treat a problem that has been detected by the sensor.

8.4.2.5. *Big data analytics and bioinformatics*

In recent decades, huge amounts of data have been generated relating to biology, plant science, agriculture and the food sector more generally. This has arisen from numerous large-scale research projects and initiatives including the Human Genome Project (HGP; Collins *et al.* 1998), the sequencing of plant, animal and pathogenic genomes, environmental monitoring data, sensor data, smart farming and traceability initiatives. In addition, there have recently been moves to enable greater data sharing, particularly that which has been generated with funding from the public purse. Indeed, many research funding bodies now have strong data sharing policies that seek to maximise the research opportunities that such large volumes of diverse data provide. However, regardless of how much data is available, it has no value or impact unless it can be understood and then this understanding used to innovate. The sheer quantity, breadth and depth of this data (often referred to as big data) means that it can often overwhelm the capacity of existing standard 'off-the-shelf' data analysis software tools to capture, store, manage, and analyse it. For example, according to Valdivia-Granda (2008), data relating to the DNA for a single organism's genome requires 10 megabytes of computer storage which increases to more than 10 gigabytes when the genome is annotated. Consequently, it is not possible to efficiently manage, interpret or use these huge data-sets without calling on high-performance computing capabilities, specialist tools and the expertise of other disciplines, in particular mathematics, statistics and computer science.

With respect to biological data, particularly that relating to genomic sequencing, this combination of disciplines, generally known as bioinformatics, is now an important strand of most biological sciences and is being used to facilitate a wide range of discoveries and innovations. This mass of genomic data has led to the development of several globally accessible databases to store, retrieve and organise the data as well as a range of computer tools to enable the data to be analysed, annotated and visualised in 3D to enable a better understanding of a wide range of different biological systems (Neerincx and Leunissen, 2005; Meyer and Mewes, 2002). These tools can be used, for example, to predict the function of different genes and the factors which affect them. They can be used to identify specific responses that plants might have to biotic or abiotic stresses which enables farming practices and cropping regimes, for example, to be modified, potentially leading to improved crop health and yields. Understanding plant genomic data using bioinformatics analytical techniques can also be used to develop new plant cultivars that offer pest and disease resistance, drought resilience, improved quality and so potentially reduced economic and environmental costs (Rajamanickam, 2012).

Rice was the first sequenced crop genome and the impact that it had on rice genetics and breeding research was immediate, leading to greatly improved rice cultivars (Eckardt, 2000). Globally, rice is the second most important crop (wheat being the first) with 50 per cent of the world's population dependent upon it for nutrition. Consequently, the sequencing of the rice genome had the potential for dramatically impacting on rice production. Indeed, the molecular understanding of the genetic basis for rice nutrient use (nitrogen and phosphorus) allowed rice researchers to engineer a low input 'green super rice' that should help meet the challenge of the growing world population and food security (Jackson, 2016; Zhang, 2007). The success of this project lead to an explosion in other similar studies. A good example is the International Tomato Genome Sequencing Project which began in 2004 and aimed to decode the genome sequence of tomato (*Solanum lycopersicum*). The project created an international bioinformatics portal for comparative Solanaceae genomics that can store and analyse the sequence data and derived information from this project and associated genomics activities in other solanaceous plants (Ranjan *et al.*, 2012). The next stage of the work is to use this knowledge to develop new cultivated tomato hybrids with improved agronomic traits (Kumar and Khurana, 2014).

It is not just food crops that bioinformatics has the potential to improve in the future. The Biofuels Feedstock Genomic Resource is an Internet-based portal for genome sequence data for plant species relevant to biofuel feedstock production. The data is currently being used to identify natural variants in maize (*Zea mays*) and switchgrass (*Panicum virgatum*) to enable new varieties to be developed that offer the potential for greater biomass production effectively increasing the amount of biofuel that can be produced from the same land area (Childs *et al.*, 2012).

These large genomic databases are also being used to aid the precise identification of a range of biological organisms. Plants, fungi, animals and insects are generally identified using their morphological features such as shape, size, texture and colour. This takes considerable experience and often a professional taxonomist is required. Nevertheless, errors are frequent, especially if the specimen sample is juvenile (so not fully developed), damaged or atypical. However, in 2003 it was proposed that very short sections of a genetic sequence could be used to identify species (Hebert *et al.*, 2003) in a manner similar to barcoding, whereby a supermarket scanner distinguishes products using the black stripes of the Universal Product Code (UPC). The gene region that is being used as the standard barcode for most animal species is a 648 base-pair region in the mitochondrial cytochrome c oxidase 1 gene ('CO1'). CO1 is proving highly effective in identifying birds, fish, insects and many other animal groups, however, it is not effective for plant identification and so two gene regions in the chloroplast, 'matK' and 'rbcL' are being used as the barcode regions for plants (Benson *et al.*, 2012).

Bioinformatics is becoming of increasing importance to agriculture, offering a wealth of opportunities. There is no doubt that new crops and varieties will, in the future, be developed based on molecular methods and bioinformatics. In a similar way, large data sets collated from monitoring, sensors, smart farming and precision agriculture are also set to influence sustainable farming in the future. Sensors in fields and crops, imagery from satellites and drones, and detailed local weather data is generating very large data sets which, especially when individual data sets are combined to cover large land areas, give a highly detailed picture of what is happening on the ground. This data can then be analysed using data mining techniques.

Data mining was first used in the marketing field and, according to Foss and Stone (2001), refers to the extraction of previously unknown information from large data sets and its use in business decision-making to support the formulation of tactical and strategic marketing initiatives and measuring their success. The term 'data mining' is now used well beyond the marketing sector and tends to refer to the process of finding new correlations and new behaviour rules in big data which can be exploited. Like bioinformatics, it is highly interdisciplinary and requires expertise in a wide range of disciplines including computers and database technology, expert systems, information theories, statistics and mathematics.

Big data and data mining has the potential to significantly alter how farming is conducted (Bauckhage and Kersting, 2013; Sharma and Mehta, 2012). It can be used to optimise farming systems in terms of the crops/varieties grown, plant nutrition and crop protection considering both sustainability and profitability (Reddy and Ankaiah, 2005; Sadok *et al.*, 2009). There are also applications for the conservation of natural resources (Berry *et al.*, 2003, 2005; Andújar *et al.*, 2006), ecosystem services (Xu *et al.*, 2008) and biodiversity, where the analysis of spatial big data can be used to improve models, identify new biological mechanisms and for developing conservation strategies (Kelling *et al.*, 2009; Hochachka *et al.*, 2007; Stockwell, 2006; Chavan and Ingwersen, 2009). However, it is not just on the farm that big data has future application. Radio-Frequency Identification (RFID)-based traceability systems can provide a constant data stream on farm products as they move through the supply chain. This can give information on shelf life, wastage and consumer preferences and be used to make more informed decisions (Ji and Tan, 2017; Ahearn *et al.*, 2016). Big data analytics also has application in the design, management and use of agricultural machinery. For example, John Deere, an international company that, amongst other things, manufactures large-scale farm machinery, has recently launched a number of innovative products that provide interconnectivity between equipment owners, operators, dealers and agricultural consultants that seeks to enhance productivity and increase efficiency. The company also uses sensors on their equipment to help farmers manage their fleet and to decrease tractor downtime and promote fuel efficiency. The information is combined with historical and real-time weather data, soil conditions, crop features and many other data sets to enable the optimisation of farming operations.

8.4.2.6. *Nanotechnology*

The concept of nanotechnology first emerged in the 1980s and at that time tended to refer to developments at the molecular level. In the intervening decades, the concept has become a reality but tends to now refer to any development less than 100 nm in size and which has a novel, exploitable property. To put this size into perspective, the human eye cannot see anything less than 10,000 nm in diameter and a human hair is typically 80,000 nm in diameter. Current successful commercial applications are still somewhat limited but the potential is considerable, particularly within medicine, biotechnology, electronics and energy industries and, not least, the agri-environment and agri-food sectors.

Nanotechnology offers opportunities to increase the sustainability of agriculture and it could undoubtedly provide the innovation needed to help address issues such as food security, climate change and a polluted environment. Within the agri-environment,

nanotechnology offers the opportunity for smart delivery of chemical inputs, significantly reducing the amounts applied and ensuring that they are provided at the point of need such that environmental losses are minimised to the extreme (Parisi *et al.*, 2015). The range of applications are summarised in Table 8.4.

In the agri-food sector, nanotechnology offers opportunities for improving the sustainability of the whole food chain (Valdes *et al.*, 2009). For example, within food processing, nanoparticles can improve the bioavailability of neutraceuticals in common, affordable ingredients such as rice and vegetable oil. They can also be used to selectively bind chemicals or pathogens from food, and antibodies can be attached to fluorescent nanoparticles and used within food packaging to identify food-borne pathogens. They can also be used as antimicrobial and antifungal surfaces for food preparation and storage (Fraceto *et al.*, 2016). Nano-barcodes have also been proposed for food traceability (Prasanna, 2007).

As an emerging technology, there is currently a distinct lack of nano-specific policy and regulation. Nevertheless, this is not far away. The EC strategy on key enabling technologies (EC, 2012) identified nano-technologies as an important innovation for the agri-environment and agri-food sectors. Ideally, as is the case with pesticides, biocides and GM developments, for example, regulation should seek to facilitate and harmonise the identification, characterisation and control of risks associated with substances and products (in this instance, nanosubstances) in order to protect human health, biodiversity and the environment. However, it should also, wherever possible, not hinder innovation or create obstacles to enhancing competitiveness.

There is, however, one significant issue: these substances often exhibit very different

Table 8.4: Examples of nanotechnology in the agri-environment

Area	Description	References
Crop protection	Both synthetic crop protection agents and bio-pesticides can be delivered in nano-formulations such as nanocapsules, nanoemulsions and nanoseed coatings. There have been a few reported experimental scale applications such as that with Neem Seed Oil nanoemulsions and, in recent years, nanosilver (nano-Ag) antimicrobial pesticides have been registered in the USA although not without controversy. Current nano-Ag products are primarily targeted for non-food applications such as protecting plastics and fabrics from fungi, moulds and mildews. As such, they do have application for protecting our cultural heritage (see Chapter 6). A number of studies with nano-Ag have demonstrated that the use of a colloidal nanosilver solution may considerably improve the growth and health of various plants. Nano-formulations of pesticides can also be exploited to improve their structure and function and enable controlled-release toward the target organism.	Nuruzzaman *et al.*, 2016; Mishra and Singh, 2015; Ghotbi *et al.*, 2014; Jolanta *et al.*, 2011.
Crop nutrition	Synthetic fertilisers can be delivered to the crop as nanoparticles, nanocapsules and viral capsids. These formulations not only deliver nutrients directly to the crop but research has shown they also are better absorbed by the crop and as such small amounts are applied the risk of nitrogen pollution is significantly reduced. Nano-fertiliser use may also help resolve the global problem of phosphorus (see Chapter 5).	Milani *et al.*, 2012; Liu and Lal, 2015.
Diagnostics	Electrochemically-active carbon formulated as nanotubes or nanofibres can be used as highly sensitive biochemical sensors for monitoring plant health, growth and environmental conditions.	Fraceto *et al.*, 2016.
Soil quality	Some nanomaterials such as hydrogels, nanozeolites and nanoclays can be used as soil enhancers by improving water retention and providing a slow release of water, helping to prevent drought damage.	Sekhon, 2014.
Water management	Nanosenors currently under development will have the potential to measure soil moisture and stress in the soil and if coupled with smart automated technology, will be able to optimise crop irrigation.	Fraceto *et al.*, 2016.
Pollution remediation	Nanoclays can filter and bind a variety of different pollutants in the soil. This offers opportunities for cleaning up contaminated soils at a much quicker rate than is currently possible and potentially cheaper.	Fraceto *et al.*, 2016.
Livestock	Nanochips and nano-barcodes have been proposed to enable the traceability of livestock. Current animal health vaccine delivery systems need to be stored at low temperatures and have a limited shelf life. Nano-formulations can overcome these limitations. Nanotechnologies also have future application in animal breeding by nanotube implant to identify optimum levels of oestradiol in the blood and thus the optimum time for insemination.	Prasanna, 2007.
Technological delivery	Nanoparticles can be used to deliver DNA to plants in a form of targeted genetic engineering. Using GM technology, plants can be engineered to produce nanoparticles for applications in other industries such as waste management and polymer production. For example, nanofibres produced from wheat straw and soybean hulls have been developed for reinforced polymer manufacturing.	Alemdar and Sain, 2008.

physical, chemical and toxicological properties compared to the same substance at a larger scale. This could be greater chemical reactivity, electrical conductivity, strength, mobility, solubility, magnetic or optical properties (Balbus *et al.*, 2005). In addition, some studies have shown that these substances do not necessarily behave in the environment as would be expected based on fate and transport studies for their larger-scale counterparts (e.g. Grillo *et al.*, 2012). Only a few toxicological studies have so far been conducted but these have demonstrated that nanomaterials may be more toxic and care is needed. Studies have demonstrated that some nanomaterials appear to have the potential to damage skin, brain and lung tissue, to be mobile or persistent in the environment or to kill microorganisms, and can be more damaging than when at the large scale (Balbus *et al.*, 2005; Dreher, 2004). As such, these novel properties may pose new risks to workers, consumers, the public, biodiversity and the environment and, consequently, the usual risk assessment approaches, as currently applied to pesticides, biocides and chemicals more generally, cannot be used and new approaches are beginning to be developed. In addition, current data for environmental fate and toxicology probably does not apply and nanotoxicology data are required. However, whilst many studies are ongoing, it is not known at this stage if the data being generated is appropriate or that it will be useful for assessment processes as these are still to be developed (Hankin *et al.*, 2011).

Within the EU, it is probable that REACH (EU Regulation on Registration, Evaluation, Authorisation and Restriction of Chemicals) will be the main control mechanism for nanotechnological substances, but in terms of food applications and crop protection it may be that the EFSA will retain governance.

8.4.3. Biology

8.4.3.1. Introduction
As mentioned above, technological responses/developments are often developed in tandem with advances in our biological knowledge. Biology, and biological understanding, is often the key restricting factor, therefore any advances that can be made to overcome restrictions can be significant. The following sections explore advances in biological knowledge that have been made recently or will be made in the future, in response to some of the pressures and drivers. In some instances, this involves the development of new techniques, based on improved biological knowledge (such as partial rootzone drying). In other instances, it involves improving the biology of plants and animals themselves (such as through genetic modification) to help tackle the challenges that lie ahead.

8.4.3.2. Partial rootzone drying (PRD)
Partial Rootzone Drying (PRD), pioneered in vineyards in Australia, is an irrigation technique that improves the water use efficiency of production without significant reduction in yield, and in some instances can increase yield (De la Hera *et al.*, 2007). It uses biochemical responses of plants to water stress to achieve a balance between vegetative and reproductive development. The technique essentially involves irrigating approximately half of the root system of a crop while the other half is left to dry. Then, after a period (5–14 days, depending on soil and climatic conditions), the dry half of the root system is irrigated, while the wet half is left to dry. Dehydrating roots send chemical signals, such as abscisic acid (ABA), to the shoots and leaves via the xylem, reducing stomatal conductance, transpiration and vegetative growth. Meanwhile, roots of the watered side maintain a favourable plant water status.

PRD has increased water use efficiency in grapevines, by up to 50 per cent or more compared to conventional irrigation, thus making PRD a potentially valuable tool to apply in water-stressed regions. Although pioneered for grapes, PRD can also be applied to other crops, including winter wheat, potatoes, maize and other vegetable and fruit crops (Sepaskhah and Ahmadi, 2010).

8.4.3.3. *Biostimulants*

Agricultural biostimulants are a new but rapidly growing approach to encouraging plant growth and all-round crop development. Biostimulants are a diverse group of substances, that includes chemical compounds and microorganisms, which, when applied to a crop, have positive effects on vigour, yield, quality and tolerance of abiotic stress. Such substances include, for example, protein hydrolysates, seaweed extracts, humic and fulvic acids, fungi and rhizobacteria. They work through a plant's life cycle offering a wide range of benefits which have been reviewed by Van Oosten *et al.* (2017) in detail and can be summarised as:

- Increased yields and crop quality by increasing the plant's metabolic efficiency.
- Enabling the crop to tolerate and recover rapidly from stresses arising from extreme weather conditions, drought, soil salinisation and pollution such as heavy metals.
- Improving the way the plant uses the available nutrients, minerals and water.
- Increasing soil fertility by encouraging strong soil microorganism populations, which in turn also encourages increased crop productivity.
- Enhancing the quality of fresh produce by, for example, increasing or reducing sugar content, enriching colour or by affecting fruit seeding.

Whilst many of these advantages are similar to the benefits seen from an efficient and effective crop nutrient strategy, agricultural biostimulants operate in a different way to nutrients and have demonstrable effects even when nutrients are short.

Much research into the use of biostimulants, particularly for helping to address food security, has been undertaken in recent years which helps to demonstrate the value of these substances. For example, the use of biostimulant compounds derived from seaweeds have been shown to significantly improve seed germination, the health and growth of plants, particularly via shoot and root elongation, as well as offering improved water and nutrient uptake, frost and saline resistance, biocontrol and resistance towards phytopathogenic organisms, remediation of pollutants of contaminated soil and fertilisation (Nabti *et al.*, 2017; Lötze and Hoffman, 2016; Rengasamy *et al.*, 2016). Microbial plant biostimulants, including certain types of *Azotobacter, Aspergillus, Lactobacillus and Bacillus,* have been shown to increase yields (Macouzet, 2016). Chitosan, which is a polysaccharide extracted from the shells of various crustaceans, has also been investigated regarding its potential as a biostimulant. A review by Qavami *et al.* (2017) demonstrated its value for crop preservation, growth enhancement and for inducing resistance to pathogens.

Biostimulants also offer other opportunities for the agri-environment, particularly by enhancing the performance of soils. For example, they can be used for the phytoremediation of contaminated soils (Arthur *et al.*, 2016; Decesaro *et al.*, 2016). Microbial inoculants have also been used successfully to improve the ability of poor soils to supply nutrients to crop plants and enhance soil structural stability (Alori *et al.*, 2017).

8.4.3.4. Biotechnology

Biotechnology is 'the application of scientific techniques to modify and improve plants, animals, and microorganisms to enhance their value' (Wieczorek, 2003), and agricultural biotechnologies therefore are intended to increase their value in relation to some element of agricultural production and/or the characteristics of the final product. To many, this implies some form of scientific intervention, often with dark connotations, but in reality we have (knowingly or not) been manipulating the genetic make-up of plants and animals in order to encourage the desirable traits society depends on, since the dawn of the earliest civilisations (Sager, 2001). Generally, this has been through a process of selecting plants for propagation and animals for breeding, as a result of their possessing some characteristic it was deemed advantageous to reproduce and, in essence, this is how much development is still done. However, as our knowledge of genetics has improved, so has our ability to guide the development process. For example, plant varieties have been improved to increase yields, raise the nutritional value of the crop, improve flavour or texture, increase their resistance to pests and diseases, and indeed for anything considered beneficial (Wieczorek, 2003; Sager, 2001), and equally, animals have been bred to produce more milk, meat, eggs, and so on (Sager, 2001). The semi-dwarf wheats, for example, were first bred in the 1940s and 50s, in order to allow crops to take advantage of fertiliser-based nitrogen enrichment to grow more vigorously and therefore produce more grain, without collapsing under the weight (logging), a problem which can be particularly prevalent in wet, windy conditions (Borojevic and Borojevic, 2005).

Traditional plant and animal breeding is, however, a slow process, in which the genetic material of 'parent' varieties/breeds combines in a relatively uncontrolled way, such that desirable traits and undesirable ones (e.g. poor yields) may both be transmitted to the following generation (Wieczorek, 2003), and achieving the combination sought may take many iterations. Nevertheless, our ability to achieve our aims in this respect has increased massively in recent decades, as our understanding of the role played by genetics has improved. Since the mid 19th century, when Gregor Mendel, a Czechoslovakian scientist, carried out his groundbreaking research into the way in which traits were inherited in peas, the science of molecular biology has advanced rapidly, so that there is now a much more detailed understanding of the genes responsible for many traits (positive and negative) in a whole range of crops and livestock. So-called gene marking techniques are now well established in animal breeding (especially in cattle) for example, with the marker density in the cow genome now at an average of about one marker per 10,000 base pairs, compared to one marker per 3 million base pairs on the first release in 2004, the change in marker density being an indicator of the expansion of knowledge of base pair sequences and their likely linkage to phenotype traits. Mapping base pair sequences is, nevertheless, a time-consuming and costly process, and only the first step in a hugely complex process (in which computer science and statistics now have a huge part to play) of working out the function of pairs and sequences within DNA systems. Descriptive science of this sort is also only part of the equation, since to be of practical benefit, traits determined to be advantageous then need to be embedded in commercially viable crop varieties and animal breeds; nevertheless, they are often the precursor to more practical discoveries and developments (Sage, 2001).

Inevitably, this technology will play a very significant role in agriculture in the future and, indeed, is already evident in current breeding programmes. For example, the American Angus Association offers a genomic profiling service, which allows the DNA of young animals to be compared to existing breed standards and predictions are made on a range of production and carcass traits. As a result, farmers can get an early indication of the likely potential of breeding animals, with benefits in terms of time, cost and performance. In terms of plants, scientists are working on a range of traits which could transform production, including those responsible for improved nitrogen efficiency, nitrogen fixing, water efficiency, salt tolerance, timing of flowering, optimising of grain shape (in cereals), pod shatter (in brassicas) and C3 versus C4 photosynthesis, to name but a few. In many cases, however, it will be at least a decade before they reach the market, and in some cases 20 to 30 years since, although science can help to speed up the process by working out which parts of the genome are likely to be involved in particular traits and to use that information to identify and select individual plants carrying those traits, other elements are less easily amended because the cross-breeding process still has within it a significant random element. Nevertheless, it allows plant and animal breeding to be a much more targeted process than was the case in the past, and it facilitates the evaluation of a greater number of options much quicker than previously. Monsanto, for example, have developed and operate a facility for the laboratory selection of maize and soya bean seeds based on their genetic make-up (Hildebrant, 2012). The so-called 'chipping' process takes a small slice of the endosperm of a seed grain (without destroying the germplasm) and analyses the DNA within it, looking for particular sequence matches, with seeds carrying the desired matches being retained, whilst

those remaining are discarded. This process, which is highly automated and can screen many thousands of individual seeds per day, replaces the much more time-consuming and costly process of growing all the seeds into mature plants and then looking for the desirable traits within those plants. Techniques like this, greatly reduce the resources and time needed to select and multiply breeding lines, and are contributing not only to speeding up the development of new lines, but also to reducing development costs, which means that breeding and selection effort for more minor crops, including vegetable and salad crops, becomes commercially viable.

Genetic engineering, however, has the potential to further speed up the process. Since the 1970s, the developing science of molecular biology has made the direct manipulation of DNA in both plants and animals a reality. As a result, it has become possible to introduce the characteristics sought, and indeed remove those seen as a problem, much quicker and in a more controlled, targeted way than in the past, thus significantly reducing the element of chance in the process (Sage, 2001; Wieczorek, 2003). It has also become possible to introduce traits from species too distant from the one of interest, to have been possible though traditional breeding programmes (Wieczorek, 2003). This sort of genetic engineering can reduce the time taken to introduce some crop modifications by a factor of 100 or even 1000 (Sage, 2001), meaning that the pace of future change is far more rapid than in previous generations. Regardless of the techniques being used (genetic engineering or plant/animal breeding), however, the goals remain more or less the same as has been the case for many years. Namely, to increase the value of crops and livestock in some way, albeit that the scope for doing so raises some interesting possibilities.

CONVENTIONAL CROPS

Breeding high-yielding crop varieties is a current priority and this is likely to continue long-term due to food security pressures. With respect to seed grains, such as wheat, there are two ways in which superior-yielding cultivars can be produced. Either the aim is to produce more seed per seed head or produce larger seeds, the latter being potentially beneficial as larger seeds tend to emerge quicker than smaller ones and so have an evolutionary advantage as they have first access to available nutrients, water and space, thus tend to produce stronger, healthier plants. However, both approaches are being considered by plant breeders and considerable success has been achieved recently with wheat varieties such as 'Oahe'. Oahe is a new winter wheat cultivar which promises high yields and excellent disease resistance to stripe rust, leaf rust and wheat streak mosaic virus, along with resistance to fusarium head blight that is comparable to other popular varieties. Research in this area is not restricted to grains. For example, a new tomato variety, Arka Rakshak, has been developed in India which, it is claimed, can yield up to 19 kg of fruit per plant, almost double typical yields, with even so-called high-yielding varieties only typically giving 12–15 kg of fruit. The Arka Rakshak tomato variety is also resistant to three commercially important diseases: tomato leaf curl virus, bacterial wilt and early blight. New rice varieties are also on the horizon that are low-input and ultra high-yielding. Research trials by the National Crops Resources Research Institute in Namulonge, Uganda, have been undertaken with several new varieties (including Okile, Wilta-9, Agoro and Komboka) that have high disease resistance and can yield up to 30 times more than traditional rice varieties. Results have confirmed excellent field performance even in marginal soils with low fertility and low water

conditions (*The Hindu*, 2016). Considering that rice is the most important crop in the developing world, these types of crop breeding varieties can, potentially, make huge differences for food security.

Other desirable crop traits can also be introduced through breeding which may also result in higher crop yields, for example, resistance to environmental stresses such as drought and heat, which are likely to be of increasing importance as climate change impacts are realised. Drought is probably the most damaging climatic condition in terms of agricultural productivity and food security. Inadequate rainfall will eventually affect plant growth and development and so reduce crop yields. Prolonged periods of drought can lead to the complete loss of the crop and can even lead to the abandonment of the land, especially in areas where severe droughts are commonplace – a situation more likely in the future due to climate change. As discussed in Chapters 4 and 5, if severe and frequent periods of drought mean that irrigation is needed to allow efficient crop growth, this can also cause other environmental problems such as soil salinisation. Hence, there is the need to develop drought- and salt-tolerant varieties. Research in this area is extensive (e.g. Levy *et al.*, 2013; Levy and Veilleux, 2007; Hanin *et al.*, 2016). For example, the winter wheat variety 'Davlatle', developed using conventional breeding techniques, is currently being evaluated for its potential to improve food security by producing high yields on saline fields. It is currently being trialled in soil with a medium-level salinity and in the harsh climate of Dashoguz in Turkmenistan, central Asia, a region characterised by low rainfall, limited irrigation and extreme heat. Should the variety be successful under these conditions, it is expected that it would do well in most situations worldwide. A Dutch research team is currently exploring a potential solution for growing potato and other halophyte crops on land considered unsuitable due to salinisation. The approach is simple in that it is not relying on the breeding of new varieties or GM, but exploring the existing 5000+ varieties systematically to identify those that will survive under saline conditions. Crops are irrigated in such a manner that the water and salinity can be carefully controlled and accurately measured (de Vos *et al.*, 2013).

Looking further into the future, as the world's population grows and resources such as water become scarcer, research is being undertaken to investigate the possibility of breeding crops with saline tolerance which would allow seawater rather than freshwater to be used for irrigation. This field of research has been ongoing for many years but has not yet had a great deal of success, however, the future looks brighter. Using seawater for irrigation is only a viable option if it is cost-effective; the value of the crop needs to be greater than the cost to produce it and pumping seawater can be expensive. Whilst early research in this area looked, with some success, towards breeding salt tolerance into crops such as wheat and barley, the amount of tolerance achieved was not sufficient to cope with the extreme growing environment caused by seawater irrigation. However, success was achieved in the late 1990s by attempting to domesticate wild halophytes that could be used as food, forage and oilseed crops. Since that date, research has continued and has shown that this approach has value. For example, various glasswort species that are common in coastal marshlands already demonstrate the ability to grow under saline conditions. One that offers most promise is *Salicornia bigelovii*, which has high value as livestock feed and its seeds contain high levels of oil (3%) and protein (35%), much like soybeans and other oilseed crops. In addition,

the salt content is less than 3 per cent, the oil is highly polyunsaturated and similar to safflower oil in fatty-acid composition. It can be extracted from the seed and refined using conventional oilseed equipment; it is also edible, with a pleasant, nut-like taste (Glen *et al.*, 1998; Lyra *et al.*, 2016).

Another area of great interest is improving the utilisation of nutrients. As discussed in Chapter 5, phosphorus is an essential nutrient for healthy and productive cropping. It is also a major constituent of the fertilisers required to sustain high crop yields. Levels of phosphate are sub-optimal in most agricultural ecosystems and when phosphate fertiliser is applied it is rapidly immobilised owing to fixation and microbial activity. Thus, cultivated plants generally use less than 30 per cent of the applied phosphate with the rest being lost to the environment. Not only is this a waste of resources and phosphorus is non-renewable, but it can also lead to environmental problems such as eutrophication (see Chapter 4). As is the case with abiotic stress, plant breeders are looking to produce phosphate-efficient crops that optimise fertiliser use as well as those that are more tolerant to low-phosphate stress. Producing crops where phosphorus is in deficit is difficult, but research has identified a number of potential options such as the use of endophytic fungi said to have an inoculation effect against P-deficiency (Iranshahi *et al.*, 2016) and, in particular, breeders are trying to exploit the fact that by increasing the density and length of root hairs, a crop's ability to acquire nutrients can be significantly improved (Vance *et al.*, 2003; Narang *et al.*, 2000). Root hairs are most effective at mining phosphorus from soil due to the large root surface area in direct contact with the soil and these types of root structures show the greatest ability to sustain high grain yields in low-P fields (Gahoonia and Nielsen, 2004). Work is ongoing globally but there have already been some notable successes, for example, new cultivars of bean that can cope with low fertility soils in Africa have been successfully introduced (Lunze *et al.*, 2012), a number of low P-tolerant rice genotypes (landraces, old improved and new improved varieties) have been identified (Aluwihare *et al.*, 2016) and new *Coffea* genotypes have been identified that are better adapted to low soil phosphorus (Neto *et al.*, 2016).

Whilst in many respects the focus of breeding new crop varieties is likely to focus on addressing food security and providing traits allowing crops to be productive in sub-optimal, potentially stressful environments, the demand for non-food crops cannot be overlooked, particularly with respect to biofuels. In Switzerland, Syngenta has released 63 new hybrid corn seeds for 2016, including two new Enogen hybrids for improving ethanol production yields. The Enogen hybrids, as well as the other new seed releases, were developed using the company's Yield Engineering System (Y.E.S.) that combines intelligent analytics, testing and technology to help get hybrids to market faster. The new hybrids include three traits meant to help farmers boost yields, including protection against specific pests such as corn ear worm and western bean cutworm and more efficient water use. Similarly, many other companies, such as Du Pont and Monsanto, are working on identifying and bringing to market corn hybrids with high ethanol yield potential.

GENETIC MODIFICATION

Modern technology may speed up conventional breeding processes, but the process can still take decades. Thus, when faced with multiple pressures in a relatively short time period, as described in the perfect storm (see Chapter 8.2), society may need everything available in its 'toolbox' to tackle them all and this may well need to include genetic modification (GM). GM is not without

considerable controversy and whilst many parts of the world are not resistant to GM, Europe has, to date, had an anti-GM stance.

The rationale behind the development of GM technologies was undoubtedly that it would be of huge benefit to mankind, particularly in terms of food security. Simply put, it would not have been in the interest of plant or animal breeders or food companies to produce a product that would ultimately cause harm. Nevertheless, there have been endless debates regarding ethics and around the risks GM and related technologies such as cloning pose to both human health and the environment. For example, in the early days of GM when Dolly the sheep was cloned (Wilmut *et al.*, 1997) there was considerable debate regarding the ethics of cloning which was exacerbated by Dolly's short lifespan due to lung disease. Scientists have since shown that Dolly's death was unrelated to cloning and the fears, in this instance, associated with GM were unfounded. Similarly, considerable controversy was caused by the introduction of terminator seeds which ultimately seriously damaged the reputation of many big plant breeders and GM technology generally. This terminator technology (also refered to as suicide seeds) is the genetic modification of plants to make them produce sterile seeds. The objective was to address concerns that GM crops would be able to spread, uncontrolled, into the environment. However, this development was not particularly successful as terminator genes were able to spread by cross-fertilisation and accidental mixing, resulting in the contamination of conventional crops such that these too would produce sterile seeds and would no longer be GM-free. There were also other concerns regarding the longer-term reduction in the crop gene pool if cross-breeding of plants in nature were restricted by sterile seed production. The ethical case was also difficult to justify as, particularly in the developing world, many farmers cannot afford to buy fresh seed each year and normal practice is for them to save their own seed. Critics saw this development as a way for companies to increase their profits at the expense of the poor (Mukherjee and Kumar, 2014).

In general, the concerns around GM crops can be summarised thus:

- Could genetic material contaminate other crops and wild plants and so result in unforeseeable environmental and food security issues?
- Could GM crops become 'super weeds' which are difficult or even impossible to control? This threat is often compared to the environmental problems caused by the introduction of alien species (e.g. giant hogweed, grey squirrel in the UK). Indeed, there are already examples of this in Canada and the US where GM herbicide-tolerant oilseed rape has pollinated other rape crops. The seed from this can germinate in following crops growing as a weed. Due to their herbicide resistance, farmers are turning to highly toxic herbicides in an attempt to control them (Wieczorek, 2003).
- Similarly, there is concern that insect pests could develop resistance to the crop protection features of GM crops, although, despite widespread use, this has not yet been observed (Wieczorek, 2003).
- Would GM crops result in greater pollution from pesticides? Whilst there is an argument that, as some GM crops would be resistant to certain pests and disease there would be less need to use pesticides, there is also the case that as some GM crops are herbicide-tolerant, this could encourage greater herbicide use. There is also the issue of superweeds mentioned above, where more toxic herbicides are needed to control them.
- Would GM crops be dangerous for

wildlife? There is some evidence to support this. For example, in the USA, a massive decline in populations of the Monarch butterfly has been reported to be caused largely by loss of habitat due to the spraying of RoundUp Ready GM crops with the weedkiller glyphosate (Pleasants and Oberhauser, 2013). In addition, researchers at Cornell University found that pollen from *Bt* maize could kill the caterpillars of the Monarch butterfly, although it also has to be said that in a field environment they are unlikely to come into contact with such pollen (Wieczorek, 2003).

- Concerns relating to possible unknown health risks such as increased antibiotic resistance, unexpected allergic reactions and the potential for new toxins to be created (Herman, 2003; Wieczorek, 2003). As yet, there is little evidence to support these concerns although there is no significant monitoring either.

The main social issues revolve around the labelling of foods containing GM-derived ingredients, with opponents saying that consumers should be given the information they need in order to make the choice for themselves as to whether to eat GM or not (Wieczorek, 2003). In the EU, foods containing GM ingredients, or ingredients derived from GM organisms, must indicate this fact on the label or where the product is sold if packaging is absent (FSA, 2013). In the US, on the other hand, this is only a requirement if the GM food is nutritionally different to its non-GM equivalent, meaning that many foods can contain ingredients from genetically engineered crops without the consumer knowing (Wieczorek, 2003).

The above concerns have led to a number of well publicised campaigns against genetic engineering in agriculture, particularly in Europe, where GM has to date received little support, with very few varieties being licenced for production. Nevertheless, worldwide, GM crops were planted by 18 million farmers in 28 countries in 2014, encompassing a total cropped area of 181.5 million ha (Huesing *et al.*, 2016), and in the US, where biotechnology has been broadly accepted (albeit that considerable numbers of opponents do exist), 69.5 million ha of GM crops were grown in 2012, accounting for 90 per cent of maize and 96 per cent of soya bean production in the country. Indeed, such extensive areas of crops developed through genetic engineering are grown, that a significant proportion (potentially as many as 60–70 per cent) of food products are thought to contain at least some genetically engineered material (Wieczorek, 2003). In contrast, only 119,000 ha was planted in Europe in 2012, 90 per cent of which was in Spain with around 0.1 million hectares of GM maize planted in 2015 (James, 2016).

Bt maize used to be grown in France, Germany and Romania, but this is no longer the case, and this is despite the fact that the EU has a shortfall in conventionally grown feed products for livestock, and therefore has to import a good deal from abroad, much of which is GM (Dunwell, 2014). This is mainly due to the political pressure brought to bear by protest groups, often including 'direct action' (Dunwell, 2014). This has meant that GM farming is seen as socially unacceptable, with entrenched positions being taken by many, regardless of any genuine scientific evidence one way or the other. In 2012, this led to BASF announcing (as others have done since) that it would shift its GM operations to the USA due the stance taken in Europe (Dunwell, 2014). This means that the European development base is greatly reduced, which could result in any crops developed being tailored to conditions found elsewhere in the world, thus presenting a competitive advantage to producers in those other areas.

There are signs things are beginning to change, in a few countries at least, with individuals who were once fervent opponents of GM production reconsidering the technology in recognition of the benefits it could provide economically, socially and environmentally (Lynas, 2013). The softening of attitudes at national level is, however, largely from outside Europe and appears to be the result of high market prices of non-GM grain forcing some Asian countries (e.g. South Korea and Japan, albeit the latter purchases are mainly for animal feed) to purchase cheaper GM grains. There are exceptions, for example, the British government is one country that appears to have softened its stance, and now seems to support the adoption of GM crop production. Indeed, in January 2017, it was one of only eight countries which voted in favour of licensing two new varieties of GM maize for use in the EU (and reapproving another), albeit the varieties in question are of little relevance to the UK, as the relevant traits target pests that are not found in the country. In addition, the UK has given the go-ahead for new field trials of a potentially high-yielding variety of wheat, that laboratory tests suggested could convert solar energy into biomass (photosynthesise) more efficiently (Rothamsted Research, 2017). This could give the impression of a change in direction, however, there are many vocal lobby groups across Europe, particularly in France and Germany, who continue to strongly oppose GM and so how much inroad GM production will make into European markets in the future is difficult to predict.

Where GM has been accepted, the focus has been largely on enhancing pest resistance and developing herbicide tolerance. Pest-resistant crops account for a significant proportion of the GM crops grown around the world. In the US, *Bt* maize and cotton, in which DNA from the insecticidal organism

Bacillus thuringiensis is included in the crop to kill common insect pests, are commonplace, and indeed, account for the majority of the maize grown (Wieczorek, 2003). Resistance to herbicides allows the use of such chemicals to remove/reduce weed populations without damaging growing crops. Herbicide-resistant varieties (e.g. soya bean, maize, cotton) accounted for around 80 per cent of the GM crops grown around the world in 2008 (Duke and Powles, 2009), with the majority of these being glyphosate-resistant. Despite the recent scientific debates as to whether this chemical is carcinogenic or not (with some groups calling for a ban), glyphosate is one of the most commonly used broad spectrum herbicides in the world, such that the ability to use it to control weeds within the crop being produced without also impacting on that crop, is seen as highly desirable (Wieczorek, 2003).

Genetic engineering has the potential for delivering other advances in areas with wide-ranging implications for sustainable development, agriculture and wider society alike. In particular, amending the nutritional content of crops can aid in supplying the needs of populations in areas which have to date suffered from inadequate supplies, including many in the developing world. Clearly, boosting the nutrient content of food products has been a goal of plant breeders for some time but with the development of new techniques, it has become possible to modify crops to include nutrients that would not otherwise be there at all (Sager, 2001). Golden Rice, for example, is a biofortified crop (also known as a nutraceutical product), intended to address vitamin A deficiencies in diets that are heavily dependent on rice, as many around the world are. Rice normally produces ß-carotene (provitamin A) in its leaves and not endosperm, meaning that traditional breeding would not have been capable of changing this situation, so genetic

engineering was used, allowing rice plants to produce endosperm supplemented with provitamin A (Ye *et al.*, 2000), in the hope of making a significant contribution to global health. Indeed, many have heralded it as a major step forward in alleviating the impacts of poverty in the developing world, although due to the concerns surrounding all genetically engineered organisms (see above), others have questioned the wisdom of such an approach.

Other substances too can be added to crops through genetic engineering, including a number with potential medical benefits. The use of genetically engineered plants as relatively economic bioreactors for the production of therapeutic molecules is known as 'biopharming', and has been suggested for the production of a range of different products (Daniell *et al.*, 2001). A number of advantages from the use of such systems have been put forward, including the fact that farm-based production is cheaper than that requiring industrial facilities. Other advantages include that the technology required for planting and harvesting is already available, the need for purification can be eliminated if the product is of food grade anyway and the risks associated with product contamination may be minimised (Daniell *et al.*, 2001). Experiments have been ongoing since the early 1990s, and although most substances are in the clinical trial stage or earlier, there have already been some successes. Mapp Biopharmaceutical Inc. of San Diego, for example, was allowed to trial its ZMapp product (produced in tobacco plants) during the 2014 Ebola outbreak in west Africa, and although the number of people treated was too small to come to any definitive conclusions, the results were very promising, encouraging the FDA to permit further work (Yao *et al.*, 2015). Others in the pipeline, include a number of antibodies, vaccines, hormones and other therapeutic substances being produced in tobacco, maize, potato and other crops. Going even further, it may be possible to produce very novel substances in agricultural plants and animals. Researchers at the Utah State University, for example, have been investigating the use of a number of other species to produce the proteins found in spider silk, including alfalfa, silkworms and goats, as a first step in producing synthetic spider silk (Teulé *et al.*, 2012.). Spider silk is one of the strongest substances known, but collecting it is somewhat problematic, so if the raw ingredients could be produced more easily (and converted into fibre) it would have significant industrial applications. Such systems are yet to be commercialised, but nevertheless, they illustrate the potential scope of genetic engineering techniques, and perhaps some of the reasons that people (rightly or wrongly) are concerned about them.

Given the multiple threats faced in terms of food security and the environment, genetically engineered agricultural products would, on the face of it, seem to be a great leap forward, offering benefits in relation to crop protection, yield and nutrition to name a few (Wieczorek, 2003). The reduced reliance on potentially harmful chemicals, could also be beneficial for the environment, through a reduction in the impact on water quality, biodiversity and the emission of greenhouse gases in their production. In addition, where yields can be increased, the amount of land needed to sustain a given level of agricultural production decreases, with implications for the extent to which natural land needs to be turned over to farming to feed the population. Despite this, however, lingering doubts remain in the minds of many, with potential problems being highlighted by environmentalists and others, and although many are based on a flawed understanding of the science, others (in theory at least) could exist. Even where this is not the case, if advanced

biotechnologies are to be widely adopted, and any environmental (and other) benefits realised, then their social acceptability has to be assured. It is therefore essential to address concerns relating to human health, the environment and their impact on social issues, head on.

8.4.4. *Management/economic*

Alongside technological, biological and biotechnological responses, there are also management and economic responses. Farms, like any other business, will respond and adapt to changing social and economic demands. When faced with economic pressures, the usual response is to try to become more efficient, and a common approach to do this is to utilise economies of scale and increase the size of production units. Bigger production units aim to reduce overheads per unit of output and thus increase efficiency. This, in itself, can of course, via competition, increase the pressure on other farms (as illustrated in Figure 8.6).

Figure 8.6 shows that pressure from consumers, in the form of demand for cheaper food, places pressure on small farms. To meet the demand for cheaper food, farming business may scale up their units of production to improve efficiency and reduce the price of the produce. This, in turn, places further pressure on the remaining smaller farms. This process has been very evident in the UK dairy industry. Figure 8.7 shows how the average herd size has been increasing and the number of holdings decreasing.

The ultimate outcome of this process has become known as the 'mega-farm', where there are units with several thousand head of livestock. The USA has had mega-farms for over ten years, for example Fair Oaks Farms in Indiana, which houses 30,000 cows and produces enough milk to supply the entire city of Chicago. Fair Oaks owns around 7,700 hectares of land and its cows

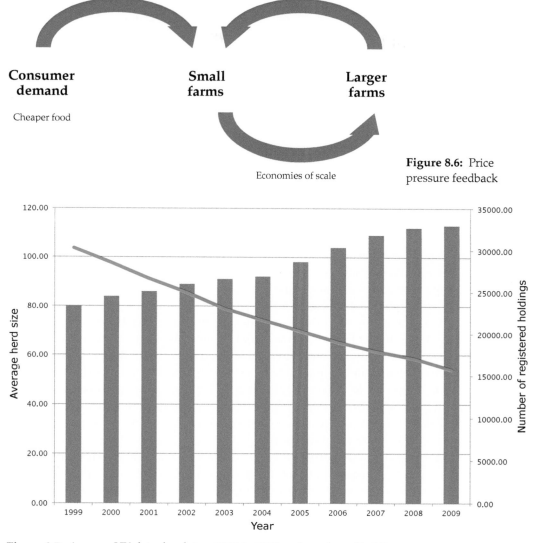

Competition: cheaper prices

Consumer demand

Cheaper food

Small farms

Larger farms

Economies of scale

Figure 8.6: Price pressure feedback

Figure 8.7: Average UK dairy herd size, 1999 to 2009 and number of holdings (Created from: Hawkins, 2011)

live in ten barns, 3,000 per facility, tended by a workforce of 400. Waste from the cows is processed in a state-of-the-art digester, producing enough methane to generate all the electricity the farm requires (Estabrook, 2010).

In 2010, plans were submitted for a mega-dairy in Nocton in Lincolnshire, UK. The plans were to have four open-sided barns housing 8,000 cows, milked 24/7 in an automated wheel-shaped rotating milking parlour. It would have been Europe's largest dairy farm. However, the plans were dropped in 2011 due to widespread opposition and a fierce media campaign against the idea, regardless of any merits it did

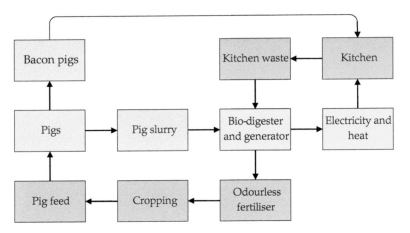

Figure 8.8: Green circle of pig production

have, and it was ultimately stopped by the Environment Agency, who were concerned about the potential impact of farm pollution. Plans for mega-dairy farms are not confined to the UK. In France, there has been strong opposition to a 1000-cow dairy farm (The Dairy Site, 2014) and in Spain, despite opposition (Green, 2017) a 20,000-cow dairy farm is currently under construction on 3,000 ha that will produce over half a million litres of milk per day and employ 300 staff (McCullough, 2017).

It is not just the dairy sector that is scaling up in size. In 2011, Midland Pig Producers proposed to build a mega-pig unit housing 25,000 sows in a state-of-the art facility. After initial objections, this was reduced to 2,500 sows, producing 1000 pigs a week for sale. The proposed plan was on 30 acres of land adjacent to Foston Hall women's prison in Derbyshire in the UK. In many respects, the pig unit was environmentally sound, following a so-called green circle of pig production (Figure 8.8). This system was very focused on containing and recycling energy and nutrients to minimise any environmental impact. The bacon pigs would provide food to the prison kitchen; the pig slurry would go into an anaerobic digester; waste from the kitchen would also go into

the digester; electricity and heat from the digester would be supplied to the kitchen; and anaerobic digestate would be used as a fertiliser to grow more pig feed.

Given the environmental credentials of the production system, the Environment Agency withdrew its objections to the proposed pig unit. However, the concept of the 'mega' pig farm was still not socially acceptable. There was a strong campaign against it and ultimately Midland Pig Producers decided to withdraw their application to build it.

Clearly, mega-farms involving livestock strike an emotive chord with the public, thus their social acceptability is in doubt. However, there are big arable farms that have not received such attention, and now there are also large greenhouse complexes such as Thanet Earth, the UK's largest greenhouse complex, spread over 90 hectares. It started producing food in October 2008 (2.5 m tomatoes, 0.75 m peppers and 0.7 m cucumbers per week ~ 2% UK consumption) and has four greenhouses so far and three more planned (to supply 4 per cent of UK consumption). It is named after its location, the Isle of Thanet in the north-east of Kent (in Roman times it was an island, but is now part of the mainland). Thanet Earth may be the largest greenhouse complex in the UK, but it is dwarfed by other areas of greenhouses elsewhere in Europe. For example, Figure 8.9 shows a satellite view of the south of Spain. The lighter area within the red box looks a similar colour to the mountains to the north, so perhaps it is a similar landscape, but actually, the light colour in this region is not mountains, but glasshouses and polytunnels covering the entire area.

Questions relating to the sustainability of this production and the social acceptance of this system need to be asked. Would the residents of the Isle of Thanet in the UK accept their environment ending up like Almeria in Spain, with a sea of plastic as far as the eye can see? This would undoubtedly raise objections on a similar scale to those seen for livestock mega-farms.

8.4.5. *Political/policy responses*
Trying to predict future agri-environmental policy even in the short term is a challenge, not least because the scope is large. Policy can change slowly and deliberately following the collation of scientific evidence, consultation and debate. However, it can also be volatile and change rapidly, taking a new direction to accommodate changed priorities of a different political power or even those of a new government minister wishing to leave his or her mark. Assuming, despite current European political and financial uncertainties, that European policy will continue broadly in the same general direction as it has over the last two decades, it is likely that there will be changes in policy for two reasons. First, to address perceived weakness and failures in the current approach and, second, to mitigate emerging agri-environmental issues.

From the previous chapters, there can be little doubt that in many instances agri-environmental policy is not producing the intended outcomes as often environmental standards are not being met. The quality of European soils is falling and, indeed, the area of fertile soil available for agricultural production is less than is needed to secure sufficient, affordable, quality food required by a rapidly expanding population. There are also major concerns regarding the loss of biodiversity exemplified by decreasing populations of pollinator species. Many of Europe's inland waters are of poor quality and oceans are also polluted. In addition, it is clear that poor air quality is affecting human health, crop productivity and climate change, and unless agriculture's GHG emissions can be further cut, it will steadily contribute more and more towards total EU emissions. There is, therefore, a pressing need to mitigate further polluting emissions from the agricultural sector.

The CAP, the foundation of all EU agri-environmental policy, is a heavy burden on EU Member States (MSs) as currently Pillar 1 direct payments account for over 70 per cent of CAP payments and almost 30 per cent of the total EU budget, therefore it is obvious that it cannot go on unchecked. There are many observers who believe that direct payments are no longer effective and that the approach does not now serve its original purpose of income support for those most in need, providing greater food security and enabling efficient resource use. Nor does it aid the delivery of rural environmental services or the move towards a more sustainable agriculture. The question is, therefore, how will it change to reduce costs and improve environmental performance?

One option on the table for the CAP post 2020 is that direct payments to farmers are no longer given as a matter of course but, instead, payments are replaced by targeted assistance to help farmers face specific challenges or provide specific public goods. Pillar 1 direct payments could be replaced, in whole or in part, with contracts for services. In

Figure 8.9: Greenhouse complex, Almeria, Spain
(Photos courtesy of: Google Maps; Jet Propulsion Laboratory [JPL]; ANE)

effect, this would mean a shift in the organisation of the CAP budget whereby Pillar 2 (i.e. the rural development programme/agri-environment schemes) becomes the main funding mechanism. Due to the potential impact this would have on the industry, the process is likely to be phased in over a period of time.

Whilst Pillar 2 funding comes with more stringent environmental demands that may help mitigate pollution problems, other changes across the policy spectrum will also be likely to address more specific issues. Pollution prevention has been the major concern across most of Europe for many, many years. Despite current policies and various emission targets (see Chapter 2.2), air quality is still a problem across many

European cities and indeed many rural areas. Although a new National Emissions Ceilings Directive came into force in December 2016 and needs to be implemented in EU MSs by June 2018, it is difficult to see how this will have the intended outcomes, considering that the previous lower emissions standards were frequently breached, unless it drives nations to take a more disciplined policy approach to transport and particularly towards reducing the use of diesel engine vehicles. With respect to ammonia from agricultural sources, the new Directive does require a reduction in ammonia emissions which will potentially require changes in some farming practices such as reducing the use of urea fertilisers. As a result, it is likely that some sort of intervention or incentive will be required to drive this forward.

There are already significant pressures on the agricultural industry to reduce GHG emissions but there is only so far the industry can go in this respect because of its constant interaction with natural cycles (see Chapter 1). Research has shown that it is the livestock industry that is responsible for the majority of agricultural GHG emissions (i.e. due to ruminant enteric fermentation) and the primary cause for the recent rise in global methane emissions (Schaefer *et al.*, 2016). There have been a number of campaigns across the developed world to raise awareness of the climate impacts, and indeed health risks, of high levels of meat consumption but these have largely been ineffective. There has also been a push to increase locally-grown food to reduce the need to transport food over long distances. However, the carbon footprint of locally-grown food is not always better than imported food. For example, Avetisyan *et al.* (2014) considered what the impact on global GHG emissions would be if imported beef and dairy products were substituted with equivalent domestically produced goods. The study

found that transport emission reductions realised were swamped by changes in global emissions due to differences in GHG emissions intensities of production. Therefore, diverting consumption to local goods only reduces global emissions when undertaken in regions with relatively low emissions intensities (Avetisyan *et al.*, 2014). There are similar studies available looking at other agricultural products. It has been shown that imported tomatoes grown in Spain or Italy have half the GHG emissions as tomatoes grown locally in Austria using heated greenhouses (Theurl *et al.*, 2014). Another study found that tomatoes grown in the UK emit more than three times as much CO_2 per ton as tomatoes imported from Spain (Webb *et al.*, 2013). Therefore, it seems that new technologies offer the best chance of reducing GHGs further; these might include, for example, feed additives that inhibit methane production (Lewis *et al.*, 2015), or even new low-emission livestock breeds (Roehe *et al.*, 2016). The problem here is to ensure sufficient uptake by the industry to make a difference; adoption on just a few farms will not be enough and there is little incentive to farm businesses to adopt technologies that do not offer sufficient financial returns to offset the implementation costs. Therefore, some form of policy intervention that involves farm subsidies is possible but is likely to be highly problematic as incentivising farmers to, for example, use feed additives might well be seen as governments giving unfair boosts to manufacturers.

The effective control of agricultural water pollution is still an unresolved issue in many parts of Europe. The key aim for future policy can only be to reduce the loss of farm-derived pollutants into water systems but this needs to be done in a way that does not damage agriculture's productivity. According to the OECD (2012) a mix of policy instruments is likely to be more effective at addressing agricultural water quality issues than a single policy instrument such as a pollution tax. There is also increasing use of innovative policy tools, such as water quality trading and agreements between water supply utilities and farmers to reduce pollution and water treatment costs and this approach may gather pace in the future. An increasing emphasis of policies is towards changing the behaviour of farmers, the agro-food chain and other stakeholders to improve water quality. Nevertheless, stricter enforcement of regulations can assist in meeting the polluter pays principle and will also lower the burden on government budgets (OECD, 2012).

A recent European task force looking at future agricultural policy requirements identified a number of priorities that included the need to set clear, strategic targets for agriculture so that farmers would better appreciate the task that confronts them; to clarify the trade-offs in reaching a low-carbon strategy whilst also paying attention to soils, water quality and biodiversity conservation targets (Buckwell, 2017). The task force also emphasised that this could not be done by the CAP alone but would require greater and more stringent regulation at national level. It was also recognised that there is a pressing need to develop a new culture that involves greater engagement of all stakeholders rather than heavy controls, inspections and sanctions. The long-term objective could well be to internalise the environmental costs of farming into food prices so that it better signals socially aware consumption patterns. This will not be achieved without the active engagement of the private food processing and retailing sectors.

However, this lighter touch approach can mean that some, perhaps less understood and so less valued, ecosystem services are neglected. Schemes that provide Payment for Ecosystem Services (PES) have been under discussion across Europe for a few years. It

is a form of market-based instrument that is increasingly used to finance nature conservation and involves payments being made to land managers in exchange for the provision of specified ecosystem services or for actions anticipated to deliver these services, over and above what would otherwise be provided in the absence of payment. Payment is given by the beneficiary, so in terms of agriculture, it might well be expected to come from the public purse. For example, PES is used as a voluntary means of securing financial sources for multifunctional and protective forest management and sustainable maintenance of ecosystem services within European forest environments. Schemes tend to be funded under Pillar 2 of the CAP (JRC, 2016).

Scientifically, payment for ecosystem services comes with a wealth of problems (Redford and Adams, 2009; Schröter *et al.*, 2014; Reed *et al.*, 2014). Markets only exist for a certain range of ecosystem services, and some services are not amenable to pricing or valuation. There are numerous examples where this holds true, such as the nutrient value of atmospheric dust from the African Sahel carried across the Atlantic. Processes such as fire, drought, disease and flooding are generally considered not to be beneficial to humans yet they are vital for ecosystem function, structuring landscapes, and providing vital services and regulatory functions to non-humans. Thus, there is the risk that a financial focus on services seen as a value to humans will see other services neglected and will create problems longer term (Redford and Adams, 2009). There is also the risk that if some ecosystem services become scarce in the future, their perceived value will increase, distorting markets such that scarcity becomes more of a governing factor than the actual benefits they deliver. In extreme situations, this could cause conflict and discord. Despite these issues,

PES appears to be gaining popularity with many European policy makers and there is the possibility that future agri-environment schemes (Pillar 2 of the CAP) are modified such that they can more effectively pay for the provision of ecosystem services that are needed, potentially on a regional basis (Reed *et al.*, 2014). This would fit well with the possibility of greater use of Pillar 2-funded contracts for services, discussed above.

There are other pressures that might drive a move away from heavy controls, inspections and sanctions; a step that would be welcomed by both governments and farmers alike. Legislation and its enforcement is becoming more and more problematic in terms of the cost of design, implementation, policing and dealing with non-compliance, with much of the burden placed on national governments and so the public purse. In addition, the European Commission wishes to reduce the administration costs for businesses by simplifying regulation. Whilst many believe that an incentive approach would be more effective than rules and regulations, it is open to debate if this would solve Europe's environmental degradation. As regulation has often failed, it is difficult to believe that a softer approach would be more effective. Therefore, it is clear that any policy amendments in this area will need to carefully considered and, perhaps, it will be the administrative side that is streamlined rather than a fundamental shift in policy and standards.

In some areas, however, it is probable, that in order to ensure the safety of humans and animals and to protect the environment, regulatory processes may need to become more stringent. One such example is that of chemical risk assessments that underpin the authorisation for a particular chemical (e.g. pesticide, biocide) to be marketed, sold and used. Current risk assessment processes are based on an estimate of exposure under a range of different scenarios and a knowledge of the toxicity of an individual substance. Each chemical is considered in isolation, however, whilst exposure to a particular chemical on its own may not cause any adverse effects, the simultaneous exposure to multiple chemicals may result in a combination effect being seen. Considering the vast range of different chemicals that any individual could be exposed to, this is a highly complex and challenging issue to deal with. There are four possible types of combination effects:

1. **The antagonistic effect:** where exposure to one particular substance cancels out the effect of another and so the overall outcome is potentially less harmful.
2. **The synergistic effect:** where exposure to more than one substance causes an effect greater than that expected from a simple addition.
3. **Accumulative effect:** where the effect is a simple addition.
4. **Cocktail effect:** where the effect seen from exposure to the mixture is unexpected and could not be predicted by considering the individual chemicals.

Approaches to deal with this are currently being developed and are still in their infancy but in the future it will probably mean that regulatory risk assessments will need to be modified to consider this issue. How that will be done is unclear, but it could possibly mean the loss of some chemicals and more stringent rules on how and when others are used.

Many of the potential policy changes discussed here are likely to be seen in the relatively short term, perhaps within the next decade. How agri-environmental policy will change in the medium- to long-term future is little short of crystal ball gazing. There are too many uncertainties related to the

European and global political landscapes, global finances and trade, and how effective current policy instruments will be at tackling environmental degradation to make predictions.

8.4.6. *Integrated responses*

The sections above have explored numerous responses that may occur in the near future to tackle a range of challenges for food production and agri-environmental management. As mentioned above, many of the responses overlap, e.g. technological advances are constrained by biological knowledge, and both have to work within the realms of what is both practical and economically viable on farms. Additionally, as outlined in Chapter 7, farms need to achieve multiple objectives to make progress towards sustainability, thus more integrated approaches are needed. Therefore, responses also need to be more integrated, if only to ensure that they do not solve one issue at the expense of another. This holistic perspective can also be broadened futher beyond the scale of individual farms to encompass policy. Policies, rules and regulations often set the 'landscape' in which farm businesses operate, thus will influence how those farms and their practices evolve. Therefore, if farms are aiming to tackle multiple objectives, then policies must do the same. For example, policies aimed at reducing greenhouse gas emissions need to be in harmony with policies (and farm-level actions) that aim to reduce other pollutants, manage water resources and enhance biodiversity. There is always scope for unintended consequences to arise from any initiatives and actions, but taking a holistic perspective and drawing upon the best knowledge and data (perhaps using the technological advances outlined above to gather data), can allow the scope for unintended consequences to be minimised, and for multiple objectives to be realised.

8.5. *A vision of the future*

This chapter has explored the pressures that exist (now and in the future) and some of the responses that are emerging to tackle the challenges that lie ahead. Some of the visions of the future of farming can be quite extreme, as illustrated in Figure 8.1, but, in reality, the future of farming is likely to be something far more familiar and the changes that do occur will be more related to aspects that are not so visible (or at least not to the lay person).

Our knowledge of agricultural production and its effects and impacts on the environment has grown substantially over the past century and that knowledge is embedded in our understanding of complex and dynamic social, political, economic and environmental systems. Taking this into account and drawing upon the content of the previous chapters in this book, there are four broad aspects to consider with respect to future farming systems:

- **Multiple objectives**: Farms now and in the future need to be multifunctional. They need to produce food (and probably other bio-resources such as fibre and fuel) as well as supporting provisioning, regulating and cultural ecosystem services. These multiple objectives demand integrated decision-making and optimal solutions. Win–win solutions are ideal but in many instances there will be trade-offs between objectives. The farms of the future will need to make decisions that balance potentially competing objectives of satisfying consumer/societal demands and their own needs to run a profitable, sustainable business.
- **Loops and leaks**: The scientific understanding that underpins agricultural production has to date largely focused on agronomy and livestock husbandry,

in order to maximise the efficiency of production. However, to meet multiple objectives, including reduced pollution, scientific understanding needs to extend to agri-environmental management. This includes understanding pollutant fate and transport in the environment, and managing and recycling flows of materials and energy. Farmers will need to understand the flows, create loops to recycle and reuse materials and energy and reduce pollutants leaking from the system.

- **Knowledge transfer**: For the farms of the future to meet multiple objectives, and manage the loops and leaks in the system, they will need robust scientific knowledge to underpin practical and profitable solutions. Therefore, it will be important that there are mechanisms in place to ensure scientific knowledge is effectively transferred into the industry. Novel and more efficient ways of delivering targeted, timely advice and support will be essential. Ensuring that farms have the most up-to-date knowledge will help to increase their adaptive capacity.
- **Adaptability**: Farms need to be adaptable to respond to changing social, economic and environmental demands. This will involve ensuring that the industry has the adaptive capacity (resilience) to cope with changes. Changes may often be gradual but on occasions they may be sudden shocks which the system must be equally able to cope with to be sustainable. Examples can include changes to prices (e.g. oil price, fertilisers, livestock feed, etc.), consumer demand (e.g. cheaper food, higher welfare standards, etc.) or environmental concerns (e.g. climate change; extreme weather events; pests and diseases, etc.).

As outlined with the example of the perfect storm (see Chapter 8.2), there are significant challenges now and in the near future for food production and agri-environmental management. Society needs to tackle all these challenges simultaneously and not just focus on a few. Sustainable solutions will be those which meet multiple objectives and those which are based on a good understanding of the system and the cause–effect mechanisms. Reliable, robust and pragmatic knowledge to realise sustainable solutions needs to reach the minds, and the hands, of those that facilitate evolution. This includes decision makers at the strategic level of policy down to those at the practical level of farm management. This will help increase innovation, capability, adaptive capacity and future resilience.

The old expression, 'Give a man a fish, and you feed him for a day. Teach a man to fish, and you feed him for a lifetime', is a neat analogy for knowledge transfer, sustainability and agri-environmental management, albeit perhaps it should be amended to 'Give a man a fish, and you feed him for a day. Teach a man to fish sustainably, and you feed him for a lifetime'. However, although there are many definitions of sustainability (Vos, 2007; White, 2013), there are many arguments about what is or is not sustainable and no agreed definition on what sustainability means (Goodland, 1995; Kates *et al.*, 2005). For some, this presents a perplexing issue: 'how do you achieve something if you do not know what it is?' This is overcome to some extent with the concept that sustainability is not necessarily about the destination, and it is the journey that is more important (Chaharbaghi and Willis, 1999; Milne, *et al.*, 2006). That said, it is important that progress is made towards more sustainable systems and the journey is not simply one of going around in a circle. Therefore, it could be argued that both the journey and direction of travel are important. A key element of both the journey and direction of travel is learning (Berkes and Turner, 2006; Espinosa and Porter, 2011), which highlights the importance of knowledge transfer, the adaptive capacity it creates and consequently the ability to change the direction of travel.

Concepts such as sustainability and resilience are emergent properties of complex adaptive systems (Fiksel, 2003; Kay *et al.*, 1999). Successful agri-environmental management is one in which the decision makers and stakeholders are able to learn, adapt and improve. It is an evolutionary process within our social, economic and environmental systems from which sustainability will either emerge or not. The challenge for society, be it policy makers or farmers, is to learn, to understand the interactions within agri-enviromental systems and innovate solutions that lead to the emergence of a sustainable food production system.

References

AeroFarms (2017) *AeroFarms website*. Available at: http://aerofarms.com/

Ahearn, M.C., Armbruster, W. and Young, R. (2016) Big data's potential to improve food supply chain environmental sustainability and food safety. *International Food and Agribusiness Management Review*, Special Issue – Volume 19 Issue A.

Alchanatis, V., Cohen, Y., Cohen, S., Moller, M., Sprinstin, M., Meron, M., Tsipris, J., Saranga, Y. and Sela, E (2010) Evaluation of different approaches for estimating and mapping crop water status in cotton with thermal imaging. *Precision Agriculture*, 11(1), 27–41. DOI: 10.1007/s11119-009-9111-7

Alemdar, A. and Sain, M. (2008) Isolation and characterisation of nanofibers from agricultural residues – wheat straw and soy hulls. *Bioresource Technology*, 99(6), 1664–1671. DOI: 10.1016/j.biortech.2007.04.029

Alori, E.T., Dare, M.O. and Babalola, O.O. (2017) Microbial inoculants for soil quality and plant health. In: Lichtfouse, E. (ed.) *Sustainable Agriculture Reviews*. Springer International Publishing, pp. 281–307.

Aluwihare, Y.C., Ishan, M., Chamikara, M.D.M., Weebadde, C.K., Sirisena, D.N., Samarasinghe, W.L.G. and Sooriyapathirana, S.D.S.S. (2016) Characterization and selection of phosphorus deficiency tolerant rice genotypes in Sri Lanka. *Rice Science*, 23(4), 184–195. DOI: 10.1016/j.rsci.2015.10.001

Andújar, J.M., Aroba, J., La de Torre, M. and Grande, J.A. (2006) Contrast of evolution models for agricultural contaminants in ground waters by means of fuzzy logic and data mining. *Environmental Geology*, 49(3):458–466. DOI: 10.1007/s00254-005-0103-2

Arthur, G.D., Aremu, A.O., Kulkarni, M.G., Okem, A., Stirk, W.A., Davies, T.C. and Van Staden, J. (2016) Can the use of natural biostimulants be a potential means of phytoremediating contaminated soils from goldmines in South Africa? *International Journal of Phytoremediation*, 18(5), 427–434. DOI: 10.1080/15226514.2015.1109602

Avetisyan, M., Hertel, T. and Sampson, G. (2014) Is local food more environmentally friendly? The GHG emissions impacts of consuming imported versus domestically produced food. *Environmental and Resource Economics*, 58(3), 415–462. DOI: 10.1007/s10640-013-9706-3

Balbus, J., Denison, R., Florini, K. and Walsh, S. (2005) Getting nanotechnology right the first time. *Issues in Science and Technology*, 21(4), 65–71.

Baranowski, P., Jedryczka, M., Mazurek, W., Babula-Skowronska, D., Siedliska, A. and Kaczmarek, J. (2015) Hyperspectral and thermal imaging of oilseed rape (Brassica napus) response to fungal species of the genus Alternaria. *PloS one*, 10(3), e0122913. DOI: 10.1371/journal.pone.0122913

Bauckhage, C. and Kersting, K. (2013) Data mining and pattern recognition in agriculture. *KI-Künstliche Intelligenz*, 27(4), 313–324. DOI: 10.1007/s13218-013-0273-0

Beddington, J. (2009) *Food security: a global challenge*. Presentation to a BBSRC workshop on food security. 19 February 2009, London.

Benson, D.A., Cavanaugh, M., Clark, K., Karsch-Mizrachi, I., Lipman, D.J., Ostell, J. and Sayers, E.W. (2012) GenBank. *Nucleic Acids Research*, 41(D1), D36–D42. DOI: 10.1093/nar/gks1195

Berkes, F. and Turner, N.J. (2006) Knowledge, learning and the evolution of conservation practice for social-ecological system resilience. *Human Ecology*, 34(4), 479–494. DOI: 10.1007/s10745-006-9008-2

Berry, J.K., Delgado, J.A., Khosla, R. and Pierce, F.J. (2003) Precision conservation for environmental sustainability. *Journal of Soil and Water Conservation*, 58(6), 332–339.

Berry, J.K., Delgado, J.A. Pierce, F.J. and Khosla, R. (2005) Applying spatial analysis for precision conservation across the landscape. *Journal of Soil and Water Conservation*, 60(6), 363–370.

Bishop-Hurley, G., Henry, D., Smith, D., Dutta, R., Hills, J., Rawnsley, R., Hellicar, A., Timms, G, Morshed, A. Rahman, A. and D'Este, C. (2014) An investigation of cow feeding behavior using motion sensors. In: Instrumentation and Measurement Technology Conference (I2MTC) Proceedings, 2014 IEEE International (pp. 1285–1290). IEEE.

Bogue, R. (2016) Robots poised to revolutionise agriculture. Robot: An International Journal, 43(5), 450–456. DOI: 10.1108/IR-05-2016-0142

Borg, E., Fichtelmann, B., Schiller, C., Kuenlenz, S., Renke, F., Jahnke, D. and Wloczyk, C. (2014) DEMMIN-Test Site for Remote Sensing in Agricultural Application. JECAM Science Meeting, 21–23 Juli 2014, Ottawa, Ontario, Canada.

Borojevic, K. and Borojevic, K. (2005) The transfer and history of 'reduced height genes' (Rht) in wheat from Japan to Europe. *Journal of Heredity*, 96(4), 455–459. DOI: 10.1093/jhered/esi060

Brown, J.F. and Pervez, M.S. (2014) Merging remote sensing data and national agricultural statistics to model change in irrigated agriculture. *Agricultural Systems*, 127, 28–40. DOI: 10.1016/j.agsy.2014.01.004

Buckwell, A. (2017) *Out of the box thinking on the CAP*. CAP2020: Debating the future of the Common Agricultural Policy. Available at: http://www.cap2020.ieep.eu/2017/1/16/out-of-the-box-thinking-on-the-cap?s=1&selected=latest

Cavender-Bares, K. and Lofgren, J. (2016) *Robotic platform and method for performing multiple functions in agricultural systems*. U.S. Patent 9 265 187, Feb. 23, 2016. Available: https://www.google.com/patents/US9265187

Chaharbaghi, K. and Willis, R. (1999) The study and practice of sustainable development. *Engineering Management Journal*, 9(1), 41–48. DOI: 10.1049/em:19990115

Chavan, V.S. and Ingwersen, P. (2009) Towards a data publishing framework for primary biodiversity data: challenges and potentials for the biodiversity informatics community. *BMC Bioinformatics*, 10(14), S2. DOI: 10.1186/1471-2105-10-S14-S2

Chen, M., Willgoose, G.R. and Saco, P.M. (2014) Spatial prediction of temporal soil moisture dynamics using HYDRUS-1D. *Hydrological Processes*, 28(2), 171–185. DOI: 10.1002/hyp.9518

Chiaradia, E.A., Facchi, A., Masseroni, D., Ferrari, D., Bischetti, G.B., Gharsallah, O., De Maria, S.C., Rienzner, M., Naldi, E., Romani, M. amd Gandolfi, C. (2015) An integrated, multisensor system for the continuous monitoring of water dynamics in rice fields under different irrigation regimes. *Environmental Monitoring and Assessment*, 187(9), 586. DOI: 10.1007/s10661-015-4796-8

Childs, K.L., Konganti, K. and Buell, C.R. (2012) The Biofuels Feedstock Genomics Resource: a web-based portal and database to enable functional genomics of plant biofuel feedstock species. *Database*, bar061.

Chuanzhong, X., Pei, W., Lina, Z., Yanhua, M. and Yanqiu, L. (2017) Compressive sensing in wireless sensor network for poultry acoustic monitoring. *International Journal of Agricultural and Biological Engineering*, 10(2), 94–102. DOI: 10.3965/j.ijabe.20171002.2148

Collins, F.S., Patrinos, A., Jordan, E., Chakravarti,

A., Gesteland, R. and Walters, L. (1998) New goals for the U.S. Human Genome Project: 1998–2003. *Science, 282*(5389), 682–689. DOI: 10.1126/science.282.5389.682

Contreras, S., Hunink, J.E. and Baille, A. (2014) *Building a Watershed Information System for the Campo de Cartagena basin (Spain) integrating hydrological modeling and remote sensing.* FutureWater Report, 125. FutureWater, Cartagena, Spain.

Daniell, H., Streatfield, S.J. and Wycoff, K. (2001). Medical molecular farming: production of antibodies, biopharmaceuticals and edible vaccines in plants. *Trends in Plant Science*, 6(5), 219–226. DOI: 10.1016/ S1360-1385(01)01922-7

Davies, I. (2016) Japanese to open giant lettuce farm run by robots. *Farmers Weekly*, 2 February 2016.

Decesaro, A., Rampel, A., Machado, T. S., Thomé, A., Reddy, K., Margarites, A.C. and Colla, L.M. (2016) Bioremediation of soil contaminated with diesel and biodiesel fuel using biostimulation with microalgae biomass. *Journal of Environmental Engineering*, 04016091

De la Hera, M.L., Romero, P., Gómez-Plaza, E. and Martinez, A. (2007) Is partial root-zone drying an effective irrigation technique to improve water use efficiency and fruit quality in field-grown wine grapes under semiarid conditions? *Agricultural Water Management*, 87(3), 261–274. DOI: 10.1016/j. agwat.2006.08.001

de Vos, A.C., Broekman, R., de Almeida Guerra, C.C., van Rijsselberghe, M. and Rozema, J. (2013) Developing and testing new halophyte crops: a case study of salt tolerance of two species of the Brassicaceae, Diplotaxis tenuifolia and Cochlearia officinalis. *Environmental and Experimental Botany*, 92, 154–164. DOI: 10.1016/j.envexpbot.2012.08.003

Doolittle, J.A. and Brevik, E.C. (2014) The use of electromagnetic induction techniques in soils studies. *Geoderma*, 223–225, 33–45. DOI: 10.1016/j.geoderma.2014.01.027

Dragg, J.L., Bizzell, R.M., Trichel, M.C., Hatch, R.E., Phinney, D.E. and Baker, T.C. (1984) *Remote sensing advances in agricultural inventories.* 17th International Symposium on Remote Sensing of Environment; May 9–13, 1983; Ann Arbor, MI.

Dreher, K.L. (2004) Health and environmental impact of nanotechnology: toxicological assessment of manufactured nanoparticles. *Toxicological Sciences*, 77(1), 3–5. DOI: 10.1093/ toxsci/kfh041

Duke, S.O. and Powles, S.B. (2009) Glyphosate-resistant crops and weeds: now and in the future. *AgBioForum*, 12(3&4), 346–357. Available at: http://www.agbioforum.org/ v12n34/v12n34a10-duke.htm

Dunwell, J.M. (2014) Genetically modified (GM) crops: European and transatlantic divisions. *Molecular Plant Pathology*, 15(2), 119–121. DOI: 10.1111/mpp.12087

EC (2012) *A European Strategy for key enabling technologies – a bridge for growth and jobs.* European Commission, communication to the European Parliament, the council and the European Economic and Social Committee of the Regions, 2012: 0341:FIN:EN:PDF

Eckardt, N.A. (2000) Sequencing the rice genome. *Plant Cell*, 12(11), 2011–2017. DOI: 10.1105/tpc.12.11.2011

Esch, T., Metz, A., Marconcini, M. and Keil, M. (2014) Combined use of multi-seasonal high and medium resolution satellite imagery for parcel-related mapping of cropland and grassland. *International Journal of Applied Earth Observation and Geoinformation*, 28, 230–237. DOI: 10.1016/j.jag.2013.12.007

Espinosa, A. and Porter, T. (2011) Sustainability, complexity and learning: insights from complex systems approaches. *The Learning Organization*, 18(1), 54–72. DOI: 10.1108/09696471111096000

Estabrook, B. (2010) *A tale of two dairy farms*. The Atlantic, 10 August 2010. Available at: https:// www.theatlantic.com

Fiksel, J. (2003) Designing resilient, sustainable systems. *Environmental Science and Technology*, 37(23), 5330–5339. DOI: 10.1021/es0344819

Flourish (2017) *Flourish project website*: Available at: http://flourish-project.eu/

Foss, B. and Stone, M. (2001) *Successful Customer Relationship Marketing: New Thinking, New*

Strategies, New Tools for Getting Closer to Your Customers. Kogan Page Limited, London.

Fraceto, L.F., Grillo, R., de Medeiros, G.A., Scognamiglio, V., Rea, G. and Bartolucci, C. (2016) Nanotechnology in agriculture: which innovation potential does it have? *Frontiers in Environmental Science*, 4(20), 1–5. DOI: 10.3389/fenvs.2016.00020

FSA (2013) *GM Labelling*. Food Standards Agency (FSA), Available at: https://www.food. gov.uk/science/novel/gm/gm-labelling

Gahoonia, T.S. and Nielsen, N.E. (2004) Barley genotypes with long root hairs sustain high grain yields in low-P field. *Plant Soil*, 262, 55–62. DOI: 10.1023/B:PLSO.0000037020.58002.ac

Gautam, K. A., Bector, V., Singh, V. and Singh, M. (2016) Spectral analysis for monitoring crop growth using tractor mounted spectroradiometer and hand held Greenseeker in cotton. *Agricultural Engineering*, 2, 21–30.

Ghotbi, R.S., Khatibzadeh, M. and Kordbacheh, S. (2014) Preparation of neem seed oil nanoemulsion. In: *Proceedings of the 5th International Conference on Nanotechnology: Fundamentals and Applications*, Prague, Czech Republic, Paper (No. 150, pp. 11–13).

Glen, E.P., Brown, J.J. and O'Leary, J.W. (1998) Irrigating crops with seawater. *Scientific America*, August, pp. 77–81.

Goodland, R. (1995) The concept of environmental sustainability. *Annual Review of Ecology and Systematics*, 26, 1–24. DOI: 10.1146/annurev.es.26.110195.000245

GOS (2010) *Land use futures: making the most of land in the 21st century*. Final Project Report. The Government Office for Science (GOS), London.

Gottschalk, R., Burgos-Artizzu, X.P., Ribeiro, A., Pajares, G. and Sanchez-Miralles, A. (2008) Real-time image processing for the guidance of a small agricultural field inspection vehicle. *International Journal of Intelligent Systems Technologies and Applications*, 8(1–4), 434–443.

Green, M. (2017) *Spanish farm union opposes 20,000 cow 'mega-dairy'*. Agra Europe. 26 January 2017. Available at: https://www. agra-net.com/agra/agra-europe/meat-livestock/dairy/spanish-farm-union-opposes-20000-cow-mega-dairy-539924.htm

Grillo, R., Dos Santos, N.Z.P., Maruyama, C.R., Rosa, A.H., De Lima, R. and Faceto, L.F. (2012) Poly(epsilon-caprolactone) nanocapsules as carrier systems for herbicides: physico-chemical characterization and genotoxity evaluation. *Journal of Hazardous Materials.* 23(1–9), 231–232. DOI: 10.1016/j.jhazmat.2012.06.019

Growing UNDERGROUND (2017) Company website. Available at: http://www.growing-underground.com/

Guarino, M., Norton, T., Berckmans, D., Vranken, E. and Berckmans, D. (2017) A blueprint for developing and applying precision livestock farming tools: A key output of the EU-PLF project. *Animal Frontiers*, **7**(1), 12–17. DOI: 10.2527/af.2017.0103

Guo, H., Chen, J., Tian, L., Leng, Q., Xi, Y. and Hu, C. (2014) Airflow-induced triboelectric nanogenerator as a self-powered sensor for detecting humidity and airflow rate. *ACS Applied Materials & Interfaces*, 6(19), 17184–17189. DOI: 10.1021/am504919w

Guo, Y., Huang, J., Shi, Z. and Li, H. (2015) Mapping spatial variability of soil salinity in a coastal paddy field based on electromagnetic sensors. *PloS one*, 10(5), e0127996. DOI: 10.1371/journal.pone.0127996

Gutiérrez, J., Villa-Medina, J.F., Nieto-Garibay, A. and Porta-Gándara, M.Á. (2014) Automated irrigation system using a wireless sensor network and GPRS module. *IEEE Transactions on Instrumentation and Measurement*, 63(1), 166–176. DOI: 10.1109/TIM.2013.2276487

Hakkim, V.A., Joseph, E.A., Gokul, A.A. and Mufeedha, K. (2016) Precision farming: the future of Indian agriculture. *Journal of Applied Biology & Biotechnology*, 4(6), 68–72. DOI: 10.7324/JABB.2016.40609

Hanin, M., Ebel, C., Ngom, M., Laplaze, L. and Masmoudi, K. (2016) New insights on plant salt tolerance mechanisms and their potential use for breeding. *Frontiers in Plant Science*, **7**, 1–17. DOI: 10.3389/fpls.2016.01787

Hankin, S., Boraschi, D., Duschl, A., Lehr, C.M. and Lichtenbeld, H. (2011) Towards nanotechnology regulation–publish the unpublishable. *Nano Today*, 6(3), 28–23. DOI: 10.1016/j.nantod.2011.03.002

Hannemann, L.L. (2016) *Design and testing of an autonomous ground robot for agricultural applications*. BioResource and Agricultural Engineering, BioResource and Agricultural Engineering Department, California Polytechnic State University, San Luis Obispo, USA.

Hawkins, O. (2011) *Dairy industry in the UK: statistics*. House of Commons Library. Standard Note: SN/SG/2721. 30 June 2011.

Hebert, P.D., Cywinska, A. and Ball, S.L. (2003) Biological identifications through DNA barcodes. *Proceedings of the Royal Society of London B: Biological Sciences*, 270(1512), 313–321. DOI: 10.1098/rspb.2002.2218

Herman, E.M. (2003) Genetically modified soybeans and food allergies. *Journal of Experimental Biology*, 54(386), 1317–1319. DOI: 10.1093/jxb/erg164

Hildebrant, D. (2012) *Seed chipper speeds up genetic progress for many crops*. Available at: http://www.minnesotafarmguide.com/news/agri-tech/seed-chipper-speeds-up-genetic-progress-for-many-crops/article_ce7a743e-0809-11e2-a638-0019bb2963f4.html

Hochachka, W.M., Caruana, R., Fink, D., Munson, A.R.T., Riedewald, M., Sorokina, D. and Kelling, S. (2007) Data-mining discovery of pattern and process in ecological systems. *Journal of Wildlife Management*, 71(7), 2427–2437. DOI: 10.2193/2006-503

Huesing, J.E., Andres, D., Braverman, M.P., Burns, A., Felsot, A.S., Harrigan, G.G., Hellmich, R.L., Reynolds, A. and Shelton, A.M. (2016) Global adoption of genetically modified (GM) crops: challenges for the public sector. *Journal of Agricultural and Food Chemistry*, 64(2), 394–402. DOI: 10.1021/acs.jafc.5b05116

Iranshahi, D.R., Sepehri, M., Khoshgoftarmanesh, A.H., Eshghizadeh, H.R. and Abadi, V.J.M. (2016) Inoculation effects of endophytic fungus (Piriformospora indica) on antioxidant enzyme activity and wheat tolerance under phosphorus deficiency in hydroponic system. *Journal of Science and Technology of Greenhouse Culture*, 6(24), 74–85. DOI: 10.18869/acadpub.ejgcst.6.4.75

Jackson, S.A. (2016). Rice: the first crop genome. *Rice*, 9(1):14. DOI: 10.1186/s12284-016-0087-4

James, C. (2016) Global status of commercialised Biotech/GM crops 2016. ISAAA Brief 52-2016 Available at: http://www.isaaa.org/resources/publications/briefs/52/executivesummary/default.asp

Ji, G. and Tan, K. (2017) A big data decision-making mechanism for food supply chain. In: *MATEC Web of Conferences* (Vol. 100, p. 02048). EDP Sciences.

Jolanta, P., Marcin, B. and Zygmunt, K. (2011) Nanosilver – making difficult decisions. *Ecological Chemistry and Engineering*, 18(2), 185–195.

JRC – Joint Research Centre (2016) *Payments for Forest Ecosystem Services: SWOT analysis and possibilities for implementation*. JRC technical Reports, JRC103176, DOI: 10.2788/957929 (online), European Commission.

Kates, R.W., Parris, T.M. and Leiserowitz, A.A. (2005) What is sustainable development? Goals, indicators, values, and practice. *Environment: Science and Policy for Sustainable Development*, 47(3), 8–21. DOI: 10.1080/00139157.2005.10524444

Kay, J.J., Regier, H.A., Boyle, M. and Francis, G. (1999) An ecosystem approach for sustainability: addressing the challenge of complexity. *Futures*, 31(7), 721–742. DOI: 10.1016/S0016-3287(99)00029-4

Kelling, S., Hochachka, W.M., Fink, D., Riedewald, M., Caruana, R., Ballard, G. and Hooker, G. (2009). Data-intensive science: a new paradigm for biodiversity studies. *BioScience*, 59(7), 613–620. DOI: 10.1525/bio.2009.59.7.12

Kim, H.J., Kim, W.K., Roh, M.Y., Kang, C.I., Park, J.M. and Sudduth, K.A. (2013) Automated sensing of hydroponic macronutrients using a computer-controlled system with an array of ion-selective electrodes. *Computers and Electronics in Agriculture*, 93, 46–54. DOI: 10.1016/j. compag.2013.01.011

Kumar, R. and Khurana, A. (2014) Functional genomics of tomato: opportunities and challenges in post-genome NGS era. *Journal of Biosciences*, 39(5), 917–929. DOI: 10.1007/ s12038-014-9480-6

Lagi, M., Bertrand, K.Z. and Bar-Yam, Y. (2011) *The food crises and political instability in North Africa and the Middle East.* Available at: http:// necsi.edu/research/social/foodcrises.html

Lee, W.S., Alchanatis, V., Yang, C., Hirafuji, M., Moshou, D. and Li, C. (2010) Sensing technologies for precision specialty crop production. *Computers and Electronics in Agriculture*, 74(1), 2–33. DOI: 10.1016/j. compag.2010.08.005

Levy, D. and Veilleux, R.E. (2007) Adaptation of potato to high temperatures and salinity–a review. *American Journal of Potato Research*, 84(6), 487–506. DOI: 10.1007/BF02987 885

Levy, D., Coleman, W.K. and Veilleux, R.E. (2013) Adaptation of potato to water shortage: irrigation management and enhancement of tolerance to drought and salinity. *American Journal of Potato Research*, 90(2), 186–206. DOI: 10.1007/s12230-012-9291-y

Lewis, K.A., Tzilivakis, J., Green, A. and Warner, D.J. (2015) The potential of feed additives to improve the environmental impact of European livestock farming: a multi-issue analysis. *International Journal of Agricultural Sustainability*, 13(1), 55–68. DOI: 10.1080/14735903.2014.936189

Liu, R.Q., and Lal, R. (2015) Potentials of engineered nanoparticles as fertilizers for increasing agronomic productions. *Science of the Total Environment.* 514, 131–139. DOI: 10.1016/j.scitotenv.2015.01.104

Lötze, E. and Hoffman, E.W. (2016) Nutrient composition and content of various biological active compounds of three South African-based commercial seaweed biostimulants. *Journal of Applied Phycology*, 28(2), 1379–1386. DOI: 10.1007/ s10811-015-0644-z

Lubaba, C.H., Hidano, A., Welburn, S.C., Revie, C.W. and Eisler, M.C. (2015) Movement behaviour of traditionally managed cattle in the eastern province of Zambia captured using two-dimensional motion sensors. *PloS one*, 10(9), e0138125. DOI: 10.1371/journal. pone.0138125

Lunze, L., Abang, M.M., Buruchara, R., Ugen, M.A., Nabahungu, N.L., Rachier, G.O., Ngongo, M. and Rao, I. (2012) Integrated soil fertility management in bean-based cropping systems of Eastern, Central and Southern Africa. In: Whalen J.K. (ed.) *Soil Fertility Improvement and Integrated Nutrient Management: A Global Perspective.* INTECH, Rijeka.

Lynas, M. (2013) Lecture to Oxford Farming Conference, 3 January 2013. Available at: http://www.marklynas.org/2013/01/lecture-to- oxford-farming-conference-3-january- 2013

Lyra, D.A., Ismail, S., Butt, K.U.R B. and Brown, J. (2016) Evaluating the growth performance of eleven Salicornia bigelovii populations under full strength seawater irrigation using multivariate analyses. *Australian Journal of Crop Science*, 10(10), 1429. DOI: 10.21475/ ajcs.2016.10.10.p7258

Macouzet, M. (2016) Critical aspects in the conception and production of microbial plant biostimulants. *Probiotic Intelligentsia*, 5(2), 29–38. DOI: 10.13140/RG.2.2.14620.69765

Mat, N.N., Rowshon, K.M., Guangnan, C. and Troy, J. (2014) Prediction of sugarcane quality parameters using visible-shortwave near infrared spectroradiometer. *Agriculture and Agricultural Science Procedia*, 2, 136–143. DOI: 10.1016/j.aaspro.2014.11.020

McCullough, C. (2017) *Spain to build Europe's largest dairy farm.* Dairy Global. 13 April 2017. Available at: http://www.dairyglobal.net/ Articles/General/2017/4/A-new-dairy-farm- that-will-hold-capacity-to-milk-20000-cows- is-currently-under-construction-in-Spain- 119810E/

Meyer K. and Mewes, H.W. (2002) How can we deliver the large plant genomes? Strategies and perspectives. *Current Opinion* in Plant Biology, 5(2), 173–177. DOI: 10.1016/S1369-5266(02)00235-2

Milani, N., McLaughlin, M.J., Stacey, S.P., Kirby, J.K., Hettiarachchi, G.M., Beak, D.G. and Cornelis, G. (2012) Dissolution kinetics of macronutrient fertilizers coated with manufactured zinc oxide nanoparticles. *Journal of Agricultural and Food Chemistry*, 60(16), 3991–3998. DOI: 10.1021/jf205191y

Milne, M.J., Kearins, K. and Walton, S. (2006) Creating adventures in Wonderland: the journey metaphor and environmental sustainability. *Organization*, 13(6), 801–839. DOI: 10.1177/1350508406068506

Mishra, S., and Singh, H. B. (2015) Biosynthesized silver nanoparticles as a nanoweapon against phytopathogens: exploring their scope and potential in agriculture. *Applied Microbiology and Biotechnology*, 99(3), 1097–1107. DOI: 10.1007/s00253-014-6296-0

Moeinian, K., Mehrasbi, M.R., Hassanzadazar, H., Kamalid, K. and Rabiei, E. (2016) Fluoride levels in soil and crops of tomato and onion farms of Zanjan. *Journal of Human Environment and Health Promotion*, 2(1), 47–51.

Mukherjee, S. and Kumar, N.S. (2014) Terminator gene technology – their mechanism and consequences. *Science Vision*, 14(1), 51–58.

Nabti, E., Jha, B. and Hartmann, A. (2017) Impact of seaweeds on agricultural crop production as biofertilizer. *International Journal of Environmental Science and Technology* 14(5), 1119–1134. DOI: 10.1007/s13762-016-1202-1

Narang, R.A., Bruene, A. and Altmann, T. (2000) Analysis of phosphate acquisition efficiency in different Arabidopsis accessions. *Plant Physiology*, 124(4), 1786–1799. DOI: 10.1104/pp.124.4.1786

Neerincx, P. and Leunissen, J. (2005) Evolution of web services in Bioinformatics. *Briefings in Bioinformatics*, 6(2), 178–188. DOI: 10.1093/bib/6.2.178

Neto, A.P., Favarin, J.L., Hammond, J.P., Tezotto, T. and Couto, H.T. (2016) Analysis of phosphorus use efficiency traits in Coffea genotypes reveals Coffea arabica and Coffea canephora have contrasting phosphorus uptake and utilization efficiencies. *Frontiers in Plant Science*, **7**, 1–10. DOI: 10.3389/fpls.2016.00408

Nuruzzaman, M., Rahman, M.M., Liu, Y. and Naidu, R. (2016) Nanoencapsulation, nanoguard for pesticides: a new window for safe application. *Journal of Agriculture and Food Chemistry*. 64(7), 1447–1483. DOI: 10.1021/acs.jafc.5b05214

OECD (1993) *OECD Core Set of Indicators for Environmental Performance Reviews*. OECD Environment Monographs No. 83. OECD, Paris.

OECD (2012) *Water Quality and Agriculture: Meeting the Policy Challenge*. OECD ISBN: 9789264168053.

Parisi, C., Vigani, M. and Rodriguez-Cerezo, E. (2015) Agricultural nanotechnologies: what are the current possibilities? *Nano Today*, 10(2), 124–127. DOI: 10.1016/j.nantod.2014.09.009

Park, Y. and Moon, J. (2016) Smart dairy management system development using biometric/environmental sensors and farm control gateway. *IEMEK Journal of Embedded Systems and Applications*, 11(1), 15–20. DOI: 10.14372/IEMEK.2016.11.1.15

Peteinatos, G.G., Korsaeth, A., Berge, T.W. and Gerhards, R. (2016) Using optical sensors to identify water deprivation, nitrogen shortage, weed presence and fungal infection in wheat. *Agriculture*, 6(2), 24. DOI: 10.3390/agriculture6020024

Pleasants, J. M. and Oberhauser, K. S. (2013) Milkweed loss in agricultural fields because of herbicide use: effect on the monarch butterfly population. *Insect Conservation and Diversity*, 6(2), 135–144. DOI: 10.1111/j.1752-4598.2012.00196.x

Prasanna, B.M. (2007) *Nanotechnology in Agriculture*. ICAR National Fellow, Division of Genetics, IARI, New Delhi.

Qavami, N., Badi, H.N., Labbafi, M.R., Mehregan, M., Tavakoli, M. and Mehrafarin, A. (2017) Overview on Chitosan as a valuable ingredient and biostimulant in pharmaceutical industries and agricultural products. *Trakia*

Journal of Sciences, 15(1), 83. DOI: 10.15547/tjs.2017.01.014

Quemada, M., Gabriel, J.L. and Zarco-Tejada, P. (2014) Airborne hyperspectral images and ground-level optical sensors as assessment tools for maize nitrogen fertilization. *Remote Sensing*, 6(4), 2940–2962. DOI: 10.3390/rs6042940

Rajamanickam, E. (2012) Application of bioinformatics in agriculture. In: K.M. Singh and M.S. Meena (eds.) *ICTs for Agricultural Development under Changing Climate*, 1st Edition. Narendra Publishing House, New Delhi, pp.163–179.

Ranjan, A., Ichihashi, Y. and Sinha, N.R. (2012) The tomato genome: implications for plant breeding, genomics and evolution. *Genome Biology*, 13(8), 167. DOI: 10.1186/gb-2012-13-8-167

Reddy, P.K. and Ankaiah, R. (2005) A framework of information technology-based agriculture information dissemination system to improve crop productivity. *Current Science*, 88(12), 1905–1913.

Redford, K.H. and Adams, W.M. (2009) Payment for ecosystem services and the challenge of saving nature. *Conservation Biology*, 23(4), 785–787. DOI: 10.1111/j.1523-1739.2009.01271.x

Reed, M.S., Moxey, A., Prager, K., Hanley, N., Skates, J., Bonn, A., Evans, C.D. Glenk, K. and Thomson, K. (2014) Improving the link between payments and the provision of ecosystem services in agri-environment schemes. *Ecosystem Services*, 9, 44–53. DOI: 10.1016/j.ecoser.2014.06.008

Rembold, F., Atzberger, C., Savin, I. and Rojas, O. (2013) Using low resolution satellite imagery for yield prediction and yield anomaly detection. *Remote Sensing*, 5(4), 1704–1733. DOI: 10.3390/rs5041704

Rengasamy, K.R., Kulkarni, M.G., Papenfus, H.B. and Van Staden, J. (2016) Quantification of plant growth biostimulants, phloroglucinol and eckol, in four commercial seaweed liquid fertilizers and some by-products. *Algal Research*, 20, 57–60. DOI: 10.1016/j.algal.2016.09.017

Rios do Amaral, L. and Molin, J.P. (2014) The effectiveness of three vegetation indices obtained from a canopy sensor in identifying sugarcane response to nitrogen. *Agronomy Journal*, 106(1), 273–280. DOI: 10.2134/agronj2012.0504

Roehe, R., Dewhurst, R.J., Duthie, C.A., Rooke, J.A., McKain, N., Ross, D.W., Hyslop, J.J., Waterhouse, A., Freeman, T.C., Watson, M. and Wallace, R.J. (2016) Bovine host genetic variation influences rumen microbial methane production with best selection criterion for low methane emitting and efficiently feed converting hosts based on metagenomic gene abundance. *PLoS Genetics*, 12(2), e1005846. DOI: 10.1371/journal.pgen.1005846

Rothamsted Research (2017) *Rothamsted research is granted permission by Defra to carry out field trial with GM wheat plants*. Available at: http://www.rothamsted.ac.uk/news-views/rothamsted-research-granted-permission-defra-carry-out-field-trial-with-gm-wheat-plants

Sadok, W., Angevin, F., Bergez, J.É., Bockstaller, C., Colomb, B., Guichard, L., Reau, R. and Doré, T. (2009) Ex ante assessment of the sustainability of alternative cropping systems: implications for using multi-criteria decision-aid methods. A review. In: Lichtfouse, E., Navarette, M., Debaeke, P., Souchere, V. and Alberola, C. (eds.) *Sustainable Agriculture*. Springer, Netherlands, pp. 753–767.

Sager, B. (2001) Scenarios on the future of biotechnology. *Technological Forecasting and Social Change*, 68(2), 109–129. DOI: 10.1016/S0040-1625(00)00107-4

Schaefer, H., Milkaloff Fletcher, S.E., Veidt, C., Lassey, K.R., Brailsford, G.W, Bromley, T.M., Dlugokencky, E.J., Michel, S.E., Miller, J.B., Levin, I., Lowe, D.C., Martin, R.J., Vaughn, B.H. and White, J.W.C. (2016) A 21st century shift from fossil-fuel to biogenic methane emissions indicated by $^{13}CH_4$. *Science*, 352(6281), 80–84. DOI: 10.1126/science.aad2705

Schröter, M., Zanden, E.H., Oudenhoven, A.P., Remme, R.P., Serna-Chavez, H.M., Groot,

R.S. and Opdam, P. (2014) Ecosystem services as a contested concept: a synthesis of critique and counter-arguments. *Conservation Letters*, **7**(6), 514–523. DOI: 10.1111/conl.12091

Sekhon, B.S. (2014) Nanotechnology in agri-food production: an overview. *Nanotechnology, Science and Application.* **7**, 31–53. DOI: 10.2147/NSA.S39406

Sepaskhah, A.R. and Ahmadi, S.H. (2010) A review on partial root-zone drying irrigation. *International Journal of Plant Production*, 4(4), 241–258. DOI: 10.22069/ijpp.2012.708

Sharma, L. and Mehta, N. (2012) Data mining techniques: a tool for knowledge management system in agriculture. *International Journal of Scientific and Technology Research*, 1(5), 67–73.

Sharma, L.K., Bu, H., Franzen, D.W. and Denton, A. (2016) Use of corn height measured with an acoustic sensor improves yield estimation with ground based active optical sensors. *Computers and Electronics in Agriculture*, 124, 254–262. DOI: 10.1016/j.compag.2016.04.016

Shaw, R., Williams, A.P., Miller, A. and Jones, D.L. (2013) Assessing the potential for ion selective electrodes and dual wavelength UV spectroscopy as a rapid on-farm measurement of soil nitrate concentration. *Agriculture*, 3(3), 327–341. DOI: 10.3390/agriculture3030327

Sheeran, J. (2012) *Public-private partnerships innovating to end malnutrition.* 22nd Annual Martin J. Forman Memorial Lecture. 4th December 2012. International Food Policy Research Institute (IFPRI), Washington, DC, USA.

Sky Greens (2017) Sky Greens website. Available at: https://www.skygreens.com/

Stockwell, D.R. (2006). Improving ecological niche models by data mining large environmental datasets for surrogate models. *Ecological Modelling*, 192(1–2), 188–196. DOI: 10.1016/j.ecolmodel.2005.05.029

Sumriddetchkajorn, S. (2013) Mobile device-based optical instruments for agriculture. In: *Proceedings Volume 8881, Sensing Technologies for Biomaterial, Food, and Agriculture 2013.* DOI: 10.1117/12.2030626

Sun, Y., Zhou, H., Qin, Y., Lammers, P. S., Berg, A., Deng, H., Cai, X., Wang, D and Jones, S B. (2014) Horizontal monitoring of soil water content using a novel automated and mobile electromagnetic access-tube sensor. *Journal of Hydrology*, 516, 50–55. DOI: 10.1016/j.jhydrol.2014.01.067

SWEEPER (2017) *SWEEPER – Sweet Pepper Harvesting Robot.* Available at: http://www.sweeper-robot.eu/

Teulé, F., Miao, Y.G., Sohn, B.H., Kim, Y.S., Hull, J.J., Fraser, M. J., Lewis, R.V. and Jarvis, D.L. (2012) Silkworms transformed with chimeric silkworm/spider silk genes spin composite silk fibers with improved mechanical properties. *Proceedings of the National Academy of Sciences*, 109(3), 923–928. DOI: 10.1073/pnas.1109420109

The Dairy Site (2014) *French mega dairy proposal meets resistance.* The Dairy Site. 24 September 2014. Available at: http://www.thedairysite.com/news/46600/french-mega-dairy-proposal-meets-resistance/

The Hindu (2016) Bengaluru's tomato variety gets researchers national award. *The Hindu*, 18.08.2016. Available at: http://agritech.tnau.ac.in/daily_events/2016/english/Aug/18_aug_16_eng.pdf

Theurl, M.C., Haberl, H., Erb, K.H. and Lindenthal, T. (2014) Contrasted greenhouse gas emissions from local versus long-range tomato production. *Agronomy for Sustainable Development*, 34(3), 593–602. DOI: 10.1007/s13593-013-0171-8

Vaccari, D.A. (2009) Phosphorus: a looming crisis. *Scientific American*, June 2009, 54–59.

Vadivambal, R. and Jayas, D. S. (2011) Applications of thermal imaging in agriculture and food industry—a review. *Food and Bioprocess Technology*, 4(2), 186–199. DOI: 10.1007/s11947-010-0333-5

Valdes, M.G., Gonzalez, A.C.V., Calzon, J.A.G. and Diaz-Garcia, M.E. (2009) Analytical nanotechnology for food analysis. *Microchimica Acta*, 166(1), 1–19. DOI: 10.1007/s00604-009-0165-z

Valdivia-Granda, W. (2008) The next

meta-challenge for bioinformatics. *Bioinformation*, 2(8), 358–362.

Van Oosten, M.J., Pepe, O., Pascale, S., Silletti, S. and Maggio, A. (2017) The role of biostimulants and bioeffectors as alleviators of abiotic stress in crop plants. *Chemical and Biological Technologies in Agriculture*, 4, 5. DOI: 10.1186/s40538-017-0089-5

Vance, C.P., Uhde-Stone, C. and Allan, D.L. (2003) Phosphorus acquisition and use: critical adaptations by plants for securing a non-renewable resource. *New Phytologist*, 157, 423–447. DOI: 10.1046/j.1469-8137.2003.00695.x

VertiCrop (2017) VertiCrop website. Available at: http://www.verticrop.com/

Vos, R.O. (2007) Defining sustainability: a conceptual orientation. *Journal of Chemical Technology and Biotechnology*, 82(4), 334–339. DOI: 10.1002/jctb.1675

Warrick, A.W., Gardner, W.R. and Wang, M. (1983) Crop yield as affected by spatial variations of soil and irrigation. *Water Resource Research*, 19(1), 181–186. DOI: 10.1029/WR019i001p00181

Webb, J., Williams, A. G., Hope, E., Evans, D. and Moorhouse, E. (2013) Do foods imported into the UK have a greater environmental impact than the same foods produced within the UK? *The International Journal of Life Cycle Assessment*, 18(7), 1325–1343. DOI: 10.1007/s11367-013-0576-2

White, M.A. (2013) Sustainability: I know it when I see it. *Ecological Economics*, 86, 213–217. DOI: 10.1016/j.ecolecon.2012.12.020

Wieczorek A. (2003) Use of biotechnology in agriculture – benefits and risks. University of Hawaii, Honolulu (HI), 6 p. (Biotechnology; BIO-3).

Wilmut, I., Schnleke, A.E., McWhir, J., Kind, A.J. and Campbell, K.H.S. (1997) Viable offspring derived from fetal and adult mammalian cells. *Nature*, 385(6619), 810–813. DOI: 10.1038/385810a0

Wójtowicz, M., Wójtowicz, A. and Piekarczyk, J.

(2016) Application of remote sensing methods in agriculture. *Communications in Biometry and Crop Science*, 11(1), 31–50.

World Bank (2013) *Food Price Watch*. Year 4, Issue 14, July 2013.

World Bank (2016) *Agricultural land (% of land area)*. Available at: http://data.worldbank.org/indicator/AG.LND.AGRI.ZS

Xu, L., Liang, N. and Gao, Q. (2008) An integrated approach for agricultural ecosystem management. *IEEE Transactions on Systems, Man, and Cybernetics, Part C (Applications and Reviews)*, 38(4), 590–599. DOI: 10.1109/TSMCC.2007.913894

Yang, C., Westbrook, J.K., Suh, C.P.C., Martin, D.E., Hoffmann, W.C., Lan, Y., Fritz, B.K. and Goolsby, J.A. (2014) An airborne multispectral imaging system based on two consumer-grade cameras for agricultural remote sensing. *Remote Sensing*, 6(6), 5257–5278. DOI: 10.3390/rs6065257

Yang, M.T., Chen, C.C. and Kuo, Y.L. (2013) Implementation of intelligent air conditioner for fine agriculture. *Energy and Buildings*, 60, 364–371. DOI: 10.1016/j.enbuild.2013.01.034

Yao, J., Weng, Y., Dickey, A. and Wang, K.Y. (2015) Plants as factories for human pharmaceuticals: applications and challenges. *International Journal of Molecular Sciences*, 16(12), 28549–28565. DOI: 10.3390/ijms161226122

Ye, X., Al-Babili, S., Klöti, A., Zhang, J., Lucca, P., Beyer, P. and Potrykus, I. (2000) Engineering the provitamin A (ß-carotene) biosynthetic pathway into (carotenoid-free) rice endosperm. *Science*, 287(5451), 303–305. DOI: 10.1126/science.287.5451.303

Zhang, Q. (2007) Strategies for developing green super rice. *Proceedings of the National Academy of Sciences of the United States of America*, 104(42):16402–16409. DOI: 10.1073/pnas.0708013104

Index